SYSTEMS THEORY RESEARCH

(Problemy Kibernetiki)

Edited by A. A. Lyapunov

Volume 23

Translated from Russian

CONSULTANTS BUREAU • NEW YORK-LONDON • 1973

The original Russian text was published by Nauka Press in Moscow in 1970 under the general direction of the Scientific Council on Complex Problems of Cybernetics of the Academy of Sciences of the USSR, Academician A. I. Berg, Chairman. The present translation is published under an agreement with Mezhdunarodnaya Kniga, the Soviet book export agency.

Library of Congress Catalog Card Number 68-15025

ISBN 978-1-4757-0081-7 ISBN 978-1-4757-0079-4 (Ebook)
DOI 10.1007/978-1-4757-0079-4

A Division of Plenum Publishing Corporation
227 West 17th Street, New York, N. Y. 10011

United Kingdom edition published by Consultants Bureau, London
A Division of Plenum Publishing Company, Ltd.
Davis House (4th Floor), 8 Scrubs Lane, Harlesden, London, NW10 6SE, England

SYSTEMS THEORY RESEARCH

ПРОБЛЕМЫ КИБЕРНЕТИКИ

PROBLEMY KIBERNETIKI

PROBLEMS OF CYBERNETICS

CONTENTS

THEORY OF CONTROL SYSTEMS

PROGRAMMING

CONTROL PROCESSES IN LIVING ORGANISMS

BRIEF COMMUNICATIONS

THEORY OF CONTROL SYSTEMS

ON THE COMPLETENESS OF FUNCTIONS HAVING DELAYS†

L. A. Biryukova and V. B. Kudryavtsev

Moscow

This paper studies completeness conditions for a certain class of automata without feedbacks – so-called functions with delays [1]. The problems considered in the paper are related to investigations begun in [1] and to an obvious degree are a continuation of them.

Assume ${}^l\hat{P}_2$ is the set of all logic-algebra functions having delays not exceeding l and $\mathfrak{M} \subseteq {}^l\hat{P}_2$. The set $\mathfrak{M} \subseteq {}^l\hat{P}_2$ is called *l-complete* if by means of "synchronous superposition" operations one may obtain any logic-algebra function having the delay l by starting from the elements of the set $\mathfrak{M}$. The paper investigates the conditions which l-complete systems must satisfy. The functional system studied, as became clear, has numerous interesting properties.

In particular, it turned out that in the general case not every class in ${}^l\hat{P}_2$ can be expanded to an l-precomplete class, and therefore the criterion of l-completeness may not be formulated solely in terms of l-precomplete classes. This fact already holds for $l = 1$. However, in this case we were nevertheless able to show that a finite system is complete when and only when it does not belong to a certain finite number of 1-precomplete classes and three rigorously increasing chains of closed classes, none of which is contained in any of the 1-precomplete classes. From this, in particular, there derives the existence of an algorithm which establishes 1-completeness of any finite system of functions having delays. It is shown that in the general case a finite l-complete subsystem cannot be isolated from just any l-complete system, and that the power of the set of closed classes in ${}^l\hat{P}_2$ is equal to the continuum. The case of $l = 1$ is investigated in particular detail, although many results may easily be carried over to the case of arbitrary l.

In the paper extensive use is made of the results of [1, 2, 3] and all notations which are not defined may be found there.

§ 1. Basic Notions

Assume $X = \{x_1, x_2, \ldots, x_n, \ldots\}$ is the set of Boolean variables that take 0 or 1 as their values. Sometimes we shall use the notation x, y, z, ... to denote letters from X. Let us use P_2 to denote the set of all logic-algebra functions which depend on variables from the set X. Hereafter logic-algebra functions shall be called simply functions for brevity. Assume τ is a parameter which takes one of the values 0, 1, ..., l. We shall call it a *delay*. Let us consider the set ${}^l\hat{P}_2$ of all pairs of the form $(f(x_1, \ldots, x_n), \tau)$, where $f(x_1, \ldots, x_n) \in P_2$ and $\tau \in \{0, 1, \ldots, l\}$. Sometimes the pair (f, τ) shall be called a *function f having*

† Original article submitted November 15, 1968.

the delay τ. We shall consider two pairs (f, τ_1) and (φ, τ_2) to be equal and denote that fact by $(f, \tau_1) = (\varphi, \tau_2)$ if $\tau_1 = \tau_2$ and the functions f and φ differ only, perhaps, by fictitious variables. We shall assume ahead of time that with the stipulation of the pair (f, τ), all pairs equal to it are stipulated simultaneously. In the set ${}^l\hat{P}_2$ we shall inductively introduce synchronous-superposition operations as follows.

Definitions. 1. Assume that we have the pair

$$(f(x_1, x_2, \ldots, x_{i-1}, x_i, x_{i+1}, \ldots, x_n), \tau) \tag{1}$$

and the variable x_j; then the pair $(f(x_1, \ldots, x_{i-1}, x_j, x_{i+1}, \ldots, x_n), \tau)$ is obtained from the pair (1) by means of synchronous-superposition operations (the rule for redesignating the variables).

2. Assume we have the pairs

$$(f(x_1, \ldots, x_n), \tau), \quad (g_{i_1}(x_1, \ldots, x_m), 0), \quad (g_{i_2}(x_1, \ldots, x_m), 0), \ldots, (g_{i_r}(x_1, \ldots, x_m), 0),$$
$$1 \leqslant i_j \leqslant n \text{ for } 1 \leqslant j \leqslant r \text{ and } i_1 < i_2 < \ldots < i_r. \tag{2}$$

Then the pair $(f(G_1, G_2, \ldots, G_n), \tau)$, where G_k denotes either a variable or a function g_k, $k = 1, \ldots, n$, is obtained from (2) by means of synchronous-superposition operations (the rule of substituting functions with zero delays into a function with delay).

3. Assume we have the pairs

$$(f(x_1, \ldots, x_n), \tau_1), \quad (g_1(x_1, \ldots, x_m), \tau_2), \quad (g_2(x_1, \ldots, x_m), \tau_2), \quad \ldots, \quad (g_n(x_1, \ldots, x_m), \tau_2), \tag{3}$$

then the pair $(f(g_1, g_2, \ldots, g_n), \tau)$, where $\tau = \tau_1 + \tau_2$ and $\tau \leq l$, is obtained from the pairs (3) by means of synchronous-superposition operations (the rule of substituting functions with delays into a function with delay).

Remark. Let us note the following two important facts affecting the synchronous-superposition operations. First, if the system $\mathfrak{M} \subseteq {}^l\hat{P}_2$ consists solely of functions having a zero delay, then in essence the operations 1-3 which we introduced for elements from $\mathfrak{M}$ are completely analogous to the "conventional" superposition operations in the class of functions P_2. Second, we wish to emphasize the fact that operation 3 together with the substitution of some pair of the form (φ, τ_2), $\tau_2 > 0$ in place of a certain variable of the function f in the pair (f, τ_1) "compels" us to replace each of the remaining variables of the functions f by a pair having a delay which is likewise equal to τ_2, the condition $\tau_1 + \tau_2 \leq l$ being valid under these conditions.

Assume $\mathfrak{M} \subseteq {}^l\hat{P}_2$. The set $[\mathfrak{M}]$ is called the closure of $\mathfrak{M}$ if it contains those and only those pairs which are obtained from the pairs of the set $\mathfrak{M}$ by means of a finite number of applications of synchronous-superposition operations.

The set $\mathfrak{M}$ is called closed if $\mathfrak{M} = [\mathfrak{M}]$.

Definition. The set $\mathfrak{M}$ is called l-complete if $[\mathfrak{M}]$ contains all pairs having delays equal to l.

Definition. The set $\mathfrak{M}$ is called l-precomplete if

1) $\mathfrak{M}$— is not l-complete;

2) $\mathfrak{M} \cup \{(g, \tau)\}$ is l-complete for any pair (g, τ) from ${}^l\hat{P}_2$ such that $(g, \tau) \notin \mathfrak{M}$. It is obvious that an l-complete class is closed.

As is well known, each function $f(x_1, \ldots, x_n)$ may be stipulated (and uniquely at that) by means of a Zhegalkin polynomial; i.e., the equation

$$f(x_1, \ldots, x_n) = \sum_{\substack{(i_1, \ldots, i_s) \\ 1 \leq i_j \leq n}} C_{i_1, i_2, \ldots, i_s} x_{i_1} x_{i_2} \ldots x_{i_s} + d$$

holds, where $C_{i_1, \ldots, i_s}$ and d are equal to zero or unity, and multiplication and addition is carried out modulo two. The *degree of the function* f is called the degree of its Zhegalkin polynomial. The *degree of linearity of the function* f is called the number of linear terms in its Zhegalkin polynomial. The *rank of the function* f is called the number of terms in its Zhegalkin polynomial.

Let us introduce the following notation for the sets which we shall require further on [2, 3]:

A	all α-functions;
B	all β-functions;
Γ	all γ-functions;
Δ	all δ-functions;
C_2	all functions $f(x_1, x_2, \ldots, x_n)$, such that $f(0, \ldots, 0) = 0$;
C_3	all functions $f(x_1, x_2, \ldots, x_n)$, such that $f(1, \ldots, 1) = 1$;
A_1	all monotonic functions;
D_3	all self-dual functions;
L_1	all linear functions;
Y	all even functions;
$\overline{A}_1$	all functions which are negations of monotonic functions;
D_1	all self-dual α-functions;
F_5^∞	all α-functions which satisfy condition $\langle A^\infty \rangle$;
F_8^∞	all functions satisfying the condition $\langle A^\infty \rangle$;
F_6^∞	all monotonic α-functions satisfying the condition $\langle A^\infty \rangle$;
F_7^∞	all monotonic functions satisfying the condition $\langle A^\infty \rangle$;
F_5^μ	all α-functions satisfying the condition $\langle A^\mu \rangle$, $\mu \geqslant 2$;
F_8^μ	all functions satisfying the conditions $\langle A^\mu \rangle$, $\mu \geqslant 2$;
F_6^μ	all monotonic α-functions satisfying the condition $\langle A^\mu \rangle$;
F_7^μ	all monotonic functions satisfying the condition $\langle A^\mu \rangle$;
F_1^∞	all α-functions satisfying the condition $\langle a^\infty \rangle$;
F_2^μ	all monotonic α-functions satisfying the conditions $\langle a^\mu \rangle$;
F_2^∞	all monotonic α-functions satisfying the condition $\langle a^\infty \rangle$;
F_1^μ	all α-functions satisfying the condition $\langle a^\mu \rangle$;
F_4^∞	all functions satisfying the condition $\langle a^\infty \rangle$;
F_4^μ	all functions satisfying the condition $\langle a^\mu \rangle$;
F_3^∞	all monotonic functions satisfying the condition $\langle a^\infty \rangle$;
F_3^μ	all monotonic functions satisfying the condition $\langle u^\mu \rangle$;
A_2	all monotonic α-functions and β-functions;
A_3	all monotonic α-functions and γ-functions;
A_4	all monotonic α-functions;
D_2	all self-dual monotonic functions;
O_1	all functions equal to x, and all functions derived from it by redesignating the variables without identification;
O_2	all functions equal to the function 1;
O_3	all functions equal to the function 0;

O_4 all functions equal to the functions x or $\bar{x}$, and all functions derived from them by redesignating the variables without identification;
O_5 all functions equal to the functions 1 or x, and all functions derived from them by redesignating the variables without identification;
O_6 all functions equal to the functions 0 or x, and all functions derived from them by redesignating the variables without identification;
O_7 all functions equal to the functions 0 or 1;
O_8 all functions equal to the functions 0, 1, or x, and all functions derived from them by redesignating the variables without identification;
O_9 all functions equal to the functions 1, 0, x, or $\bar{x}$, and all functions derived from them by redesignating the variables without identification;
S_1 all logical sums;
S_3 all logical sums and all functions equal to 1;
S_5 all logical sums and all functions equal to 0;
S_6 all logical sums and all functions equal to 0 or 1;
P_1 all logical products;
P_3 all logical products and all functions equal to 0;
P_5 all logical products and all functions equal to 1;
P_6 all logical products and all functions equal to 0 or 1;
L_2 all linear α- and β-functions;
L_3 all linear α- and γ-functions;
L_4 all linear α-functions;
L_5 all linear self-dual functions.

From the basic Post theorems [2] it follows that each of the sets enumerated above, with the exception of the sets B, Γ, Δ, Y, $\bar{A}_1$, is closed relative to superposition operations, and no closed sets exist which are different from them.

Note that the notation for the sets, with the exception of the sets A, B, Γ, Δ, Y, $\bar{A}_1$, coincides with the notation in [2]. The sets A, B, Γ, Δ are denoted in the same way as they are in [1].

The notation which has been introduced shall be used in a certain modified form to denote the subsets from ${}^l\hat{P}_2$. Thus, assume $\mathfrak{N}$ is any of the sets of functions just examined; then we use ${}^\tau\mathfrak{N}$ to denote the set of all pairs (f, τ) such that $f \in \mathfrak{N}$.

We shall be interested in the following problem. Assume there is a finite system of pairs $\mathfrak{B} \subset {}^l\hat{P}_2$. It is required to clarify the conditions under which this system has the property of l-completeness.

The paper will carry out a detailed consideration of the case in which $l = 1$. Namely, a criterion of 1-completeness will be obtained which is formulated in terms of 1-precomplete classes and certain expanding chains of closed classes.

Assume the set $\mathfrak{M} \subseteq {}^lP_2$, and assume

$$(f_1, 0), \quad (f_2, 0), \ldots, (\varphi_1, 1), \quad (\varphi_2, 1), \ldots, (\psi_1, l), \quad (\psi_2, l), \ldots$$

are all functions with 0, unitary, etc., delays, respectively (certain of these sets may turn out to be empty), which are contained in $\mathfrak{M}$. In order to denote $\mathfrak{M}$ it will sometimes be convenient for us to use the following notation:

$$\{(f_i, 0), (\varphi_j, 1), \ldots, (\psi_k, l)\},$$

where the parameters i, j, ..., k run the gamut of values from a natural series.

§ 2. Certain Necessary Conditions of 1-Completeness of Finite Systems

Lemma 1. If $\mathfrak{B}$ is a finite 1-complete system, then it necessarily contains the following pairs: $(f, 0)$ and $(\varphi, 1)$, where f and φ are not constants.

Proof. In accordance with the conditions of the theorem, let us consider two cases.

a) Assume condition 1) is not fulfilled. Then it is obvious that by using pairs from $\mathfrak{B}$, one cannot obtain any pair of the form (f, τ), where f depends essentially on a larger number of variables than any of the functions appearing in the pairs of the system $\mathfrak{B}$, by means of synchronous-superposition operations.

b) Assume that condition 2) has not been fulfilled. It is obvious that a pair of the form $(\varphi, 1)$ must appear in $\mathfrak{B}$. Assume φ = const. Since functions having a unitary delay may not be substituted into one another, it follows that in order to obtain functions with unitary delays, one may substitute only constants with a unitary delay in place of all variables in a function having a zero delay, or vice versa. It is obvious that as a result we again obtain a constant having a unitary delay; i.e., $\mathfrak{B}$ is not a 1-complete system, which is what it has been required to prove.

Lemma 2. If $\mathfrak{B}$ is a finite 1-complete system, then the pair $(f, 0) \notin {}^0L_1$ is contained in it.

Proof. Let us assume that $\mathfrak{B}$, having zero delays, contains only linear functions.

Further, it is not difficult to see that the substitution of a linear function having a zero delay into the pair $(\psi, 1)$ does not raise the degree of ψ; furthermore, the substitution of functions having a unitary delay into a linear function having a zero delay yields a function with a unitary delay whose degree does not exceed the maximum degree out of the degrees of the substituted functions. From this it follows that the degree of any function having delay from $\mathfrak{B}$ has a degree no larger than r, which it was required to prove.

Lemma 3. If $[\mathfrak{B}]$ is a finite 1-complete system, then $\mathfrak{B}$ contains the pairs $(f, 0) \notin {}^0P_6$ and $(g, 0) \notin {}^0S_6$.

Proof. By virtue of the duality principle it is sufficient to prove, for example, that $\mathfrak{B}$ contains the pair $(f, 0) \notin {}^0P_6$. Let us assume that the lemma is invalid (i.e., a 1-complete finite system $\mathfrak{B}$, exists which contains only elements of the set ${}^0P_6$ as functions having zero delay). We shall show that in this case the system $\mathfrak{B}$ cannot be 1-complete.

The set $[\mathfrak{B}]$ may be partitioned into two subsets which may, in general, cross. In the first subset $[\mathfrak{B}]$, we shall include all functions having delays from $[\mathfrak{B}]'$ which are obtained by the substitution of a function having zero delay into a function having unitary delay from $\mathfrak{B}$. Assume the highest degree of linearity and the highest rank of functions appearing in pairs of the set $\mathfrak{B}$, are equal to $m \geq 1$ and $r \geq 1$ $(r \geq m)$, respectively. It is obvious that each function contained in a pair from the set $[\mathfrak{B}]'$, has a degree of linearity no higher than r.

In the second subset $[\mathfrak{B}]''$ we shall include all functions having delays from $[\mathfrak{B}]$, which are obtained by substitution of the functions having unitary delays from $[\mathfrak{B}]'$ into functions having zero delay from $\mathfrak{B}$. It is obvious that $[\mathfrak{B}]' \cup [\mathfrak{B}]'' = [\mathfrak{B}]$. We shall show that the set $[\mathfrak{B}]''$ cannot contain the pair (f, τ), where f is an α-function whose degree of linearity is higher than r. Note that the set $[\mathfrak{B}]$ contains only the pairs (0, 0), (1, 0), and pairs of the form $(x_1 \cdot x_2 \cdot \ldots \cdot x_n, 0)$, whose degree of linearity is obviously no higher than r, as functions having zero delay. By virtue of this it is sufficient to consider functions having a unitary delay solely from $[\mathfrak{B}]''$. Note that each pair $(\varphi, 1) \in [\mathfrak{B}]''$, where φ differs from a constant, may be represented in the form $(f_1 \cdot f_2 \cdot \ldots f_s, 1)$, where $(f_i, 1) \in [\mathfrak{B}]'$, $i = 1, \ldots, s$. But proof of the fact that each α-function φ having delay 1 which belongs to $[\mathfrak{B}]''$, has a degree of linearity no higher than r will be carried out by induction with respect to the number s. For s = 1 our statement is obvious. Assume that it is true for all $s \le k$, $k \ge 1$. We shall show that our statement is true for s = k + 1. Let us consider the pair $(f_1 \cdot f_2 \cdot \ldots \cdot f_k \cdot f_{k+1}, \tau)$. We denote the product of the first k functions in this pair by g. Relative to the functions g and f_{k+1} the following distribution of values of free terms c and d is possible in their Zhegalkin polynomials.

1) c = 0, d = 0. In this case, $g \cdot f_{k+1}$ is an α-function only when g and f_{k+1} are themselves α-functions. The latter, according to the induction proposition, have a degree of linearity $\le$ r. From this it follows that the degree of linearity of $g \cdot f_{k+1}$ is no higher than r.

2) c = 0, d = 1. In this case, $g \cdot f_{k+1}$ is an α-function only when g is an α-function and f_{k+1} is a β-function. Since, according to the induction proposition, the degree of linearity of g is no higher than r, it follows that the degree of linearity of $g \cdot f_{k+1}$ is no higher than r.

3) c = 1, d = 0. This case is completely analogous to case 2).

4) c = 1, d = 0. In this case $g \cdot f_{k+1}$ cannot be an α-function.

Thus, $[\mathfrak{B}]$ does not contain a single pair (f, τ) such that f is an α-function whose degree of linearity is higher than r. This contradicts the postulated 1-completeness of the system $\mathfrak{B}$, which is what it was required to prove.

Let us consider the system N containing the following 19 sets of functions with delays (the symbol ${}^iQ$ denotes the set of all pairs of the form (f, i), where

$$\begin{aligned}
&\mathfrak{N}_1 = ({}^0P_2 \cup {}^1C), \text{ where } C = \{0, 1\}, && \mathfrak{N}_{11} = ({}^0F_5^2 \cup {}^1\mathrm{A} \cup {}^1\Gamma \cup {}^1\Delta),\\
&\mathfrak{N}_2 = ({}^0\mathrm{A} \cup {}^1\mathrm{B} \cup {}^1\Gamma), && \mathfrak{N}_{12} = ({}^0\mathrm{A} \cup {}^1\Delta \cup {}^1\Gamma),\\
&\mathfrak{N}_3 = ({}^0C_3 \cup {}^1C_3), && \mathfrak{N}_{13} = ({}^0A_2 \cup {}^1\Delta \cup {}^1\Gamma \cup (1, 1)),\\
&\mathfrak{N}_4 = ({}^0A_1 \cup {}^1A_1), && \mathfrak{N}_{14} = ({}^0A_2 \cup {}^1\mathrm{A} \cup {}^1\mathrm{B} \cup (0, 1)),\\
&\mathfrak{N}_5 = ({}^0A_4 \cup {}^1A \cup {}^1\mathrm{B} \cup {}^1\Gamma), && \mathfrak{N}_{15} = ({}^0C_2 \cup {}^1C_2),\\
&\mathfrak{N}_6 = ({}^0A_4 \cup {}^1\Delta \cup {}^1\mathrm{B} \cup {}^1\Gamma), && \mathfrak{N}_{16} = ({}^0F_1^2 \cup {}^1\mathrm{A} \cup {}^1\mathrm{B} \cup {}^1\Delta),\\
&\mathfrak{N}_7 = ({}^0A_1 \cup {}^1\bar{A}_1), && \mathfrak{N}_{17} = ({}^0\mathrm{A} \cup {}^1\Delta \cup {}^1\mathrm{B}),\\
&\mathfrak{N}_8 = ({}^0D_3 \cup {}^1D_3), && \mathfrak{N}_{18} = ({}^0A_3 \cup {}^1\mathrm{A} \cup {}^1\Gamma \cup (1, 1)),\\
&\mathfrak{N}_9 = ({}^0D_3 \cup {}^1Y), && \mathfrak{N}_{19} = ({}^0A_3 \cup {}^1\Delta \cup {}^1\mathrm{B} \cup (0, 1)).\\
&\mathfrak{N}_{10} = ({}^0D_1 \cup {}^1A \cup {}^1\Delta), &&
\end{aligned}$$

In §3 it will be proved that each of the sets of the system N is 1-precomplete.

§ 3. Proof of 1-Precompleteness of Classes from the System N

The closure of each of the classes of the system N easily derives from the definition of these classes. Let us show that for any $\mathfrak{N}_i$ and any pair $(f, \tau) \in {}^1\hat{P}_2$, such that $(f, \tau) \notin \mathfrak{N}_i$, the set $\mathfrak{N}_i \cup \{(f, \tau)\}$ is 1-complete. Let us consider each of the cases separately.

Lemma 4. The set $\mathfrak{N}_1 = ({}^0P_2 \cup {}^1C)$ is 1-precomplete.

Proof. The lemma will be proved if we show that the set $\mathfrak{M} = \mathfrak{N}_1 \cup \{(f, 1)\}$, where $f \not\equiv c$ is 1-complete. For this purpose it is obviously sufficient to show that $[\mathfrak{M}] \supseteq {}^0P_2 \cup \{(x, 1)\}$. Since in the pair $(f, 1)$ the function $f(x_1, \ldots, x_n)$ is different from a constant, it follows that one can find a variable (without loss of generality it may be assumed that this is x_1) on which the function depends essentially (i.e., for certain collections $\tilde{\alpha} = (0, \alpha_2, ..., \alpha_n)$ and $\tilde{\alpha}' = (1, \alpha_2, ..., \alpha_n)$ the relation $f(\tilde{\alpha}) \neq f(\tilde{\alpha}')$. holds). Let us substitute the pair $(\alpha_i, 0) \in ({}^0P_2 \cup {}^1C)$ into the pair $(f, 1)$ in place of the variable x_i, $i \neq 1$, while the pair $(x, 0)$ is substituted in place of the variable x_1 if $f(\tilde{\alpha}) = 0$, and the pair $(\bar{x}, 0)$ if $f(\tilde{\alpha}) = 1$ (obviously, the pairs $(x, 0)$ and $(\bar{x}, 0)$ belong to $({}^0P_2 \cup {}^1C)$). It can easily be seen that as a result of this substitution we obtain a pair equal to the pair $(x, 1)$, which is what it was required to prove.

Lemma 5. The set $\mathfrak{R}_2 = ({}^0\mathrm{A} \cup {}^1\mathrm{B} \cup {}^1\Gamma)$ is 1-precomplete.

Proof. Let us show that the set $\mathfrak{M} = \mathfrak{R}_2 \cup \{(f, \tau)\}$, where $(f, \tau) \notin ({}^0\mathrm{A} \cup {}^1\mathrm{B} \cup {}^1\Gamma)$ is 1-complete. First we show that in the case in which the pair (f, τ) is equal to the pair $(x, 1)$ the system $\mathfrak{M}$ will be 1-complete. The sorting of all remaining possible pairs (f, τ) shall be reduced to this basic case. Thus, let us show that the system $\mathfrak{M}' = ({}^0\mathrm{A} \cup {}^1\mathrm{B} \cup {}^1\Gamma) \cup \{(x, 1)\}$ is 1-complete. For the proof we note that for any function $\varphi(x_1, ..., x_n)$ one can find an α-function $\psi(y_1, y_2, x_1, ..., x_n)$ of the form $y_1 \vee \varphi \cdot y_2$. Let us consider the pair $(\psi, 0) \in ({}^0\mathrm{A} \cup {}^1\mathrm{B} \cup {}^1\Gamma)$. Let us substitute the pair $(0, 1)$ into this pair instead of the variable y_1, the pair $(1, 1)$ instead of the variable y_2, and the pair $(x_i, 1)$ instead of the variable x_i. Each of the enumerated pairs belongs to $[\mathfrak{M}']$. As a result we obviously obtain the pair $(\varphi, 1)$ (i.e., the system $\mathfrak{M}'$ is 1-complete).

Let us consider the logically possible cases for the pair (f, τ).

1) (f, τ) is an α-function having unitary delay. Since the pair $(f(x, ..., x), 1)$ is equal tc the pair $(x, 1)$, then everything reduces to the basis case.

2) (f, τ) is an δ-function having unitary delay. Note that $({}^0\mathrm{A} \cup {}^1\mathrm{B} \cup {}^1\Gamma)$ contains the pair $(y_1 \vee y_2\bar{z}, 0)$. Let us substitute the pair $(0, 1)$ into this pair instead of the variable y_1, the pair $(1, 1)$ instead of the variable y_2, and the pair $(f(x, ..., x), 1)$, which is obviously equal to the pair $(\bar{x}, 1)$, instead of the variable z. It can easily be seen that as a result we obtain the pair $(x, 1)$. Thus, this case has been reduced to the basic one.

It remains for us to consider the case in which (f, τ) is either a β-function having zero delay, a γ-function having zero delay, or a δ-function having zero delay. The proof of 1-completeness in each of these cases derives from the following concepts. The set $\mathfrak{R}_2$ contains the pair $(x \vee \bar{y}, 1)$. Let us substitute the pair $(f(x, ..., x), 0)$ into this pair instead of the variable y, where the function f is either a β- or a δ-function; as a result we obtain the pair $(x, 1)$. If the function f is a γ-function, then we substitute the pair $(f(x, ..., x), 0)$ into the pair $(x \cdot \bar{y}, 1) \in ({}^0\mathrm{A} \cup {}^1\mathrm{B} \cup {}^1\Gamma)$ in place of the variable y; as a result we obtain the pair $(x, 1)$. Therefore, the consideration of these cases also can be reduced to the basic case, which is what it was required to prove.

Lemma 6. The set $\mathfrak{R}_3 = ({}^0C_3 \cup {}^1C_3)$ is 1-precomplete.

Proof. Let us show that the set $\mathfrak{M} = \mathfrak{R}_3 \cup \{(f, \tau)\}$, where $(f, \tau) \notin \mathfrak{R}({}^0C_3, {}^1C_3)$ is 1-complete. Since it is obvious that ${}^1C_3 \cup {}^1\bar{C}_3$, where ${}^1\bar{C}_3$ denotes the set of all pairs $(f, 1)$ such that $(\bar{f}, 1) \in {}^1C_3$ is a 1-complete set, it follows that for proof of 1-completeness of the system it is sufficient to show that either the pair $(\bar{x}, 1) \in [\mathfrak{M}]$ or the pair $(\bar{x}, 0) \in [\mathfrak{M}]$.

Let us consider the logically possible cases for the pair (f, τ).

1) (f, τ) is a β-function with a zero delay.

The pair $(f(x, ..., x), 0)$ is obviously equal to the pair $(1, 0)$.

Note that the set $\mathfrak{R}$ contains the pair $(x \cdot \bar{y}, 1)$. Let us replace the variables x by the pair $(1, 0)$ in this pair, and the variable y by the pair $(x, 0)$. It can easily be seen that as a result we obtain the pair $(\bar{x}, 1)$.

2) (f, τ) is a δ-function having a zero delay. It is obvious that the pair $(f(x, \ldots, x), 0)$ coincides with the pair $(\bar{x}, 0)$.

3) (f, τ) is a β-function with a unitary delay. The pair $(f(x, \ldots, x), 0)$ is equal to the pair (1, 1). Note that the set $\mathfrak{N}$ contains the pair $(x \cdot \bar{y}, 0)$. In place of the variable x we substitute the pair (1, 1) in this pair, and in place of the variable y the pair (x, 1). As a result we obtain the pair $(\bar{x}, 1)$.

4) (f, τ) is a δ-function having a unitary delay. It is obvious that the pair $(f(x, \ldots, x), 0)$ coincides with the pair $(\bar{x}, 1)$, which is what it was required to prove.

Lemma 7. The set $\mathfrak{N}_4({}^0A_1 \cup {}^1A_1)$ is 1-precomplete.

Proof. Let us show that the set $\mathfrak{M} = \mathfrak{N}_4 \cup \{(f, \tau)\}$, where $(f, \tau) \notin ({}^0A_1 \cup {}^1A_1)$ is 1-complete. First we shall show that for the case in which the pair (f, τ) is equal to the pair $(\bar{x}, 1)$, the system $\mathfrak{M}$ will be 1-complete. But the sorting of all remaining possible pairs (f, τ) shall be reduced to this basic case. Thus, let us show that the system $\mathfrak{M}' = \mathfrak{N}_4 \cup (\bar{x}, 1)$ is 1-complete. Since the pairs $(x \cdot y, 0)$, $(x \vee y, 0)$, $(x, 1)$ and $(x, 1)$ belong to $\mathfrak{M}'$, it follows that by applying synchronous-superposition operations to them we may obviously obtain the perfect disjunctive normal form with a unitary delay for any function $\varphi(x_1, \ldots, x_n)$. Therefore, the system $\mathfrak{M}'$ is 1-complete.

Let us consider the logically possible cases for the pair (f, τ).

1) (f, τ) is a nonmonotonic function having a zero delay. Since the function $f(x_1, \ldots, x_n)$ is nonmonotonic, it follows that, as is well known, one can find collections $\tilde{\alpha}(\alpha_1, \alpha_2, \ldots, \alpha_{i-1}, 0, \alpha_{i+1}, \ldots, \alpha_n)$ and $\tilde{\alpha}'(\alpha_1, \alpha_2, \ldots, \alpha_{i-1}, 1, \alpha_{i+1}, \ldots, \alpha_n)$ which are adjacent in the i-th component and are such that $f(\tilde{\alpha}) = 1$, $f(\tilde{\alpha}') = 0$ hold. In the pair (f, τ) let us replace the variable x_j, $j \neq i$ by the pair $(\alpha_j, 1) \in \mathfrak{N}_4$ and the variable x_i by the pair $(x, 1) \in \mathfrak{N}_4$. It can easily be seen that as a result of the substitution we obtain a pair equal to the pair $(\bar{x}, 1)$. This case has been reduced to the basic one.

2) (f, τ) is a nonmonotonic function having a unitary delay. This case obviously can be reduced to the basic one also, which is what it was required to prove.

Lemma 8. The set $\mathfrak{N}_5 = ({}^0A_4 \cup {}^1\mathrm{A} \cup {}^1\mathrm{B} \cup {}^1\Gamma)$ is 1-precomplete.

Proof. Let us show that $\mathfrak{M} = \mathfrak{N}_5 \cup \{(f, \tau)\}$, where $(f, \tau) \notin \mathfrak{N}_5$ is 1-complete. It is obvious that for this it is sufficient to show that either the pair $(\bar{x}, 0)$ or $(\bar{x}, 1)$ belongs to $[\mathfrak{M}]$.

Let us consider the logically possible cases for the pair (f, τ).

1) (f, τ) is a nonmonotonic function having a zero delay. From it, as has already been shown in Lemma 7, one can obtain the pair $(\bar{x}, 1)$ by means of the pairs (x, 1), (1, 1), $(0\ 1) \in \mathfrak{M}$.

2) (f, τ) is a function identically equal to unity and having a zero delay. Substituting the pair (1, 0) for the variable x and the pair (x, 0) for the variable y in the pair $(x \cdot \bar{y}, 1) \in \mathfrak{M}$, we obtain the pair $(\bar{x}, 1)$.

3) (f, τ) is a function which is identically equal to zero and has a zero delay. Substituting the pair (0, 0) for the variable x and the pair (x, 0) for the variable y in the pair $(x \vee \bar{y}, 1) \in \mathfrak{M}$, we obtain the pair $(\bar{x}, 1)$.

4) (f, τ) is a δ-function having a unitary delay. It is obvious that the pair $(f(x, \ldots, x), 1)$ coincides with the pair $(\bar{x}, 1)$, which is what was required to prove.

The validity of the following statement is similarly established.

Lemma 9. The set $\mathfrak{R}_6 = ({}^0A_4 \cup {}^1\Delta \cup {}^1B \cup {}^1\Gamma)$ is 1-precomplete.

Lemma 10. The set $\mathfrak{R}_7 = ({}^0A_1 \cup {}^1\overline{A}_1)$ is 1-precomplete.

Proof. Let us show that the set $\mathfrak{M} = \mathfrak{R}_7 \cup (f, \tau)$, where $(f, \tau) \notin \mathfrak{R}_7$, is 1-complete. For the case in which the pair (f, τ) is equal to the pair (x, 1), the system $\mathfrak{M}$ will obviously be 1-complete. The sorting of all remaining possible pairs (f, τ) shall be reduced to this basic case.

Let us consider the logically possible cases for the pair (f, τ).

1) (f, τ) is a nonmonotonic function having a zero delay. From it one may obviously obtain the pair (x, 1) using the pairs (0, 1), (1, 1), $(\overline{x}, 1) \in \mathfrak{M}$.

2) (f, τ) is not a negation of a monotonic function having a unitary delay. If the function $f(x_1, \ldots, x_n)$ is either a β-function, an α-function, or a γ-function, then it is obvious that $[\mathfrak{M}]$ contains the pair (x, 1). If the function $f(x_1, \ldots, x_n)$ is a δ-function, then one can obviously find two collections $\tilde{\alpha} = (\alpha_1, \ldots, \alpha_n)$ and $\tilde{\alpha}' = (\alpha_1', \ldots, \alpha_n')$ such that $\tilde{\alpha} < \tilde{\alpha}'$ and $f(\tilde{\alpha}) < f(\tilde{\alpha}')$. Let us partition the variables $(x_1, \ldots, x_n)$ by collections $\tilde{\alpha}$ and $\tilde{\alpha}'$ into no more than three groups. To the first group we assign the variables x_i for which $\alpha_i = \alpha_i' = 0$, while for the second group we assign the variables x_j for which $\alpha_j = \alpha_j' = 1$, and to the third group we assign the variables x_k for which $\alpha_i = 0$, $\alpha_i' = 1$. In this case in the pair (f, τ) let us replace the variables from the first group by the pair (0, 0), the variables from the second group by the pair (1, 0), and the variables from the third group by the pair (x, 0). It is obvious that as a result we obtain the pair (x, 1), which is what it was required to prove.

Lemma 11. The set $\mathfrak{R}_8 = ({}^0D_3 \cup {}^1D_3)$ is 1-precomplete.

Proof. Let us show that the set $\mathfrak{M} = \mathfrak{R}_8 \cup \{(f, \tau)\}$ is 1-complete, where $(f, \tau) \notin \mathfrak{R}_8$.

Let us consider the logically possible cases for the pair (f, τ).

1) (f, τ) is a nonself-dual function having a zero delay. Since the set D_3 is precomplete in P_2, it is obvious that the system $\mathfrak{M}$ is 1-complete.

2) (f, τ) is a nonself-dual function having a unitary delay. Since from any nonself-dual function one may obtain 1 or 0 by substituting x and $\overline{x}$ for certain variables, it is obvious that $[\mathfrak{M}]$ contains the pairs (1, 1), and (0, 1). Note that for any function $\varphi(x_1, \ldots, x_n)$ one can find a self-dual function $f(x, x_1, \ldots, x_n)$ in the form $(x \vee \varphi(x_1, \ldots, x_n)) \cdot (\overline{x} \vee \overline{\varphi}(\overline{x}_1, \ldots, \overline{x}_n))$. Let us consider the pair $(f, 0) \in \mathfrak{R}_8$. In this pair let us replace the variable x by the pair (0, 1), and the variable x_i by the pair $(x_i, 1)$. As a result it is obvious that we obtain the pair $(\varphi, 1)$, which is what it was required to prove.

Lemma 12. The set $\mathfrak{R}_9 = ({}^0D_3 \cup {}^1Y)$ is 1-precomplete.

Proof. Let us show that the set $\mathfrak{M} = \mathfrak{R}_9 \cup (f, \tau)$, where $(f, \tau) \notin \mathfrak{R}_9$ is 1-complete. For the case in which the pair (f, τ) coincides with the pair (x, 1), the system $\mathfrak{M}$, will obviously be 1-complete.

Let us enumerate the logically possible cases for the pair (f, τ).

1) (f, τ) is a nonself-dual function having a zero delay. In this case the system $\mathfrak{M}$, is obviously 1-complete.

2) (f, τ) is an odd function having a unitary delay. Since the function $f(x_1, \ldots, x_n)$ is not even, it follows, as is well known, that one can find at least two collections $\tilde{\alpha} = (\alpha_1, \ldots, \alpha_n)$ and $\tilde{\alpha}' = (\overline{\alpha}_1, \ldots, \overline{\alpha}_n)$ on which the function $f(x_1, \ldots, x_n)$ takes opposite values c and $\overline{c}$. Let us partition the variables $x_1, \ldots, x_n$ into two groups in accordance with the collection $\tilde{\alpha}$. To the

first group let us assign the variables x_i for $\alpha_i = 0$, while to the second group let us assign the variables x_j for which $\alpha_j = 1$. In the pair (f, τ) let us replace the variables from the first group by the pair $(x, 0)$, while the variables from the second group are replaced by the pair $(\bar{x}, 0)$. As a result we obtain the pair $(g(x), 1)$, where $g(x)$ is either x or $\bar{x}$; i.e., $[\mathfrak{M}]$ contains the pair $(x, 1)$. Thus, this case has been reduced to the one considered above, which is what it was required to prove.

Lemma 13. The set $\mathfrak{R}_{10} = ({}^0D_1 \cup {}^1A \cup {}^1\Delta)$ is 1-precomplete.

Proof. Let us show that the set $\mathfrak{M} = \mathfrak{R}_{10} \cup (f, \tau)$, where $(f, \tau) \notin \mathfrak{R}_{10}$ is 1-complete. First let us establish the fact that for the case in which $[\mathfrak{M}]$ contains the pairs $(0, 1)$ and $(1, 1)$, the system $\mathfrak{M}$ is 1-complete. Note that for any function $\varphi(x_1, \ldots, x_n)$ one can find a self-dual α-function $f(x, y, x_1, \ldots, x_n)$ in the form $(\varphi(x_1, \ldots, x_n) \vee x) y \vee (\overline{\varphi}(\bar{x}_1, \ldots, \bar{x}_n) \vee y) x$. In the pair $(f(x, y, x_1, \ldots, x_n), 0)$ let us replace the variable x by the pair $(0, 1)$, the variable y by the pair $(1, 1)$, and the variable x_i by the pair $(x_i, 1)$. Each of these pairs belongs to $[\mathfrak{M}]$. As a result we obviously obtain the pair $(\varphi, 1)$; i.e., the system $\mathfrak{M}$ is 1-complete in this basic case.

Let us consider the logically possible cases for the pair (f, τ), and let us show that each of them will reduce to the basic case.

1) (f, τ) is a nonself-dual function having a zero delay. From this pair one can obviously obtain either the pair $(0, 1)$ or the pair $(1, 1)$ by means of the pairs $(\bar{x}, 1)$ and $(x, 1)$. Since the pair $(x + y + z, 0) \in {}^0D_1$, then, by replacing the variable x by the pair $(1, 1)$, the variable y by the pair $(x, 1)$, and the variable z by the pair $(\bar{x}, 1)$ in it we obtain the pair $(0, 1)$; or, by replacing the variable x by the pair $(0, 1)$, the variable y by the pair $(x, 1)$, and the variable z by the pair $(\bar{x}, 1)$ in it we obtain the pair $(1, 1)$.

2) (f, τ) is a δ-function having a zero delay. In this case the system $\mathfrak{M}$ is 1-complete, since it contains the 1-precomplete class $\mathfrak{R}_8$ and does not coincide with it.

3) (f, τ) is a γ-function having a unitary delay.

4) (f, τ) is a β-function having a unitary delay.

Cases 3) and 4) can immediately be reduced to the first case. The lemma has been proved.

Lemma 14. The set $\mathfrak{R}_{11} = ({}^0F_5^2 \cup {}^1A \cup {}^1\Gamma \cup {}^1\Delta)$ is 1-precomplete.

Let us show that the set $\mathfrak{M} = \mathfrak{R}_{11} \cup (f, \tau)$, where $(f, \tau) \notin \mathfrak{R}_{11}$ is 1-complete. Let us first establish the fact that for the case in which the pair (f, τ) is equal to the pair $(1, 1)$ the system $\mathfrak{M}$ will be 1-complete. Sorting of all the remaining logically possible cases for the pair (f, τ) will be reduced to this basic case. Let us show that the system $\mathfrak{M}' = \mathfrak{R}_{11} \cup (1, 1)$ is 1-complete. Note that for any function $\varphi(x_1, \ldots, x_n)$ one can find a β-function of the form $\varphi(x_1, \ldots, x_n) \vee (z + y + 1)$ and a function $f(x, y, z, x_1, \ldots, x_n) \in F_5^2$ of the form $x(\varphi(x_1, \ldots, x_n) \vee (z + y + 1))$. In the pair $(f(x, y, z, x_1, \ldots, x_n), 0) \in \mathfrak{R}_{11}$ let us replace the variable x by the pair $(1, 1)$, the variable y by the pair $(\bar{x}, 1)$, the variable z by the pair $(x, 1)$, and the variable x_j by the pair $(x_j, 1)$. It is obvious that as a result we obtain the pair $(\varphi, 1)$; therefore, in this basic case the system is 1-complete.

Let us consider the logically possible cases for the pair (f, τ):

1) (f, τ) is a β-function having a unitary delay,

2) (f, τ) is a γ-function having a zero delay,

3) (f, τ) is a β-function having a zero delay,

4) (f, τ) is a δ-function having a zero delay.

Each of these cases can obviously be reduced immediately to the basic case.

5) (f, τ) is an α-function having a zero delay, which does not belong to ${}^0F_5^2$. In this case it is obvious that α-functions belong to $[\mathfrak{M}]$, which have a zero delay and among which there are α-functions which do not satisfy the condition $\langle A^2 \rangle$, a nonself-dual α-function, and, consequently [2], a α-function which does not satisfy the condition $\langle a^2 \rangle$; then [2] $[\mathfrak{M}]$ contains the pair $(x \vee y, 0)$, and from it we obtain the pair (1, 1) by substituting the pairs (x, 1) and $(\bar{x}, 1)$. Consequently, this case also can be reduced to the basic one, which is what it was required to prove.

Lemma 15. The set $\mathfrak{R}_{12} = ({}^0A \cup {}^1\Delta \cup {}^1\Gamma)$ is 1-precomplete.

Proof. Let us show that the set $\mathfrak{M} = \mathfrak{R}_{11} \cup (f, \tau)$, where $(f, \tau) \notin \mathfrak{R}$, is 1-complete. First let us establish the fact that in the case in which the pair (f, τ) is equal to the pair (1, 1), the system $\mathfrak{M}$ will be 1-complete. This case is called the basic one.

Thus, let us show that the system $\mathfrak{M}' = \mathfrak{R}_{12} \cup (1, 1)$ is 1-complete. Note that for any function $\varphi(x_1, \ldots, x_n)$ one can find an α-function $f(x, y, \bar{x}_1, \ldots, \bar{x}_n)$ of the form $x \vee y\varphi(\bar{x}_1, \ldots, \bar{x}_n)$. In the pair $(f, 0)$ let us replace the variable x by the pair (0, 1), the variable y by the pair (1, 1), and the variable x_i by the pair $(\bar{x}_i, 1)$. All these pairs belong to $[\mathfrak{M}']$. It can easily be seen that as a result we obtain the pair $(\varphi, 1)$.

Let us consider the logically possible cases for the pair (f, τ).

1) (f, τ) is a β-function having a unitary delay,

2) (f, τ) is a δ-function having a zero delay,

3) (f, τ) is a γ-function having a zero delay,

4) (f, τ) is a β-function having a zero delay.

It can easily be seen that $[\mathfrak{M}']$ contains the pair (1, 1) in each of these cases.

5) (f, τ) is an α-function having a unitary delay. In this case the set $[\mathfrak{M}]$ is 1-complete, since $\mathfrak{M}$ contains the 1-precomplete set $\mathfrak{R}_{10}$ and does not coincide with it; this is what it was required to prove.

Lemma 16. The set $\mathfrak{R}_{13} = ({}^0A_2 \cup {}^1\Delta \cup {}^1\Gamma \cup (1, 1))$ is 1-precomplete.

Proof. Let us show that the set $\mathfrak{M} = \mathfrak{R}_{13} \cup (f, \tau)$, where $(f, \tau) \notin \mathfrak{R}_{13}$, is 1-complete.

Let us consider the logically possible cases for the pair (f, τ).

1) (f, τ) is a β-function having a unitary delay, which differs from a constant. In this case it is obvious that the pairs $(x \cdot y, 0)$, $(x \vee y, 0)$, $(x, 1)$, $(x, 1)$ belong to $[\mathfrak{M}]$; consequently, for any function $\varphi(x_1, \ldots, x_n)$ one may obtain a perfect disjunctive normal form having a unitary delay (i.e., the system $\mathfrak{M}$ is 1-complete).

2) (f, τ) is an α-function having a unitary delay. This case is analogous to case 1).

3) (f, τ) is a nonmonotonic function having a zero delay. In this case it is obvious that the pair $(x, 1) \in [\mathfrak{M}]$, and consequently it is analogous to case 1).

4) (f, τ) is a function which is identically equal to zero and has a zero delay. In this case the system $\mathfrak{M}$ is 1-complete, since it contains the 1-precomplete set $\mathfrak{R}_7$ and does not coincide with it; this is what it was required to prove.

The validity of the following postulate is established similarly.

Lemma 17. The set $\mathfrak{R}_{14} = ({}^0A_2 \cup {}^1A \cup {}^1B \cup (0, 1))$ is 1-precomplete.

Since the classes $\mathfrak{R}_{15}-\mathfrak{R}_{19}$ are dual relative to the classes $\mathfrak{R}_3$, $\mathfrak{R}_{11}$, $\mathfrak{R}_{12}$, $\mathfrak{R}_{14}$, $\mathfrak{R}_{13}$, it follows that by virtue of the duality principle its 1-precompleteness derives from Lemmas 6, 14, 15, 17, 16.

§ 4. The Basic Theorem on 1-Completeness of Finite Systems of Functions Having Delays

We shall show that the following basic theorem on 1-completeness holds.

Theorem 1. In order for the finite system $\mathfrak{B} \subseteq {}^1\hat{P}_2$ to be 1-complete it is necessary and sufficient that it does not belong entirely to any 1-precomplete class from the system N, and among the functions having a zero delay there are contained the pairs

$$(g_1, 0) \notin {}^0L_1, \quad (g_2, 0) \notin {}^0P_6, \quad (g_3, 0) \notin {}^0S_6.$$

The necessity of the statements of the theorem derive from Lemmas 2 and 3 and the fact that the classes of the system N are not 1-complete. Let us show that the conditions of the theorem are sufficient for the system $\mathfrak{B}$ to be 1-complete. Thus, assume the system $\mathfrak{B}$ has the form $\{(f_i, 0), (\varphi_j, 1)\}$ and satisfies the conditions of the theorem.

All such systems $\mathfrak{B}$ will be partitioned into several groups as a function of the properties of the functions having zero delay which appear in $\mathfrak{B}$.

I. The system $\mathfrak{B}$ includes a δ-function having zero delay.

II. The system $\mathfrak{B}$ includes only δ-functions having zero delay.

III. The system $\mathfrak{B}$ contains only α- and β-functions having zero delay.

IV. The system $\mathfrak{B}$ includes only α- and γ-functions having a zero delay.

V. The system $\mathfrak{B}$ includes only α-, β-, and γ-functions having zero delay.

These five groups completely define all possible forms of the system $\mathfrak{B}$. The closure of the system of functions having zero delay from $\mathfrak{B}$ coincides with some specified closed class ${}^0P_2$. In each of these specific cases it will be shown that $\mathfrak{B}$ is a 1-complete system.

The proof of sufficiency will derive from a series of lemmas.

Let us consider each of the groups I-V separately.

I. The system $\mathfrak{B}$ includes a δ-function with zero delay. From a consideration of 1-precomplete classes of the system N it follows that in order for the system $\mathfrak{B}$ not to belong to any 1-precomplete class it must satisfy one of the following two conditions:

1) among the functions having zero delay in $\mathfrak{B}$ one can find a nonself-dual function, while among the functions having a unitary delay one can find a function which differs from a constant;

2) among the functions having a unitary delay in $\mathfrak{B}$ one can find an odd function and a nonself-dual function.

Lemma 18. If the system $\mathfrak{B}$ contains a δ-function having a zero delay and does not belong to any 1-precomplete class from the system N, then $\mathfrak{B}$ is a 1-complete system.

Proof. It is sufficient to consider separately the two cases in each of which the system $\mathfrak{B}$ satisfies one of the conditions 1), 2).

A. The system $\mathfrak{B}$ contains a nonself-dual function having a zero delay. It is obvious that in this case $[\{(f_i, 0)\}]$ coincides with the class $^0P_2$; further, having the pair $(\varphi, 1)$, where φ differs from a constant, and the pairs (0, 0), (1, 0), $(\bar{x}, 0)$, one may obtain the pair (x, 1) by means of synchronous-superposition operations, and this in turn involves the relation $[\mathfrak{B}] \supseteq {}^1P_2$.

B. The system $\mathfrak{B}$ contains only self-dual functions having a zero delay. There exist only three closed classes of logic-algebra functions of such a type: O_4, L_5, and D_3 [4]. It is obvious that $[\{(f_i, 0)\}]$ may coincide only with $^0D_3$. The pair (x, 1) can be obtained from an odd function having a unitary delay by means of synchronous-superposition operations using the pairs (x, 0), and $(\bar{x}, 0)$, which evidently belong to $^0D_3$. Therefore, it may be assumed that the set $^1D_3 \subset [\mathfrak{B}]$. Thus, the system $\mathfrak{B}$ is 1-complete, since it contains the 1-precomplete class $\mathfrak{R}_8$ and does not coincide with it.

II. The system $\mathfrak{B}$ includes only α-functions having a zero delay. From a consideration of the 1-precomplete classes of the system N it follows that in order for the system $\mathfrak{B}$ not to belong to any 1-precomplete class the system must satisfy one of six conditions:

1) among the functions having a zero delay in $\mathfrak{B}$ one can find a function which does not satisfy the condition $\langle A^2 \rangle$, a function which does not satisfy the condition $\langle a^2 \rangle$, and a function which does not belong to the class D_1; among the functions having a unitary delay one can find a δ-function and an α-function;

2) among the functions having a zero delay in $\mathfrak{B}$ one can find a function which does not satisfy the condition $\langle a^2 \rangle$; among the functions having a unitary delay one can find a δ-, α-, and β-function;

3) among the functions having a zero delay in $\mathfrak{B}$ one can find a function which does not satisfy the condition $\langle A^2 \rangle$; among the functions having a unitary delay one can find a δ-, α-, and γ-function;

4) among the functions having a zero delay in $\mathfrak{B}$ one can find a function which does not belong to the class A_4; among the functions having a unitary delay one can find an α-, β-, and γ-function;

5) among the functions having a zero delay in $\mathfrak{B}$ one can find a function which does not belong to the class A_4; among the functions having a unitary delay one find a δ-, β-, and γ-function;

6) among the functions having a unitary delay in $\mathfrak{B}$ one can find an α-, β-, γ-, and δ-function.

Lemma 19. *If the system $\mathfrak{B}$ contains only an α-function having a zero delay and does not belong to any 1-precomplete class $\mathfrak{R}_1$—$\mathfrak{R}_{19}$, then $\mathfrak{B}$ is a 1-complete system.*

Proof. Let us consider six cases separately, in each of which the system $\mathfrak{B}$ satisfies one of the conditions 1)-6).

A. The system $\mathfrak{B}$ contains a nonself-dual function having a zero delay, a function which does not satisfy the condition $\langle A^2 \rangle$, and a function which does not satisfy the condition $\langle a^2 \rangle$. In this case (see [2]) the pairs $(x \cdot y, 0)$ and $(x \vee y, 0)$ belong to $[\mathfrak{B}]$, and $[\{(f_i, 0)\}]$ coincides with the class $^0A_4$ or ^0A. It is sufficient to consider the case in which $[\{(f_i, 0)\}]$ coincides with the class $^0A_4$. In this case the pairs (x, 1), $(\bar{x}, 1)$, (0, 1), (1, 1) belong to $[\mathfrak{B}]$, and consequently for any function $\varphi(x_1, \ldots, x_n) \in P_2$ one may use synchronous-superposition operations (just as in Lemma 7) to obtain the perfect disjunctive normal form having a unitary delay. Thus, the system $\mathfrak{B}$ is 1-complete.

B. The system $\mathfrak{B}$ contains a function having a zero delay which does not satisfy the condition $\langle a^2\rangle$. In this case $[\{(f_i, 0)\}]$ may coincide with one of the classes: ${}^0D_1$, ${}^0F_5^\mu$, ${}^0F_6^\mu$ ($\mu = 2, 3, \ldots$).

Let us consider each of these three cases.

a) $[\{(f_i, 0)\}]$ coincides with ${}^0D_1$. Note that for any function $\varphi(x_1, \ldots, x_n)$ one can find a self-dual α-function $f(x, y, x_1, \ldots, x_n)$ and $(\varphi(x_1, x_2, \ldots, x_n) \vee x)\cdot y \vee (\overline{\varphi}(\overline{x}_1, \overline{x}_2, \ldots, \overline{x}_n) \vee y)\,x$. In the pair $(f(x, y, x_1, \ldots, x_n), 0)$ let us replace the variable x by the pair (0, 1), the variable y by the pair (1, 1), and the variable x_i by the pair $(x_i, 1)$. Each of these pairs belongs to $[\mathfrak{B}]$. As a result we obviously obtain the pair $(\varphi, 1)$, i.e., the system $\mathfrak{B}$ is 1-complete.

b) $[\{(f_i, 0)\}]$ coincides with one of the classes ${}^0F_5^\mu$ ($\mu = 2, 3, \ldots$), for example, with the narrower class ${}^0F_5^\infty$. Note that for any function $\varphi(x_1, \ldots, x_n)$ one can find a function $f(x, y, z, x_1, \ldots, x_n)$ from the class F_5^∞ of the form $x\cdot(\varphi(x_1, \ldots, x_n) \vee (z+y+1))$. In the pair $(f(x, y, z, x_1, \ldots, x_n), 0)$ let us replace the variable x by the pair (1, 1), the variable y by the pair $(\overline{x}, 1)$, the variable z by the pair (x, 1), and the variable x_i by the pair $(x_i, 1)$. Each of these pairs belongs to $[\mathfrak{B}]$. As a result we obviously obtain the pair $(\varphi, 1)$, i.e., the system $\mathfrak{B}$ is 1-complete.

c) $[\{(f_i, 0)\}]$ coincides with one of the classes ${}^0F_6^\mu$ ($\mu = 2, 3, \ldots$), for example, with the narrower class ${}^0F_6^\infty$. Note that for any monotonic function $\varphi(x_1, \ldots, x_n)$ one can find a function $f(x, x_1, \ldots, x_n)$ from the class F_6^∞ of the form $x \cdot \varphi(x_1, \ldots, x_n)$.

Substituting the pair (1, 1) for the variable x, and the pair $(\overline{x}_i, 1)$ or $(x_i, 1)$ for the variables x_i in the pair $(f(x, x_1, \ldots, x_n), 0)$, one may obtain the perfect disjunctive normal form for any function having a unitary delay; i.e., the system $\mathfrak{B}$ is 1-complete.

C. The system $\mathfrak{B}$ contains a nonmonotonic function having a zero delay. In this case $[\{(f_i, 0)\}]$ obviously coincides with one of the classes: ${}^0A$, ${}^0D_1$, ${}^0F_5^\mu$, ${}^0F_1^\mu$ ($\mu = 2, 3, \ldots$).

Let us consider each of these four cases.

a) $[\{(f_i, 0)\}]$ coincides with the class ${}^0A$. Since the pairs (x, 1), $(\overline{x}, 1)$, (0, 1), (1, 1) obviously belong to the class $[\mathfrak{B}]$, it follows that the system $\mathfrak{B}$ is 1-complete.

b) $[\{(f_i, 0)\}]$ coincides with ${}^0D_1$. A system $\mathfrak{B}$ of this form, as was shown in **B**, is 1-complete.

c) $[\{(f_i, 0)\}]$ coincides with one of the classes ${}^0F_5^\mu$ ($\mu = 2, 3, \ldots$). A system $\mathfrak{B}$ of this form (as was shown in **B**) is 1-complete.

d) $[\{(f_i, 0)\}]$ coincides with one of the classes ${}^0F_1^\mu$ ($\mu = 2, 3, \ldots$). Since the classes ${}^0F_1^\mu$ are respectively dual with respect to the classes ${}^0F_5^\mu$ ($\mu = 2, 3, \ldots$), it follows that by virtue of the quality principle the system $\mathfrak{B}$ is 1-complete.

D. In this case the system $\mathfrak{B}$ contains functions of the type α, β, γ, and δ having a unitary delay. It is evident that $[\{(f_i, 0)\}]$ [2] coincides with one of the classes: ${}^0A$, ${}^0A_4$, ${}^0D_1$, ${}^0F_1^\mu$, ${}^0F_6^\mu$, ${}^0F_5^\mu$, ${}^0F_2^\mu$ ($\mu = 2, 3, \ldots$), ${}^0D_2$.

All of these subcases, with the exception of the two latter ones, have already been considered above; therefore in each of them the system $\mathfrak{B}$ is 1-complete. Since the classes ${}^0F_2^\mu$ are respectively dual with the classes ${}^0F_6^\mu$ ($\mu = 2, 3, \ldots$), it follows that by virtue of the duality principle the system $\mathfrak{B}$ is 1-complete in this case also. If $[\{(f_i, 0)\}]$ coincides with ${}^0D_2$, then since for any monotonic functions $\varphi(x_1, x_2, \ldots, x_n)$ the pair $(x\cdot y \vee \overline{x}\cdot y\varphi(x_1, \ldots, x_n) \vee x\cdot\overline{y}\cdot\overline{\varphi}(\overline{x}_1, \ldots, \overline{x}_n), 0) \in {}^0D_2$, one may obviously obtain the perfect disjunctive normal form for any logic-algebra function having a unitary delay (i.e., the system $\mathfrak{B}$ is 1-complete in this case also).

The cases in which the system $\mathfrak{B}$ satisfies conditions 3) and 5) can be considered in dual fashion, which is what it was required to prove.

III. The system $\mathfrak{B}$ includes only α- and β-functions having a zero delay. From a consideration of 1-precomplete classes of the system N, it follows that in order for the system not to belong to any 1-precomplete class from the system N it must satisfy one of eight conditions:

1) among the functions having a zero delay in $\mathfrak{B}$ one can find a nonmonotonic function; among the functions having a unitary delay one can find a γ-function which is nonzero;

2) among the functions having a zero delay in $\mathfrak{B}$ one can find a nonmonotonic functions; among the functions having a unitary delay one can find a δ-function;

3) among the functions having a zero delay in $\mathfrak{B}$ one can find a nonmonotonic function; among the functions having a unitary delay one can find a β-function which differs from 1 and the constant 0;

4) among the functions having a zero delay in $\mathfrak{B}$ one can find a nonmonotonic function; among the functions having a unitary delay one can find an α-function and the constant 0;

5) among the functions having a zero delay in $\mathfrak{B}$ one can find only monotonic functions; among functions having a unitary delay one can find β- and γ-functions which are respectively different from 1 and 0;

6) among functions having a zero delay in $\mathfrak{B}$ one can find only monotonic functions; among the functions having a unitary delay one can find α- and γ-functions, the γ-function being nonzero;

7) among the functions having a zero delay in $\mathfrak{B}$ one can find only monotonic functions; among the functions having a unitary delay one can find β- and δ-functions, the β-functions being different from 1;

8) among the functions having a zero delay in $\mathfrak{B}$ one can find only monotonic functions; among the functions having a unitary delay one can find only α- and δ-functions.

Lemma 20. *If the system $\mathfrak{B}$ contains only α- and β-functions having a zero delay and does not belong to any 1-precomplete class from the system N, then $\mathfrak{B}$ is a 1-complete system.*

Proof. Let us consider separately the cases in each of which the system $\mathfrak{B}$ satisfies one of conditions 1-8).

The system $\mathfrak{B}$, satisfying one of the conditions 1-4) contains a nonmonotonic function having a zero delay, and therefore $[\{(f_i, 0)\}]$ coincides with the class ${}^0C_2$ or with one of the classes ${}^0F_4^{\mu}$ ($\mu = 2, 3, \ldots$) (for example, with a narrower one of them – ${}^0F_4^{\infty}$). It can easily be seen that the pairs $(x, 1)$, $(\bar{x}, 1)$, $(0, 1)$, and $(1, 1)$ belong to the set $[\mathfrak{B}]$. If $[\{(f_i, 0)\}]$ coincides with ${}^0C_2$, then the system $\mathfrak{B}$, is obviously 1-complete.

Assume $[\{(f_i, 0)\}]$ coincides with ${}^0F_4^{\infty}$. Note that for any function $\varphi(x_1, \ldots, x_n) \in P_2$ one can find a function $f(x, x_1, \ldots, x_n)$ of the form $x \vee \varphi(x_1, \ldots, x_n)$, which satisfies the condition $\langle a^{\infty} \rangle$. Substituting in the pair $(f(x, x_1, \ldots, x_n), 0)$ the pair $(0, 1) \in [\mathfrak{B}]$, for the variable x and the pair $(x_i, 1) \in [\mathfrak{B}]$, for the variable x_i, we obtain the pair $(\varphi, 1)$ (i.e., the system $\mathfrak{B}$ is 1-complete).

The system $\mathfrak{B}$, satisfying one of the conditions 5-8) contains only monotonic functions having zero delay, and therefore $[\{(f_i, 0)\}]$ coincides with the class ${}^0A_2$ or with one of the classes ${}^0F_3^{\mu}$ ($\mu = 2, 3, \ldots$) (for example, with a narrower one of them, ${}^0F_3^{\infty}$). It can easily be seen that the pairs $(x, 1)$, $(\bar{x}, 1)$, $(0, 1)$, and $(1, 1)$ belong to the class $[\mathfrak{B}]$.

If the system $[\{(f_i, 0)\}]$ coincides with ${}^0A_2$, then for any function $\varphi(x_1, \ldots, x_n) \in P_2$ one may use synchronous-superposition operations (just as in Lemma 7) to obtain its perfect disjunctive normal form having a unitary delay.

Assume $[\{(f_i, 0)\}]$ coincides with ${}^0F_3^\infty$. Note that for any monotonic function $\varphi(x_1, \ldots, x_n)$ which differs from a constant, one can find a function $f(x, x_1, \ldots, x_n) \in F_3^\infty$ of the form $x \vee \varphi(x_1, \ldots, x_n)$. Substituting the pair (0, 1) for the variable x, and the pair $(x_i, 1)$ or $(\bar{x}_i, 1)$ for the variable x_i in the pair $f(x, x_1, \ldots, x_n)$, one may obtain the perfect disjunctive normal form or a function from P_2 having a unitary delay. Thus, the system $\mathfrak{B}$ is 1-complete.

By virtue of the duality principle the following lemma is valid:

Lemma 21. *If the system $\mathfrak{B}$ contains only α- and γ-functions having zero delay and does not belong to any 1-precomplete class from the system N, then $\mathfrak{B}$ is a 1-complete system.*

Thereby the case IV has been considered fully.

V. The system $\mathfrak{B}$ includes only α-, β-, and γ-functions having zero delay. From the consideration of 1-precomplete classes of the system N it follows that in order for the system $\mathfrak{B}$ not to belong to any 1-precomplete class it must satisfy one of the following two conditions:

1) among the functions having a zero delay in $\mathfrak{B}$ one can find a nonmonotonic function; among the functions having a unitary delay one can find a function which differs from a constant;

2) among the functions having a unitary delay in $\mathfrak{B}$ one can find a nonmonotonic function and a function which is not a negation of a monotonic function.

Lemma 22. *If the system $\mathfrak{B}$ contains only α-, β-, and γ-functions having zero delay and does not belong to any 1-precomplete class of the system N, then $\mathfrak{B}$ is a 1-complete system.*

Proof. If the system $\mathfrak{B}$ does not belong to any 1-precomplete class, then it is obvious that $[\{(f_i, 0)\}]$ coincides either with ${}^0P_2$ or with ${}^0A_1$. Obviously, the pairs (0, 1), (1, 1), (x, 1), $(\bar{x}, 1)$ belong to $[\mathfrak{B}]$, and now it is not difficult to show that the system $[\mathfrak{B}]$ will be 1-complete in each of these cases, which is what it was required to prove.

This completes the proof of the sufficiency of the conditions of Theorem 1.

From the proof of the basic Theorem 1 the following statement derives.

Theorem 2. *From any finite 1-complete system $\mathfrak{B}$ one may isolate a 1-complete subsystem $\mathfrak{B}'$ containing no more than five functions having delays.*

As will be shown below, the requirement of finiteness of the complete system in Theorem 2 is essential.

§ 5. Certain Properties of Systems of Pairs from ${}^1\hat{P}_2$

In this section the following proposition, in particular, will be established.

Theorem 3. *In the set of pairs from ${}^1\hat{P}_2$ there do not exist 1-precomplete classes which differ from the classes $\mathfrak{R}_1$—$\mathfrak{R}_{19}$.*

For proof of this theorem we shall need certain auxiliary statements.

Assume $\mathfrak{R} = \{(f_i, 0), (\varphi_j, 1)\}$ is a closed class from ${}^1\hat{P}_2$, which is not contained in any of the classes $\mathfrak{R}_1, \ldots, \mathfrak{R}_{19}$. Then the following statements hold.

Lemma 23. If in $\mathfrak{R}$ the system $\{(f_i, 0)\}$ coincides with some one of the classes ${}^0P_1$, ${}^3P_3$, ${}^0P_5$, ${}^0P_6$ or with some one of the classes ${}^0S_1$, ${}^0S_3$, ${}^0S_5$, ${}^0S_6$, then $\mathfrak{R}$ is not a 1-precomplete class.

Proof. By virtue of the duality principle it is sufficient to consider, for example, the case in which $\{(f_i, 0)\}$ coincides with some one of the classes ${}^0P_1$, ${}^0P_3$, ${}^0P_5$, ${}^0P_6$. In order to be specific let us assume that the system $\mathfrak{R}$ is a 1-precomplete class, and (again to be specific) that the set $\{(f_i, 0)\}$ coincides with the class ${}^0P_6$ (the remaining cases may be considered to be analogous). The function $g = x_1^{\sigma_1} \vee x_2^{\sigma_2} \vee \ldots \vee x_n^{\sigma_n}$ shall be called an (m, n)-function, where m denotes the number of ones in the collection $(\sigma_1, \ldots, \sigma_n)$. From the fact that $\mathfrak{R}$ is not a 1-complete system and $\mathfrak{R} \supset {}^0P_6$, it obviously follows that there exists a constant k such that for any (m, n)-function for which $m < n$, the condition $m < k$ holds, and the function $g_1 = x_1 \vee x_2 \vee \ldots \vee x_k \vee x_{k+1} \vee \bar{x}_{k+3}$ does not belong to $\mathfrak{R}$. Let us consider the system $\mathfrak{R} \cup \{g_1\}$. By definition it must be 1-complete. We shall show, however, that the function $g_2 = x_1 \vee \vee x_2 \vee \ldots \vee x_{k+1} \vee x_{k+2} \vee \bar{x}_{k+3}$ cannot be contained in $[\mathfrak{R} \cup \{g_1\}]$, from which the validity of the statement of the lemma will follow.

Let us assume that $g_2 \in [\mathfrak{R} \cup \{g_1\}]$. Then $g_2 = f_1 \cdot f_2 \cdot \ldots \cdot f_s$ holds, where a certain function $f_i \not\equiv c$ is derived from the function g_1 by substituting functions from P_6 into it. After transformations, f_i may obviously be represented in the form $\mathfrak{A}_1 \vee \mathfrak{A}_2 \vee \ldots \vee \mathfrak{A}_q \vee \bar{x}_{j_1} \vee \bar{x}_{j_2} \vee \ldots \vee \bar{x}_{j_p}$, where $q < k + 2$, and each $\mathfrak{A}_u$ is not a cofactor of any other $\mathfrak{A}_v$ and is independent of the variables $x_{j_1}, \ldots, x_{j_p}$.

Assume $q = 0$. Then $f_i(1, \ldots, 1) = 0$, which would involve the vanishing of the function g_2 on the set $(1, 1, \ldots, 1)$; however, $g_2(1, \ldots, 1) = 1$.

Assume $q > 0$. Since $q < k + 2$, it follows that one can obviously find two collections from which f_i will vanish, and thereby the function g_2 must vanish on these same collections; however, this is impossible, since g_2 vanishes only on one collection. Thus, $g_2 \notin [\mathfrak{R} \cup \{g_1\}]$, which is what it was required to prove.

Lemma 24. If in $\mathfrak{R}$ the system $\{(f_i, 0)\}$ coincides with some one of the closed classes of linear functions and the degrees of functions having a unitary delay from the system $\mathfrak{R}$ are bounded in aggregate, then the system $\mathfrak{R}$ is not a 1-precomplete class.

Proof. Let us suppose that the system $\mathfrak{R}$ is a 1-precomplete class. Let us use r to denote the highest degree of functions from the system $\{(\varphi_j, 1)\}$. Assume that the pair $(\psi(x_1, \ldots, x_n), 1)$ does not belong to the set $\mathfrak{R}$, and that the degree of ψ is equal to $r + 1$. Then, obviously, the set $\mathfrak{R}$ does not belong to the sequence of functions having a unitary delay and degrees $r + 2$, $r + 3$, ..., from which the function $\psi(x_1, \ldots, x_n)$ can be derived by identification of variables. From the supposition of 1-precompleteness of the system $\mathfrak{R}$ it follows that the system $\{\mathfrak{R} \cup (\psi(x_1, \ldots, x_n), 1)\}$ must be 1-complete; however, this is not so because from pairs of this system we obviously may not obtain functions having a unitary delay which are of degree $r + 2$ and higher using synchronous-superposition operations; this is what we were required to prove.

Below we shall consider only those systems $\mathfrak{R}$, in which the degrees of functions having a unitary delay are not bounded in aggregate.

Lemma 25. If in $\mathfrak{R}$ the system $\{(f_i, 0)\}$ coincides with some one of the classes ${}^0O_1$, ${}^0O_2$, ..., ${}^0O_0$, then $\mathfrak{R}$ is not a 1-precomplete class.

The proof of this lemma is completely analogous to the proof of Lemma 23.

Lemma 26. If in $\mathfrak{R}$ the system $\{(f_i, 0)\}$ coincides with the class L_3, then $\mathfrak{R}$ is not a 1-precomplete class.

Proof. More precisely, we shall establish that in the case considered the set $\mathfrak{R}$ is 1-complete. For proof of this fact it is sufficient to show that $\mathfrak{R}$ contains a pair (1, 1) and all pairs of the form $(x_1 \cdot x_2 \cdot \ldots \cdot x_r, 1)$, $r = 1, 2, \ldots$.

Actually, having these pairs and the pair $(x + y, 0)$, which by stipulation belongs to the set $\mathfrak{R}$, one may use synchronous-superposition operations to obtain the pair $(P_f, 1)$ for any pair $(f, 1)$, where P_f is the Zhegalkin polynomial of the function f.

Let us begin by showing that $(1, 1) \in \mathfrak{R}$. From the fact that $\mathfrak{R}$ does not belong to the set $\mathfrak{R}_1, \ldots, \mathfrak{R}_{19}$ it follows that $\mathfrak{R}$ must include the pair $(\varphi, 1)$, where φ is either a β- or a δ-function.

Let us identify the variables in the pair $(\varphi, 1)$; we obtain either (1, 1) (if φ is a β-function) or $(\bar{x}, 1)$ (if φ is a δ-function). Substituting the pair (0, 0) from $\mathfrak{R}$, into $(\bar{x}, 1)$, we likewise obtain the pair (1, 1).

Let us now show that any pair $(x_1 \cdot x_2 \cdot \ldots \cdot x_r, 1) \in \mathfrak{R}$. Assume the pair $(\psi(x_1, \ldots, x_n), 1) \in \mathfrak{R}$ and the degree of ψ is equal to $k \geq 1$. Without loss of generality it may be assumed that the Zhegalkin polynomial of the function ψ has the form

$$x_1 \cdot x_2 \cdot \ldots \cdot x_k + \sum_{i=1}^{s} \mathfrak{A}_i + \sum_{j=1}^{n} c_j x_j + c, \tag{1}$$

where $\mathfrak{A}_i$ are conjunctions of degree no higher than k, while $c, c_j \in \{0, 1\}$. Let us show how, starting from the pair $(\psi, 1)$, one can use synchronous-superposition operations to obtain the pair $(x_1 \cdot x_2 \cdot \ldots \cdot x_k, 1)$. In the pair $(\psi, 1)$ let us replace the variables $x_{k+1}, \ldots, x_n$ by the pair (0, 0). As a result we obtain the pair $(\psi', 1)$. The Zhegalkin polynomial of the function ψ' is obviously the sum of all those terms of the polynomial (1) which do not contain variables differing from $x_1, \ldots, x_k$, as cofactors, plus a free term. In the pair $(\psi', 1)$ we replace the variable x_1 by the pair (0, 0). As a result we obtain the pair $(\psi^{(1)}, 1)$. In the pair $(y_1 + y_2, 0) \in \mathfrak{R}$ let us replace the variables y_1 by the pair $(\psi', 1)$, and the variable y_2 by the pair $(\psi^{(1)}, 1)$. As a result we obtain the pair $(\psi^{(2)}, 1)$. The Zhegalkin polynomial of the function $\psi^{(2)}$ is obviously the sum of all those terms of the polynomial (1) which contain x_1 as a cofactor. In the pair $(\psi^{(2)}, 1)$ let us now replace the variable x_2 by the pair (0, 0). As a result we obtain the pair $(\psi^{(3)}, 1)$. In the pair $((y_1 + y_2), 0)$ let us replace the variable y_1 by the pair $(\psi^{(2)}, 1)$, and the variable y_2 by the pair $(\psi^{(3)}, 1)$. As a result we obtain the pair $(\psi^{(4)}, 1)$. The Zhegalkin polynomial of the function $\psi^{(4)}$ is obviously the sum of all those terms of the polynomial (1) which contain x_1 and x_2 as cofactors. Having carried out analogous constructions for each x_i, $i = 1, \ldots, k$, we obviously obtain the pair $(x_1 \cdot \ldots \cdot x_k, 1)$.

Let us now note that since the set of degrees of functions having delays which are included in $\mathfrak{R}$, is not bounded in aggregate, one can find a sequence

$$(f_1, 1), (f_2, 1), \ldots, (f_n, 1), \ldots$$

$(f_i, 1) \in \mathfrak{R}$, such that the condition $i = 1, 2, \ldots$, applies to the degrees $s_1, s_2, \ldots$ of the functions $f_1, f_2, \ldots$. From this, by virtue of the constructions described above, we derive the fact that all pairs of the form $(x_1 \cdot \ldots \cdot x_{s_i}, 1)$. belong to the set $\mathfrak{R}$. Since, obviously, any pair $(x_1, \cdot \ldots x_r, 1)$ may be obtained from a pair $(x_1 \cdot \ldots \cdot x_{s_j}, 1)$, where $j > r$, by identifying certain variables, it follows that all pairs $(x_1 \cdot \ldots \cdot x_r, 1)$, $r = 1, 2, \ldots$ belong to $\mathfrak{R}$, which is what it was required to prove.

By virtue of the duality principle the following statement holds:

Lemma 27. If in $\mathfrak{R}$ the system $\{(f_i, 0)\}$ coincides with the class L_2, then $\mathfrak{R}$ is not a 1-precomplete class.

Lemma 28. If in $\mathfrak{R}$ the system $\{(f_i, 0)\}$ coincides with the class L_1, then $\mathfrak{R}$ is not a 1-precomplete class.

The proof of this statement is completely analogous to the proof of Lemma 26.

Lemma 29. If in $\mathfrak{R}$ the system $\{(f_i, 0)\}$ coincides with the class L_4, then $\mathfrak{R}$ is not a 1-precomplete class.

Proof. Let us assume that $\mathfrak{R}$ is a 1-precomplete class. If $\mathfrak{R}$ is not contained in any of the functions $\mathfrak{R}_1, \mathfrak{R}_2, \ldots, \mathfrak{R}_{19}$, it follows that $\mathfrak{R}$ contains either α-, β-, and γ-functions or β-, γ-, and δ-functions, or α- and δ-functions having a unitary delay. In the first and second cases one can evidently obtain the pairs $(\bar{x}, 1)$ and $(x, 1)$, respectively, by identifying the variables and substituting the pairs obtained into the pair $((x+y+z), 0) \in \mathfrak{R}$. For the case in which the system $\mathfrak{R}$ contains δ- and α-functions having a unitary delay, it follows that by identifying the variables and substituting the previously obtained pairs $(\bar{x}, 1)$ and $(x, 1)$ in special fashion [2] into the pair $(f, 0) \in L_4$, where f is a nonself-dual function, we obtain the pair $(c, 1)$, where $c = \{0, 1\}$. Replacing the variables by the pairs $(c, 1)$, $(x, 1)$, $(\bar{x}, 1)$, respectively, in the pair $((x + y + z), 0)$, we obtain the pair $(\bar{c}, 1)$ as a result. Consequently, the system $\mathfrak{R}$ contains the pairs $(0, 1)$, $(1, 1)$, $(x, 1)$, $(\bar{x}, 1)$.

From the supposition that $\mathfrak{R}$ is a 1-precomplete class it derives that $\mathfrak{R}$ is not 1-complete. By virtue of this there must exist a natural number n_0, which is such that for any pair from $\mathfrak{R}$ of the form $(x_1, \ldots, x_r, 1)$ the condition $r \le n_0$ holds. Otherwise, starting with these pairs and also with the pairs $(\bar{x}, 1)$, $(x + y + z, 0)$, $(1, 1)$, and $(0, 1)$, we could construct the pair $(P_f, 1)$, where P_f is the Zhegalkin polynomial for f.

Assume n_0 is such that in $\mathfrak{R}$ there exists a pair $(x_1, \ldots, x_{n_0}, 1)$ but not a pair $(x_1, \ldots, x_{n_0+1}, 1)$. It is obvious that $n_0 \ge 1$. By virtue of what has been said above, it may be assumed that $\mathfrak{R}$ contains all functions of degree no higher than n_0 having a unitary delay. From the assumed 1-precompleteness of $\mathfrak{R}$ it follows that the set $\mathfrak{M} = \mathfrak{R} \cup \{(x_1 \ldots x_{n_0+1}, 1)\}$ must be 1-complete. We shall show that if $\mathfrak{M}$ is complete, then $\mathfrak{R}$ must necessarily contain a pair $(x_1, \ldots, x_{n_0+1}, 1)$. This will contradict the choice of the number n_0, and thereby the lemma will be proved.

Thus, assume $[\mathfrak{M}] = {}^1\hat{P}_2$. Then, in particular, $[\mathfrak{M}]$ must contain a pair $(x_1 \cdot \ldots \cdot x_{n_0+2}, 1)$. It is evident that one can find functions f of degree $n_0 + 2$ and g of degree $n_0 + 1$ which are such that $x_1 \cdot x_2 \cdot \ldots \cdot x_{n+2} = f + g$, it being true that the pair $(f, 1) \in \mathfrak{R}$. Let us consider the function f in greater detail. Assume f' is a function stipulated by a Zhegalkin polynomial which is obtained from the Zhegalkin polynomial of the function f by eliminating all terms of degree $n_0 + 2$ and $n_0 + 1$. It is obvious that $(f', 1) \in \mathfrak{R}$. Further, it is obvious that the pair $(f + f', 1) \in \mathfrak{R}$, it being possible to assume without loss of generality that the Zhegalkin polynomial of the function $f + f'$ has the form

$$x_1 \cdot x_2 \ldots x_{n_0+2} + \sum_{i=1}^{s} \mathfrak{A}_i,$$

where each term of $\mathfrak{A}_i$, that is the product of certain variables from the set $\{x_1, x_2, \ldots, x_{n_0+2}\}$ has the degree $n_0 + 1$ and the number of terms $s \ge 1$. Let us consider two cases.

1. $s > 1$. Assume that the sum $\sum_{i=1}^{s} \mathfrak{A}_i$ contains the terms $x_1 x_2 \cdot \ldots \cdot x_{u-1} x_{u+1} \cdot \ldots \cdot x_{n_0+2}$ and $x_1 x_2 \cdot \ldots \cdot x_{v-1} x_{v+1} \cdot \ldots \cdot x_{n_0+2}$, $u < v$. Let us identify the variables x_u and x_v in the function f'.

Assume f'' is the function which is obtained under these conditions. It is obvious that the Zhegalkin polynomial of this function has the form

$$x_1x_2\cdot\ldots\cdot x_{u-1}x_ux_{u+1}\cdot\ldots\cdot x_{v-1}x_{v+1}\cdot\ldots\cdot x_{n_0+2}+\sum_{j=1}^{r} b_j,$$

where the degrees of all terms $b_1, \ldots, b_r$ are less than $n_0 + 1$. Since $\{(f'', 1), (\sum_{j=1}^{r} b_j, 1)\} \subseteq \mathfrak{R}$ and since $f''+\sum_{j=1}^{r} b_j = x_1x_2 \ldots x_{u-1}x_ux_{u+1}\cdot\ldots\cdot x_{v-1}x_{v+1}+\ldots+x_{n_0+2}$, it follows that the pair $(x_1\cdot x_2\cdot\ldots\cdot x_{u-1}x_ux_{u+1}\cdot\ldots\cdot x_{v-1}x_{v+1}\cdot\ldots\cdot x_{n_0+2}, 1)\in\mathfrak{R}$.

2. $s = 1$. Assume $\mathfrak{A}_1$ does not contain the variable x_j. Let us identify any pair of variables x_u and x_v, $u \neq j$, $v \neq j$, $u < v$ in the function f'. As a result we obtain a certain function f''. Obviously, the Zhegalkin polynomial of this function is the sum of two terms, one of which has the degree $n_0 + 1$, while the second has the degree n_0. In exactly the same way as in case 1 we construct a pair $(x_1\cdot\ldots\cdot x_{n_0+1}, 1)$, which must belong to $\mathfrak{R}$.

Thus, $\mathfrak{R}$ contains the pair $(x_1\cdot\ldots\cdot x_{n_0+1}, 1)$, and this contradicts the choice of n_0; this is what it was required to be proved.

Lemma 30. If in $\mathfrak{R}$ the system $\{(f_i, 0)\}$ coincides with the class L_5, then $\mathfrak{R}$ is not a 1-precomplete class.

The proof of this statement is analogous to the proof of Lemma 29.

Proof of Theorem 3. Assume $\mathfrak{R}$ is a 1-precomplete class. Let us show that $\mathfrak{R}$ coincides with one of the classes of the system N. Let us assume that the latter is not fulfilled; more precisely, $\mathfrak{R}$ does not coincide with any of the classes indicated and is not contained in any of them. Let us consider the possible cases for a system $\{f_i, 0\}$, $\mathfrak{R}$. If the system $\{f_i, 0\}$ were to contain the pairs $(f, 0)$, $(\varphi, 0)$, and $(\psi, 0)$, such that $f\notin L_1$, $\varphi\notin P_6$, and $\psi\in S_6$, then from the fact that the set $\mathfrak{R}$ does not belong to the class of the system N it would follow by virtue of Theorem 1 that $\mathfrak{R}$ is a 1-complete system. Thus, (see [2]), the system $\{(f_i, 0)\}$ may only coincide with one of the following classes: S_i, P_i, L_j, O_k, $i = 1, 3, 5, 6$, $j = 1, 2, 3, 4, 5$, $k = 1, 2, \ldots, 9$. However, by virtue of Lemmas 23-29, $\mathfrak{R}$ is not 1-precomplete in any of these cases. The theorem has been proved.

Theorem 4. There is a continuum of closed classes in ${}^1\hat{P}_2$.

Proof. Let us construct an infinite sequence of functions having unitary delays, in which any pair does not belong to the closure of the set of the remaining pairs. Thereby different subsequences of our sequence will correspond to different closed classes which are closures of these subsequences. In order to complete the proof it remains for us to note that one may choose a continuum of different subsequences.

Following [2], let us consider the function

$$h_\mu(x_1, \ldots, x_{\mu+1}) = x_2\cdot x_3\cdot\ldots\cdot x_{\mu+1} \vee x_1x_3\cdot\ldots\cdot x_{\mu+1} \vee \ldots$$
$$\ldots \vee x_1\cdot\ldots\cdot x_{i-1}x_{i+1}\cdot\ldots\cdot x_{\mu+1} \vee \ldots \vee x_1\cdot\ldots\cdot x_\mu, \quad \mu \geqslant 2.$$

We choose

$$(h_2, 1)(h_3, 1), \ldots, (h_\mu, 1), \ldots$$

as the sequence about which we spoke above. Only the operation of identification of variables is applicable to the terms of this sequence.

Let us show that none of the pairs of this sequence can be expressed in terms of the remaining ones by means of the operation of identification of variables. The latter fact derives from the fact that the function h_μ has the property $\langle A^\mu \rangle$ but does not have the property $\langle A^{\mu+1} \rangle$ (see [2]), while at the same time in the identification of any pair of variables of the function h_μ a function is obtained which has the property $\langle A^\infty \rangle$; this is what it was required to prove.

As was established in §4, from each 1-complete finite system one may isolate a subsystem consisting of no more than five functions having delays, as well as a 1-complete function. However, in the general case this statement turns out to be untrue. Moreover, the following statement holds.

Theorem 5. In ${}^1P_2$ there exists a countable 1-complete system no finite subsystem of which is 1-complete.

As such a system one may, for example, choose the set ${}^1P_2$.

The set ${}^1P_2$ likewise has one more interesting property. Assume $\mathfrak{M} \subseteq {}^1P_2$ and $[\mathfrak{M}] = {}^1P_2$; assume further that $\mathfrak{M}' \subset \mathfrak{M}$ and the difference $\mathfrak{M} \setminus \mathfrak{M}'$ contains only a finite number of functions having delays. Then $[\mathfrak{M}'] = {}^1P_2$. Actually, assume $\mathfrak{M} \setminus \mathfrak{M}'$ consists of the pairs

$$(f_1(x_1, \ldots, x_{r_1}), 1), (f_2(x_1, \ldots, x_{r_2}), 1), \ldots, (f_s(x_1, \ldots, x_{r_s}), 1),$$

where $r_i \leq r_{i+1}$ for $i = 1, s-1$. Let us consider the pairs $(\varphi_1, 1), \ldots, (\varphi_s, 1)$, where $\varphi_i = f_i(x_1, \ldots, x_{r_i}) + \sum_{j=1}^{2r_s} x_{r_{i+j}}$. Obviously, $(\varphi_i, 1) \in [\mathfrak{M}']$. Moreover,

$$(f_i(x_1, \ldots, x_{r_i}), 1) = (\varphi_i(x_1, \ldots, x_{r_i}, x_{r_i}, \ldots, x_{r_i}), 1).$$

From this it follows that $[\mathfrak{M}'] = {}^1P_2$.

A 1-complete infinite system of functions having delays from ${}^1P_2$, for which no intrinsic subsystem is 1-complete is called a basis. Constructions analogous to those carried out in the proofs of Lemmas 23-30 allow us to establish the validity of the following statement which is a generalization of the property just considered for infinite 1-complete systems.

Theorem 6. No bases exist in ${}^1\hat{P}_2$.

The validity of the following statement likewise derives from a description of the system N of 1-precomplete classes from Theorem 1.

Theorem 7. In ${}^1P_2$ there exist closed classes which are not 1-complete and are not contained in any 1-complete class.

As such a closed class one may take, for example, the class ${}^0L_1 \cup \mathfrak{N}^r$, where $\mathfrak{N}^r$ is the set of all pairs of the form $(f, 1)$, while the function f has a degree no higher than $r \geq 1$.

The statements proved above allow another formulation of the basic Theorem 1 to be presented.

Theorem 8. In order for a finite system $\mathfrak{B} \subseteq {}^1\hat{P}_2$ to be 1-complete, it is necessary and sufficient that it does not belong to the classes of the system N or to three rigorously increasing chains of closed classes

$$[{}^0P_6 \cup \mathfrak{M}^1] \subset [{}^0P_6 \cup \mathfrak{M}^2] \subset \ldots \subset [{}^0P_6 \cup \mathfrak{M}^r] \subset \ldots,$$
$$[{}^0S_6 \cup \mathfrak{M}^1] \subset [{}^0S_6 \cup \mathfrak{M}^2] \subset \ldots \subset [{}^0S_6 \cup \mathfrak{M}^r] \subset \ldots,$$
$$[{}^0L_1 \cup \mathfrak{N}^1] \subset [{}^0L_1 \cup \mathfrak{N}^2] \subset \ldots \subset [{}^0L_1 \cup \mathfrak{N}^r] \subset \ldots,$$

where $\mathfrak{M}^r$ is the set of all pairs of the form $(f, 1)$ such that the rank of f does not exceed r, r = 1, 2, ..., any element of any chain not being contained in any 1-precomplete class and the union of all elements of any of the chains being 1-complete.

The requirement of finiteness in this theorem is essential, since the system ${}^1\mathrm{A} \cup {}^1\mathrm{B} \cup {}^1\Gamma \cup \{(\bar{x}, 1)\}$ obviously satisfies the conditions of nonappearance in the sets indicated in the theorem, but, as is easily seen, is not 1-complete.

Literature Cited

1. V. B. Kudryavtsev, "Completeness theorem for a certain class of automata without feedbacks," in: Problemy Kibernetiki, No. 8, Fizmatgiz, Moscow (1962).
2. S. V. Yablonskii, G. P. Gavrilov, and V. B. Kudryavtsev, Logic Algebra Functions and Post Classes, Nauka, Moscow (1966).
3. S. V. Yablonskii, Functional Constructions in k-Valued Logic, Transactions of the V. A. Steklov Mathematics Institute of the Academy of Sciences, Vol. 51 (1958).
4. E. L. Post, Two-Valued Interactive Systems of Mathematical Logic, Princeton (1941).

ASYMPTOTICALLY STABLE DISTRIBUTIONS OF CHARGE ON VERTICES OF AN n-DIMENSIONAL CUBE†

V. K. Leont'ev

Novosibirsk

We consider the set E^n of binary sequences of length n. Let $M = \{A_1, A_2, \ldots, A_s\}$ be any s-subset from E^n. Consider the number

$$H(M) = \sum_{1 \leq i < j \leq s} \frac{1}{\rho(A_i, A_j)}, \tag{1}$$

where $\rho(A_i, A_j)$ is the Hemming distance in E^n.‡ S. V. Yablonskii has posed the problem of finding an s-subset $M \subseteq E^n$, in which the functional H(M) has a minimum. Physically the set M can be interpreted as a stable position of s like charged particles placed on vertices in E^n. In [1] this problem was completely solved for the case $s(n) = 2^{n-1}$. In this case it turned out that there exist two extremal sets, both of even parity. For other s, however, the question of the structure of the sets remained open. In [2] an asymptotic formulation of the problem was considered. It consists of the following. Suppose $H_s(n) = \min_{M \subseteq E^n} H(M)$, it is required to find a sequence $\{M_n\}$ of s-subsets from E^n such that

$$\lim_{n\to\infty} \frac{H_s(M_n)}{H_s(n)} = 1.$$

It was also remarked in that paper that since $\log s(n) \sim n$, all s-subsets from E^n have asymptotically identical energy § and hence in this case the asymptotic formulation of the problem turns out to be devoid of interest.

---

† Original article submitted February 5, 1968.

‡ The so-called Hemming distance between points $A = (\alpha_1, \alpha_2, \ldots, \alpha_n)$ and $B = (\beta_1, \beta_2, \ldots, \beta_n)$ is the number

$$\rho(A, B) = \sum_{i=1}^{n} |\alpha_i - \beta_i|.$$

§ The function $H(M) = \sum_{1 \leq i < j \leq s} \frac{1}{\rho(A_i, A_j)}$, proceeding from the physical analogy introduced above, is called the energy of the set M.

The basic results of the present work consist of the following:

1) the discovery of an asymptotic expression for $H_s(n)$, where $s = s(n)$ is an arbitrary increasing function;

2) a proof that "almost all" s-subsets of E^n have asymptotically minimum energy and the discovery of an estimate of the deviation of $H(M)$ from $H_s(n)$ for almost all s-subsets M from E^n;

3) the construction, in the case of functions of the form $s(n) = 2^{\varphi(n)}$ and with $\varphi(n)$ an arbitrary increasing integral function, of s-subsets of asymptotically minimal energy.

Besides, it is shown that the MacDonald equidistant code [3] has absolutely minimal energy. That is, instances of sets with absolutely minimal energy – even parity and MacDonald code – are at the same time sets with maximal minimum distance among all sets of the same cardinality.

§ 1

In this section the asymptotic minimum energy is found and the distribution studied for sets with asymptotically minimal energy. Here it is convenient to consider immediately more general functionals than energy, which introduce no additional difficulty in the obtaining of the needed results and which give supplementary information about the distribution of the minima of other functionals.

Let $\{M_1, M_2, \ldots, M_{C^s_{2^n}}\}$ be all s-subsets of vertices of the n-dimensional unit cube E^n and let $\varphi(x)$ be an arbitrary function defined on the set of natural numbers. Let $M = \{A_1, A_2, \ldots, A_s\}$ and

$$\xi(M) = \sum_{1\le i<j\le s} \varphi[\rho(A_i A_j)].$$

In the sequel we shall consider a probability space whose points are the $C^s_{2^n}$ subsets of cardinality s, and to each subset we assign the probability $p = \frac{1}{C^s_{2^n}}$. The function ξ will be considered a random quantity defined in this space.

The values of the mathematical expectation and the variance of the random quantity ξ are set out in the following lemmas.

Lemma 1.

$$M\xi = \frac{s(s-1)}{2(2^n-1)} \sum_{k=1}^{n} C^k_n \varphi(k).$$

Proof. Suppose $M_1, M_2, \ldots, M_{C^s_{2^n}}$ are all the subsets of cardinality s from E^n. Introduce the function

$$\xi^k_{ij} = \begin{cases} 1, & \text{if} \quad (A_i, A_j) \in M_k, \\ 0, & \text{if} \quad (A_i, A_j) \bar{\in} M_k. \end{cases}$$

Obviously, with the help of the function ξ^k_{ij} we can express $M\xi$ as follows:

$$M\xi = \frac{1}{2C^s_{2^n}} \sum_{k=1}^{C^s_{2^n}} \sum_{i\neq j} \xi^k_{ij} \varphi[\rho(A_i, A_j)] \tag{2}$$

(the summation extends over all pairs consisting of distinct points of the set E^n). Changing the order of summation in (2) we set

$$M\xi = \frac{1}{2C^s_{2^n}} \sum_{i \neq j} \varphi[\rho(A_i, A_j)] \sum_{k=1}^{C^s_{2^n}} \xi^k_{ij}. \tag{3}$$

It is obvious that $A_{ij} = \sum_{k=1}^{C^s_{2^n}} \xi^k_{ij}$ is equal to the number of s-subsets from E^n containing the pair (A_i, A_j), i.e., is equal to $C^{s-2}_{2^n-2}$ independently of the points A_i and A_j. Hence:

$$M\xi = \frac{1}{2} \frac{C^{s-2}_{2^n-2}}{C^s_{2^n}} \sum_{i \neq j} \varphi[\rho(A_i, A_j)]. \tag{4}$$

By the symmetry in E^n we can write

$$S = \sum_{i \neq j} \varphi[\rho(A_i, A_j)] = 2^n \sum_{\substack{j \\ j \neq i}} \varphi[\rho(A_i, A_j)], \tag{5}$$

where A_i is an arbitrary point of E^n. Choose as A_i the origin. Then $\rho(A_i, A_j) = \| A_j \|$. Considering that in E^n there are exactly C^k_n points with norm k, we get from (5)

$$S = 2^n \sum_{k=1}^{n} C^k_n \varphi(k). \tag{6}$$

Finally, from (4) and (6) we get

$$M\xi = \frac{s(s-1)}{2(2^n-1)} \sum_{k=1}^{n} C^k_n \varphi(k).$$

Lemma 1 is proved.

Let

$$\Psi(n) = \sum_{k=1}^{n} C^k_n \varphi(k); \quad W(n) = \sum_{k=1}^{n} C^k_n \varphi^2(k); \quad \omega(s, n) = 2s^2 - 2s(2^{n+1}-1) + 2^{n+1}(2^n-1).$$

Lemma 2.

$$D\xi = \frac{\omega(s, n)\, s(s-1)}{4(2^n-1)(2^n-2)(2^n-3)} \left[W(n) - \frac{\Psi^2(n)}{2^n-1}\right].$$

Proof. By the relation

$$D\xi = M\xi^2 - (M\xi)^2$$

and by Lemma (1), it is enough, in calculating variance, to find

$$M\xi^2 = \frac{1}{C^s_{2^n}} \sum_{k=1}^{C^s_{2^n}} [\xi(M_k)]^2.$$

As in the previous case, we bring $M\xi^2$ to a more suitable form:

$$M\xi^2 = \frac{1}{C^s_{2^n}} \sum_{k=1}^{C^s_{2^n}} \left\{\frac{1}{2} \sum_{i \neq j} \xi^k_{ij} \varphi[\rho(A_i, A_j)]\right\}^2.$$

(Henceforward we shall abbreviate $\varphi[\rho(A_i, A_j)]$ simply by φ_{ij}.) Further,

$$M\xi^2=\frac{1}{4C^s_{2^n}}\left[\sum_{k=1}^{C^s_{2^n}}\sum_{i=j}\xi^k_{ij}\varphi^2_{ij}+\sum_{k=1}^{C^s_{2^n}}\sum_{i\neq j}\xi^k_{ij}\xi^k_{ji}\varphi_{ij}\varphi_{ji}+\sum_{k=1}^{C^s_{2^n}}\sum_{(i,\,j)\neq(s,\,r)}\xi^k_{ij}\xi^k_{s,\,r}\varphi_{ij}\varphi_{sr}\right].$$

Noting that $\varphi_{ij} = \varphi_{ji}$, we can write:

$$M\xi^2=\frac{1}{2C^s_{2^n}}\sum_{k=1}^{C^s_{2^n}}\sum_{i\neq j}\xi^k_{ij}\varphi^2_{ij}+\frac{1}{4C^s_{2^n}}\sum_{k=1}^{C^s_{2^n}}\sum_{(i,\,j)\neq(s,\,r)}\xi^k_{ij}\xi^k_{sr}\varphi_{ij}\varphi_{sr}.$$

From Lemma 1 there follows

$$S_1=\frac{1}{2C^s_{2^n}}\sum_{k=1}^{C^s_{2^n}}\sum_{i\neq j}\xi^k_{ij}\varphi^2_{ij}=\frac{s\,(s-1)}{2\,(2^n-1)}W\,(n).$$

Now we calculate the sum

$$S_2=\frac{1}{4C^s_{2^n}}\sum_{k=1}^{C^s_{2^n}}\sum_{(i,\,j)\neq(s,\,r)}\xi^k_{ij}\xi^k_{sr}\varphi_{ij}\varphi_{sr}.$$

Changing the order of summation, we get

$$S_2=\frac{1}{4C^s_{2^n}}\sum_{(i,\,j)\neq(s,\,r)}\varphi_{ij}\varphi_{sr}\sum_{k=1}^{C^s_{2^n}}\xi^k_{ij}\xi^k_{sr}.$$

Let $R\,(A_i,\ A_j,\ A_s,\ A_r)=\sum_{k=1}^{C^s_{2^n}}\xi^k_{ij}\xi^k_{sr}$. Clearly, this sum depends only on the number of different points among A_i, A_j, A_r, A_s. This number can be equal to 3 or 4. Denote the cardinality of the set M by $|M|$. Then

$$S_2=\frac{1}{4C^s_{2^n}}\sum_{|A_i\cup A_j\cup A_s\cup A_r|=4}\varphi_{ij}\varphi_{sr}R\,(A_i,\ A_j,\ A_s,\ A_r)+\frac{1}{4C^s_{2^n}}\sum_{|A_i\cup A_j\cup A_s\cup A_r|=3}\varphi_{ij}\varphi_{sr}R\,(A_i,\ A_j,\ A_sA_r).$$

Clearly, in the first sum $R\,(A_i,\ A_j,\ A_s,\ A_r)$ is equal to the number of s-subsets containing the four points A_i, A_j, A_s, A_r, i.e., is equal to $C^{s-4}_{2^n-4}$, while in the second sum $R\,(A_i,\,A_j,\,A_s,\,A_r)=C^{s-3}_{2^n-3}$. Thus

$$S_2=\frac{C^{s-4}_{2^n-4}}{4C^s_{2^n}}\sum_{|A_i\cup A_jA_s\cup A_r|=4}\varphi_{ij}\varphi_{sr}+\frac{C^{s-3}_{2^n-3}}{4C^s_{2^n}}\sum_{|A_i\cup A_j\cup A_s\cup A_r|=3}\varphi_{ij}\varphi_{sr}.$$

We start by calculating the second sum. Since $|A_i\cup A_j\cup A_s\cup A_r|=3$, there are the four possibilities: $A_i = A_s$; $A_i = A_r$; $A_j = A_s$; $A_j = A_r$. Corresponding with this, the second sum breaks up into four equal sums. We calculate one of them, for example, the first. We have

$$S^{(2)}_{2,\,1}=\sum_{i,\,j,\,r}\varphi_{ij}\varphi_{ir}.$$

Clearly

$$S^{(2)}_{2,1}=\sum\varphi_{ij}\varphi_{ir}=\sum_{i,j}\varphi_{ij}\sum_{r}\varphi_{ir}-\sum_{i,j}\varphi^2_{ij}.$$

We have

$$\sum_r\varphi_{ir}=\sum_{k=1}^{n}C^k_n\varphi(k)=\Psi(n)$$

(as remarked above, this sum does not depend on the point A_i because of the symmetry of the set E^n).

Analogously,

$$\sum_{i,j}\varphi_{ij}=2^n\Psi(n),\qquad\sum_{i,j}\varphi^2_{ij}=2^nW(n).$$

Accordingly we get for the second sum in (7) the closed expression

$$S^{(2)}_2=\frac{C^{s-3}_{2^n-3}}{4C^s_{2^n}}\sum_{|A_i\cup A_j\cup A_s\cup A_r|=3}\varphi_{ij}\varphi_{sr}=2^n\frac{C^{s-3}_{2^n-3}}{C^s_{2^n}}[\Psi^2(n)-W(n)].\tag{8}$$

We now calculate the first sum in (7). We get

$$S^{(1)}_2=\sum_{|A_i\cup A_j\cup A_s\cup A_r|=4}\varphi_{ij}\varphi_{sr}=\sum_{i,j}\varphi_{ij}\sum_{s,r}\varphi_{sr}-\sum_{i,j,s}\varphi_{ij}\varphi_{si}-$$
$$-\sum_{i,j,s}\varphi_{ij}\varphi_{sj}-\sum_{i,j,r}\varphi_{ij}\varphi_{ir}-\sum_{i,j,r}\varphi_{ij}\varphi_{jr}-\sum_{i,j}\varphi_{ij}\varphi_{ji}-\sum_{i,j}\varphi^2_{ij}.$$

The sum of the second, third, fourth, and fifth were calculated above. The sum of the first, sixth, and seventh will also be calculated according to the demonstration of Lemma 1. Thus we get

$$S^{(1)}_2=2^n(2^n-4)\Psi^2(n)+2^{n+1}W(n).\tag{9}$$

Since

$$\frac{C^{s-3}_{2^n-3}}{C^s_{2^n}}=\frac{s(s-1)(s-2)}{2^n(2^n-1)(2^n-2)}\text{ and }\frac{C^{s-4}_{2^n-4}}{C^s_{2^n}}=\frac{s(s-1)(s-2)(s-3)}{2^n(2-1)(2^n-2)(2^n-3)},\tag{10}$$

it follows, taking account of (7)-(10), after simple transformations, that we have

$$D\xi=\frac{\omega(s,n)s(s-1)}{4(2^n-1)(2^n-2)(2^n-3)}\left[W(n)-\frac{\Psi^2(n)}{2^n-1}\right].$$

We now apply the result just obtained to find the asymptotic minimum of the random quantity ξ. Let

$$H^s_\varphi(n)=\min_{M\subseteq E^n}\xi(M),$$

where $|M|=s$. We shall suppose that $\varphi(x)$ lies in a class of functions $R=\{\varphi(x)\}$ with the following properties:

1) $\varphi(x)$ is convex in $[0, +\infty]$,

2) $\varphi(x)$ is decreasing in $[0, +\infty]$,

3) there exists a $\lambda > 0$ and a constant c such that

$$\lim_{n\to\infty} \varphi(n) n^{\lambda} = c.$$

Theorem 1. If $\varphi(x) \in R$ and $s(n) \leq 2^n$ is an arbitrary increasing function, then

$$H_{\varphi}^{s}(n) \sim \frac{s^2}{2}\varphi\left(\frac{n}{2}\right).$$

Theorem 1 shows that a minimum of the random variable ξ for the indicated class of functions asymptotically coincides with its mathematical expectation.

Remark. The class of functions $R = \{\varphi(x)\}$ unquestionably does not exhaust the functions for which the assertion of Theorem 1 holds. In particular, Condition 3) obviously can be significantly weakened.

It would probably be of interest to find the conditions on the class of functions R under which Theorem 1 holds.

We introduce two lemmas for the proof of Theorem 1.

Lemma 3.

$$H_{\varphi}^{s}(n) \leqslant \frac{s^2}{2}\varphi\left(\frac{n}{2}\right).$$

Proof. Clearly

$$H_{\varphi}^{s}(n) \leqslant M\xi.$$

From Lemma 1 we get

$$H_{\varphi}^{s}(n) \leqslant \frac{s(s-1)}{2(2^n-1)} \sum_{k=1}^{n} C_n^k \varphi(k). \tag{11}$$

We proceed to find the asymptotic expression

$$L(n) = \frac{\sum_{k=1}^{n} C_n^k \varphi(k)}{2^n - 1}.$$

To do this we first show that for any $p > 0$ the asymptotic equality

$$T_p(n) = \frac{1}{2^n} \sum_{k=1}^{n} \frac{C_n^k}{k^p} \sim \frac{2^p}{n^p} \tag{12}$$

holds. This follows without difficulty from noting that the Bernshtein polynomials $B_n(x) \sum_{k=1}^{n} C_n^k \left(\frac{n}{k}\right)^p x^k (1-x)^{n-k}$ approximate the function $f(x) = 1/x^{\rho}$ at the point $x = 1/2$.

Further, using Condition (3) we find $\lambda > 0$ and c such that

$$\lim_{n\to\infty} \varphi(n) n^{\lambda} = c.$$

Consider the equality

$$S(n) = \left| \frac{1}{2^n} \sum_{k=1}^{n} C_n^k \varphi(k) - \frac{c}{2^n} \sum_{k=1}^{n} \frac{C_n^k}{k^\lambda} \right|.$$

We have

$$S(n) \leqslant \frac{1}{2^n} \sum_{n=1}^{k} C_n^k \left| \varphi(k) - \frac{c}{k^\lambda} \right| = \frac{1}{2^n} \sum_{k=1}^{n} C_n^k \, r(k),$$

where $r(k) = \left| \varphi(k) - \frac{c}{k^\lambda} \right|$. We have $r(k) = o\left(\frac{1}{k^\lambda}\right)$, i.e., there exists an $\varepsilon > 0$ such that

$$r(k) < \frac{1}{k^{\lambda+\varepsilon}}.$$

Hence we get, using (12),

$$S(n) = o(T_\lambda(n)).$$

That is,

$$L(n) \sim \frac{2^\lambda}{n^\lambda} \sim \varphi\left(\frac{n}{2}\right). \qquad (13)$$

Substituting (13) in (11) we get

$$H_\varphi^s(h) \leqslant \frac{s^2}{2} \varphi\left(\frac{n}{2}\right),$$

as was to be proved.

Lemma 4.

$$H_\varphi^s(n) \geqslant \frac{s^2}{2} \varphi\left(\frac{n}{2}\right).$$

Proof. Since, by Condition 1), φ is convex downwards on $[0, +\infty]$, it follows, using Jensen's inequality, that

$$\sum_{1 \leq i < j \leq s} \varphi[\rho(A_i, A_j)] \geqslant \frac{s(s-1)}{2} \varphi\left[\frac{2 \sum_{1 \leq i < j \leq s} \rho(A_i, A_j)}{s(s-1)}\right]. \qquad (14)$$

It remains to provide an upper bound for

$$E_s(n) = \sum_{1 \leq i < j \leq s} \rho(A_i, A_j).$$

This was done in [5], where it is shown that

$$E_s(n) \leqslant \frac{ns^2}{4}. \qquad (15)$$

From (14), (15), and Condition (2) we get

$$H_\varphi^s(n) \geqslant \frac{s(s-1)}{2} \varphi\left[\frac{ns}{2(s-1)}\right]. \qquad (16)$$

Using the condition on the increase of the function S(n) and Condition 3) we get

$$H_\varphi^s(n) \geqslant \frac{s^2}{2}\varphi\left(\frac{n}{2}\right),$$

as was to be proved.

The following theorem gives useful information about the character of the distribution of the minimum of the random quantity ξ for the indicated class of functions.

Theorem 2. Let $\varphi(x) \in R$. Then for almost all s-subsets M of E^n the inequality

$$\left|\xi(M) - \frac{s^2}{2}\varphi\left(\frac{n}{2}\right)\right| \leqslant \frac{s}{2}\varphi\left(\frac{n}{2}\right)$$

holds.

Proof. From Lemmas 1 and 2 and the definition of the class $R = \{\varphi(x)\}$ it is not difficult to obtain

$$M\xi \sim \frac{s^2}{2}\varphi\left(\frac{n}{2}\right), \tag{17}$$

$$D\xi \sim \frac{s^2}{2}\frac{\varphi^2\left(\frac{n}{2}\right)}{\Delta(n)}, \tag{18}$$

where $\lim_{n\to\infty}\Delta(n) = \infty$. Substituting now in Chebyshev's inequality

$$P\{|\xi - M\xi| \geqslant t\} < \frac{D\xi}{t^2},$$

$t = \frac{s}{2}\varphi\left(\frac{n}{2}\right)$, and noting (17) and (18), we get what is required.

For the case $\varphi(x) = 1/x$, Theorem 3 can be formulated as follows. Almost all positions of s identically charged particles at vertices of a unit n-dimensional cube are asymptotically stable. Here, for almost all distributions M there holds the following estimate of the deviation of the energy H(M) from the minimum:

$$\left|H(M) - \frac{s^2}{n}\right| \leqslant \frac{s}{n}.$$

§ 2

In this section, sets are constructed having asymptotically minimal energy. For the construction of such sets there serves certain simple information from the theory of group codes. The entirety of this information can be found in [3].

The subset $G \subseteq E^n$ † is called a group code if G is a subgroup of the group E^n. It is at the same time clear that G is a linear space over the field GF(2). If the dimension of the code G as a linear space is equal to k, then G is called a group (n, k) code. The metric structure of a group (n, k) code is completely determined by the set of weights of its code points, i.e., by the vector $M(G) = (a_1(n), a_2(n), \ldots, a_n(n))$, where $a_i(n)$ is the number of code points of weight i. The vector $M(G) = (a_1(n), a_2(n), \ldots, a_n(n))$ is called the spectrum of the code G. A group code G* is called dual with respect to the group code G if the subspaces G

† The set E^n is an Abelian group for the operation of positional addition mod 2.

and G* are orthogonal. It is obvious that the dimension of the code G* dual to the (n, k)-code G is equal to n – k, i.e., the code G* is an (n, n – k) code.

Let M(G) = (a_1(n), a_2(n), ..., a_n(n)) be a spectrum of the (n, k) code G and M(G*) = (b_1(n), b_2(n), ..., b_n(n)) be a spectrum of the dual code G*. Suppose also the dimension of the code G satisfies the following inequality

$$k(n) > \Delta_\omega(n) = \log n + \log\log n + \omega(n),$$

where ω(n) is an arbitrary increasing function of n. Under these circumstances the following assertion holds.

Lemma 5. *For the energy of the (n, k(n))-code G_n there holds the following asymptotic equality*

$$H(G_n) \sim \frac{2^{2k}}{n} + \frac{2^{2k}}{n} \sum_{i=1}^{[\frac{n}{2}]} \frac{b_i}{C_n^i \left(1 - \frac{2_i}{n}\right)}.$$

Proof. We use the MacWilliams formula [6] which connects the spectrum of the code G_n with the spectrum of the dual code

$$\sum_{i=0}^{n} b_i (1+x)^{n-i} (1-x)^i = 2^{n-k} \sum_{i=0}^{n} a_i x^i. \tag{19}$$

From (19) there follows

$$\sum_{i=0}^{n} b_i \int \frac{(1+x)^{n-i}(1-x)^i}{x} dx = 2^{n-k} \sum_{i=0}^{n} \frac{a_i}{i} x^i. \tag{20}$$

Transforming the integral on the left into two parts, we have

$$u_i(x) = \int \frac{(1+x)^{n-i}(1-x)^i}{x} dx = \int \frac{(1-x)^i}{x} dx + \int \frac{\sum_{s=1}^{n-i} C_{n-i}^s x^s}{x} (1-x)^i dx. \tag{21}$$

The first and the second integrals in (21) we take individually:

$$u_i^1(x) = \int \frac{(1-x)^i}{x} dx = \ln x + \int \left(\sum_{r=1}^{i} (-1)^r C_i^r x^{r-1}\right) dx = \ln x + \sum_{r=1}^{i} (-1)^r C_i^r \frac{x^r}{r};$$

$$u_i^2(x) = \int \left[\sum_{s=1}^{n-i} C_{n-i}^s x^{s-1} (1-x)^i\right] dx = \sum_{s=1}^{n-i} C_{n-i}^s \int x^{s-1}(1-x)^i dx.$$

Further:

$$v_{i,s}(x) = \int x^{s-1}(1-x)^i dx = \sum_{m=0}^{i} (-1)^m C_i^m \frac{x^{m+s}}{m+s}.$$

It is known [4] that

$$u_i^{(1)} = \sum_{r=1}^{i} (-1)^r C_i^r \frac{1}{r} = \sum_{r=1}^{i} \frac{1}{r}. \tag{22}$$

We calculate $v_{i,s}$ (1). We have

$$v_{i,s}(1)=\int_0^1 x^{s-1}(1-x)^i\,dx=B(s-1,\ i),$$

where B(s − 1, i) is the Euler integral. It is equal to

$$B(s-1,\ i)=\frac{(s-1)!\,i!}{(s+i)!}. \tag{23}$$

Using (20)-(23) we get

$$2^{n-k}\sum_{i=1}^{n}\frac{a_i}{i}=\sum_{i=1}^{n}b_i\sum_{r=1}^{i}\frac{1}{r}+\sum_{i=1}^{n}b_i\sum_{s=1}^{n-i}C_{n-i}^{s}\frac{(s-1)!\,i!}{(s+i)!}. \tag{24}$$

Further:

$$C_{n-i}^{s}\frac{(s-1)!\,i!}{(s+i)!}=\frac{1}{C_n^i}\,\frac{C_n^{s+i}}{s}. \tag{25}$$

From (24) and (25), setting s + i = r, we get

$$2^{n-k}\sum_{i=1}^{n}\frac{a_i}{i}=\sum_{i=1}^{n}b_i\sum_{r=1}^{i}\frac{1}{r}+\sum_{i=0}^{n}\frac{b_i}{C_n^i}\sum_{r=i+1}^{n}\frac{C_n^r}{r-i}. \tag{26}$$

Since

$$H(G_n)=2^{k-1}\sum_{i=1}^{n}\frac{a_i}{i},$$

it follows that

$$H(G_n)=\frac{2^{2k-1}}{2^n}\sum_{i=1}^{n}b_i\sum_{r=1}^{i}\frac{1}{r}+\frac{2^{2k-1}}{2^n}\sum_{i=0}^{n}\frac{b_i}{C_n^i}\sum_{r=i+1}^{n}\frac{C_n^r}{r-i}. \tag{27}$$

We now find the asymptotic expression for the second sum in (27). Changing order of summation we have

$$\Phi_2(n)=\frac{2^{2k-1}}{2^n}\sum_{i=0}^{n}\frac{b_i}{C_n^i}\sum_{r=i+1}^{n}\frac{C_n^r}{r-i}=\frac{2^{k-1}}{2^n}\sum_{r=1}^{n}C_n^r\sum_{i=0}^{r-1}\frac{b_i}{C_n^i(r-i)}.$$

Let

$$\varphi(n,\ r)=\sum_{i=0}^{r-1}\frac{b_i}{C_n^i(r-i)}.$$

Then

$$\Phi_2(n)=\frac{2^{2k-1}}{2^n}\sum_{r=1}^{n}C_n^r\varphi(n,\ r).$$

Since $b_i/C_n^i \leq 1$ it follows that

$$\varphi(n, r) < \ln r < \ln n.$$

Further, for $\delta(n) = \ln(n)/\sqrt{n}$ we have

$$\frac{1}{2^n}\sum_{r=1}^{n} C_n^r \varphi(n, r) \sim \frac{1}{2^n}\sum_{|2r-n|<n\delta(n)} C_n^r \varphi(n, r). \tag{28}$$

Indeed, if $|2r - n| > n\delta(n)$, then $\frac{(2r-n)^2}{n^2\delta^2(n)} > 1$, so that

$$\frac{1}{2^n}\sum_{|2r-n|>n\delta(n)} C_n^r(n, r) < \frac{1}{2^n}\frac{(2r-n)^2}{n^2\delta^2(n)}\sum_{|2r-n|>n\delta(n)} C_n^r\varphi(n, r) <$$

$$< \frac{4\max\varphi(n, r)}{n^2\delta^2(n)}\frac{1}{2^n}\sum_{r=0}^{n} C_n^2\left(r-\frac{n}{2}\right)^2 < \frac{4\ln n}{n^2\delta^2(n)}\cdot\frac{n}{4} = \frac{\ln n}{n^2\delta^2(n)}, \tag{29}$$

since

$$\frac{1}{2^n}\sum_{r=0}^{n} C_n^r\left(r-\frac{n}{2}\right)^2 = \frac{n}{4}.$$

Now in (29), putting $\delta(n) = \ln n/\sqrt{n}$, we get (28), as was to be proved.

We note, moreover, that for $\varphi(n) \leq \sqrt{n}\ln n$

$$\lim_{n\to\infty}\sum_{i=\left[\frac{n}{2}\right]}^{\left[\frac{n}{2}\right]+\varphi(n)}\frac{b_i}{C_n^i} = 0. \tag{30}$$

Hence,

$$\sum_{i=\left[\frac{n}{2}\right]}^{\left[\frac{n}{2}\right]+\varphi(n)}\frac{b_i}{C_n^i} < \varphi(n)\max_{\left[\frac{n}{2}\right]<i<\left[\frac{n}{2}\right]+\varphi(n)}\frac{b_i}{C_n^i} < \varphi(n)\frac{|G_n^*|\sqrt{n}}{2^n},$$

since $b_i < |G_n^*|$. Further, from the condition $k(n) > \Delta_\omega(n)$ it follows that

$$|G_n^*| = 2^{n-k} < \frac{2^n}{h\ln n\Delta(n)},$$

where $\Delta(n) = 2^{\omega(n)}$. Hence

$$\varphi(n)\frac{|\varphi_n^*|\sqrt{n}}{2^n} < \sqrt{n}\ln n\frac{2^n\sqrt{n}}{2^n n\ln n\Delta(n)} = \frac{1}{\Delta(n)}.$$

Since $\lim_{n\to\infty}\omega(n) = \infty$, we have proved (30).

Moreover,

$$\varphi\left(n, \left[\frac{n}{2}\right]+\varphi(n)\right)=\sum_{i=0}^{\left[\frac{n}{2}\right]} \frac{b_i}{C_n^i\left[\frac{n}{2}+\varphi(n)-i\right]}+\sum_{i=\left[\frac{n}{2}\right]}^{\left[\frac{n}{2}\right]+\varphi(n)} \frac{b_i}{C_n^i\left[\frac{n}{2}+\varphi(n)-i\right]}$$

Using (30) it is easy to show that the second sum tends to zero as n goes to infinity. Hence

$$\varphi\left(n, \left[\frac{n}{2}\right]+\varphi(n)\right) \sim \sum_{i=0}^{\left[\frac{n}{2}\right]} \frac{b_i}{C_n^i\left[\frac{n}{2}+\varphi(n)-i\right]}=\frac{2}{n} \sum_{i=0}^{\left[\frac{n}{2}\right]} \frac{b_i}{C_n^i\left[1-\frac{2i}{n}+\frac{2\varphi(n)}{n}\right]} \sim \frac{2}{n} \sum_{i=0}^{\left[\frac{n}{2}\right]} \frac{b_i}{C_n^i\left[1-\frac{2i}{n}\right]},$$

i.e.,

$$\varphi\left(n, \left[\frac{n}{2}\right]+\varphi(n)\right) \sim \varphi\left(n, \left[\frac{n}{2}\right]\right). \tag{31}$$

It follows now at once from (31) that

$$\frac{1}{2^n} \sum_{|2r-n|<n\delta(n)} C_n^r \varphi(n, r) \sim \varphi\left(n, \frac{n}{2}\right).$$

Hence

$$H(G_n) \sim \frac{2^{2k-1}}{2^n} \sum_{i=1}^{n} b_i \sum_{r=1}^{i} \frac{1}{r}+\frac{2^{2k}}{n} \sum_{i=0}^{\left[\frac{n}{2}\right]} \frac{b_i}{C_n^i\left[1-\frac{2i}{n}\right]}. \tag{32}$$

Using the condition $k(n) > \Delta_\omega(n)$, we get

$$\frac{2^{2k-1}}{2^n} \sum_{i=1}^{n} b_i \sum_{r=1}^{i} \frac{1}{r}<\frac{2^{2k-1}}{2^n} \sum_{i=1}^{n} b_i \ln i<\frac{\ln n}{2^{n-2k+1}} 2^{n-k}=2^{k-1} \ln n=o\left(\frac{2^{2k}}{n}\right). \tag{33}$$

From (32) and (33) we get finally

$$H(G_n) \sim \frac{2^{2k}}{n}+\frac{2^{2k}}{n} \sum_{i=1}^{\left[\frac{n}{2}\right]} \frac{b_i}{C_n^i\left(1-\frac{2i}{n}\right)},$$

as was to be proved.

The following theorem shows that the presence in the code U_n of a code G_n of sufficiently large dimension having asymptotically minimal energy is a sufficient criterion that the code U_n too has asymptotically minimal energy.

Theorem 3. Let the group $(n, k(n))$-code G_n with $k(n) \geq \Delta_\omega(n)$, is an arbitrary increasing function, have asymptotically minimal energy, and let G belong to some other group code U where $\omega(n)$ is an arbitrary increasing function, have asymptotically minimal energy, and let G_n belong to some other group code U_n, i.e., let it be a subgroup of U_n. Then the code U_n also has asymptotically minimal energy.

Proof. It follows from Lemma 4 that

$$H(G_n) \geqslant \frac{2^{2k}}{n}. \tag{34}$$

We get from Lemma 6

$$H(G_n) \sim \frac{2^{2k}}{n} + \frac{2^{2k}}{n} \sum_{i=1}^{\left[\frac{n}{2}\right]} \frac{b_i}{C_n^i\left(1-\frac{2i}{n}\right)} . \tag{35}$$

That is, it follows from (35) and (34) that the group (n, k(n))-code G_n with $k(n) \geq \Delta_\omega(n)$ has asymptotically minimal energy if and only if

$$\lim_{n\to\infty} W(G_n^*) = 0, \tag{36}$$

where

$$W(G_n^*) = \sum_{i=1}^{\left[\frac{n}{2}\right]} \frac{b_i}{C_n^i\left(1-\frac{2i}{n}\right)} .$$

Since by hypothesis the code G_n has asymptotically minimal energy, it follows that (35) is fulfilled. From the direct inclusion $G_n \subseteq U_n$ follows the inverse inclusion for the dual codes, i.e., $G_n^* \supseteq U_n^*$. Hence

$$b_i(U_n^*) \leqslant b_i(G_n)$$

which means

$$W(U_n^*) < W(G_n^*).$$

Now from (35) we get

$$\lim_{n\to\infty} W(U_n^*) = 0,$$

i.e., the code U_n has asymptotically minimal energy, as was to be proved.

Theorem 3 shows that in learning to construct an asymptotically extremal (n, k(n))-code G_n with $k(n) \geq \Delta_\omega(n)$, we could construct an asymptotically extremal (n, φ(n))-code G_n for any function $\varphi(n) \geq k(n)$. This theorem illustrates the huge diversity in methods of constructing these extremal codes.

It is worth noting another interesting fact about the "universality" of a group (n, k(n))-code G_n with asymptotically minimal energy.

Theorem 4. For any increasing function $s(n) \leq 2^{k(n)}$ in the code G_n there exists a set of cardinality s(n) with asymptotically minimal energy.

Proof. Choose as a probability space the set of all subsets of cardinality s(n) from the code G_n and consider the random variable

$$\xi(M) = \sum_{1\leq i<j\leq s} \frac{1}{\rho(A_iA_j)} ,$$

where $M = \{A_1, A_2, \ldots, A_s\} \subseteq G_n$. As in Lemma 1, using symmetry of the code G_n, the mathematical expectation of ξ is readily calculated:

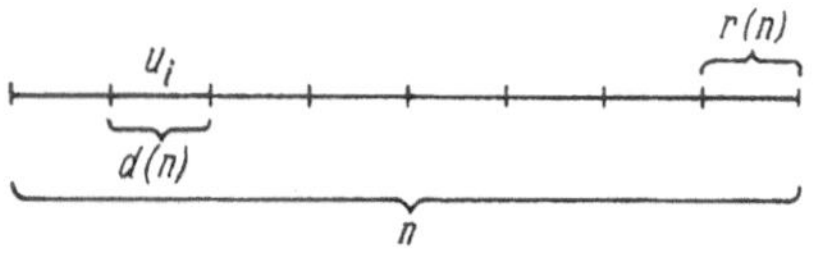

Fig. 1

$$M\xi = \frac{s(s-1)}{2(2^{k(n)}-1)} \sum_{k=1}^{n} \frac{a_k}{k},$$

where $\{a_1(n), a_2(n), \ldots, a_n(n)\}$ is the spectrum of the code G_n. Since by hypothesis

$$H(G_n) \sim \frac{2^{2k}}{n},$$

it follows that

$$\sum_{k=1}^{n} \frac{a_k}{k} \sim \frac{2^{k+1}}{n}.$$

Hence

$$M\xi \sim \frac{s^2}{n}, \qquad (37)$$

which proves the theorem, since from (37) there follows that among s-subsets of the code G_n there must exist at least one whose energy does not surpass $M\xi$.

Accordingly, in the construction of sets of any "increasing cardinality" with asymptotically minimal energy, it effectively suffices to have only one group extremal (n, k(n))-code with $k(n) \geq \Delta_\omega(n)$, where $\omega(n)$ is an arbitrary increasing integral function with natural argument.

We use this circumstance in §3 to construct sets with asymptotically minimal energy.

§ 3

Now consider a special class of group codes in which we can successively construct an extremal set.

We let n be the length of a binary sequence and let d(n) be a natural number: $2 \leq d(n) < n$. We divide n by d(n):

$$n = d(n)\, s(n) + r(n),$$

and correspondingly we separate our sequence into s(n) blocks of length d(n) and a block of length r(n) (see Fig. 1).

Each block u_i is filled either entirely with single zeros or entirely with single ones. It is clear that the thus obtained set of "block sequences" (points) is a group (n, s + 1)-code. The spectrum is readily calculated of the code G_n. Actually each point of the code G_n has weight either vd (v = 0, 1, ..., s) or vd + r (v = 0, 1, ..., r). Here it is obvious that $a_{vd} = a_{vd+r} = C_s^v$ and $a_r = 1$ (for $r \neq 0$). We have the following result.

Theorem 5. *The code G_n has an asymptotically minimal energy for $S(n) \geq \log n + \omega(n)$, where $\omega(n)$ is any increasing function.*

Proof. We calculate the energy of the code G_n. We have

$$H(G_n) = 2^s \sum_{v=1}^{s} \frac{C_s^v}{vd} + 2^s \sum_{v=0}^{s} \frac{C_s^v}{vd+r} + \frac{2s}{r}.$$

Consider the asymptotic behavior of $H(G_n)$ as $n \to \infty$ and $s(n) \geq \log n + \omega(n)$, where $\omega(n)$ is some increasing function. We have

$$2^s \sum_{v=1}^{s} \frac{C_s^v}{vd} = \frac{2^s}{d} \sum_{v=1}^{s} \frac{C_s^v}{v} \sim \frac{2^{2s+1}}{sd}$$

and

$$2^s \sum_{v=0}^{s} \frac{C_s^v}{vd+r} = \frac{2^s}{d} \sum_{v=0}^{s} \frac{C_s^v}{v+\frac{r}{d}} \sim \frac{2^s}{d} \sum_{v=1}^{s} \frac{C_s^v}{v} \sim \frac{2^{2s+1}}{sd},$$

since $0 < r/d < 1$. Hence $H(G_n) \sim \frac{2^{2s+2}}{s(n)\,d(n)} + \frac{2^s}{r(n)}$; however, it is clear that $n = s(n)\,d(n) \left[1 + \frac{r(n)}{s(n)\,d(n)}\right] \sim s(n)\,d(n)$, since $\lim_{n\to\infty} s(n) = \infty$. Hence,

$$H(G_n) \sim \frac{(2^{s+1})^2}{n} + \frac{2^s}{r(n)}; \tag{38}$$

however,

$$\frac{2^s}{r(n)} < 2^s = o\left(\frac{2^{2s}}{n}\right)$$

on the strength of the condition $s(n) \geq \log n + \omega(n)$. Thus we get finally

$$H(G_n) \sim \frac{(2^{s+1})^2}{n},$$

i.e., the code G_n has asymptotically minimal energy.

With the aid of Theorem 5 we can construct an asymptotically extremal $(n, s(n))$-code for the function $s(n)$ having the form

$$s(n) = \left[\frac{n}{d(n)}\right], \tag{39}$$

where $2 \leqslant d(n) \leqslant \frac{n}{\log n + \omega(n)}$ for arbitrary increasing function $\omega(n)$. There are two unsatisfactory things about this method. First, the sequence $s(n) = [n/d(n)]$ does not represent all integral functions $s(n) \leq n$ and second, there are bounds on the growth of the function $s(n)$, i.e.,

$$\log n + \omega(n) \leqslant S(n) \leqslant \frac{n}{2}.$$

Our approximative problem will be free of these defects. Theorem 3 shows that in the construction of asymptotically extremal $(n, k(n))$-codes with an arbitrary function $(kn) \geq \Delta_\omega(n)$ it suffices to construct only one $(n, k(n))$-code with any particular function $k(n)$ satisfying the inequality

$$k(n) \geqslant \Delta_\omega(n).$$

By Theorem 5 we obviously are able to do this. Thus only the case $k(n) < \Delta_\omega(n)$ remains to be cleared up.

This case can conveniently be broken down into three subcases:

I. $\log n+\omega(n)<k(n)<\Delta_\omega(n)$.
II. $\log n<k(n)<\log(n)+\omega(n)$.
III. $k(n)\leqslant \log n$.

The first subcase presents no difficulty. We need only follow the preceding construction using, as d(n), the function $\left[\frac{n}{k(n)-1}\right]$. All the foregoing arguments remain in force.

In the second case we try to change the construction in such a way that in (38), r(n) will be an increasing function, which we then choose in the capacity of $\omega(n)$. [It is easy to imagine that $r(n) \equiv 0$; the above introduced construction would need no changes.] That is to say, we let

$$\log n<k(n)<\log n+\omega(n).$$

As above, we divide sequences of length n into blocks of length d(n) of such a form that the number of all the blocks will be k(n). If, meanwhile, the unique block of length $r(n) = n - [k(n) - 1]d(n)$ increases with growing n, then all the previous constructions, in order, proceed without change. If this is not the case then we introduce the following point of view: an increasing sequence of blocks in some growing length is counted as a sequence of blocks, but of such form that in that sequence of blocks the next to the last block had a growing length. Obviously this can be done. Clearly, the dimension of the code does not change because of this. Its spectrum will emerge as follows:

C_{k-2}^{v} points of weights vd, v = 0, 1, ..., k − 2;

C_{k-2}^{v} points of weights $vd + r_1$, v = 0, 1, ..., k − 2 (r_1 is the length of the last block);

C_{k-2}^{v} points of weights $vd + d_1$, v = 0, 1, ..., k − 2 (d_1 is the length of the next to last block);

C_{k-2}^{v} points of weights $vd + d_1 + r_1$, v = 0, 1, ..., k − 2;

one point of weight r_1;

one point of weight d_1.

We now calculate the energy of this code. We have

$$H(G_n)=2^{k-1}\sum_{v=1}^{k-2}\frac{C_{k-2}^{v}}{vd}+2^{k-1}\sum_{v=0}^{k-2}\frac{C_{k-2}^{v}}{vd+d_1}+2^{k-1}\sum_{v=0}^{k-2}\frac{C_{k-2}^{v}}{vd+r_1}+2^{k-1}\sum_{v=0}^{k-2}\frac{C_{k-2}^{v}}{vd+r_1+d_1}+\frac{2^{k-1}}{r_1}+\frac{2^{k-1}}{d_1}.$$

As in the preceding case, it is not difficult to get an asymptotic expression for $H(G_n)$:

$$H(G_n)\sim 4\cdot\frac{2^{2k-2}}{n}+\frac{2^k}{r_1}+\frac{2^k}{r_2}, \tag{40}$$

where now $\lim_{n\to\infty} r_1(n)=\infty$ and $\lim_{n\to\infty} d_1(n)=\infty$. On the strength of the condition

$$k(n)>\log n$$

it follows from (40) that

$$H(G_n)\sim\frac{2^{2k}}{n},$$

i.e., the code G_n has asymptotically minimal energy, as was to be proved.

In case III we need only use the equidistance of the MacDonald [3] or the Plotkin [3] codes, which, between arbitrary pairs of points, have distance asymptotically equal to n/2. Any of the subsets of these codes with increasing cardinality is an asymptotic extremal subset, as required.

We can, incidentally, easily get from Lemma 5 the following precise inequality of the energy of s-subsets

$$H(M) \geqslant \frac{(s-1)^2}{n}. \tag{41}$$

Choosing s = n + 1 and, in the role of M, that MacDonald code all code points of which have weight d = (n + 1)/2, we set

$$H(M) = C_{n+1}^2 \cdot \frac{1}{d} = n. \tag{42}$$

From (41) and (42) it follows that the code M has absolutely minimal energy.

Literature Cited

1. B. S. Zil'berman, "On the distribution of charge in the vertices of the unit n-dimensional cube," Dokl. Akad. Nauk, Vol. 149, No. 3 (1963).
2. T. N. Kruglova, "On asymptotic methods of solving problem on charges," in: Problemy Kibernetiki, Vol. 13, Moscow (1965).
3. W. Peterson, Error Correcting Codes, Wiley, New York (1961).
4. G. Polya and G. Szego, Aufgaben und Lehrsätze aus der Analysis, Springer, Leipzig (1935).
5. D. D. Joshi, "A note on upper bounds for minimum distance codes," Inf. Control, 1(3):289-295 (1958).
6. J. MacWilliams, "The structure and properties of binary cyclic alphabets," Bell. Syst. Techn. Journ., 44(2):303-332 (1965).

ON NETWORKS CONSISTING OF FUNCTIONAL ELEMENTS WITH DELAYS†

O. B. Lupanov

Moscow

§ 1. Statement of the Problem and Formulation of Results

As is well known, in "traditional" methods for the realization of logic-algebra functions (by contact networks, Π-networks, networks consisting of the functional elements, formulas) the so-called "Shannon effect" holds: "almost all functions" of n arguments have "an almost identical" complexity which is asymptotically equal to the complexity of the most complex function of n arguments. The hypothesis on this effect was stated by C. E. Shannon in 1949 (see [11]) and was subsequently proved by the author of the present paper (see, for example, [5, 3]). In certain cases (for example, for disjunctive normal forms) the "weakened Shannon effect" holds – "almost all functions of n arguments have almost identical complexity"; true, this complexity is less than the complexity of the most complex function [1, 7, 8].

In the present paper a certain natural class of supervisory systems is considered – proper networks consisting of functional elements with delays, for which these effects generally do not hold.

We shall consider networks consisting of elements of the following form. Each element E_i is juxtaposed with a certain logic-algebra function $\varphi_i(x_1, \ldots, x_{k_i})$ (which depends essentially on k_i arguments) and two positive numbers: P_i, the "weight of the element," and T_i, the "delay of the element"; these numbers are not assumed to be integers. If $k_i \geq 2$, then the numbers $P_i/(k_i - 1)$ and $T_i/\log k_i$ ‡ shall be called the *reduced weight* and the *reduced delay* of the element E_i, respectively. It is assumed that the number of inputs of the element E_i for $k_i \geq 1$ is equal to k_i, while for $k_i = 0$ (i.e., if the element realizes a constant) it is equal to 1.

The networks are constructed over a finite set $\mathscr{E} = \{E_1, \ldots, E_r\}$ of elements of the indicated form in accordance with the rules of constructing networks from functional elements [5, 4].

---

† Original article submitted December 30, 1968.

‡ Here we have in mind binary logarithms throughout.

We shall call the chain between the elements $E^{(1)}$ and $E^{(2)}$ the sequence of elements $E_{i_1}, \ldots, E_{i_s}$, having the properties:

1) a certain input of the element E_{i_j} is connected to the output of the element $(2 \leq j \leq s)$;

2) $E_{i_{j-1}} (2 \leqslant j \leqslant s)$;
2) $E^{(1)} = E_{i_1}$, $E^{(2)} = E_{i_s}$.

We shall call the number $T_{i_1} + \ldots + T_{i_s}$ the delay of this chain. We shall call a chain principal if a certain input of its first element E_{i_1} is an input of the network, while the output of its last element E_{i_s} is an output of the network. The network S shall be called proper if the delays of all of its principal chains are equal.† This common number shall be called the delay of the network and denoted by the symbol T(S). The function realized by the network and the complexity L(S) of the network are defined in the same way as they are for networks consisting of functional elements.

We shall assume that the system $\mathscr{E}$ of original elements is complete in the sense that for each logic-algebra function f one may construct a proper network which realizes f (with a certain delay); i.e., we have in view "completeness in the second sense" in the terminology of V. B. Kudryavtsev [2]; however, here commensurateness of the delays of the basis elements is not assumed.

A complete basis shall be called regular if networks in this basis may be used to realize all four functions of one argument x, $\bar{x}$, 0, 1,‡ having one and the same delay. Otherwise, the basis shall be called irregular. From the results obtained by V. B. Kudryavtsev [2] it follows that in the case of commensurate delays (for example, integer delays) the basis is regular.§

The logic-algebra function $f(x_1, \ldots, x_n)$ shall be called an x-function (correspondingly an $\bar{x}$-function, a 0-function, a 1-function) if the function $f(x, \ldots, x)$ is equal to x (correspondingly $\bar{x}$, 0, 1)¶ ; x- and $\bar{x}$-functions shall likewise be called ϕ-functions, while 0- and 1-functions shall be called c-functions.

Let us consider the following Shannon functions:

$L(f)$ — the minimum of the complexities of (proper) networks which realize the function f;

$T(f)$ — the minimum of the delays of networks which realize the function f. Assume $\mathfrak{N}_n$ ($\mathfrak{N}_n^c$, $\mathfrak{N}_n^\Phi$), respectively) is the set of all functions (correspondingly, c-functions, ϕ-functions) $f(x_1, \ldots, x_n)$, and assume

$$L(n) = \max_{f \in \mathfrak{N}_n} L(f), \quad T(n) = \max_{f \in \mathfrak{N}_n} T(f);$$
$$L^c(n) = \max_{f \in \mathfrak{N}_n^c} L(f), \quad T^c(n) = \max_{f \in \mathfrak{N}_n^c} T(f);$$
$$L^\Phi(n) = \max_{f \in \mathfrak{N}_n^\Phi} L(f), \quad T^\Phi(n) = \max_{f \in \mathfrak{N}_n^\Phi} T(f).$$

† From this definition it follows that for any network elements the delays of all chains traveling from the inputs of the network to the inputs of this element are equal.

‡ Below (§2) another (equivalent) definition will be given. The existence of irregular basis will be established in §4 (Lemma 14).

§ See likewise below, p. 47.

¶ In E. Post's terminology (see [10, 9]) these are correspondingly α-, δ-, γ-, β-functions.

Assume Φ is the set of all basis functions $\varphi_i(x_1, \ldots, x_{k_i})$, which depend essentially on at least two arguments, while Φ^* is its subset consisting of all Φ-functions. Assume, finally, that

$$\rho = \min_{\varphi_i \in \Phi} \frac{P_i}{k_i - 1}, \qquad \tau = \min_{\varphi_i \in \Phi} \frac{T_i}{\log k_i};$$

$$\rho^* = \min_{\varphi_i \in \Phi^*} \frac{P_i}{k_i - 1}, \qquad \tau^* = \min_{\varphi_i \in \Phi^*} \frac{T_i}{\log k_i}.$$

It is obvious that if the numbers ρ^* and τ^* are defined,† then

$$\rho^* \geqslant \rho, \qquad \tau^* \geqslant \tau.$$

The following statements hold.

Theorem 1. *If the basis $\mathscr{E}$ is regular, then*

1) $L(n) \sim L^c(n) \sim L^\Phi(n) \sim \rho \frac{2^n}{n}$;

2) $T(n) \sim T^c(n) \sim T^\Phi(n) \sim \tau n$‡);

3) *moreover, for any $\varepsilon > 0$ and any function $f(x_1, \ldots, x_n)$ of a fairly large number of arguments there exists a network S which realizes f and is such that*

$$L(S) < (1+\varepsilon)\rho\frac{2^n}{n}, \qquad T(S) < (1+\varepsilon)\tau n.$$

Theorem 2. *If the basis $\mathscr{E}$ is irregular, then*

1a) $L^c(n) \sim \rho \frac{2^n}{n}$;

1b) $L(n) \sim L^\Phi(n) \sim \rho^* \frac{2^n}{n}$;

2a) $T^c(n) \sim \tau n$;

2b) $T(n) \sim T^\Phi(n) \sim \tau^* n$ §);

3) *moreover, for any $\varepsilon > 0$ and any c-function (correspondingly, Φ-function) $f(x_1, \ldots, x_n)$ of a fairly large number of arguments there exists a network S which realizes f and is such that*

$$L(S) < (1+\varepsilon)\rho\frac{2^n}{n}, \qquad T(S) < (1+\varepsilon)\tau n$$

(respectively, $L(S) < (1+\varepsilon)\rho^*\frac{2^n}{n}$, $T(S) < (1+\varepsilon)\tau^* n$).

These theorems show that although ρ and τ (and likewise ρ^* and τ^*) may be attained on various elements, it is nevertheless true that for almost all functions it is possible to construct networks which are asymptotically optimal simultaneously with respect to both complexity and delay. Theorem 2, moreover, indicates that in the case of an irregular basis and

† From Lemma 2 proved below it follows that in the case of an irregular basis Φ is non-empty.

‡ Actually, for "almost all" functions $f(x_1, \ldots, x_n)$ of n arguments, $L(f) \sim \rho(2^n/n)$ and $T(f) \sim \tau n$.

§ For "almost all" c-functions (correspondingly, Φ-functions) $f(x_1, \ldots, x_n)$ of n arguments, $L(f) \sim \rho(2^n/n)$ and $T(n) \sim \tau(n)$ [$L(f) \sim \rho^*(2^n/n)$ and $T(f) \sim \tau^* n$, respectively].

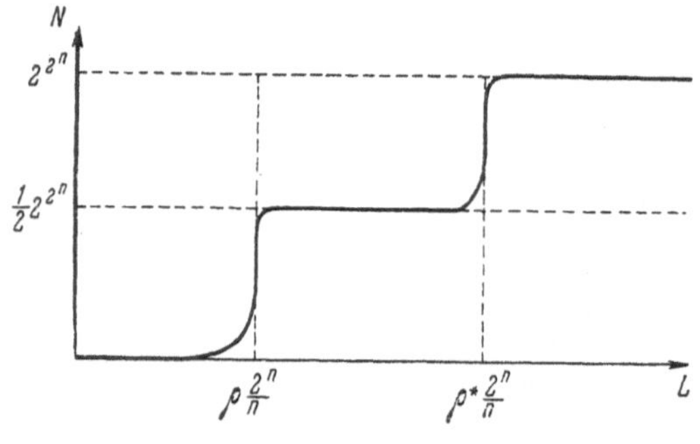

Fig. 1

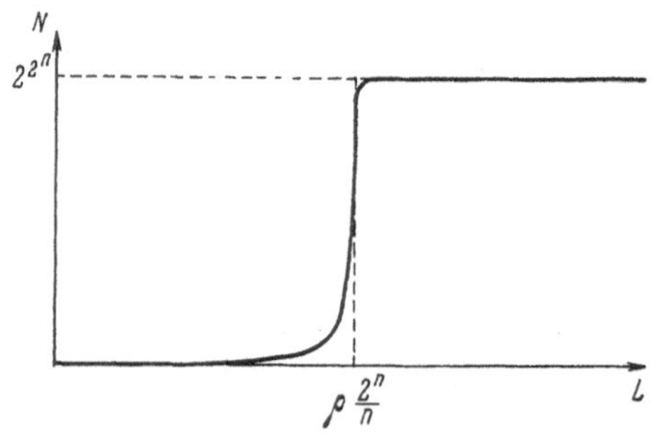

Fig. 2

struct networks which are asymptotically optimal simultaneously with respect to both complexity and delay. Theorem 2, moreover, indicates that in the case of an irregular basis and $\rho^* > \rho$ "almost all" functions can be partitioned into two subsets, each containing "almost half" of all the functions; these subsets are realized with different complexities. This phenomenon may be more conveniently expressed as follows. Let us use $N_n(L)$ to denote the number of functions $f(x_1, \ldots, x_n)$ for which $L(f) \leq L$. The behavior of the functions $N_n(L)$ is depicted approximately in Fig. 1. In the case of "traditional" methods of realization, the approximate behavior of the corresponding function $N_n(L)$ is depicted in Fig. 2. The increase of the bounds for $L^\Phi(n)$ and $T^\Phi(n)$ in the case of an irregular basis occurs due to the fact that in this case networks for Φ-functions may consist of elements belonging only to a part of the basis (" Φ-elements" – Lemma 2).

The first statement of Theorem 1 (for the condition that the delays of the elements are integers) is an almost trivial corollary of the theorem on the asymptotic behavior of the Shannon function for networks consisting of functional elements ([5]; see also [6], Theorem D.12). The second statement (for the same condition) is likewise almost obvious. The third statement is somewhat less trivial (if ρ and τ are attained on different elements). It turns out to be possible due to the fact that the "complexity of the network" and the "level of the network" are created by different parts of the network, and these parts of the network may be constructed from elements of different kinds.

Plan for Proving This Theorem. In §3 the lower bound of the functions L and T will be established. In obtaining the bounds for $L^\Phi(n)$ in the case of an irregular basis the property mentioned above of such basis (§2) is used. The lower bounds for $T(n)$ and $T^c(n)$ are likewise actually known. Their proof is presented solely for completeness of the picture.

The methods of network synthesis mentioned in the third statements of both theorems (these methods give the upper bounds for the asymptotic relationships in the first two statements) are described in §§8-10 (§§4-7 contain auxiliary propositions).

Three synthesis methods will be presented:

1) for realization of any functions in an arbitrary regular basis;

2) for realization of 0-functions in an arbitrary irregular basis;

3) for realization of x-functions in an arbitrary irregular basis.

It is obvious that special methods of network synthesis (in an irregular basis) for 1-functions and for $\bar{x}$-functions are not required – the corresponding networks can be obtained from networks for 0-functions and x-functions as a result of attaching networks realizing negation to their outputs.

§ 2. Certain Properties of Regular Basis

We shall call the set of functions $\mathfrak{M}$ joint (relative to the basis $\mathscr{E}$) if all functions from $\mathfrak{M}$ can be realized by networks with identical delays. By virtue of the completeness of the basis, the following statement is valid.

(*) Assume $\mathfrak{M}$ is a joint set and $\varphi(x_1, \ldots, x_m)$ is an arbitrary function. Then the set of functions of the form $\varphi(\varphi_1, \ldots, \varphi_m)$, where $\varphi_i \in \mathfrak{M}$, is joint.

Lemma 1. In order for the system of functions $\{x, \bar{x}, 0, 1\}$ to be joint (i.e., in order for the basis to be regular) it is necessary and sufficient that just one of the pairs

$$\{x, \bar{x}\}, \{x, 0\}, \{x, 1\}, \{\bar{x}, 0\}, \{\bar{x}, 1\} \tag{1}$$

be joint.

Proof. The necessity is obvious.

Sufficiency. Assume that the pair $\{x, \bar{x}\}$ is joint. Let us consider the function $g(x, y, z)$ which satisfies the conditions

$$g(0, 0, 0) = g(0, 0, 1) = g(1, 1, 1) = 0,$$
$$g(0, 1, 1) = g(1, 0, 0) = g(1, 1, 0) = 1.$$

Then

$$g(x, x, x) = 0, \quad g(x, x, \bar{x}) = x, \quad g(\bar{x}, \bar{x}, x) = \bar{x}, \quad g(x, \bar{x}, \bar{x}) = 1,$$

and by virtue of (*) the system $\{x, \bar{x}, 0, 1\}$ is joint.

Assume that the pair $\{x \oplus \alpha, \beta\}$ † is joint (any of the four latter pairs in (1)). Let us consider the function $\varphi_1(x) = x \oplus \alpha \oplus 1$. By virtue of (*) the functions $\varphi_2(x) = \varphi_1(x \oplus \alpha) = \bar{x}$ and $\varphi_3(x) = \varphi_1(\beta) = \alpha \oplus \beta \oplus 1$. are joint.

Further, the functions

$$\varphi_2(\varphi_2(x)) = x, \quad \varphi_2(\varphi_3(x)) = \alpha \oplus \beta, \quad \varphi_3(\varphi_3(x)) = \alpha \oplus \beta \oplus 1$$

are joint; i.e., x, 0, 1 are joint.

Finally, let us consider the function $h(x_1, x_2, x_3, x_4)$ which satisfies the conditions

$$h(0, 0, 0, 0) = h(0, 0, 0, 1) = h(1, 1, 1, 1) = 0,$$
$$h(0, 0, 1, 1) = h(0, 1, 1, 1) = 1.$$

Then the functions

$$h(0, 0, 0, x) = 0, \quad h(0, 0, x, 1) = x, \quad h(0, x, 1, 1) = 1, \quad h(x, 1, 1, 1) = \bar{x}$$

are joint.

The lemma has been proved.

Corollary (see [2]). If all elements of the basis have rational delays, then the basis is regular.

† The symbol $\oplus$ implies addition modulo 2.

Actually, networks S_1 and S_2 exist which realize x and 0, respectively; assume that their (rational) delays are equal to p_1/q_1 and p_2/q_2. Having connected q_1p_2 copies of the network S_1 in cascade, we obtain the network for x having the delay p_1p_2. Analogously, from p_1q_2 copies of the network S_2 we obtain the network for 0 having the delay p_1p_2. In this way the pair $\{x, 0\}$ is joint, and by virtue of Lemma 1 the basis is regular.

Assume $* \in \{x, \bar{x}, 0, 1, \phi, c\}$. The basis element realizing the $*$-function is called a $*$-element.

Lemma 2. *Assume $\mathscr{E}$ is an irregular basis and S is a network in $\mathscr{E}$, which contains at least one c-element. Then S realizes a certain c-function.*

Proof. Assume S' is the network obtained from S as a result of identification of the inputs. Let us consider the principal chain $\Delta = [E_{i_1}, \ldots, E_{i_s}]$, passing through a certain c-element E_{i_j}. Then either constants are applied to all of the inputs of this element, or the function x is applied to all of its inputs, or the function $\bar{x}$ is applied to all of its inputs (since otherwise one of the pairs (1) would be joint, which is impossible for an irregular basis). In all three cases a constant is realized at the output of this c-element. Therefore (again by virtue of the irregularity of the basis) constants are likewise applied to the inputs of all the next elements of the chain $E_{i_{j+1}}, \ldots, E_{i_s}$, and the network S' realizes the constants.

The lemma has been proved.

Let us consider two properties of a basis.

I. Two chains with identical delays exist which consist solely of ϕ-elements, one of which contains an even number of $\bar{x}$-elements, while the other contains an odd number of $\bar{x}$-elements.

II. Two chains exist with identical delays, one of which consists solely of ϕ-elements, while the other contains at least one c-element.

The chain $E_{i_1}, \ldots, E_{i_s}$ shall be called *special* if all inputs of the element E_{i_1} are attached to one input of the network, while all inputs of the element E_{i_l} $(2 \leqslant l \leqslant s)$ are attached to the output of the element $E_{i_{l-1}}$ (Fig. 3). Each special chain (treated as a network) realizes a certain function of one argument, say of x.

Let us now give a different definition of a regular basis, which is more convenient for checking.

Lemma 3. *In order for a basis to be regular it is necessary and sufficient that it have one of the properties (N_I) and (N_{II}).*

Proof. Sufficiency. If property I is fulfilled (property II, respectively), then having formed the corresponding special chain we find that the pair $\{x, \bar{x}\}$ (a certain pair $\{x \oplus \alpha, \beta\}$), respectively) is joint; therefore, the basis is regular by virtue of Lemma 1.

Necessity. Assume that the basis is regular, and assume that T* is the minimal positive delay with which functions of a certain pair (1) may be realized (see Lemma 1). Assume S_1 and S_2 are two networks having the delay T* which realize the functions indicated. By virtue of minimality of the number T*, one of the following possibilities† holds for each element E of these networks:

† Compare with the proof of Lemma 2; there an analogous situation arose due to the incompatibility of any pairs at all from (1).

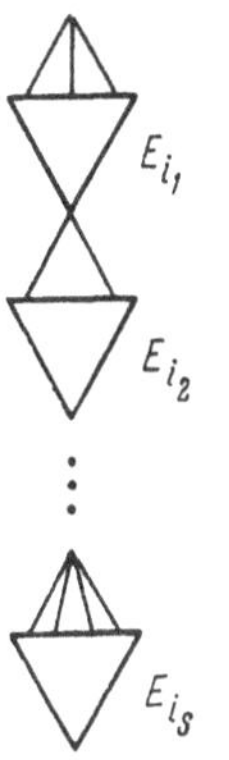

Fig. 3

a) constants are applies to all inputs of this element;

b) the function x is applied to all inputs of this element;

c) the function $\bar{x}$ is applied to all inputs of this element.

Assume S is one of those networks which realize the function $x \oplus \alpha$, and $E_{i_1}, \ldots, E_{i_s}$ is its principal chain. Then for all elements of this chain b) or c) is fulfilled (since if constants are applied to the inputs of the element E_{i_j} a constants will appear at its output; therefore, constants are applied to all inputs of the element $E_{i_{j+1}}$, etc.). Therefore, the chain consists only of ϕ-elements.

Let us now consider two cases.

1) The networks S_1 and S_2 realize x and $\bar{x}$. In this case their principal chains consist of ϕ-elements, an even number of $\bar{x}$ elements being contained in the first and an odd number in the second (i.e., the property I is fulfilled).

2) The networks S_1 and S_2 realize $x \oplus \alpha$ and β. In this case the principal chain Δ_1 of the network S_1 consists of ϕ-elements. Let us consider the principal chain $\Delta_2 = [E_{i_1}, \ldots$ of the network S_2. Assume E_{i_j} is the first element of this chain, which produces a constant (since E_{i_s} produces a constant, it follows that such an element exists). Then b) or c) is fulfilled for it. Therefore, E_{i_j} is a c-element. For the chains Δ_1 and Δ_2 the property II is fuilfilled.

The lemma has been proved.

§ 3. Lower Bounds of Shannon Functions

A. Complexity Estimates

Lemma 4. The following relationships are valid:

$$L(n) \geqslant \rho \frac{2^n}{n}, \quad L^c(n) \geqslant \rho \frac{2^n}{n}, \quad L^\Phi(n) \geqslant \rho \frac{2^n}{n},$$

while in the case of an irregular basis,

$$L^\Phi(n) \geqslant \rho^* \frac{2^n}{n}.$$

Actually, it is well known (see [5, 6]) that for "conventional" networks consisting of functional elements (without matching of the delays of the circuits) the Shannon function $L^0(n)$ satisfies the condition

$$L^0(n) \geqslant \rho \frac{2^n}{n},$$

where ρ is the minimum of the reduced weights of the basis elements. In the case when only proper networks are used the Shannon functions may only increase. Therefore, in our case we have

$$L(n) \geqslant \rho \frac{2^n}{n}.$$

This same inequality is also valid separately for Φ-functions and c-functions (since the logarithms of the numbers of functions from $\mathfrak{N}_n^\Phi$ and from $\mathfrak{N}_n^c$ are asymptotically equal to 2^n).

In the case of an irregular basis the lower bound for $L^{\Phi}(n)$ is raised: by virtue of Lemma 2 one may use networks over only a portion of the basis consisting of Φ-elements in the realization of Φ-functions; therefore, the constant ρ in the lower bound may be replaced by ρ^*.

B. Delay Estimates. Let us use M(T) to denote the maximum number of elements in irreducible† proper networks in the basis $\mathscr{E}$, which have one output and satisfy the condition

$$T(S) \leqslant T.$$

Lemma 5. $M(T) \leqslant C_1 T 2^{\frac{T}{\tau}}$, where τ is the minimum of the reduced delays of the basis elements.‡

Proof. Assume T_0 is the least of the delays of the basis elements. Let us partition the domain of variation of T into half-intervals $J_i = [(i-1)T_0, iT_0)$. We shall use induction with respect to i – the number of the half-interval to which T belongs – to prove the inequality

$$M(T) \leqslant \frac{T}{T_0} 2^{\frac{T}{\tau}} \tag{2}$$

If i = 1, i.e., $0 \leq T < T_0$, then M(T) = 0, and the inequality (2) is obvious.

The Inductive Step. Assume $T \in J_i$, $i \geqslant 2$. In this case $T \geq T_0$, and M(T) is attained on the network S having at least one element. Assume E is an output element with k inputs and the delay T^*.

If k = 1, then we have, on the basis of using the inductive proposition and the inequalities $T \geq T_0$ and $T^* \geq T_0$,

$$M(S) \leqslant 1 + M(T-T^*) \leqslant 1 + \frac{T-T^*}{T_0} 2^{\frac{T-T^*}{\tau}} \leqslant 1 + \frac{T-T_0}{T_0} 2^{\frac{T}{\tau}} = 1 + \frac{T}{T_0} 2^{\frac{T}{\tau}} - 2^{\frac{T}{\tau}} < \frac{T}{T_0} 2^{\frac{T}{\tau}}.$$

If $k \geq 2$, then $\tau \leqslant \frac{T^*}{\log k}$, $\log k \leqslant \frac{T^*}{\tau}$, $k \leqslant 2^{\frac{T^*}{\tau}}$, and by analogy we have

$$M(S) \leqslant 1 + kM(T-T^*) \leqslant 1 + k\frac{T-T^*}{T_0} 2^{\frac{T-T^*}{\tau}} \leqslant 1 + \frac{T-T^*}{T_0} 2^{\frac{T}{\tau}} = 1 + \frac{T}{T_0} 2^{\frac{T}{\tau}} - \frac{T^*}{T_0} 2^{\frac{T}{\tau}} \leqslant \frac{T}{T_0} 2^{\frac{T}{\tau}}.$$

The lemma has been proved.

Lemma 6. The relationships

$$T(n) \geqslant \tau n, \quad T^{C}(n) \geqslant \tau n, \quad T^{\Phi}(n) \geqslant \tau n$$

are valid, while in the case of an irregular basis,

$$T^{\Phi}(n) \geqslant \tau^* n.$$

† A network is called irreducible if for each element the following condition is fulfilled: its output is either an output of the network, or it is attached to the input of a certain element. It is obvious that for each proper network S one can indicate an irreducible proper network S' which realizes the same function as S does and is such that $L(S') \leq L(S)$ and $T(S') = T(S)$.

‡ Throughout this paper the letter C with its subscripts is used to denote constants (in general, constants which depend on the basis).

Proof. Assume ε is an arbitrary positive number. Let us consider networks whose delay does not exceed $(1-\varepsilon)\tau n$. By virtue of Lemma 5 the number of elements in each such network does not exceed

$$C_1(1-\varepsilon)\tau n 2^{(1-\varepsilon)n} = C_2 \frac{n2^n}{2^{\varepsilon n}} = o\left(\frac{2^n}{n}\right).$$

Therefore, for $n > n_\varepsilon$ it follows by virtue of Lemma 4 that not just any function of n arguments may be realized by such a network. Thus, for $n > n_\varepsilon$ we have

$$T(n) > (1-\varepsilon)\tau n.$$

In exactly the same way the following inequalities are valid for fairly large n:

$$T^c(n) > (1-\varepsilon)\tau n, \qquad T^\Phi(n) > (1-\varepsilon)\tau n.$$

Finally, in the case of an irregular basis the constant τ may be replaced by τ^* in the lower bound for $T^\Phi(n)$ (for the same reason as it is in the proof of Lemma 4).

The lemma has been proved.

§ 4. Certain Auxiliary Statements

Lemma 7. There exist networks S_1, S_2, S_3, which have one and the same delay $T^{(1)}$ and realize the functions x, xy, $x\vee y$, respectively.

Proof. By virtue of the completeness of the basis, a proper network S_0 exists having a certain delay T_0, which realizes the function $\delta(x, y) = \bar{x}\vee\bar{y}$. Therefore, the function $\delta(\delta(x_1, x_2), \delta(x_3, x_4)) = x_1x_2\vee x_3x_4$ may be realized with a delay $2T_0$.

Identifying the inputs for this function in the network, we obtain the required networks:

when $x_1=x,\ x_2=x,\ x_3=x,\ x_4=x$ — function for x;
when $x_1=x,\ x_2=y,\ x_3=x\ \ x_4=y$ — function for xy;
when $x_1=x,\ x_2=x,\ x_3=y,\ x_4=y$ — function for $x\vee y$.

Lemma 8. For any r and m, $r \le m$, there exists a network $K_{r,m}$ which realizes $x_1x_2 \ldots x_r$, and is such that

$$L(K_{r,m}) \leqslant C_3 m, \qquad T(K_{r,m}) =]\log m[\, T^{(1)}.$$

Proof. Assume $\mu =]\log m[$. The network $K_{r,m}$ is obtained from a μ-story dichotomic tree which is constructed from the network S_2 (and realizes a conjunction of length 2^μ) as a result of identifying certain inputs. Let us note also that $m \le 2^\mu < 2m$.

The following analogous lemma is valid.

Lemma 9. For any r and m, $r \le m$, there exists a network $D_{r,m}$ which realizes $x_1\vee x_2\vee\ldots\vee x_r$, and is such that

$$L(D_{r,m}) \leqslant C_4 m, \qquad T(D_{r,m}) =]\log m[\, T^{(1)}.$$

Lemma 10. There exist networks S_4, S_5, S_6, S_7 having one and the same delay $T^{(2)}$ and realizing the functions 1, 0, $\bar{x}y$, $\bar{x}\vee y$, respectively.

Proof. By virtue of the completeness of the basis, proper networks S_α and S_β exist which realize the functions $\alpha(x, y) = x \vee \bar{y}$ and $\beta(x, y) = \bar{x}y$, respectively, having the delays T_α and T_β. Therefore, the following functions may be realized with a delay $T_\alpha + T_\beta$:

$$\alpha(\beta(x, x), \beta(x, x)) = 1, \quad \beta(\alpha(x, x), \alpha(x, x)) = 0,$$
$$\beta(\alpha(x, x), \alpha(y, x)) = x\bar{y}, \quad \alpha(\beta(x, x), \beta(y, x)) = x \vee \bar{y}.$$

Assume $K_m^0(x_1, \ldots, x_m)$ is the system of all conjunctions $x_1^{\sigma_1} \ldots x_m^{\sigma_m}$ which are such that $(\sigma_1, \ldots, \sigma_m) \neq (0, \ldots, 0)$, $(\sigma_1, \ldots, \sigma_m) \neq (1, \ldots, 1)$.

Lemma 11. For any conjunction from K_m^0 there exists a network S which realizes this conjunction and is such that

$$L(S) \leqslant C_5 m, \quad T(S) = T^{(2)} +]\log m[\, T^{(1)}.$$

Proof. The network S is obtained from the network $K_{m,m}$ as a result of attaching m S_6 networks for $\bar{x}y$ to its inputs (under these conditions a conjunction of the form $u_1 \ldots u_m \bar{v}_1 \ldots \bar{v}_m$ is realized) and identifying certain inputs.

Along with the networks S_1-S_7 the network S_8 (which exists by virtue of the completeness of the basis) will be used further on; this network has a certain delay $T^{(3)}$ and realizes the function $\xi(x, y, z, w) = x(yz \vee \bar{w})$.

The functions $x_1 x_2 \ldots x_m \vee x_{q_1}^{\sigma_{q_1}} x_{q_2}^{\sigma_{q_2}} \ldots x_{q_l}^{\sigma_{q_l}}$, where $l \leq m$, $1 \leq q_j \leq m$, and the collection $(\sigma_{q_1}, \ldots, \sigma_{q_l})$ differs from the zero collection, shall be called A-functions. From this definition it follows that any A-function is an x-function.

Lemma 12. For any A-function $\alpha(x_1, \ldots, x_m)$ there exists a proper network S which realizes this function and is such that

$$L(S) \leqslant C_6 m, \quad T(S) = T^{(3)} +]\log m[\, T^{(1)}.$$

Proof. It is easy to check the fact that

$$\xi(u_1, v_1, w_1, v_2)\, \xi(u_2, v_2, w_2, v_3) \ldots \xi(u_{m-1}, v_{m-1}, \omega_{m-1}, v_m)\, \xi(u_m, v_m, w_m, v_1) =$$
$$= u_1 u_2 \ldots u_m (v_1 v_2 \ldots v_m w_1 w_2 \ldots w_m \vee \bar{v}_1 \bar{v}_2 \ldots \bar{v}_m).$$

We shall denote this function by $\omega(u_1, \ldots, u_m, v_1, \ldots, v_m, w_1, \ldots, w_m)$.

Assume now that $\alpha = x_1 \ldots x_m \vee x_{q_1}^{\sigma_{q_1}} \ldots x_{q_l}^{\sigma_{q_l}}$ is an arbitrary A-function. Its variables can be partitioned into three groups:

1) variables appearing in the second conjunction without negations (i.e., appearing in both conjunctions); assume these are $x_{i_1}, \ldots, x_{i_a}$;

2) variables appearing in the second conjunction with negations; assume these are $x_{j_1}, \ldots, x_{j_b}$;

3) variables which do not appear in the second conjunction; assume these are $x_{h_1}, \ldots, x_{h_c}$. It is clear that if $1 \leq a \leq m$, $0 \leq b \leq m - 1$, $0 \leq c \leq m - 1$. Therefore, the function α may be derived from ω as a result of a certain substitution of variables (if $a = m$, then we assume that $v_1 = \ldots v_m = w_1 = \ldots = w_m = x_1$; if $c = 0$, then we assume, for example, as a preliminary step, that $w_1 = \ldots = w_m = v_1$).

The network S for α is derived from the network $K_{m,m}$ and m S_7 networks attached to inputs of the network $K_{m,m}$ as a result of the identification of inputs described above.

Finally, a proper network S_9 exists having a certain delay $T^{(4)}$, which realizes the function $\zeta(x, y, z) = \bar{x}y \vee xz$.

The networks S_1-S_9 shall be called canonical.

Remark. Instead of the function $\xi(x, y, z, w)$ one could take a "simpler" function $\eta(x, y, w) = x(y \vee w)$, since $\eta(x, y, w) \& \eta(x, z, w) = \xi(x, y, z, w)$, or the function $\zeta(x, y, z)$ just introduced above, since $\zeta(w, x, y) \& \zeta(y, x, x) = \eta(x, y, w)$.

Lemma 13. The system of elements realizing the functions $\delta(x, y) = \bar{x} \vee \bar{y}$, $\beta(x, y) = x\bar{y}$, $\xi(x, y, z, w) = x(yz \vee \bar{w})$ having any delays (in particular, noncommensurate delays) is complete.

Actually, any 0-function $f(x_1, \ldots, x_n)$ may be realized as a disjunction of conjunctions from K_n^0; under these conditions only elements for β and δ are used (see Proof of Lemmas 7, 8, 9, 11). Any 1-function can be realized as a negation of a 0-function (negation realizes one additional element for δ with identified inputs). Any x-function can be realized as a disjunction of A-functions; under these conditions the elements for δ and ξ are used. Any $\bar{x}$-function can be realized as a negation of an x-function.

Lemma 14. Irregular bases exist.

Actually, an example of such a basis may consist of the system of elements for the functions δ, β, ξ (Lemma 13) with the delays 1, $\sqrt{2}$, $\sqrt{3}$, respectively. Each of the conditions I and II is fulfilled, since the numbers 1, $\sqrt{2}$, $\sqrt{3}$ are linearly independent over the field of rational numbers.

Lemma 15. Assume E is an arbitrary element of a basis which realizes a function that differs from a constant and has a delay T. Then the networks Z_E and C_E exist which have the properties:

1) Z_E has three inputs and one output; its function g(x, y, z) satisfies the condition $g(0, 1, z) = z \oplus \sigma$ (for a certain σ);

$$L(Z_E) = C_7, \quad T(Z_E) = T.$$

2) C_E has two inputs and two outputs; the functions $g_0(x, y)$ and $g_1(x, y)$ which realize it satisfy the condition $g_0(0, 1) = 0$, $g_1(0, 1) = 1$;

$$L(C_E) = C_8, \quad T(C_E) = T.$$

Proof. Assume E realizes the function $\varphi(x_1, \ldots, x_k)$. Constants† exist: $\alpha_1, \ldots, \alpha_{k-1}, \sigma$ which are such that $\varphi(\alpha_1, \ldots, \alpha_{k-1}, z) = z \oplus \sigma$. Assume $i_1, \ldots, i_s$ are the numbers of the zero components of the collection $(\alpha_1, \ldots, \alpha_{k-1})$, while $j_1, \ldots, j_t$ are the numbers of its unitary components. The network Z_E is obtained as a result of attaching the inputs of the element E having the numbers $i_1, \ldots, i_s$ to the first input of the network, the inputs having the numbers $j_1, \ldots, j_t$ to the second input of the network; and the k-th input to the third input of the network.‡

The network C_E is obtained from two Z_E networks, in one of which the input z is identified with x, while in the other the input z is identified with y.

† For k = 1 the constants $\alpha_1, \ldots, \alpha_{k-1}$ are missing.

‡ If a certain constant is not used, then the network input corresponding to it is not attached to the inputs of the element E.

Remark. A chain of t cascade connected networks C_E realizes the constants 0 and 1 with the delays 0, T, ..., tT if the constants 0 and 1 are applied to the inputs of the chain. A chain of Z_E networks of even length t (in the presence of constants produced by the chain of networks C_E) realizes the function $f(z) = z$ with a delay tT.

§ 5. Pieces and Principal Pieces

We shall consider collections of zeros and ones having a certain length m. Let us juxtapose each collection $\tilde{\sigma} = (\sigma_1, \ldots, \sigma_m)$ with a number $|\tilde{\sigma}| = \sum_{i=1}^{m} \sigma_i 2^{m-i}$. The set of whole numbers l which satisfy the condition $l_1 \leq l < l_2$ will be denoted by $[l_1, l_2)$ and called a piece. A piece of the form $[\lambda 2^i, (\lambda+1)2^i)$, where λ and i are nonnegative integers, will be called a principal piece. The terms "piece" and "principal piece" shall likewise be used to denote sets of corresponding collections.

Lemma 16. Any piece I (of collections of length m or of their corresponding numbers) is a union of no more than 2m principal pieces.

Proof. Let us use $\delta(l)$ to denote the largest number δ such that 2^δ divides l [if $l = 0$, then $\delta(l) = \infty$]. Assume $I = [l_1, l_2), \delta(I) = \max_{l_1 \leq l \leq l_2} \delta(l)$,† and assume l_0 is the number for which $\delta(l_0) = \delta(I)$. The piece $[l_1, l_2)$ shall be called left if $l_0 = l_1$, right if $l_0 = l_2$, and middle if $l_1 < l_0 < l_2$. It is clear that the middle piece $[l_1, l_2)$ is the union of the right piece $[l_1, l_0)$ and the left piece $[l_0, l_2)$.

Let us now show that any left (right) piece is a union of no more than m principal pieces. Let us begin by considering the particular case $I = [0, l)$. Assume $l = 2^{i_1} + 2^{i_2} + \ldots + 2^{i_s}$ (i_j are nonnegative integers; $i_1 > i_2 > \ldots > i_s$; $s \leq m$). Then

$$I = [0, 2^{i_1}) \cup [2^{i_1}, 2^{i_1} + 2^{i_2}) \cup \ldots \cup [2^{i_1} + \ldots + 2^{i_{s-1}}, 2^{i_1} + \ldots + 2^{i_{s-1}} + 2^{i_s}) =$$
$$= [0 \cdot 2^{i_1}, 1 \cdot 2^{i}) \cup [2^{i_1 - i_2} \cdot 2^{i_2}, (2^{i_1 - i_2} + 1) 2^{i_2}) \cup \ldots \cup [(2^{i_1 - i_s} + \ldots + 2^{i_{s-1} - i_s}) 2^{i_s}, (2^{i_1 - i_s} + \ldots + 2^{i_{s-1} - i_s} + 1) 2^{i_s}),$$

i.e., I is a union of no more than m principal pieces.

Assume now that $l_1 > 0$ and $I = [l_1, l_2)$ is a left piece. We have $l_1 = (2s+1) 2^{\delta(l_1)}$. Further,

$$l_2 - l_1 < 2^{\delta(l_1)}, \tag{3}$$

since otherwise $l^* = l_1 + 2^{\delta(l_1)} \in [l_1, l_2]$, on the one hand, while $l^* = (2s+2) 2^{\delta(l_1)} = (s+1) 2^{\delta(l_1)+1}$, on the other hand; i.e., $\delta(l^*) > \delta(l_1)$, and this contradicts the fact that $\max_{l_1 \leq l \leq l_2} \delta(l)$ is attained for $l = l_1$. From (3) it follows that from the representation of the piece $[0, l_2 - l_1)$ in the form of a sum of (no more than m) principal pieces one obtains the corresponding representation of the piece $[l, l_2)$ (it is sufficient to attach l_1 to all ends).

Analogous statements are also valid for right segments.

Thereby the lemma has been proved.

Remark. The collection corresponding to numbers from the principal piece $I = [\lambda 2^i, (\lambda+1)2^i)$ constitutes all possible collections $(\sigma_1, \ldots, \sigma_{m-i}, \sigma_{m-i+1}, \ldots, \sigma_m)$ for which

† The right end is taken into account!

a portion $(\sigma_1, \ldots, \sigma_{m-i})$ is specified and $\lambda = \sum_{j=1}^{m-i} \sigma_j 2^{m-i-j}$. Therefore, the characteristic function $\varphi_I(x_1, \ldots, x_m)$ of this piece is the conjunction $x_1^{\sigma_1} \ldots x_{m-i}^{\sigma_{m-i}}$.

Assume $I(f)$ [$I_0(f)$, respectively] is the minimal number of pieces (principal pieces, respectively) whose union makes up the set of collections on which f goes to unity.

Lemma 17. Assume the function f depends on m arguments, while the functions f^+ and f^- are derived from it by adding and eliminating, respectively, one conjunction in the perfect disjunctive normal form. Then

$$I_0(f) \leqslant 2mI(f), \tag{4}$$

$$I_0(f^+) \leqslant I_0(f) + 1, \tag{5}$$

$$I_0(f^-) \leqslant I_0(f) + 2m. \tag{6}$$

Proof. The inequality (4) derives from Lemma 16. The inequality (5) is obvious. The inequality (6) derives from the following concepts. For elimination of one collection, the principal piece containing this collection is partitioned into no more than two pieces, one left and one right. Each of them is a union of no more than m principal pieces (see proof of Lemma 16).

§ 6. Principal Blocks and Their Properties

A. An (E, t)-block. Assume E is an element of a basis having at least two inputs, and that t is a nonnegative integer. We shall call an (E, t)-block a proper network constructed from elements E and constituting a t-story tree (Fig. 4). Assume E realizes the function $\varphi(x_1, \ldots, x_\varkappa)$ and has the weight P and the delay T. Then an (E, t)-block consists of $(\varkappa^t - 1)/(\varkappa - 1)$ elements, has a delay tT, a complexity $[(\varkappa^t - 1)/(\varkappa - 1)]P$, $N = \varkappa^t$ inputs, and realizes a certain function $F(y_0, \ldots, y_{N-1})$ which depends essentially on N arguments (the arguments are numbered in natural order). In particular, a network consisting of one pole is an (E, 0)-block. (E, l)-blocks $(0 \leq l \leq t)$ which are entered in the composition of an (E, t)-block and are attached to its inputs shall be called its principal subnetworks.

The set of variables assigned to the inputs of any (E, l)-block entered into the composition of an (E, t)-block (and likewise the set of numbers of these variables) shall be called an interval (having a height $t - l$) (of variables and numbers of variables, respectively). It is obvious that an interval of numbers of variables is a piece in the sense of §5.

Lemma 18. Assume S is an (E, l)-block; $S_1, \ldots, S_\varkappa$ are its principal subnetworks, which are (E, $l-1$)-blocks; $J_1, \ldots, J_\varkappa$ are the intervals of the variables of these subnetworks, and F', $F_1, \ldots, F_\varkappa$ are functions realized by the networks S, $S_1, \ldots, S_\varkappa$, respectively. Then for any i $(1 \leq i \leq \varkappa)$ one can find a collection of values of arguments which do not appear in J_i and which when substituted into the appropriate place in the function F' cause the latter to go over to the function $F_i \oplus a_i$, where a_i is equal to 0 or 1.

Proof. Since $\varphi(y_1, \ldots, y_\varkappa)$ depends essentially on all of its arguments, there exists a collection of constants $a_1, \ldots, a_{i-1}, a_i, a_{i+1}, \ldots, a_\varkappa$ which is such that

$$\varphi(a_1, \ldots, a_{i-1}, y_i, a_{i+1}, \ldots, a_\varkappa) = y_i \oplus a_i. \tag{7}$$

Further, for each function F_j $(j \neq i)$ there exists a collection of values of its arguments (i.e., arguments from J_j) which converts it into a_j. The statement of the lemma derives from this and from (7), since $F' = \varphi(F_1, \ldots, F_\varkappa)$.

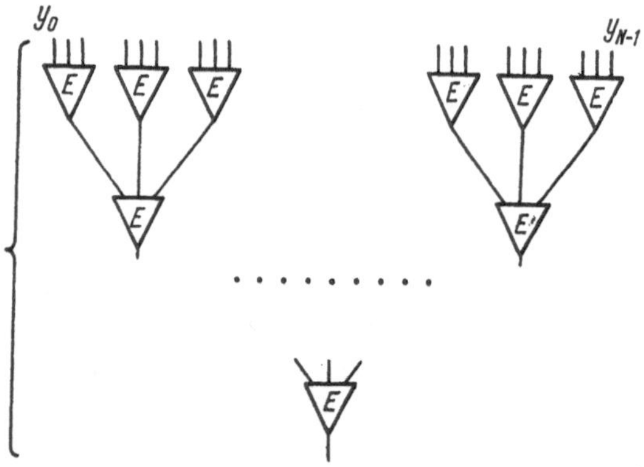

Fig. 4

Lemma 19. Assume $F(y_0, \ldots, y_{N-1})$ is the function realized by the (E, t)-block S. Then there exist constants $c_{\eta,\xi}$ $(0 \leq \eta, \xi \leq N-1)$ which have the following properties.

1) Assume I is an arbitrary interval of numbers of variables, and $\alpha \notin I$. Then all constants $c_{\beta,\alpha}$ for fixed α and any β from I are equal to one another.

2) $F(c_{\eta,0}, c_{\eta,1}, \ldots, c_{\eta,\eta-1}, y_\eta, c_{\eta,\eta+1}, \ldots, c_{\eta,N-1}) = y_\eta \oplus c_{\eta,\eta}$ $(\eta = 0, \ldots, N-1)$.

Proof. Assume $S_1, \ldots, S_\varkappa$ are principal subnetworks of the network S, which are (E, t − 1)-blocks; $F_1, \ldots, F_\varkappa$ are the functions realized by them; and $J_1, \ldots, J_\varkappa$ ($I_1, \ldots, I_\varkappa$, respectively) are the intervals of variables (the numbers of variables, respectively) of these subnetworks. By virtue of Lemma 18 there exists for any i $(1 \leq i \leq \varkappa)$ a collection of constants $c_\gamma^{(i)}$ $(\gamma \notin I_i$; i.e., $\gamma \notin [(i-1)\varkappa^{t-1}\ i\varkappa^{t-1}))$, which when substituted in place of the variables x_γ cause the function F to go over to $F_i \oplus a_i$ (a_i is a certain constant). Let us now assume that $c_{\eta,\zeta} = c_\zeta^{(i)}$. for all pairs (η, ζ) which are such that $\eta \in I_i$, $\zeta \in I_j$ and $i \neq j$. It is obvious that these constants $c_{\eta,\zeta}$ (these are part of all required constants) satisfy condition 1).

We deal analogously with "smaller" intervals. Assume, for example, that $S_{11}, \ldots, S_{1\varkappa}$ are principal subnetworks of the networks S_1, which are (E, t − 2) blocks, while $J_{11}, \ldots, J_{1\varkappa}$ ($I_{11}, \ldots, I_{1\varkappa}$, respectively) are intervals of variables (numbers of variables, respectively) of these subnetworks. Applying Lemma 18 to the network S_1, we obtain the result that for any i $(1 \leq i \leq \varkappa)$ there exist constants $c_\gamma^{(1,i)}$ $(\gamma \in I_1, \gamma \notin I_{1i})$, which when substituted in place of x_γ cause the function Φ_1 to go over into $F_{1i} \oplus a_{1i}$ (F_{1i} is the function realized by the network S_{1i}; a_{1i} is a certain constant). Let us now assume for all (η, ζ) such that $\eta \in I_{1i}$, $\zeta \in I_{1j}$ $i \neq j$, that $c_{\eta,\zeta} = c_\zeta^{(1,i)}$, etc. In the final analysis for all pairs (η, ζ) such that $\eta \neq \zeta$ there will be defined constants $c_{\eta,\zeta}$ which satisfy condition 1) and are such that for any η

$$F(c_{\eta,0}, \ldots, c_{\eta,\eta-1}, y_\eta, c_{\eta,\eta+1}, \ldots, c_{\eta,N-1}) = y_\eta \oplus c_{\eta,\eta},$$

where $c_{\eta,\eta}$ is a certain constant.

The lemma has been proved.

B. Partitioning of Collections

Lemma 20.† Assume that the element E has $\varkappa$ inputs; $F(y_0, \ldots, y_{N-1})$ is a function realized by an (E, t)-block; m is an integer and $M = \left]\frac{2^m}{N}\right[$. Then there exist

the numeration Σ of all sets of length n consisting of zeros and ones by pairs of subscripts: $\tilde{\sigma}_{l,\xi}$ $(0 \leqslant l \leqslant M-1;\ 0 \leqslant \xi \leqslant N-1)$‡;

† This lemma characterizes the construction of generalized expansion functions ψ for the case in which "an external expansion function" is a "homogeneous superposition" of functions which depend on a limited number of arguments.

‡ $0 \leq \xi \leq 2^m - 1 - (M-1)N$ for $l = M - 1$.

the system of functions $\psi_{lh}(y_h; x_1, \ldots, x_m)$, $0 \leqslant l \leqslant M-1$, $0 \leqslant h \leqslant N-1$, which have the properties:

1) $F(\psi_{l,0}(y_0, \tilde{\sigma}_{l,\xi}), \psi_{l,1}(y_1, \tilde{\sigma}_{l,\xi}), \ldots, \psi_{l,N-1}(y_{N-1}, \tilde{\sigma}_{l,\xi})) = y_\xi$; if $\tilde{\sigma} \notin \Sigma_l$, then

$$F(\psi_{l,0}(y_0, \tilde{\sigma}), \psi_{l,1}(y_1, \tilde{\sigma}), \ldots, \psi_{l,N-1}(y_{N-1}, \tilde{\sigma})) = 0,$$

2) any function ψ from the set $\Psi = \{\psi_{l,h}(\delta, x_1, \ldots, x_m)\}$ $(0 \leqslant l \leqslant M-1,\ 0 \leqslant h \leqslant N-1,\ \delta = 0, 1)$ satisfies the condition

$$I(\psi) \leqslant \varkappa t + 2.$$

Proof. Let us juxtapose each collection $\tilde{\sigma}$ of length m consisting of zeros and ones with a pair of numbers (λ, μ) which are defined as follows: $|\tilde{\sigma}| = \lambda N + \mu$; $0 \leqslant \lambda$; $0 \leqslant \mu < N$. It is obvious that λ and μ are defined uniquely according to the collection σ. Let us number the collection $\tilde{\sigma}$ by means of the indicated pairs of subscripts. Thereby the partitioning of the set of all collections $\tilde{\sigma}$ into ordered subsets Σ_l will be defined. Let us note that the collections from Σ_l form a piece.

Assume now that $c_{\eta,\zeta}$ $(0 \leq \eta, \zeta \leq N-1)$ is a collection of constants which have the properties 1) and 2) (Lemma 19) and that $(\varepsilon_0, \ldots, \varepsilon_{N-1})$ is a collection such that

$$F(\varepsilon_0, \varepsilon_1, \ldots, \varepsilon_{N-1}) = 0. \tag{8}$$

Let us now define the functions $\psi_{lh}(y_h, x_1, \ldots, x_m)$ as follows:

$$\psi_{lh}(y_h, \tilde{\sigma}_{\lambda,\mu}) = \begin{cases} \varepsilon_h, & \text{if } \lambda \neq l; \\ c_{\mu,h}, & \text{if } \lambda = l \text{ and } \mu \neq h; \\ y_h \oplus c_{h,h}, & \text{if } \lambda = l \text{ and } \mu = h. \end{cases}$$

Then the property (1) of the functions $\psi_{l,h}$ derives from the definition of these functions, the property 2) of the constants (see the formulation of Lemma 19), and (8).

Let us establish the property 2). The partitioning of the numbers of variables 0, 1, ..., N − 1 into intervals, which is defined by an (E, t)-block, induces the partitioning of each of the sets Σ_l into subsets which shall likewise be called intervals†; in particular, the sets Σ_l themselves will be intervals. It is clear that each of the intervals of the sets is a piece. Let us now consider the following system of partitionings of the sets of all collections $\tilde{\sigma}$ "relative to the collection $\tilde{\sigma}_{l,h}$":

0) $\Sigma_0, \Sigma_1, \ldots, \Sigma_{M-1}$;

1) Σ_l is partitioned into the intervals $\Sigma_{l1}, \ldots, \Sigma_{l\varkappa}$ having the length 1;

2) assume $\tilde{\sigma}_{l,h} \in \Sigma_{li_1}$. The interval Σ_{li_1} is partitioned into the intervals $\Sigma_{li_11}, \ldots, \Sigma_{li_1\varkappa}$ having a height 2, etc. Finally,

5) assume $\tilde{\sigma}_{lh} \in \Sigma_{l,i_1,\ldots,i_{t-1}}$. The interval $\Sigma_{l,i_1,\ldots,i_{t-1}}$ is partitioned into intervals $\Sigma_{l,i_1,\ldots,i_{t-1},1}, \ldots, \Sigma_{l,i_1,\ldots,i_{t-1},\varkappa}$ having the height t. Each of these intervals contains exactly one collection; assume $\Sigma_{l,i_1,\ldots,i_t} = \{\tilde{\sigma}_{lh}\}$.

From the definition of the function $\psi_{l,h}(y_h, x_1, \ldots, x_m)$ and the properties of the constants $c_{\eta,\zeta}$ it follows that the function $\psi_{l,h}(\delta, x_1, \ldots, x_m)$ is constant on each of the intervals

† In the set Σ_{M-1} the images of certain intervals may be shorter or may be absent altogether.

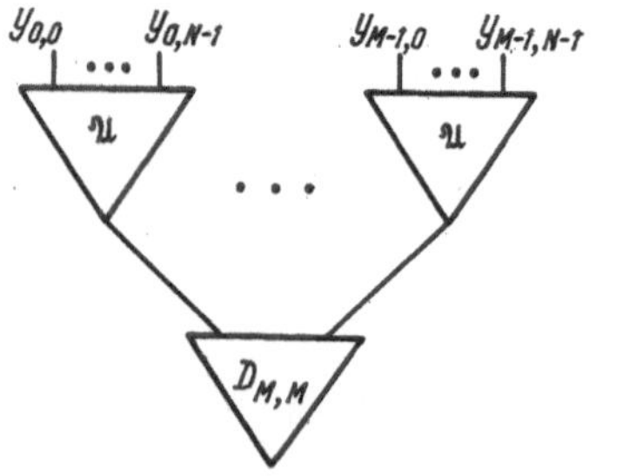

Fig. 5

Σ_i $(i \neq l)$, Σ_{li} $(i \neq i_1)$, Σ_{li_1i} $(i \neq i_2)$, etc., it being equal to ε_h on all intervals Σ_i $(i \neq l)$. Thus, $\psi_{lh}(\delta, x_1, \ldots, x_m)$ is equal to the disjunction of the characteristic functions of those of the enumerated intervals on which it is equal to 1. The number of such intervals, whose height is at least 1, does not exceed $\varkappa t$. Intervals of height 0 make up no more than two pieces $\Sigma_0 \cup \ldots \cup \Sigma_{l-1}$ and $\Sigma_{l+1} \cup \ldots \cup \Sigma_{\mu-1}$ (or are not needed at all). Finally, we have

$$I(\psi_{lh}(\delta, x_1, \ldots, x_m)) \leqslant \varkappa t + 2.$$

The lemma has been proved completely.

From the second statement of Lemma 20 and from Lemma 16 the following corollary derives directly.

Corollary.

C. An (E, t, m)-Block and the Expansion of Functions Which Are Associated with It. Assume $\mathfrak{A}$ is an (E, t)-block, and $M = \left]\frac{2^m}{N}\right[$. The (E, t, m)-block $\mathfrak{B}$ consists of M copies of the block $\mathfrak{A}$, which are attached to the inputs of the network $D_{M,M}$ (Fig. 5); the inputs of the l-th $\mathfrak{A}$ block are assigned the variables

$$y_{l,0}, \ldots, y_{l,N-1} \qquad (l = 0, 1, \ldots, M-1).$$

It is obvious that the function $\Phi(y_{0,0}, \ldots, y_{M-1,N-1})$ realized by an (E, t, m)-block is equal to $F(y_{0,0}, \ldots, y_{0,N-1}) \vee \ldots \vee F(y_{M-1,0}, \ldots, y_{M-1,N-1})$. From Lemma 20 the following property of the function Φ derives.

Assume $\tilde{\sigma}$ is an arbitrary collection of length m, and $\tilde{\sigma} = \tilde{\sigma}_{l,h}$. Then

$$\Phi(\psi_{0,0}(y_{0,0}, \tilde{\sigma}), \ldots, \psi_{0,N-1}(y_{0,N-1}, \tilde{\sigma}), \ldots, \psi_{M-1,0}(y_{M-1,0}, \tilde{\sigma}), \\ \ldots, \psi_{M-1,N-1}(y_{M-1,N-1}, \tilde{\sigma})) = y_{lh}. \tag{9}$$

It is clear that

$$L(\mathfrak{B}) = ML(\mathfrak{A}) + L(D_{M,M}) < \frac{\varkappa^t M}{\varkappa - 1} P + C_9 M, \tag{10}$$

$$T(\mathfrak{B}) = T(\mathfrak{A}) + T(D_{M,M}) < tT + C_{10} \log M. \tag{11}$$

Assume Φ is a function realized by an (E, t, m)-block, while Σ and Ψ are the corresponding numerations of the collections of length m and the system of functions $\psi_{l,h}(\delta, \tilde{z})$ (Lemma 20). Let us introduce a certain abridged notation which will often be used hereafter. Assume G is the system of functions $g_{l,h}(v_1, \ldots, v_k)$ $(0 \leq l \leq M-1,\ 0 \leq h \leq N-1)$ and that the function $g_{G,\Sigma}(\tilde{z}, \tilde{v})$ is defined in the following manner:

$$g_{G,\Sigma}(\tilde{\sigma}_{l,h}, v_1, \ldots, v_k) = g_{l,h}(v_1, \ldots, v_k).$$

The function

$$\Phi(\psi_{0,0}(g_{0,0}(\tilde{v}), \tilde{z}), \psi_{0,1}(g_{0,1}(\tilde{v})\tilde{z}), \ldots, \psi_{M-1,N-1}(g_{M-1,N-1}\tilde{v}), \tilde{z}))$$

is denoted by $[\Phi, \Sigma, \Psi, G]$.

Lemma 21.

$$g_{G,\Sigma}(\tilde{z}, \tilde{v}) = [\Phi, \Sigma, \Psi, G].$$

The statement of the lemma derives directly from the definitions of the functions $g_{G,\Sigma}$; $[\Phi, \Sigma, \Psi, G]$, and from (9).

D. A Certain Generalization. We shall consider logic-algebra functions which depend on $\tilde{x} = (x_1, \ldots, x_w)$. Assume $\mathfrak{M}$ is a certain set of collections of the values $\tilde{x}$. Two functions $f(\tilde{x})$ and $g(\tilde{x})$ shall be called equal on $\mathfrak{M}$, if $f(\tilde{\alpha}) = g(\tilde{\alpha})$ for any collection α from $\mathfrak{M}$ (use the notation $f(\tilde{x}) \underset{\mathfrak{M}}{=} g(\tilde{x})$). For certain sets equality on them will be denoted in a special manner. For the set $\mathfrak{M}_0$, consisting of all collections with the exception of $\tilde{0} = (0, \ldots, 0)$ and $\tilde{1} = (1, \ldots, 1)$ we shall likewise use the symbol $\approx$ in place of the symbol $\underset{\mathfrak{M}_0}{=}$. For the set $\mathfrak{M}_{a,b}$ of collections $(\tilde{\alpha}, \tilde{\beta})$ having the length $a+b$ ($\tilde{\alpha}$ has the length a, $\tilde{\beta}$ has the length b) which are such that $\tilde{\alpha} \neq \tilde{0}$, $\tilde{\alpha} \neq \tilde{1}$, $\tilde{\beta} \neq \tilde{0}$, $\tilde{\beta} \neq \tilde{1}$, we shall likewise use the symbol $\underset{a,b}{=}$ in place of the symbol $\underset{\mathfrak{M}_{a,b}}{=}$.

The following analog of Lemma 21 holds.

Lemma 22. Assume $\mathfrak{M}$ is a certain subset of the collections of values $(\tilde{u}, \tilde{v})$, $\tilde{u} = (u_1, \ldots, u_m)$, $v = (v_1, \ldots, v_k)$. Assume the systems of functions $\Psi^* = \{\psi^*_{l,h}(y_{l,h}, \tilde{u})\}$ and $G^* = \{g^*(\tilde{\sigma}, \tilde{v})\}$ satisfy the conditions

$$\Psi^*_{l,h}(\delta, \tilde{u}) \underset{\mathfrak{M}}{=} \Psi_{l,h}(\delta, \tilde{u}); \quad g^*(\tilde{\sigma}, \tilde{v}) \underset{\mathfrak{M}}{=} g(\tilde{\sigma}, \tilde{v}).$$

Then

$$[\Phi, \Sigma, \Psi^*, G^*] \underset{\mathfrak{M}}{=} [\Phi, \Sigma, \Psi, G].\dagger$$

Remark 1. If

$$\psi^*_{l,h}(\delta, \tilde{u}) \approx \psi_{l,h}(\delta, \tilde{u}), \quad g^*(\tilde{\sigma}, \tilde{v}) \approx g(\tilde{\sigma}, \tilde{v}),$$

then

$$[\Phi, \Sigma, \Psi^*, G^*] \underset{m,k}{=} [\Phi, \Sigma, \Psi, G].$$

Remark 2. If

$$\Psi^*_{l,h}(\delta, \tilde{u}) \approx \Psi_{l,h}(\delta, \tilde{u}), \quad g^*(\tilde{\sigma}, \tilde{u}) \approx g(\tilde{\sigma}, \tilde{u})$$

[the functions $g^*(\tilde{\sigma}, \tilde{u})$ depend on the same variables as those on which the functions $\psi_{l,h}(\delta, \tilde{u})$ depend!), it follows that

$$[\Phi, \Sigma, \Psi^*, G] \approx g(\tilde{u}, \tilde{u}).$$

§ 7. The Expansion of Functions

Assume $f(x_1, \ldots, x_n)$ is an arbitrary logic-algebra function. We shall partition its variables into three groups:

† Since on collections from $\mathfrak{M}$ all functions "corresponding to one another" which are entered in $[\Phi, \Sigma, \Psi^*, G^*]$ and in $[\Phi, \Sigma, \Psi, G]$ take equal values.

TABLE 1

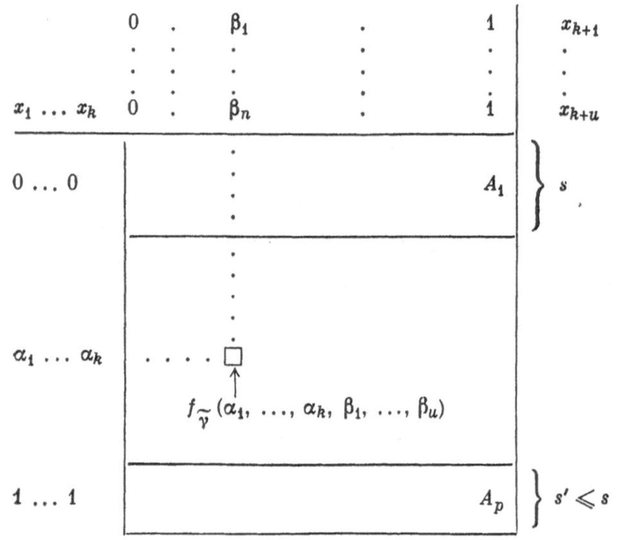

$$\widetilde{x}=(x_1, \ldots, x_k),\ \widetilde{y}=(x_{k+1}, \ldots, x_{k+u}),$$
$$\widetilde{z}=(x_{k+u+1}, \ldots, x_n);$$

let us denote $n-k-u$ by m. Let us use $f_{\widetilde{\gamma}}(\widetilde{x}, \widetilde{y})$ to denote the function $f(\widetilde{x}, \widetilde{y}, \widetilde{\gamma})$.

Assume E' is an element of a basis on which τ is attained. Assume this element has the weight P', the delay T', and a number of inputs equal to k'. Then $T' = \tau \log k'$. Assume t' is a certain parameter which satisfies the condition (for $n \to \infty$)

$$\frac{(k')^{t'}}{2^m} \to 0. \tag{12}$$

Let us now consider an (E', t')-block (and the corresponding partitioning Σ' of the collections $\widetilde{\gamma}$ of length m – Lemma 20), the (E', t', m)-block $\mathfrak{B}'$ (and the function Φ' and system of functions Ψ' which correspond to it), and the system F' of functions $f(\widetilde{x}, \widetilde{y}, \widetilde{\gamma}_{l', h'})$. Then the function $f(\widetilde{x}, \widetilde{y}, \widetilde{z})$ allows the representation (Lemma 21)

$$f(\widetilde{x}, \widetilde{y}, \widetilde{z})=[\Phi', \Sigma', \Psi', F']. \tag{13}$$

Assume $M' = \left]\frac{2^m}{(k')^{t'}}\right[$. Then by virtue of (12) we have

$$M' \sim \frac{2^m}{(k')^{t'}} \tag{14}$$

and for the block $\mathfrak{B}'$ we have [see (10), 11)]

$$\left.\begin{array}{l} L(\mathfrak{B}') < \frac{(k')^{t'} M'}{k'-1} P' + C_9 M' \leqslant C_{11} 2^m, \\ T(\mathfrak{B}') = t'T' +]\log M'[\, T^{(1)} = \tau t' \log k' +]\log M'[\, T^{(1)}. \end{array}\right\} \tag{15}$$

Each function $f_{\widetilde{\gamma}}(\widetilde{x}, \widetilde{y})$ may be stipulated by a table having two inputs (Table 1).

In this table the value of the function on the collection $(\alpha_1, \ldots, \alpha_k, \beta_1, \ldots, \beta_n)$ is placed at the intersection of the row corresponding to $(\alpha_1, \ldots, \alpha_k)$, and the column corresponding to $(\beta_1, \ldots, \beta_n)$. Let us partition the matrix which determines the values of the function $f_{\widetilde{\gamma}}(\widetilde{x}, \widetilde{y})$ into strips $A_1, \ldots, A_p$, each of which (except, perhaps, for the last) contains s rows, while the last contains s', $s' \leq s$ rows. The number p of strips satisfies the condition

$$p \leqslant \frac{2^k}{s} + 1. \tag{16}$$

Let us use $f_{\widetilde{\gamma}_{l', h'}, i}$ to denote the function which coincides with $f_{\widetilde{\gamma}_{l', h'}}$ on the strip A_i and is equal to 0 outside the strip A_i. It is clear that

$$f_{\widetilde{\gamma}_{l', h'}}(\widetilde{x}, \widetilde{y}) = \bigvee_{i=1}^{p} f_{\widetilde{\gamma}_{l', h'}, i}(\widetilde{x}, \widetilde{y}). \tag{17}$$

Assume E" is an element of the basis on which ρ is attained. Assume that this element has the weight P", the delay T", and a number of inputs equal to k". Then $P'' = \rho(k'' - 1)$. Assume t" is a certain parameter which satisfies the conditions

$$t'' \to \infty, \tag{18}$$

$$\frac{(k'')^{t''}}{2^u} \to 0. \tag{19}$$

Let us now consider an (E", t")-block (and the corresponding partitioning Σ'' of steps of length u – Lemma 20), the (E", t", u)-block $\mathfrak{B}''$ (and the function Φ'' and system of functions Ψ'' corresponding to it), and the system $F''_{l',h',i}$ of functions $f_{\tilde{\gamma}_{l',h',i}}(\tilde{x}, \tilde{\beta}_{l'',h''})$. Then the function $f_{\tilde{\gamma}_{l',h',i}}(\tilde{x}, \tilde{y})$ allows the representation

$$f_{\tilde{\gamma}_{l',h',i}}(\tilde{x}, \tilde{y}) = [\Phi'', \Sigma'', \Psi'', F''_{l',h',i}]. \tag{20}$$

Assume $M'' = \left]\frac{2^u}{(k'')^{t''}}\right[$. Then by virtue of (19), (18) we have

$$M'' \sim \frac{2^u}{(k'')^{t''}}, \quad M'' = o(2^u) \tag{21}$$

and for the block $\mathfrak{B}''$ we have

$$\left.\begin{aligned} L(\mathfrak{B}'') &< \frac{(k'')^{t''}M''}{k''-1}P'' + C_9 M'' \leqslant \rho 2^u, \\ T(\mathfrak{B}'') &= t''T'' +]\log M''[\,T^{(1)} \leqslant C_{12}(t''+u). \end{aligned}\right\} \tag{22}$$

§ 8. The Method of Synthesizing Networks in the Case of a Regular Basis

A network will have several "layers" (separated from one another by the horizontal lines in Fig. 6). Within each layer the network consists either of identical elements or of canonical networks having identical delay. The network will consist of blocks of two kinds:

Blocks which calculate "intentional" functions (denoted by the letter A with subscripts);

Matching blocks which are used to match the delays of various parts of the network. Blocks which transmit constants (for the production of delays in accordance with Lemma 15) will likewise be associated with matching blocks. Matching blocks will be denoted by the letter Z with subscripts.

In describing blocks their complexity, delay, and (in certain cases) number of outputs (denoted by the letter Ω will be indicated.

In descriptions of this and the two subsequent methods of synthesis, blocks which realize identical (or similar) functions have identical numbers. This explains the gaps in the numeration of the blocks.

If a certain block is situated inside a layer in which the elements (canonical networks) have a delay T, while the entire layer has a delay tT, it follows that the expression "the block realizes φ with any delay" implies that for any i $(0 \le i \le t)$ there exists an output of this block on which φ is realized with the delay iT (relative to the inputs of this block).

Let us now go over directly to the description of the blocks of the network (see Fig. 6).

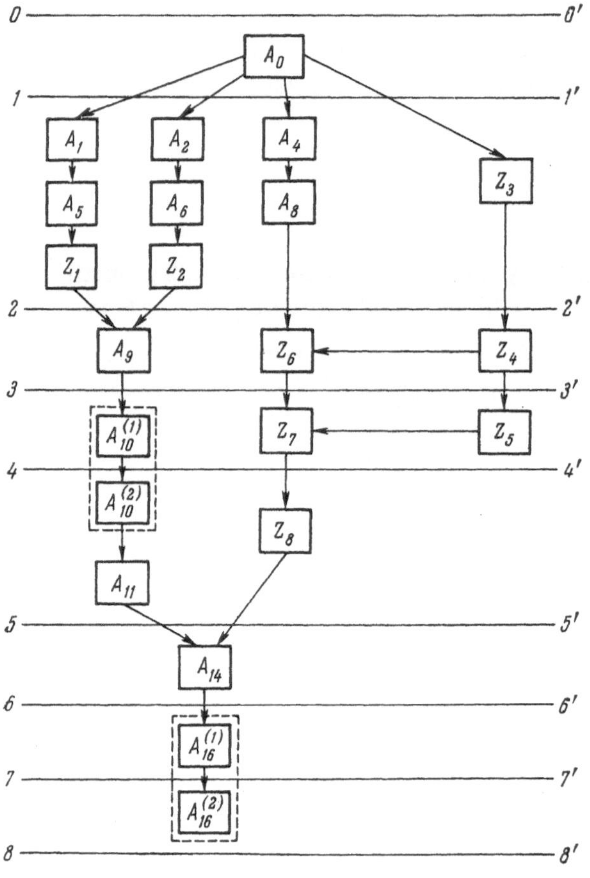

Fig. 6

The block A_0 realizes the constants 0, 1, the variables $x_1, \ldots, x_n$, and their negations $\bar{x}_1, \ldots, \bar{x}_n$. Since the basis is regular, this can be accomplished with one and the same delay T_0:

$$L(A_0) \leqslant C_{100}n, \quad T(A_0) = T_0.$$

The block A_1 realizes the conjunctions $x_1^{\sigma_1} \ldots x_k^{\sigma_k}$ (2^k of them) and the constant 0 on the basis of the functions $0, x_1, \ldots, x_n, \bar{x}_1, \ldots, \bar{x}_n$ already realized by the block A_0. Each conjunction is realized by a $K_{k,k}$ network (Lemma 8); the constant 0 is transmitted to the output of the network by means of a chain of canonical networks S_1. Therefore,

$$L(A_1) \leqslant C_{101}k2^k, \quad T(A_1) =]\log k[T^{(1)}.$$

The block A_2 realizes all conjunctions $x_{k+1}^{\sigma_{k+1}} \ldots x_{k+u}^{\sigma_{k+u}}$ and the constant 0,

$$L(A_2) \leqslant C_{102}u2^u, \quad T(A_2) =]\log u[T^{(1)}.$$

The block A_4 realizes all conjunctions $x_{n-m+1}^{\sigma_{n-m+1}} \ldots x_{n-m+j}^{\sigma_{n-m+j}}$, $0 \leq j \leq m$ (for $j = 0$ the conjunction is empty and identically equal to 1; the number of all functions is equal to 2^{m+1}) and the constant 0;

$$L(A_4) \leqslant C_{103}m2^m, \quad T(A_4) =]\log m[T^{(1)}.$$

The block A_5 realizes all possible different functions $f_{\tilde{\gamma}_{l', h'}, i}(\tilde{x}, \tilde{\beta}_{l'', h''})$. Each of them is obtained as a disjunction of no more than s conjunctions which have been realized by the block A_1 using one network $D_{r,s}$ for a certain r. We have

$$\Omega(A_5) \leqslant p2^s;$$

further, by virtue of Lemma 9,

$$L(A_5) \leqslant C_{104} ps2^s; \qquad T(A_5) =]\log s[T^{(1)}.$$

The block A_6 realizes all possible functions $\psi''_{l'', h''}(\delta, \tilde{y})$, which are obtained from the functions of the system Ψ'' as a result of substituting the constants 0 and 1 in place of the first argument. Each of them may be obtained as a disjunction of no more than 2^u conjunctions,† which are realized by the block A_2 using one network $D_{r, 2^u}$ for a certain r. We have, taking account of (19),

$$\Omega(A_6) = 2M''(k'')^{t''} \leqslant C_{105} 2^u;$$
$$L(A_6) \leqslant C_{106} M''(k'')^{t''} 2^u \leqslant C_{107} 2^{2u}; \qquad T(A_6) = uT^{(1)}.$$

The block A_8 realizes all possible functions $\psi'_{l', h'}(\delta, \tilde{z})$. By virtue of the corollary from Lemma 20, each of these functions is a disjunction of no more than (k't' + 2)2m conjunctions realized by the block A_4, and may be obtained using one network $D_{r, (k't'+2)2m}$. We have, taking account of (14),

$$\Omega(A_8) = 2M'(k')^{t_1} \leqslant C_{108} 2^m.$$

Further, since k' is a constant,

$$\left.\begin{aligned} m &\leqslant n, \\ t' \leqslant C_{109} m &\leqslant C_{109} n, \end{aligned}\right\} \tag{23}$$

it follows that

$$(k't' + 2)\, 2m \leqslant C_{110} n^2. \tag{24}$$

Therefore,

$$L(A_8) \leqslant C_{111} 2^m n^2; \qquad T(A_8) =]\log((k't'+2)\, 2m)[T^{(1)}.$$

Let us introduce the notation

$$T^* =]\log m[+]\log((k't'+2)\, 2m)[.$$

We shall assume that the conditions (for fairly large n)

$$]\log k[+]\log s[\leqslant T^*, \qquad]\log u[+ u \leqslant T^*, \tag{25}$$

$$k = O(\log n) \tag{26}$$

are fulfilled. From (24) it follows that

$$T^* \leqslant C_{112} \log n. \tag{27}$$

† Here it is sufficient to use a very rough bound for the complexity.

In order to equalize the delays of the networks A_1-A_5, A_2-A_6, A_4-A_8, we attach systems of chains of canonical networks S_1 of appropriate length to the outputs of the blocks A_5 and A_6. Assume Z_1 and Z_2 are blocks consisting of these systems of delays:

$$L(Z_1) \leqslant C_{113} p 2^s \log n,$$
$$L(Z_2) \leqslant C_{114} 2^u \log n.$$

The block A_9 realizes all possible different superpositions† $\psi''(f_{\widetilde{\gamma},i}(\widetilde{x}, \widetilde{\beta}), \widetilde{y})$ on the basis of the relationship

$$\psi''(f_{\widetilde{\gamma},i}(\widetilde{x}, \widetilde{\beta}), \widetilde{y}) = \zeta(f_{\widetilde{\gamma},i}(\widetilde{x}, \widetilde{\beta}), \psi''(0, \widetilde{y}), \psi''(1, \widetilde{y})).$$

One network S_9 is used for the realization of each of them:

$$L(A_9) < C_{115} 2^u p 2^s, \quad T(A_9) = T^{(4)}.$$

The block A_{10} realizes all functions $f_{\widetilde{\gamma}_{l'}, h', i}(\widetilde{x}, \widetilde{y})$ in accordance with the representation (20) (see likewise p. 59). One (E", t", u)-block $\mathfrak{B}''$. is used in the realization of each of these functions. Therefore [see (22)]

$$L(A_{10}) = M'(k')^{t'} pL(\mathfrak{B}''),$$
$$T(A_{10}) = T(\mathfrak{B}'') = t''T'' +]\log M''[T^{(1)}.$$

Let us require that the condition

$$\frac{2^k}{s} \to \infty \tag{28}$$

be satisfied. From (14), (16), (28), (22) we have

$$L(A_{10}) \lesssim \rho \frac{2^n}{s}.$$

For convenience we shall assume that the block A_{10} consists of two parts: The block $A_{10}^{(1)}$ containing all elements E"

$$T(A_{10}^{(1)}) = t''T'',$$

and the block $A_{10}^{(2)}$ containing the remaining elements

$$T(A_{10}^{(2)}) =]\log M''[T^{(1)}.$$

From (19), (25), (21), (27) it follows that

$$t'' = O(\log n), \tag{29}$$

$$\log M'' = O(\log n). \tag{30}$$

The block A_{11} realizes the functions $f(\widetilde{x}, \widetilde{y}, \widetilde{\gamma}_{l', h'})$ in accordance with (17). One network $D_{p,p}$ is used for one function. Therefore [see (25), (26), (27)]

$$L(A_{11}) = 2^m L(D_{p,p}) \leqslant C_{116} p 2^m, \tag{31}$$
$$T(A_{11}) =]\log p[T^{(1)} \leqslant C_{117} \log n.$$

† We drop the indices β', h', β'', h".

The block Z_3 "transmits the constants 0 and 1" to the 2-2' level. It consists of two chains of S_1 networks. From (27) it follows that

$$L(Z_3) \leqslant C_{118} \log n.$$

The block Z_4 transmits the constants 0 and 1 to the 3-3' level (see Lemma 15)

$$L(Z_4) \leqslant C_{119}.$$

The block Z_5 realizes the constants 0 and 1 with all the delays between 3-3' and 4-4' (see remark of Lemma 15). We have, taking account of (29),

$$L(Z_5) \leqslant C_{120} t'' \leqslant C_{121} \log n.$$

We shall assume the following condition is satisfied:

$$t' \text{ and } t'' \text{ are even numbers.} \tag{32}$$

The block Z_6 transmits the functions of the block A_8 to the 3-3' level (see remark of Lemma 15). Taking account of (14), we have

$$L(Z_6) \leqslant C_{122} 2^m.$$

The block Z_7 transmits these functions to the 4-4' level

$$L(Z_7) \leqslant C_{123} t'' 2^m \leqslant C_{124} 2^m \log n.$$

The block Z_8 transmits these functions to the output level of the block A_{11}. This block consists of chains of S_1 networks. We have [see (30), (31)]

$$L(Z_8) \leqslant C_{125} 2^m \log n.$$

The block A_{14} realizes the functions $\psi'_{l', h'}(f(\tilde{x}, \tilde{y}, \tilde{\gamma}_{l', h'}), \tilde{z})$ on the basis of the relationship

$$\psi'_{l', h'}(f(\tilde{x}, \tilde{y}, \tilde{\gamma}_{l', h'}), \tilde{z}) = \zeta(f(\tilde{x}, \tilde{y}, \tilde{\gamma}_{l', h'}), \psi'(0, \tilde{z}), \psi'(1, \tilde{z}))$$

and is arranged similarly to the block A_9

$$L(A_{14}) \leqslant C_{126} 2^m, \qquad T(A_{14}) = T^{(4)}.$$

The block A_{16} realizes the function $f(\tilde{x}, \tilde{y}, \tilde{z})$ in accordance with the representation (13). This block is the (E', t', m)-block $\mathfrak{B}'$. Therefore [see (15)]

$$L(A_{16}) \leqslant C_{127} 2^m, \qquad T(A_{16}) = \tau t' \log k' +] \log M' [T^{(1)}.$$

Let us now assume that

$$\left. \begin{array}{l} k = [2 \log n], \qquad u = [2 \log n], \qquad s = [n - 5 \log n], \\ t' = 2\left[\dfrac{1}{2 \log k'}(n - 5 \log n)\right], \quad t'' = 2\left[\dfrac{\log n}{2 \log k''}\right]. \end{array} \right\} . \tag{33}$$

Then the conditions (12), (18), (19), (25), (26), (28), (32) are fulfilled. It is likewise easy to check the fact that in this case $L(A_{10}) \lesssim \rho(2^n/n)$ and $L(A_i) = o(2^n/n)$ for $i \neq 10$; $L(Z_i) = o(2^n/n)$. Further, for $i \neq 16$, we have $T(A_i) = O(\log n)$;

$$\log M' \leqslant m - t' \log k' + O(1) = \log n + O(1).$$

Therefore,

$$T(A_{16}) \leqslant \tau t' \log k' + O(\log n) \sim \tau n.$$

Thus, for the entire network S we have

$$L(S) \leqslant \rho \frac{2^n}{n}, \quad T(S) \leqslant \tau n.$$

Thereby the upper bound of Shannon function is established in the case of a regular basis.

Remark. For the particular case of a regular basis, for which all of the elements have integer delays and there is a unitary delay in the basis, the synthesis method is somewhat simpler than in the general case. In this particular case:

1) It is not necessary to equalize the delays in each layer (equalization before union of the networks only is sufficient);

2) Instead of the networks based on Lemma 15 which simulate chains of delays one may use "conventional" delays. Therefore, instead of eight blocks which in one way or another are associated with the execution of the delay function (Z_1-Z_8) it is sufficient to have only two (consisting of chains of unitary delays);

one for equalizing the delays in the networks A_1-A_5 and A_2-A_6,

the second for equalizing the delays in the networks (A_1, A_2, A_{11}) and A_4-A_8.

§ 9. Method of Synthesizing Networks for 0-Functions in the Case of an Irregular Basis†

A. Auxiliary Functions and the Additional Expansion of 0-Functions. Let us introduce the notation:

$$\varkappa_0(u_1, \ldots, u_a) = \bar{u}_1\bar{u}_2 \ldots \bar{u}_a, \qquad \varkappa_1(u_1, \ldots, u_a) = u_1 u_2 \ldots u_a,$$
$$\lambda_a(u_1, \ldots, u_a) = \bar{\varkappa}_0(u_1, \ldots, u_a)\,\bar{\varkappa}_1(u_1, \ldots, u_a),$$
$$\mu_{a,b}(u_1, \ldots, u_a, v_1, \ldots, v_b) = \lambda_a(u_1, \ldots, u_a)\,\lambda_b(v_1, \ldots, v_b). ‡$$

It can easily be seen that

$$\lambda_a(u_1, \ldots, u_a) = (u_1 \vee \ldots \vee u_a)(\bar{u}_1 \vee \ldots \vee \bar{u}_a) = u_1\bar{u}_2 \vee u_2\bar{u}_3 \vee \ldots \vee u_{a-1}\bar{u}_a \vee u_a\bar{u}_1. \tag{34}$$

Let us determine the following functions of the variables $x_1, \ldots, x_n$ (the variables will be assumed to be partitioned into three groups $\tilde{x}, \tilde{y}, \tilde{z}$ which contain k, u, and m variables, respectively; see §7).

1. The functions of $\tilde{x}, \tilde{y}, \tilde{z}$ (Table 2):

$$e_1(\tilde{x}, \tilde{y}, \tilde{z}) = \mu_{k+u,m}(\tilde{x}, \tilde{y}, \tilde{z}) = \lambda_{k+u}(\tilde{x}, \tilde{y})\,\lambda_m(\tilde{z}),$$
$$e_2(\tilde{x}, \tilde{y}, \tilde{z}) = \varkappa_0(\tilde{x}, \tilde{y})\,\lambda_m(\tilde{z}), \qquad e_3(\tilde{x}, \tilde{y}, \tilde{z}) = \varkappa_1(\tilde{x}, \tilde{y})\,\lambda_m(\tilde{z}),$$
$$e_4(\tilde{x}, \tilde{y}, \tilde{z}) = \varkappa_0(\tilde{z})\,\lambda_{k+u}(\tilde{x}, \tilde{y}), \qquad e_5(\tilde{x}, \tilde{y}, \tilde{z}) = \varkappa_1(\tilde{z})\,\lambda_{k+u}(\tilde{x}, \tilde{y}),$$
$$e_6(\tilde{x}, \tilde{y}, \tilde{z}) = \varkappa_1(\tilde{x}, \tilde{y})\,\varkappa_0(\tilde{z}) = x_1 \ldots x_{k+u}\bar{x}_{k+u+1} \ldots \bar{x}_n,$$
$$e_7(\tilde{x}, \tilde{y}, \tilde{z}) = \varkappa_0(\tilde{x}, \tilde{y})\,\varkappa_1(\tilde{z}) = \bar{x}_1 \ldots \bar{x}_{k+u}x_{k+u+1} \ldots x_n.$$

† This method is also applicable for a regular basis; however, it is intentional only in the case of an irregular basis.

‡ This function is the characteristic function of the set $\mathfrak{M}_{a,b}$, introduced in §7.

TABLE 2

$\tilde{x}, \tilde{y}$ \ $\tilde{z}$	$\tilde{0}$		$\tilde{1}$
$\tilde{0}, \tilde{0}$		e_2	e_7
	e_4	e_1	e_5
$\tilde{1}, \tilde{1}$	e_6	e_3	

TABLE 3

$\tilde{x}$ \ $\tilde{y}$	$\tilde{0}$		$\tilde{1}$
$\tilde{0}$			d_2
		d_1	
$\tilde{1}$	d_2		

2. Functions of $\tilde{x}, \tilde{y}$ (Table 3):

$$d_1(\tilde{x}, \tilde{y}) = \mu_{k,u}(\tilde{x}, \tilde{y}) = \lambda_k(\tilde{x})\lambda_u(\tilde{y}),$$
$$d_2(\tilde{x}, \tilde{y}) = \varkappa_0(\tilde{x})\bar{\varkappa}_0(\tilde{y}) \vee \varkappa_0(\tilde{y})\bar{\varkappa}_0(\tilde{x}) \vee \varkappa_1(\tilde{x})\bar{\varkappa}_1(\tilde{y}) \vee \varkappa_1(\tilde{y})\bar{\varkappa}_1(\tilde{x}).$$

Assume further for the arbitrary O-function $f(\tilde{x}, \tilde{y}, \tilde{z})$ that

$$f^{(i)}(\tilde{x}, \tilde{y}\ \tilde{z}) = f(\tilde{x}, \tilde{y}, \tilde{z})\, e_i(\tilde{x}, \tilde{y}, \tilde{z}), \quad 1 \leqslant i \leqslant 7.$$

Since $f(\tilde{0}, \tilde{0}, \tilde{0}) = f(\tilde{1}, \tilde{1}, \tilde{1}) = 0$, it follows that

$$f(\tilde{x}, \tilde{y}, \tilde{z}) = \bigvee_{i=1}^{7} f^{(i)}(\tilde{x}, \tilde{y}, \tilde{z}).$$

For each function $f_{\tilde{\gamma}}(\tilde{x}, \tilde{y}) = f(\tilde{x}, \tilde{y}, \tilde{\gamma})$ (see §7) we define the functions

$$f_{\tilde{\gamma}}^{(i)}(\tilde{x}, \tilde{y}) = f_{\tilde{\gamma}}(\tilde{x}, \tilde{y})\, d_i(\tilde{x}, \tilde{y}), \quad i = 1, 2,$$
$$f_{\tilde{\gamma}}^{(0)}(\tilde{x}, \tilde{y}) = f_{\tilde{\gamma}}^{(1)}(\tilde{x}, \tilde{y}) \vee f_{\tilde{\gamma}}^{(2)}(\tilde{x}, \tilde{y}).$$

The following properties derive from the definitions given above:

1) $\gamma_{\tilde{\gamma}}^{(0)}(\tilde{x}, \tilde{y}) \approx f_{\tilde{\gamma}}(\tilde{x}, \tilde{y})$;

2) if a certain function $g(\tilde{x}, \tilde{y})$ satisfies the condition

$$g(\tilde{x}, \tilde{y}) \underset{k,\,u}{=\!=} f_{\tilde{\gamma}}(\tilde{x}, \tilde{y}), \text{ then } f_{\tilde{\gamma}}^{(1)}(\tilde{x}, \tilde{y}) = g(\tilde{x}, \tilde{y})\, d_1(\tilde{x}, \tilde{y});$$

2') analogously, if a certain function $g(\tilde{x}, \tilde{y}, \tilde{z})$ satisfies the condition $g(\tilde{x}, \tilde{y}, \tilde{z}) \underset{k+u,\,m}{=\!=} f(\tilde{x}, \tilde{y}, \tilde{z})$, then $f^{(1)}(\tilde{x}, \tilde{y}, \tilde{z}) = g(\tilde{x}, \tilde{y}, \tilde{z})\, e_1(\tilde{x}, \tilde{y}, \tilde{z})$;

3)
$$f^{(4)}(\tilde{x}, \tilde{y}, \tilde{z}) = f_{\tilde{0}}^{(0)}(\tilde{x}, \tilde{y})\, e_4(\tilde{x}, \tilde{y}, \tilde{z}),$$
$$f^{(5)}(\tilde{x}, \tilde{y}, \tilde{z}) = f_{\tilde{1}}^{(0)}(\tilde{x}, \tilde{y})\, e_5(\tilde{x}, \tilde{y}, \tilde{z}).$$

B. The Method of Synthesizing Networks. This method is more cumbersome than it is in the case of a regular basis. In order to simplify its description we shall indicate only the blocks which execute intentional functions. "Delay blocks" used to match delays are depicted by thick lines in Fig. 7. The complexity of delay blocks will be estimated

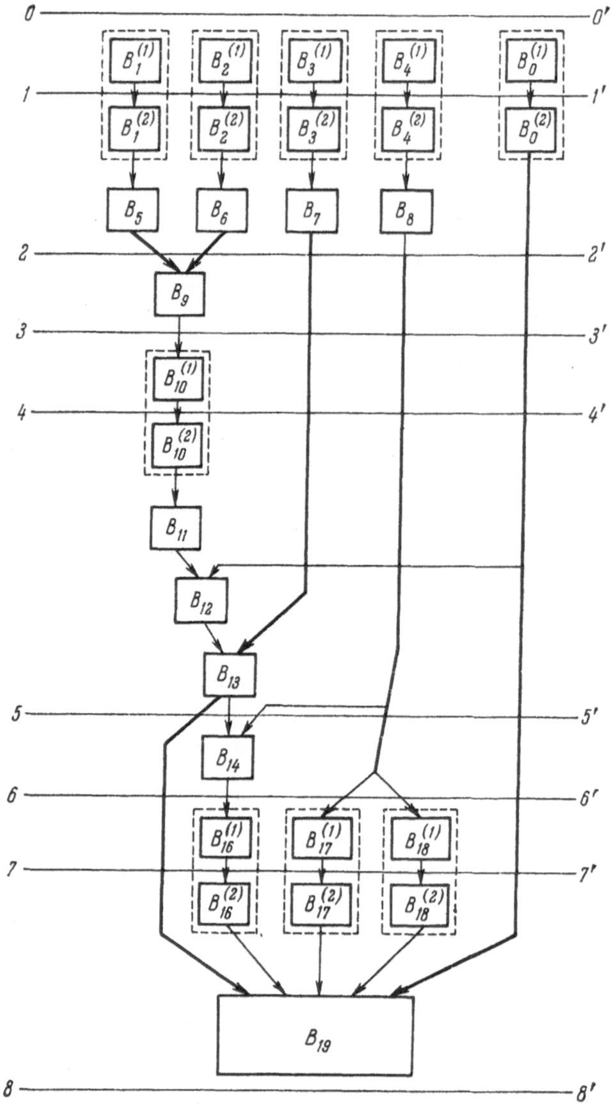

Fig. 7

at the end. In describing the method the notation from §7 will be used. The values of the parameters here are the same as those in §8. Therefore, the conditions (12), (18), (19), (25), (26), (28), (32) are fulfilled.

Let us go over to a description of the blocks.

The block B_0 (Fig. 8) realizes the auxiliary functions: 0, 1, e_i ($1 \leq i \leq 7$), d_1. It consists of the blocks B_{00}–B_{04}.

The block B_{00} realizes constants (for their use in delay blocks). It consists of the canonical S_4 and S_5 networks (Lemma 10)

$$L(B_{00}) = C_{200}, \quad T(B_{00}) = T^{(2)}.$$

The block B_{01} realizes the functions $\lambda_k(\tilde{x})$, $\lambda_u(\tilde{y})$, $\lambda_{k+u}(\tilde{x}, \tilde{y})$ and $\lambda_m(\tilde{z})$ in accordance with (34)

$$L(B_{01}) \leqslant C_{201} n,$$

$$T(B_{01}) = T^{(2)} +]\log n[\, T^{(1)}.$$

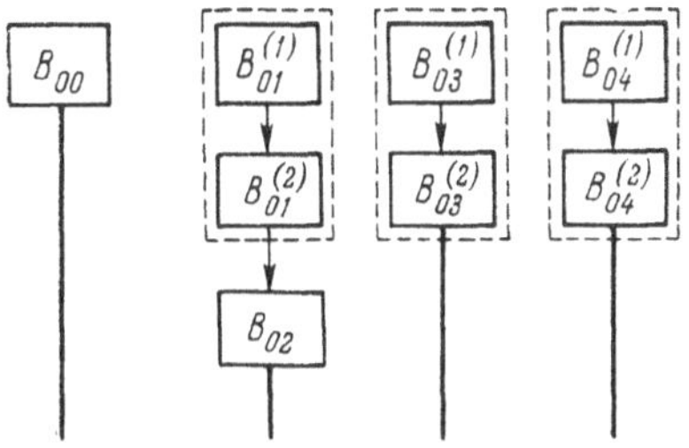

Fig. 8

The block B_{02} realizes $d_1(\tilde{x}, \tilde{y}) = \lambda_k(\tilde{x})\lambda_u(\tilde{y})$ and $e_1(\tilde{x}, \tilde{y}, \tilde{z}) = \lambda_{k+u}(\tilde{x}, \tilde{y})\lambda_m(\tilde{z})$,

$$L(B_{02}) = C_{202}, \quad T(B_{02}) = T^{(1)}.$$

The block B_{03} realizes the functions e_2-e_5 on the basis of the representation

$$e_2(\tilde{x}, \tilde{y}, \tilde{z}) = \bar{x}_1 \ldots \bar{x}_{n-m}x_{n-m+1}\bar{x}_{n-m+2} \vee \bar{x}_1 \ldots \bar{x}_{n-m}x_{n-m+2}\bar{x}_{n-m+3} \vee \ldots$$
$$\ldots \vee \bar{x}_1 \ldots \bar{x}_{n-m}x_{n-1}\bar{x}_n \vee \bar{x}_1 \ldots \bar{x}_{n-m}x_n\bar{x}_{n-m+1} \quad (35)$$

and analogous representations using Lemma 11 and networks

$$L(B_{03}) \leqslant C_{203}n^2, \quad T(B_{03}) = T^{(2)} + 2]\log n[T^{(1)}.$$

The block BO_4 realizes the functions e_6 and e_7

$$L(B_{04}) \leqslant C_{204}n, \quad T(B_{04}) = T^{(2)} +]\log n[T^{(1)}.$$

Thus,

$$L(B_0) \leqslant C_{205}n^2,$$
$$\max(T(B_{00}), T(B_{01} - B_{02}), T(B_{03}), T(B_{04})) \leqslant 2T^{(1)}\log n.$$

The blocks B_1 and B_2 realize the systems of conjunctions $K^0(\tilde{x})$ and $K^0(\tilde{y})$, respectively, and likewise the constant 0. We have (see Lemma 11)

$$L(B_1) \leqslant C_{206}k2^k, \quad T(B_1) = T^{(2)} +]\log k[T^{(1)},$$
$$L(B_2) \leqslant C_{207}u2^u, \quad T(B_2) = T^{(2)} +]\log u[T^{(1)}.$$

The block B_3 realizes the constant 0 and all possible conjunctions $x_1^{\alpha_1} \ldots x_k^{\alpha_k}x_{k+1}^{\beta_1} \ldots x_{k+u}^{\beta_u}$, which correspond to the collections $(\tilde{\alpha}, \tilde{\beta}) = (\alpha_1, \ldots, \alpha_k, \beta_1, \ldots, \beta_u)$, satisfying the condition

$$(\tilde{\alpha} = \tilde{0} \,\&\, \tilde{\beta} \neq \tilde{0}) \vee (\tilde{\alpha} = \tilde{1} \,\&\, \tilde{\beta} \neq 1) \vee (\tilde{\alpha} \neq \tilde{0} \,\&\, \tilde{\beta} = \tilde{0}) \vee (\tilde{\alpha} \neq \tilde{1} \,\&\, \tilde{\beta} = \tilde{1}).$$

The number of such conjunctions is smaller than [see (32)] $2(2^k + 2^u) = 2^{u+2}$. Therefore (Lemma 11),

$$L(B_3) \leqslant C_{208}u2^u,$$
$$T(B_3) = T^{(2)} +]\log(k+u)[T^{(1)}.$$

The block B_4 realizes all conjunctions $x_{n-m+1}^{\sigma_{n-m+1}} \ldots x_{n-m+j}^{\sigma_{n-m+j}}$, $1 \leqslant j \leqslant m$, containing at least one variable with negation, at least one variable without negation, and the constant 0:

$$L(B_4) \leqslant C_{209}m2^m,$$
$$T(B_4) = T^{(2)} +]\log m[T^{(1)}.$$

The block B_5 realizes the set of all functions $h(\tilde{x})$ which have the properties:

a) $h(\tilde{0}) = 0$, $h(\tilde{1}) = 0$;

b) $h(\tilde{x})$ outside of a certain band (see Table 1) is equal to 0. It is obvious that for each function $f_{\tilde{\nu}_{l',h'},i}(\tilde{x}, \tilde{\beta}_{l'',h''})$ there exists a function $h(\tilde{x})$, which is such that $f_{\tilde{\nu}_{l',h'},i}(\tilde{x}, \tilde{\beta}_{l'',h''}) \approx h(\tilde{x})$; this function $h(\tilde{x})$ will likewise be denoted by $f^0_{\tilde{\nu}_{l',h'},i}(\tilde{x}, \tilde{\beta}_{l'',h''})$. The system of all functions $f^0_{\tilde{\nu}_{l',h'},i}(\tilde{x}, \tilde{\beta}_{l'',h''})$ for fixed l', h', i shall be denoted by $F''^0_{l',h',i}$. It is clear that each function $h(\tilde{x})$ is a disjunction of no more than s conjunctions from $K^0(\tilde{x})$. By virtue of Lemma 9,

$$L(B_5) \leqslant C_{210} ps2^s, \quad T(B_5) =]\log s[\, T^{(1)}, \quad \Omega(B_5) \leqslant p2^s.$$

The block B_6 realizes the functions $\psi''^0_{l'',h''}(\delta, \tilde{y}) = \psi''_{l'',h''}(\delta, \tilde{y})\, \bar{\varkappa}_0(\tilde{y})\, \bar{\varkappa}_1(\tilde{y})$ as disjunctions of conjunctions from $K^0_u(\tilde{y})$

$$L(B_6) \leqslant C_{211} 2^{2u}, \quad T(B_6) = uT^{(1)}, \quad \Omega(B_6) \leqslant C_{212} 2^u.$$

It is obvious that

$$\psi''^0_{l'',h''}(\delta, \tilde{y}) \approx \psi''_{l'',h''}(\delta, \tilde{y}).$$

The system of functions

$$\psi''^0_{l'',h''}(0, \tilde{y})\, \bar{y}_{l'',h''} \vee \psi''^0_{l'',h''}(1, \tilde{y})\, y_{l'',h''} = \zeta(y_{l'',h''}, \psi''^0_{l'',h''}(0, \tilde{y}), \psi''^0_{l'',h''}(1, \tilde{y}))$$

shall be denoted by Ψ''^0.

The block B_8 realizes the functions $\psi'^0_{l',h'}(\delta, \tilde{z}) = \psi'_{l',h'}(\delta, \tilde{z})\, \bar{\varkappa}_0(\tilde{z})\, \bar{\varkappa}_1(\tilde{z})$, which are analogous to the functions $\psi''^0_{l'',h''}(\delta, \tilde{y})$. Therefore,

$$\psi'^0_{l',h'}(\delta, \tilde{z}) \approx \psi'_{l',h'}(\delta, \tilde{z}). \tag{36}$$

The function $\psi'^0_{l',h'}(\delta, \tilde{z})$ differs from $\psi'_{b',h'}(\delta, \tilde{z})$ by no more than two collections. Thereby, by virtue of Lemma 17 and the corollary of Lemma 20

$$I_0(\psi'^0_{l',h'}(\delta, \tilde{z})) \leqslant I_0(\psi'_{l',h'}(\delta, \tilde{z})) + 4m \leqslant (k't' + 4)\, 2m.$$

Since $\psi'^0_{l',h'}(\delta, \tilde{0}) = \psi'^0_{l',h'}(\delta, \tilde{1}) = 0$, it follows that for the formation of $\psi'^0_{l',h'}(\delta, \tilde{z})$ only conjunctions realized by the block B_4 are used. Therefore [see (14)]

$$L(B_8) \leqslant C_{213} M' (k')^{t'} (k't' + 4)\, 2m \leqslant C_{214} 2^m n^2,$$
$$T(B_8) =]\log((k't' + 4)\, 2m)[\, T^{(1)}, \quad \Omega(B_8) = C_{215} 2^m.$$

The system of functions

$$\psi'^0_{l',h'}(0, \tilde{z})\, \bar{z}_{l',h'} \vee \psi'^0_{l',h'}(1, \tilde{z})\, z_{l',h'} = \zeta(z_{l',h'}, \psi'^0_{l',h'}(0, \tilde{z}), \psi'^0_{l',h'}(1, \tilde{z}))$$

is denoted by Ψ'^0.

The block B_7 realizes the function $f^{(2)}_{\tilde{\nu}}(\tilde{x}, \tilde{y})$ (see p. 67) on the basis of the conjunctions realized by the block B_3,

$$L(B_7) \leqslant C_{216} 2^{u+m}, \quad T(B_7) = (u + 2)\, T^{(1)}, \quad \Omega(B_7) = 2^m.$$

It is easy to check the fact that of the networks B_0, B_1-B_5, B_2-B_6, B_3-B_7, B_4-B_8 the network B_4-B_8 has the largest delay:

$$T^{(2)}+(]\log m[+]\log((k't'+4)\,2m)[)\,T^{(1)}\sim 3T^{(1)}\log n.$$

This quantity determines the delay of the entire network between 1-1' and 2-2'.

The block B_9 is analogous to the block A_9. It realizes the functions†

$$\zeta(f^{0}_{\tilde{\gamma},\,i}(\tilde{x},\tilde{\beta}),\ \psi''^{0}(0,\tilde{y}),\ \psi''^{0}(1,\tilde{y})),$$
$$L(B_9)\leqslant C_{217}2^{u}p2^{z},\quad T(B_9)=T^{(4)}.$$

The block B_{10} is analogous to the block A_{10}. It realizes the system of functions

$$L(B_{10})\leqslant\rho\frac{2^n}{n},\quad T(B_{10})=t''T''+]\log M''[T^{(1)}.$$

Since $f^{0}_{\tilde{\gamma}_{l'},\,h',\,i}(\tilde{x},\tilde{\beta}_{l'',\,h''})\approx f_{\tilde{\gamma}_{l'},\,h',\,i}(\tilde{x},\tilde{\beta}_{l'',\,h''})$ and $\psi''^{0}_{l'',\,h''}(\delta,\tilde{y})\approx\psi''_{l'',\,h''}(\delta,\tilde{y})$, it follows that by virtue of Lemma 22

$$[\Phi'',\ \Sigma'',\ \Psi''^{0},\ F''^{0}_{l',\,h',\,i}]\underset{k,\,u}{=\!=}f_{\tilde{\gamma}_{l'},\,h',\,i}(\tilde{x},\tilde{y}).\tag{37}$$

The block B_{11} realizes the functions

$$L(B_{11})\leqslant C_{218}p2^{m},\quad T(B_{11})=]\log p[T^{(1)}$$

From (37) it follows that

$$f^{*}_{\tilde{\gamma}_{l'},\,h'}(\tilde{x},\tilde{y})\underset{k,\,u}{=\!=}f_{\tilde{\gamma}_{l'},\,h'}(\tilde{x},\tilde{y}).\tag{38}$$

The block B_{12} realizes the functions $f^{(1)}_{\tilde{\gamma}}(\tilde{x},\tilde{y})$ on the basis of the representation [see (38) and property 2) on p. 67] $f^{(1)}_{\tilde{\gamma}}(\tilde{x},\tilde{y})=f^{*}_{\tilde{\gamma}}(\tilde{x},\tilde{y})\,d_1(\tilde{x},\tilde{y})$;

$$L(B_{12})\leqslant C_{219}2^{m},\quad T(B_{12})=T^{(1)}.$$

The block B_{13} realizes the functions $f^{(0)}_{\tilde{\gamma}}(\tilde{x},\tilde{y})=f^{(1)}_{\tilde{\gamma}}(\tilde{x},\tilde{y})\vee f^{(2)}_{\tilde{\gamma}}(\tilde{x},\tilde{y})$;

$$L(B_{13})\leqslant C_{220}2^{m},\quad T(B_{13})=T^{(1)}.$$

It is clear that

$$f^{(0)}_{\tilde{\gamma}}(\tilde{x},\tilde{y})\approx f_{\tilde{\gamma}}(\tilde{x},\tilde{y}).\tag{39}$$

The system of functions $f^{(0)}_{\tilde{\gamma}_{l'},\,h'}(\tilde{x},\tilde{y})$ shall be denoted by F'^{0}.

The block B_{14} realizes the function $\zeta(f^{0}_{\tilde{\gamma}_{l'},\,h'}(\tilde{x},\tilde{y}),\ \psi'^{0}_{l',\,h'}(0,\tilde{z}),\ \psi'^{0}_{l',\,h'}(1,\tilde{z}))$;

$$L(B_{14})\leqslant C_{221}2^{m},\quad T(B_{14})=T^{(4)}.$$

† We drop the indices l', h', l'', h''.

The block B_{16} realizes the function $f^*(\tilde{x},\tilde{y},\tilde{z})=[\Phi',\Sigma',\Psi'^0,F'^0]$ by means of one block $\mathfrak{B}'$. We have [see (15)]

$$L(B_{16})\leqslant C_{222}2^m,\qquad T(B_{16})=\tau t'\log k'+]\log M'[T^{(1)}.$$

By virtue of Lemma 22, (36), and (39)

$$f^*(\tilde{x},\tilde{y},\tilde{z})\underset{k+u,\,m}{=\!=\!=}f(\tilde{x},\tilde{y},\tilde{z}).$$

Let us use F'_ε (ε = 0, 1) to denote the system of functions $f(\tilde{\varepsilon},\tilde{\varepsilon},\tilde{\gamma})$ (this is a system of constants; the collection $\tilde{\gamma}$ runs the gamut of all collections of length m).

The block B_{17} realizes the function $f_{\tilde{0}}^*(\tilde{z})=[\Phi',\Sigma',\Psi'^0,F'_0]$ (by means of one $\mathfrak{B}'$ block; we note that

$$\zeta(f(\tilde{\varepsilon},\tilde{\varepsilon},\tilde{\gamma}_{l',h'}),\ \psi'^0_{l',h'}(0,\tilde{z}),\ \psi'^0_{l',h'}(1,\tilde{z}))=\psi'^0_{l',h'}(f(\tilde{\varepsilon},\tilde{\varepsilon},\tilde{\gamma}_{l',h'}),\tilde{z}),$$

i.e., it coincides with a certain function

$$L(B_{17})\leqslant C_{223}2^m.$$

By virtue of Lemma 22,

$$f_{\tilde{0}}^*(\tilde{z})\approx f(\tilde{0},\tilde{0},\tilde{z}).$$

The block B_{18} realizes $f_{\tilde{1}}^*(\tilde{z})=[\Phi',\Sigma',\Psi'^0,F'_1]$ analogously;

$$f_{\tilde{1}}^*(\tilde{z})\approx f(\tilde{1},\tilde{1},\tilde{z});$$
$$L(B_{18})\leqslant C_{224}2^m.$$

The block B_{19} realizes $f(\tilde{x},\tilde{y},\tilde{z})$ in accordance with the representation

$$f^{(1)}(\tilde{x},\tilde{y},\tilde{z})=f^*(\tilde{x},\tilde{y},\tilde{z})e_1(\tilde{x},\tilde{y},\tilde{z}),\qquad f^{(2)}(\tilde{x},\tilde{y},\tilde{z})=f_{\tilde{0}}^*(\tilde{z})e_2(\tilde{x},\tilde{y},\tilde{z}),$$
$$f^{(3)}(\tilde{x},\tilde{y},\tilde{z})=f_{\tilde{1}}^*(\tilde{z})e_3(\tilde{x},\tilde{y},\tilde{z}),\qquad f^{(4)}(\tilde{x},\tilde{y},\tilde{z})=f_{\tilde{0}}(\tilde{x},\tilde{y})e_4(\tilde{x},\tilde{y},\tilde{z}),$$
$$f^{(5)}(\tilde{x},\tilde{y},\tilde{z})=f_{\tilde{1}}(\tilde{x},\tilde{y})e_5(\tilde{x},\tilde{y},\tilde{z}),\qquad f^{(6)}(\tilde{x},\tilde{y},\tilde{z})=f(\tilde{1},\tilde{1},\tilde{0})e_6(\tilde{x},\tilde{y},\tilde{z}),$$
$$f^{(7)}(\tilde{x},\tilde{y},\tilde{z})=f(\tilde{0},\tilde{0},\tilde{1})e_7(\tilde{x},\tilde{y},\tilde{z}),\qquad f(\tilde{x},\tilde{y},\tilde{z})=\bigvee_{i=1}^{7}f^{(i)}(\tilde{x},\tilde{y},\tilde{z}).$$

The functions $f^*(\tilde{x},\tilde{y},\tilde{z})$, $f_{\tilde{0}}^*(\tilde{z})$, $f_{\tilde{1}}^*(\tilde{z})$ are realized by the block B_{16}, B_{17}, B_{18}; the functions $f_{\tilde{0}}(\tilde{x},\tilde{y})$, $f_{\tilde{1}}(\tilde{x},\tilde{y})$ are realized by the block B_{13} (together with other functions $f_{\tilde{\gamma}}(\tilde{x},\tilde{y})$); the constants $f(\tilde{1},\tilde{1},\tilde{0})$ and $f(\tilde{0},\tilde{0},\tilde{1})$ are realized by the block B_0; the functions $e_i(\tilde{x},\tilde{y},\tilde{z})$ are realized by the block B_0;

$$L(B_{19})\leqslant C_{225},\qquad T(B_{19})=4T^{(1)}$$

(one story of the S_2 networks and the network $D_{7,8}$).

It is easy to check the fact (just as it is for the synthesis method from §8) that for i ≠ 10, $L(B_i)=o(2^n/n)$ and $L(B_{10})\leqslant\rho(2^n/n)$;

$$T(B_j) = O(\log n) \quad \text{for} \quad j \neq 16,\ 17,\ 18,$$
$$T(B_{16}) = T(B_{17}) = T(B_{18}) \sim \tau n.$$

Thus, the delay of the entire network does not exceed $C_{226}n$; therefore, the sum of the complexities of all the delay networks (depicted by the thick lines) does not exceed

$$C_{227}(\Omega(B_0) + \Omega(B_5) + \Omega(B_6) + \Omega(B_7) + \Omega(B_8) + \Omega(B_{13}))\,n \leqslant C_{228}(p2^s + 2^u + 2^m + 1)\,n + o\left(\frac{2^n}{n}\right).$$

Thus, for the proper network S which has been obtained we have

$$L(S) \leqslant \rho\frac{2^n}{n}, \qquad T(S) \sim \tau n.$$

§ 10. Method of Synthesis of Networks for x-Functions in the Case of an Irregular Basis

This method is analogous to the method of synthesizing networks for 0-functions; however there are the following basic differences.

I. Instead of the elements E' and E" (on which τ and ρ, respectively, are obtained) the elements E‴ and E^{IV} are used (on which τ^* and ρ^* are attained; see § 1).

II. Instead of the auxiliary functions e_i and d_i, their analogs $e_i(\tilde{x}, \tilde{y}, \tilde{z}) \vee \varkappa_1(\tilde{x}, \tilde{y}, \tilde{z})$ and $d_i(\tilde{x}, \tilde{y}) \vee \varkappa_1(\tilde{x}, \tilde{y})$.

III. The construction of Lemma 15 for forming delays is not used.

Let us consider these differences in greater detail.

I. Assume E‴ is an element of the basis on which τ^* is attained. This is a Φ-element. Assume this element has the weight P‴, the delay T‴, and a number of inputs equal to k‴. Then T‴ = τ^* log k‴. In the method described we shall use the (E‴, t‴, m)-block $\mathfrak{B}'''$ (instead of the block $\mathfrak{B}'$), the numeration Σ''' of collections of length m defined by it, the function Φ''' and the system of functions Ψ''' corresponding to it, as well as the system of function F‴ (actually the "old" system of functions $f(\tilde{x}, \tilde{y}, \tilde{z})$, only the collections $\tilde{\gamma}$ in it are newly numbered by pairs of indices), and the representation

$$f(\tilde{x}, \tilde{y}, z) = [\Phi''', \Sigma''', \Psi''', F'''] \tag{40}$$

based on them. The system of functions $f_{\tilde{\gamma}_{l''', h'''}, i}(\tilde{x}, \tilde{y})$. is defined in accordance with the new numeration of collections of length m.

Assume E^{IV} is an element of the basis on which ρ^* is attained. Assume this element has the weight P^{IV}, the delay T^{IV}, and a number of inputs equal to k^{IV}. Then $P^{IV} = \rho^*(k^{IV} - 1)$. In the method described we shall make use of the (E^{IV}, t^{IV}, u)-block $\mathfrak{B}^{IV}$ (instead of the block $\mathfrak{B}''$), the numeration Σ^{IV} specified by it, the function Φ^{IV} and the system of functions Ψ^{IV} corresponding to it, as well as the system of functions $F^{IV}_{l''', h''', i}$ (actually they coincide with the system $F''_{l', h', i}$) and the representations

$$f_{\tilde{\gamma}_{l''', h'''}, i}(\tilde{x}, \tilde{y}) = [\Phi^{IV}, \Sigma^{IV}, \Psi^{IV}, F^{IV}_{l', h''', i}] \tag{41}$$

based on them.

The numbers

$$M''' = \left]\frac{2^m}{(k''')^{t'''}}\right[\text{ and } M^{IV} = \left]\frac{2^u}{(k^{IV})^{t^{IV}}}\right[$$

are defined analogously.

The values of the parameters are chosen in a similar fashion:

$$k = [2 \log n], \quad u = [2 \log n], \quad s = [n - 5 \log n],$$
$$t''' = 2\left[\frac{1}{2 \log k'''}(n - 5 \log n)\right], \quad t^{IV} = 2\left[\frac{\log n}{2 \log k^{IV}}\right].$$

This time,

$$\left.\begin{aligned} &L(\mathfrak{B}''') \leqslant C_{300} 2^m = o\left(\frac{2^n}{n}\right), \\ &T(\mathfrak{B}''') = \tau^* t''' \log k''' +]\log M''' [T^{(1)} \sim \tau^* n, \\ &L(\mathfrak{B}^{IV}) \leqslant \rho^* 2^u, \\ &T(\mathfrak{B}^{IV}) = t^{IV} T^{IV} +]\log M^{IV} [T^{(1)} = O(\log n). \end{aligned}\right\} \tag{42}$$

II. Let us introduce the functions (see §9.A)

$$e_{10+i}(\tilde{x}, \tilde{y}, \tilde{z}) = e_i(\tilde{x}, \tilde{y}, \tilde{z}) \vee \varkappa_1(\tilde{x}, \tilde{y}, \tilde{z}), \quad 1 \leqslant i \leqslant 7,$$
$$d_{10+i}(\tilde{x}, \tilde{y}) = d_i(\tilde{x}, \tilde{y}) \vee \varkappa_1(\tilde{x}, \tilde{y}), \quad 1 \leqslant i \leqslant 2.$$

Remark. Assume a_{ij} $(x_1, \ldots, x_a)$ are A-functions $x_1 x_2 \ldots x_a \vee x_i \bar{x}_j$. From the definition of the function d_{11} and from (34) we derive the equation

$$\begin{aligned} d_{11}(\tilde{x}, \tilde{y}) &= (\lambda_k(\tilde{x}) \vee \varkappa_1(\tilde{x}, \tilde{y})) \,\&\, (\lambda_u(\tilde{y}) \vee \varkappa_1(\tilde{x}, \tilde{y})) = \\ &= (a_{12}(\tilde{x}, \tilde{y}) \vee a_{23}(\tilde{x}, \tilde{y}) \vee \ldots \vee a_{k-1,k}(\tilde{x}, \tilde{y}) \vee a_{k,1}(\tilde{x}, \tilde{y})) \,\& \\ &\& \,(a_{k+1,k+2}(\tilde{x}, \tilde{y}) \vee a_{k+2,k+3}(\tilde{x}, \tilde{y}) \vee \ldots \vee a_{k+u-1,k+u}(\tilde{x}, \tilde{y}) \vee a_{k+u,k+1}(\tilde{x}, \tilde{y})). \end{aligned} \tag{43}$$

An analogous representation is valid for the function $e_{11}(\tilde{x}, \tilde{y}, \tilde{z})$.

For an arbitrary x-function $f(\tilde{x}, \tilde{y}, \tilde{z})$ we place

$$f^{(j)}(\tilde{x}, \tilde{y}, \tilde{z}) = f(\tilde{x}, \tilde{y}, \tilde{z})\, e_j(\tilde{x}, \tilde{y}, \tilde{z}), \quad 11 \leqslant j \leqslant 17.$$

Since $f(\tilde{0}, \tilde{0}, \tilde{0}) = 0$, it follows that

$$f(\tilde{x}, \tilde{y}, \tilde{z}) = \bigvee_{j=11}^{17} f^{(j)}(\tilde{x}, \tilde{y}, \tilde{z}).$$

Note in addition that

$$f^{(j)}(\tilde{1}, \tilde{1}, \tilde{1}) = 1. \tag{44}$$

For each function $f_{\tilde{\gamma}}(\tilde{x}, \tilde{y}) = f(\tilde{x}, \tilde{y}, \tilde{\gamma})$ let us place

$$\left.\begin{aligned} &f_{\tilde{\gamma}}^{(j)}(\tilde{x}, \tilde{y}) = f_{\tilde{\gamma}}(\tilde{x}, \tilde{y})\, d_j(\tilde{x}, \tilde{y}), \quad j = 11, 12, \\ &f_{\tilde{\gamma}}^{(10)}(\tilde{x}, \tilde{y}) = f_{\tilde{\gamma}}^{(11)}(\tilde{x}, \tilde{y}) \vee f_{\tilde{\gamma}}^{(12)}(\tilde{x}, \tilde{y}). \end{aligned}\right\} \tag{45}$$

The following properties hold (which are analogous to the properties from §9.A)

1) $f_{\tilde{\gamma}}^{(10)}(\tilde{x}, \tilde{y}) \approx f_{\tilde{\gamma}}(\tilde{x}, \tilde{y})$;

2) if a certain function $g(\tilde{x}, \tilde{y})$ satisfies the condition $g(\tilde{x}, \tilde{y}) \underset{k,u}{=\!=} f_{\tilde{\gamma}}(\tilde{x}, \tilde{y})$, then

$$f_{\tilde{\gamma}}^{(11)}(\tilde{x}, \tilde{y}) \approx g(\tilde{x}, \tilde{y})\, d_{11}(\tilde{x}, \tilde{y})$$

(note that here the conventional equation may not hold: on the collection $\tilde{1}$ the left and right parts may take different values);

2') if $f(\tilde{x}, \tilde{y}, \tilde{z})$ is an x-function and if a certain function $g(\tilde{x}, \tilde{y}, \tilde{z})$ satisfies the condition $g(\tilde{x}, \tilde{y}, \tilde{z}) \underset{k+u,\, m}{=\!=\!=} f(\tilde{x}, \tilde{y}, \tilde{z})$, then

$$f^{(11)}(\tilde{x}, \tilde{y}, \tilde{z}) = g(\tilde{x}, \tilde{y}, \tilde{z})\, e_{11}(\tilde{x}, \tilde{y}, \tilde{z}) \vee \varkappa_1(\tilde{x}, \tilde{y}, \tilde{z})$$

[see (44)];

3)
$$f^{(14)}(\tilde{x}, \tilde{y}, \tilde{z}) = f_{\tilde{0}}^{(10)}(\tilde{x}, \tilde{y})\, e_{14}(\tilde{x}, \tilde{y}, \tilde{z}) \vee \varkappa_1(\tilde{x}, \tilde{y}, z),$$
$$f^{(15)}(\tilde{x}, \tilde{y}, \tilde{z}) = f_{\tilde{1}}^{(10)}(\tilde{x}, \tilde{y})\, e_{15}(\tilde{x}, \tilde{y}, \tilde{z})$$

(since $f_{\tilde{1}}^{(10)}(\tilde{1}, \tilde{1}) = 1$).

III. Note in addition that circuits of the elements E^{III} and E^{IV} with identified inputs having the length t^{III} and t^{IV}, respectively, realize the function $f(x) = x$. The network S_8 (for the function ξ) likewise realizes x for identification of the inputs. This allows the necessary delays to be realized in all layers without using the corollary from Lemma 15.

Let us now go over to a description of the blocks (Fig. 9).

The block D_0 realizes the "conditional constants" $e^{(0)}(\tilde{z}) = \varkappa_1(\tilde{z})$, $e^{(00)}(\tilde{x}, \tilde{y}, \tilde{z}) = \varkappa_1(\tilde{x}, \tilde{y}, \tilde{z})$, and $e^{(1)}(\tilde{z}) = x_{k+u+1} \vee \ldots \vee x_n$, and the functions e_i $(11 \le i \le 17)$ and d_{11}. It consists of the blocks D_{00}, D_{01}, D_{02}, D_{03}, D_{04}. These blocks are connected to each other similarly to the blocks B_{00}-B_{04} (see Fig. 8).

The block D_{00} realizes conditional constants. Since $\xi(x, x, x, x) = x$, it follows that D_{00} may be chosen so that

$$L(D_{00}) \leqslant C_{301} n, \quad T(D_{00}) = T^{(3)} +]\log n[\, T^{(1)}.$$

It is obvious that

$$e^{(0)}(\tilde{z}) \approx 0, \quad e^{(1)}(\tilde{z}) \approx 1. \tag{46}$$

The block D_{01} realizes the A-functions $a_{ij}(\tilde{x}, \tilde{y})$ and $a_{ij}(\tilde{x}, \tilde{y}, \tilde{z})$ (see remark) which are used for the realization of d_{11} and e_{11}. Their number is smaller than 2n. We have (Lemma 12)

$$L(D_{01}) \leqslant C_{302} n^2, \quad T(D_{01}) = T^{(3)} +]\log n[\, T^{(1)}$$

(the delays of the network for various functions a_{ij} are equalized by means of circuits from S_1 networks).

The block D_{02} realizes d_{11} in accordance with the representation (43) (p. 74), and e_{11} in accordance with an analogous representation;

$$L(D_{02}) \leqslant C_{303} n^2, \quad T(D_{02}) = (]\log n[+ 1)\, T^{(1)}.$$

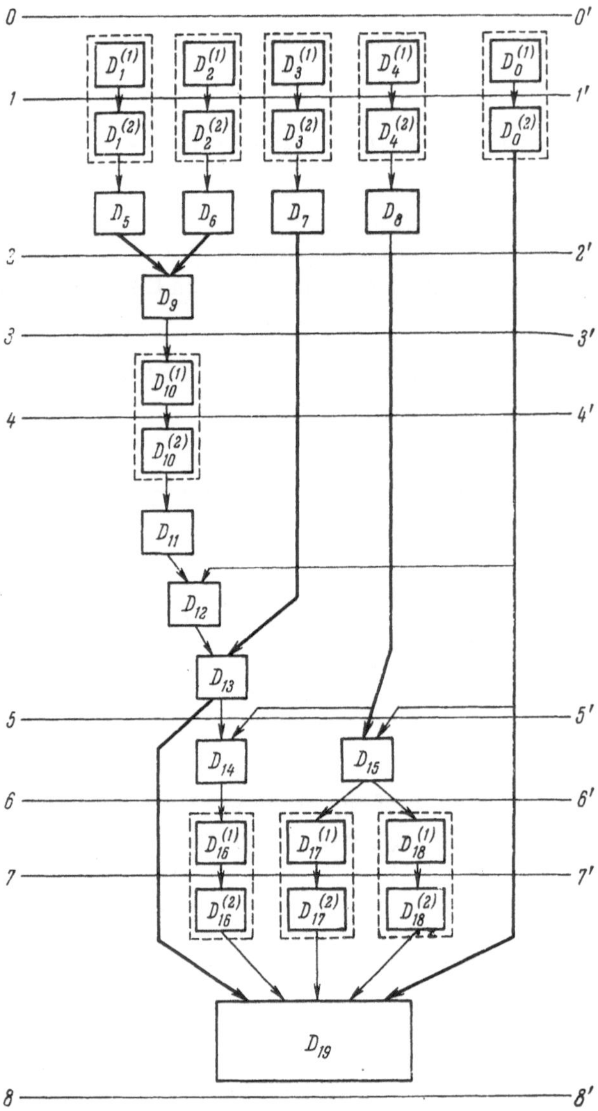

Fig. 9

The block D_{03} realizes e_{12}-e_{15}. We have [see (35)]

$$e_{12}(\tilde{x}, \tilde{y}, \tilde{z}) = (x_1 \ldots x_n \vee \bar{x}_1 \ldots \bar{x}_{n-m} x_{n-m+1} \bar{x}_{n-m+2}) \vee (x_1 \ldots x_n \vee \bar{x}_1 \ldots \bar{x}_{n-m} x_{n-m+2} x_{n-m+3}) \vee$$
$$\vee (x_1 \ldots x_n \vee \bar{x}_1 \ldots \bar{x}_{n-m} x_{n-1} \bar{x}_n) \vee (x_1 \ldots x_n \vee \bar{x}_1 \ldots \bar{x}_{n-m} x_n \bar{x}_{n-m+1}).$$

For e_{13}-e_{15} the analogous representations hold. Therefore,

$$L(D_{03}) \leqslant C_{304} n^2, \quad T(D_{03}) = T^{(3)} + 2] \log n [T^{(1)}.$$

The block D_{04} realizes the functions e_{16} and e_{17} (these are A-functions);

$$L(D_{04}) \leqslant C_{305} n, \quad T(D_{04}) = T^{(3)} +] \log n [T^{(1)}.$$

The blocks D_1 and D_2 realize A-functions of the form

$$x_1 \ldots x_k \vee x_1^{\sigma_1} \ldots x_k^{\sigma_k} \text{ and } x_{k+1} \ldots x_{k+u} \vee x_{k+1}^{\sigma_{k+1}} \ldots x_{k+u}^{\sigma_{k+u}}$$

respectively (analogs of conventional conjunctions). We have (Lemma 12)

$$L(D_1) \leqslant C_{306} k 2^k, \quad T(D_1) = T^{(3)} +]\log k[\, T^{(1)},$$
$$L(D_2) \leqslant C_{307} u 2^u, \quad T(D_2) = T^{(3)} +]\log u[\, T^{(1)}.$$

The block D_3 realizes A-functions $x_1 \ldots x_{k+u} \vee x_1^{\alpha_1} \ldots x_k^{\alpha_k} x_{k+1}^{\beta_1} \ldots x_{k+u}^{\beta_u}$, whose collections $\tilde{\alpha} = (\alpha_1, \ldots, \alpha_k)$, $\tilde{\beta} = (\beta_1, \ldots, \beta_u)$ satisfy the condition

$$(\tilde{\alpha} = \tilde{0} \,\&\, \tilde{\beta} \neq \tilde{0}) \vee (\tilde{\beta} = \tilde{0} \,\&\, \tilde{\alpha} \neq 0) \vee (\tilde{\alpha} = \tilde{1}) \vee (\tilde{\beta} = \tilde{1});$$
$$L(D_3) \leqslant C_{308} u 2^u, \quad T(D_3) = T^{(3)} +]\log(k+u)[\, T^{(1)} = O(\log n).$$

The block D_4 realizes all A-functions $x_{n-m+1} \ldots x_n \vee x_{n-m+1}^{\sigma_{n-m+1}} \ldots x_{n-m+j}^{\sigma_{n-m+j}}$ $(i \leq j \leq m)$;

$$L(D_4) \leqslant C_{309} m 2^m; \quad T(D_4) = T^{(3)} +]\log m[\, T^{(1)}.$$

The block D_5 realizes the set of all functions $h^1(\tilde{x})$ which have the properties:

a) $h^1(\tilde{0}) = 0$, $h^1(\tilde{1}) = 1$;

b) $h^1(\tilde{x})$ outside of a certain strip is equal to 0 (with the exception of the collection $\tilde{1}$ on which, as has already been said, h^1 is equal to 1). It is obvious that for each function $f_{\tilde{\gamma}_{l'''}, h''', i}(\tilde{x}, \tilde{\beta}_{l^{IV}, h^{IV}})$ there exists a function $h^1(\tilde{x})$ such that $f_{\tilde{\gamma}_{l'''}, h''', i}(\tilde{x}, \tilde{\beta}_{l^{IV}, h^{IV}}) \approx h^1(\tilde{x})$. This function $h^1(\tilde{x})$ shall likewise be denoted by $f^1_{\tilde{\gamma}_{l'''}, h''', i}(\tilde{x}, \tilde{\beta}_{l^{IV}, h^{IV}})$. The system of all functions $f^1_{\tilde{\gamma}_{l'''}, h''', i}(\tilde{x}, \tilde{\beta}_{l^{IV}, h^{IV}})$ for specified l''', h''', i shall be denoted by l''', h'''. It is clear that each function $h^1(\tilde{x})$ is a disjunction of no more than s functions realized by the block D_1. By virtue of Lemma 9,

$$L(D_5) \leqslant C_{310} p s 2^s, \quad T(D_5) =]\log s[\, T^{(1)}, \ \Omega(D_5) \leqslant p 2^s.$$

The block D_6 realizes the functions $\psi^{IV, 1}_{l^{IV}, h^{IV}}(\delta, \tilde{y}) = \psi^{IV}_{l^{IV}, h^{IV}}(\delta, \tilde{y}) \bar{\varkappa}_0(\tilde{y}) \vee \varkappa_1(\tilde{y})$ as disjunctions of the functions realized by the block D_2,

$$L(D_6) \leqslant C_{311} 2^{2u}, \quad T(D_6) = u T^{(1)}, \quad \Omega(D_6) \leqslant C_{312} 2^u.$$

It is obvious that

$$\psi^{IV, 1}_{l^{IV}, h^{IV}}(\delta, \tilde{y}) \approx \psi^{IV}_{l^{IV}, h^{IV}}(\delta, \tilde{y}).$$

The system of functions

$$\psi^{IV, 1}_{l^{IV}, h^{IV}}(0, \tilde{y}) \, \bar{y}_{l^{IV}, h^{IV}} \vee \psi^{IV, 1}_{l^{IV}, h^{IV}}(1, \tilde{y}) \, y_{l^{IV}, h^{IV}} = \zeta(y_{l^{IV}, h^{IV}}, \psi^{IV, 1}_{l^{IV}, h^{IV}}(0, \tilde{y}), \psi^{IV, 1}_{l^{IV}, h^{IV}}(1, \tilde{y}))$$

shall be denoted by $\Psi^{IV, 1}$.

The Block D_8 realizes the functions $\psi'''^{, 1}_{l''', h'''}(\delta, \tilde{z}) = \psi'''_{l''', h'''}(\delta, \tilde{z}) \bar{\varkappa}_0(\tilde{z}) \vee \varkappa_1(\tilde{z})$, which are analogous to the functions $\psi^{IV, 1}_{l^{IV}, h^{IV}}(\delta, \tilde{y})$. Therefore,

$$\psi'''^{, 1}_{l''', h'''}(\delta, \tilde{z}) \approx \psi'''_{l''', h'''}(\delta, \tilde{z}). \qquad (47)$$

By analogy with the situation prevailing for the block B_8, we have here (using the functions realized by the block D_4)

$$L(D_8) \leqslant C_{313} 2^m n^2, \quad T(D_8) =]\log((k'''t''' + 4)\, 2m)[\, T^{(1)}, \ \Omega(D_8) \leqslant C_{314} 2^m.$$

The system of functions

$$\psi^{\prime\prime\prime, 1}_{l^{\prime\prime\prime}, h^{\prime\prime\prime}}(0, \tilde{z})\, \bar{z}_{l^{\prime\prime\prime}, h^{\prime\prime\prime}} \vee \psi^{\prime\prime\prime, 1}_{l^{\prime\prime\prime}, h^{\prime\prime\prime}}(1, \tilde{z})\, z_{l^{\prime\prime\prime}, h^{\prime\prime\prime}} = \zeta(z_{l^{\prime\prime\prime}, h^{\prime\prime\prime}}, \psi^{\prime\prime\prime, 1}_{l^{\prime\prime\prime}, h^{\prime\prime\prime}}(0, \tilde{z}), \psi^{\prime\prime\prime, 1}_{l^{\prime\prime\prime}, h^{\prime\prime\prime}}(1, \tilde{z}))$$

shall be denoted by $\Psi^{\prime\prime\prime, 1}$.

The block D_7 realizes the functions $f^{(12)}_{\tilde{\gamma}}(\tilde{x}, \tilde{y})$ as disjunctions of the functions realized by the block D_3

$$L(D_7) \leqslant C_{315} 2^{u+m}, \quad T(D_7) = (u+2)\, T^{(1)}, \quad \Omega(D_7) = 2^m.$$

As in the case of the method of synthesizing networks for 0-functions, it is easy to verify the fact that of the networks D_0, D_1-D_5, D_2-D_6, D_3-D_7, D_4-D_8 the network D_4-D_8 has the largest delay:

$$T^{(3)} + (]\log m[+]\log((k^{\prime\prime\prime} t^{\prime\prime\prime} + 4)\, 2m)[)\, T^{(1)}.$$

The block D_9 realizes the functions† $\zeta(f^1_{\tilde{\gamma}, i}(\tilde{x}, \tilde{\beta}), \psi^{IV, 1}(0, \tilde{y}), \psi^{IV, 1}(1, \tilde{y}))$;

$$L(D_9) \leqslant C_{316} 2^u p 2^s, \quad T(D_9) = T^{(4)}.$$

The block D_{10} realizes the functions $[\Phi^{IV}, \Sigma^{IV}, \Psi^{IV, 1}, F^{IV, 1}_{l^{IV}, h^{IV}, i}]$. We have [see (42)]

$$L(D_{10}) = p 2^m L(\mathfrak{B}^{IV}) \leqslant \rho \frac{2^n}{n},$$
$$T(D_{10}) = T(\mathfrak{B}^{IV}) = O(\log n).$$

By virtue of the fact that $f^1_{\tilde{\gamma}_{l^{\prime\prime\prime}, h^{\prime\prime\prime}, i}}(\tilde{x}, \tilde{\beta}_{l^{IV}, h^{IV}}) \approx f_{\tilde{\gamma}_{l^{\prime\prime\prime}, h^{\prime\prime\prime}, i}}(\tilde{x}, \tilde{\beta}_{l^{IV}, h^{IV}})$ and $\psi^{IV, 1}_{l^{IV}, h^{IV}}(\delta, \tilde{y}) \approx \psi^{IV}_{l^{IV}, h^{IV}}(\delta, \tilde{y})$, it follows that by virtue of Lemma 22 and (41)

$$[\Phi^{IV}, \Sigma^{IV}, \Psi^{IV, 1}, F^{IV, 1}_{l^{\prime\prime\prime}, h^{\prime\prime\prime}, i}] \underset{k, u}{=\!=} f_{\tilde{\gamma}_{l^{\prime\prime\prime}, h^{\prime\prime\prime}, i}}(\tilde{x}, \tilde{y}). \tag{48}$$

The block D_{11} realizes the functions

$$f^{**}_{\tilde{\gamma}_{l^{\prime\prime\prime}, h^{\prime\prime\prime}}}(\tilde{x}, \tilde{y}) = \bigvee_i [\Phi^{IV}, \Sigma^{IV}, \Psi^{IV, 1}, F^{IV, 1}_{l^{\prime\prime\prime}, h^{\prime\prime\prime}, i}],$$
$$L(D_{11}) \leqslant C_{317} p 2^m, \quad T(D_{11}) =]\log p[\, T^{(1)}.$$

From (48) it follows that

$$f^{**}_{\tilde{\gamma}_{l^{\prime\prime\prime}, h^{\prime\prime\prime}}}(\tilde{x}, \tilde{y}) \underset{k, u}{=\!=} f_{\tilde{\gamma}_{l^{\prime\prime\prime}, h^{\prime\prime\prime}}}(\tilde{x}, \tilde{y}). \tag{49}$$

The block D_{12} realizes the functions $f^{***}_{\tilde{\gamma}}(\tilde{x}, \tilde{y}) = f^{**}_{\tilde{\gamma}}(\tilde{x}, \tilde{y})\, d_{11}(\tilde{x}, \tilde{y})$,

$$L(D_{12}) \leqslant C_{318} 2^m, \quad T(D_{12}) = T^{(1)}.$$

By virtue of the properties 2), (45), and (49),

$$f^{***}_{\tilde{\gamma}}(\tilde{x}, \tilde{y}) \approx f^{(11)}_{\tilde{\gamma}}(\tilde{x}, \tilde{y}). \tag{50}$$

† We drop the indices $l^{\prime\prime\prime}$, $h^{\prime\prime\prime}$, l^{IV}, h^{IV}.

The block D_{13} realizes the functions $f^{****}_{\tilde{\gamma}}(\tilde{x}, \tilde{y}) = f^{***}_{\tilde{\gamma}}(\tilde{x}, \tilde{y}) \vee f^{(12)}_{\tilde{\gamma}}(\tilde{x}, \tilde{y})$,

$$L(D_{13}) \leqslant C_{319} 2^m, \quad T(D_{13}) = T^{(1)}.$$

From (45), the property (1), and (50) it follows that

$$f^{****}_{\tilde{\gamma}}(\tilde{x}, \tilde{y}) \approx f_{\tilde{\gamma}}(\tilde{x}, \tilde{y}). \tag{51}$$

The system of functions $f^{****}_{\tilde{\gamma}_{l'''}, h'''}(\tilde{x}, \tilde{y})$ shall be denoted by $\mathfrak{F}'''^*$.

The block D_{14} realizes the system of functions

$$\zeta(f^{****}_{\tilde{\gamma}_{l'''}, h'''}(\tilde{x}, \tilde{y}), \psi'''^{;1}_{l''', h'''}(0, \tilde{z}), \psi'''^{;1}_{l''', h'''}(1, \tilde{z})).$$
$$L(D_{14}) \leqslant C_{320} 2^m, \quad T(D_{14}) = T^{(4)}$$

The block D_{16} realizes the function $f^{**}(\tilde{x}, \tilde{y}, \tilde{z}) = [\Phi''', \Sigma''', \Psi'''^{,1}, F'''^*]$ by means of one block $\mathfrak{B}'''$. We have [see (42)]

$$L(D_{16}) = o\left(\frac{2^n}{n}\right), \quad T(D_{16}) \sim \tau^* n.$$

By virtue of Lemma 22, (47), (51), (40)

$$f^{**}(\tilde{x}, \tilde{y}, \tilde{z}) \underset{k+u,\, m}{=\!=\!=} f(\tilde{x}, \tilde{y}, \tilde{z}). \tag{52}$$

We use $\mathfrak{F}^{\mathfrak{m}}_{\varepsilon}$ to denote the system of functions $e^{(f(\tilde{\varepsilon}, \tilde{\varepsilon}, \tilde{\gamma}))}(\tilde{z})$ (the collection $\tilde{\gamma}$ runs the gamut of the set of all collections of length m).

The block D_{15} realizes the functions $\zeta(e^{(e)}(\tilde{z}), \psi'''^{;1}_{l''', h'''}(0, \tilde{z}), \psi'''^{;1}_{l''', h'''}(1, \tilde{z}))$,

$$L(D_{15}) \leqslant C_{321} 2^m, \quad T(D_{15}) = T^{(4)}.$$

The block D_{17} realizes the functions $f^{**}_0(\tilde{z}) = [\Phi''', \Sigma''', \Psi'''^{,1}, F'''_0]$ by means of one block $\mathfrak{B}'''$. We have

$$L(D_{17}) = o\left(\frac{2^n}{n}\right).$$

By virtue of Lemma 22, (47), (46),

$$f^{**}_0(\tilde{z}) \approx f(\tilde{0}, \tilde{0}, \tilde{z}).$$

The block D_{18} realizes $f^{**}_1(\tilde{z}) = [\Phi''', \Sigma''', \Psi'''^{,1}, F'''_1]$, in analogous fashion,

$$f^{**}_1(\tilde{z}) \approx f(\tilde{1}, \tilde{1}, \tilde{z}),$$
$$L(D_{18}) = o\left(\frac{2^n}{n}\right).$$

Finally, the block D_{19} realizes the function $f(\tilde{x}, \tilde{y}, \tilde{z})$ in accordance with the representation

$$f^{(11)}(\tilde{x}, \tilde{y}, \tilde{z}) = f^{**}(\tilde{x}, \tilde{y}, \tilde{z})\, e_{11}(\tilde{x}, \tilde{y}, \tilde{z}) \vee \varkappa_1(\tilde{x}, \tilde{y}, \tilde{z})$$

[see property 2'), (52)]

$$f^{(12)}(\tilde{x},\tilde{y},\tilde{z}) = f_0^{**}(\tilde{z})\, e_{12}(\tilde{x},\tilde{y},\tilde{z}) \vee \varkappa_1(\tilde{x},\tilde{y},\tilde{z}),$$
$$f^{(13)}(\tilde{x},\tilde{y},\tilde{z}) = f_1^{**}(\tilde{z})\, e_{13}(\tilde{x},\tilde{y},\tilde{z}) \vee \varkappa_1(\tilde{x},\tilde{y},\tilde{z}),$$
$$f^{(14)}(\tilde{x},\tilde{y},\tilde{z}) = f_{\tilde{0}}^{****}(\tilde{x},\tilde{y})\, e_{14}(\tilde{x},\tilde{y},\tilde{z}) \vee \varkappa_1(\tilde{x},\tilde{y},\tilde{z}) \quad \text{(see (51))},$$
$$f^{(15)}(\tilde{x},\tilde{y},\tilde{z}) = f_{\tilde{1}}^{****}(\tilde{x},\tilde{y})\, e_{15}(\tilde{x},\tilde{y},\tilde{z}) \vee \varkappa_1(\tilde{x},\tilde{y},\tilde{z}),$$
$$f^{(16)}(\tilde{x},\tilde{y},\tilde{z}) = \begin{cases} \varkappa_1(\tilde{x},\tilde{y},\tilde{z}), & \text{if } f(\tilde{1},\tilde{1},\tilde{0}) = 0, \\ e_{16}(\tilde{x},\tilde{y},\tilde{z}), & \text{if } f(\tilde{1},\tilde{1},\tilde{0}) = 1, \end{cases}$$
$$f^{(17)}(\tilde{x},\tilde{y},\tilde{z}) = \begin{cases} \varkappa_1(\tilde{x},\tilde{y},\tilde{z}), & \text{if } f(\tilde{0},\tilde{0},\tilde{1}) = 0, \\ e_{17}(\tilde{x},\tilde{y},\tilde{z}), & \text{if } f(\tilde{0},\tilde{0},\tilde{1}) = 1, \end{cases}$$
$$f(\tilde{x},\tilde{y},\tilde{z}) = \bigvee_{j=11}^{17} f^{(j)}(\tilde{x},\tilde{y},\tilde{z}),$$
$$L(D_{19}) = C_{322}, \qquad T(D_{19}) = 5T^{(1)}$$

(two stories for the realization of $xy \vee z$ and the network $D_{7,8}$).

The complexity of all (complementary) delay networks is of the order of $o(2^n/n)$.

Finally, for the entire network S we have

$$L(S) \leqslant \rho^* \frac{2^n}{n}, \qquad T(S) \leqslant \tau^* n.$$

Appendix

Assume $\tilde{\sigma} = (\sigma_1, \ldots, \sigma_n)$ is a collection of zeros and ones, and assume $|\tilde{\sigma}| = \sum_{i=1}^{n} 2^{n-i}\sigma_i$. Let us introduce the notation:

$\mathfrak{F}^{n,q}$ is the class of all vector-functions (i.e., systems of functions)

$$\tilde{f}(x_1, \ldots, x_n) = (f_1(x_1, \ldots, x_n), \ldots, f_q(x_1, \ldots, x_n))$$

(in [6] such vector-functions were called (n, q)-operators); $\mathfrak{F}_c^{n,q}$ ($\mathfrak{F}_x^{n,q}$, $\mathfrak{F}_{\bar{x}}^{n,q}$), respectively) is the class of vector-functions from $\mathfrak{F}^{n,q}$, for which all components are c-functions (x-functions, $\bar{x}$-functions, respectively).

From what has been said above (see §§1-2) it follows that in the case of a regular basis it is possible for all vector-functions from $\mathfrak{F}^{n,q}$, to be realized (by a proper network), while in the case of an irregular basis it is only possible for vector-functions from $\mathfrak{F}_c^{n,q} \cup \mathfrak{F}_x^{n,q} \cup \mathfrak{F}_{\bar{x}}^{n,q}$ to be realized.

Assume $r \leq 2^n$ and

$\mathfrak{F}^{n,q,r}$ is the class of all vector-functions $\tilde{f}$ from $\mathfrak{F}^{n,q}$, which satisfy the condition

$$\text{if } |\tilde{\sigma}| \geqslant r, \text{ then } \tilde{f}(\tilde{\sigma}) = (0, \ldots, 0);$$

$\mathfrak{F}_c^{n,q,r}$ ($\mathfrak{F}_x^{n,q,r}$, $\mathfrak{F}_{\bar{x}}^{n,q,r}$), respectively) is the class of all vector-functions $\tilde{f}$ from $\mathfrak{F}_c^{n,q}$ ($\mathfrak{F}_x^{n,q}$, $\mathfrak{F}_{\bar{x}}^{n,q}$, respectively) which satisfy the condition

$$\text{if } r \leqslant |\tilde{\sigma}| \leqslant 2^n - 2, \text{ then } \tilde{f}(\tilde{\sigma}) = (0, \ldots, 0).$$

As above, for the vector-function $\tilde{f}$ (realized by a proper network in the basis considered) let us use $L(\tilde{f})$ to denote the least of the complexities, while $T(\tilde{f})$ denotes the least of the delays of the proper networks realized in $\tilde{f}$. Assume $\mathfrak{N}$ is a certain class of vector-functions realizable by proper networks. Assume

$$L(\mathfrak{N}) = \max_{\tilde{f} \in \mathfrak{N}} L(\tilde{f}), \qquad T(\mathfrak{N}) = \max_{\tilde{f} \in \mathfrak{N}} T(\tilde{f}).$$

The following statements hold which generalize Theorems 1 and 2 (see §1) and are analogs of Theorems A.10 and A.12 from [6].

Theorem A.1. *If the basis $\mathscr{E}$ is regular and $\frac{\log q_n}{2^n} \to 0$, then*

$$L(\mathfrak{F}^{n,\,q_n}) \sim L(\mathfrak{F}_c^{n,\,q_n}) \sim L(\mathfrak{F}_x^{n,\,q_n}) \sim L(\mathfrak{F}_{\bar{x}}^{n,\,q_n}) \sim \rho \frac{q_n 2^n}{n + \log q_n},$$
$$T(\mathfrak{F}^{n,\,q_n}) \sim T(\mathfrak{F}_c^{n,\,q_n}) \sim T(\mathfrak{F}_x^{n,\,q_n}) \sim T(\mathfrak{F}_{\bar{x}}^{n,\,q_n}) \sim \tau n.$$

And what is more, for any $\varepsilon > 0$ and any vector-function $\tilde{f}$ from $\mathfrak{F}^{n,\,q_n}$ there exists a network S which for fairly large n realizes the function f and is such that

$$L(S) \leqslant (1+\varepsilon)\rho \frac{q_n 2^n}{n + \log q_n}, \qquad T(S) \leqslant (1+\varepsilon)\tau n.$$

Theorem A.2. *If the basis $\mathscr{E}$ is regular,*

$$2^{n-1} < r_n \leqslant 2^n$$

and

$$\frac{\log q_n}{2^n} \to 0,$$

then

$$L(\mathfrak{F}^{n,\,q_n,\,r_n}) \sim L(\mathfrak{F}_c^{n,\,q_n,\,r_n}) \sim L(\mathfrak{F}_x^{n,\,q_n,\,r_n}) \sim L(\mathfrak{F}_{\bar{x}}^{n,\,q_n,\,r_n}) \sim \rho \frac{q_n r_n}{\log (q_n r_n)},$$
$$T(\mathfrak{F}^{n,\,q_n,\,r_n}) \sim T(\mathfrak{F}_c^{n,\,q_n,\,r_n}) \sim T(\mathfrak{F}_x^{n,\,q_n,\,r_n}) \sim T(\mathfrak{F}_{\bar{x}}^{n,\,q_n,\,r_n}) \sim \tau n.$$

And what is more, the bounds of the complexity and the delay are simultaneously attained asymptotically (i.e., on one network; see the end of the formulation of Theorem A.1).

Theorem A.3. *If the basis $\mathscr{E}$ is regular and $\frac{\log q_n}{2^n} \to 0$, then*

$$L(\mathfrak{F}_c^{n,\,q_n}) \sim \rho \frac{q_n 2^n}{n + \log q_n}, \quad L(\mathfrak{F}_x^{n,\,q_n}) \sim L(\mathfrak{F}_{\bar{x}}^{n,\,q_n}) \sim \rho^* \frac{q_n 2^n}{n + \log q_n},$$
$$T(\mathfrak{F}_c^{n,\,q_n}) \sim \tau n, \; T(\mathfrak{F}_x^{n,\,q_n}) \sim T(\mathfrak{F}_{\bar{x}}^{n,\,q_n}) \sim \tau^* n.$$

And what is more, the bounds of the complexity and the delay are simultaneously attained asymptotically.

Theorem A.4. *If the basis $\mathscr{E}$ is irregular,*

$$2^{n-1} < r_n \leqslant 2^n$$

and

$$\frac{\log q_n}{2^n} \to 0,$$

then

$$L(\mathfrak{F}_c^{n,q_n,r_n}) \sim \rho \frac{q_n r_n}{\log (q_n r_n)}, \quad L(\mathfrak{F}_x^{n,q_n,r_n}) \sim L(\mathfrak{F}_{\bar{x}}^{n,q_n,r_n}) \sim \rho^* \frac{q_n r_n}{\log (q_n r_n)},$$
$$T(\mathfrak{F}_c^{n,q_n,r_n}) \sim \tau n, \quad T(\mathfrak{F}_x^{n,q_n,r_n}) \sim T(\mathfrak{F}_{\bar{x}}^{n,q_n,r_n}) \sim \tau^* n.$$

And what is more, the bounds of the complexity and delay are simultaneously attained asymptotically.

The lower bounds of the complexity and delay (in all four theorems) are established by analogy with the corresponding bounds of Theorems 1 and 2. A certain difference resides in the fact that the lower bound of the delay of the system of functions is determined by the complexity of one function of the system – in the case given, of order $2^n/n$.

The synthesis methods which yield simultaneous attainment of asymptotically minimal complexity and asymptotically minimal delay are likewise analogous to the corresponding methods for one function. The basic difference resides in the fact that the functions $f_{\tilde{\gamma}_{l',h'},i}(\tilde{x}, \tilde{\beta}_{l'',h''})$ are realized directly as functions of k arguments with an asymptotically minimal delay ($\sim \tau k$).† The number of strips p satisfies the condition $p \leqslant \frac{r}{2^{n-k_s}} + 2$ (compare with [6], p. 102). The values of the parameters may be chosen, for example, in the following way.‡ Assume $\mu = \log (q_n r_n)$, $\lambda = \min \left(\log n, \frac{n - \log \mu}{5}\right)$. Then

$$k = [\log \mu + \lambda], \quad u = [2\lambda], \quad s = [\mu - 5 \log \mu],$$
$$t' = 2\left[\frac{1}{2 \log k'}(m - \lambda)\right], \quad t'' = 2\left[\frac{\lambda}{2 \log k'''}\right].$$

In the case of an irregular basis the block B_0 in realizing vector-functions from $\mathfrak{F}_c^{n,q}$ additionally realizes the function e = $x_1 x_2 \ldots x_n \vee \bar{x}_1 \bar{x}_2 \ldots \bar{x}_n$ in accordance with the representation e = $(x_1 \vee \bar{x}_2)(x_2 \vee \bar{x}_3) \ldots (x_{n-1} \vee \bar{x}_n)(x_n \vee \bar{x}_1)$ and using S_7 networks. Then 0-functions coinciding with the components of the original vector-function on all collections with the exception of $\tilde{0}$ and $\tilde{1}$ are realized. Finally, either 0 (if the corresponding component is a 0-function) or e (if the corresponding component is a 1-function) is added to each of the functions obtained.

Literature Cited

1. V. V. Glagolev, "Some bounds for disjunctive normal forms of functions of the algebra of logic," in: Systems Theory Research, Vol. 19, Consultants Bureau, New York (1970), p. 74.
2. V. B. Kudryavtsev, "Completeness theorem for a certain class of automata without feedbacks," in: Problemy Kibernetiki, Vol. 8, Fizmatgiz, Moscow (1962), pp. 91-115.
3. O. B. Lupanov, "On the synthesis of certain classes of supervisory systems," in: Problemy Kibernetiki, Vol. 10, Fizmatgiz, Moscow (1963), pp. 63-97.

† One may indicate a direct (but more complex) construction which does not use Theorem 1.
‡ In realizing the vector-functions from $\mathfrak{F}_x^{n,q_n}$ and from $\mathfrak{F}_{\bar{x}}^{n,q_n}$ we take k‴ and k^{IV}, respectively, instead of k′ and k″, in the expression for t′ and t″.

4. O. B. Lupanov, "On a certain class of networks consisting of functional elements," in: Problemy Kibernetiki, Vol. 7, Fizmatgiz, Moscow (1962), pp. 61-114.
5. O. B. Lupanov, "On a certain method of network synthesis," Izvestiya Vuzov, Radiofizika, 1(1):120-140 (1958).
6. O. B. Lupanov, "On a certain approach to the synthesis of supervisory systems – the principle of local coding," in: Problemy Kibernetiki, Vol. 14, Nauka, Moscow (1965), pp. 31-110.
7. S. V. Makarov, "The upper bound of the average length of a disjunctive normal form," in: Discrete Analysis (Transactions of the Mathematics Institute, Siberian Branch, Academy of Sciences of the USSR), No. 3 (1964), pp. 78-80.
8. R. G. Nigmatullin, "The variational principle in logic algebra," in: Discrete Analysis (Transactions of the Mathematics Institute, Siberian Branch, Academy of Sciences of the USSR), No. 10 (1967), pp. 69-89.
9. S. V. Yablonskii, G. P. Gavrilov, and V. B. Kudryavtsev, Logic-Algebra Functions and Post Classes, Nauka, Moscow (1966).
10. E. L. Post, "Two-valued iterative systems in mathematical logic," Princeton Ann. of Math. Studies, Vol. 5 (1941).
11. C. E. Shannon, "The synthesis of two-terminal switching circuits," Bell System Technical Journal, 28(1):59-98 (1949).

PROOF OF MINIMALITY OF CIRCUITS CONSISTING OF FUNCTIONAL ELEMENTS†

N. P. Red'kin

Moscow

Introduction

In the synthesis of circuits realizing Boolean functions it is important to construct minimal circuits. Some results in this direction were obtained by Cardot [5], who proved the minimality of relay contact circuits for a sum of n variables modulo-2, whereas Soprunenko [3] obtained a minimal realization of conjunctions and disjunctions with the aid of circuits of functional elements in a base consisting of Sheffer's stroke.

In this paper we consider the realization of Boolean functions by circuits of functional elements in a base consisting of functions realizing conjunction, disjunction, and negation (the notations are: $E^{\&}$, a conjunctor; $E^{\vee}$, a disjunctor; and E^{-}, an inverter).‡ We present circuits for the realization of a linear function, and of the comparison operator and coincidence operator. We also prove the minimality of these circuits.

The upper bounds follow directly from these circuits realizing the functions just mentioned.

The idea of the proof of the lower bounds is as follows: For any minimal circuit realizing a Boolean function (or operator), we establish the possibility of removing a certain number of elements, thus obtaining (perhaps after changing the configuration of the remaining elements) a new circuit realizing a similar function (operator), but with a smaller number of variables. We find that it is not sufficient to consider circuits with a limited number of inputs (as is done, for example, in [5]), our approach being based on the separation of bounded "pieces" from the circuits under consideration (in the sense of the number of elements they contain).

Prior to a detailed analysis, let us introduce the concepts of complexity and minimality of circuits.

Let S be a circuit of functional elements realizing conjunction, disjunction, and negation. The *complexity of a circuit* [denoted by L(S)] is defined as the number of elements occurring in this circuit. The smallest of the complexities of circuits realizing f (or the oper-

† Original article submitted March 15, 1969.

‡ The definitions of certain often-encountered concepts and terms, not given in this paper, can be found in [1], [2], and [4].

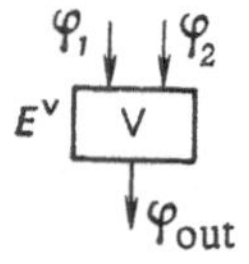

Fig. 1

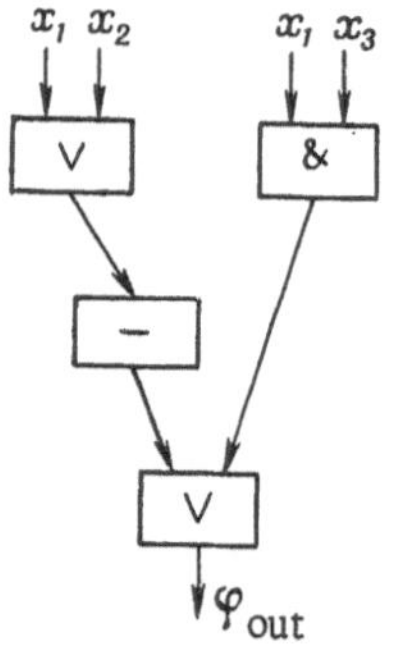

Fig. 2

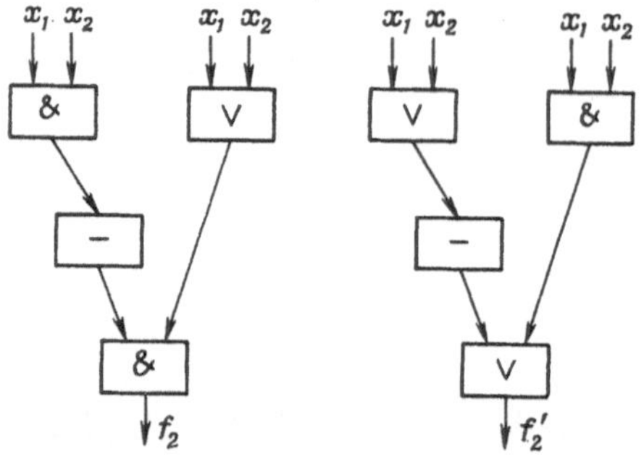

Fig. 3

ator F) will be called the complexity of a function f (or operator F) and denoted by $L(f)$ ($L(F)$). A circuit S that realizes a function f (or operator F) is said to be minimal if $L(S) = L(f)$ (or $L(S) = L(F)$).

§ 1. Two Properties of Circuits of Functional Elements $E^{\&}$, $E^{\vee}$, E^{-}

Let S be a circuit of functional elements $E^{\&}$, $E^{\vee}$, E^{-}.

Property 1. Suppose that together with the variables $x_1, \ldots, x_n$ we have also constants 0 and 1 that can be applied to the inputs of the circuit S. If we apply to the input of an element of S the identical zero or the identical unity, this element can be removed from the circuit in such a way that the function realized by this circuit does not change.†

Indeed, suppose for example that the circuit has a disjunctor $E^{\vee}$ (Fig. 1) to one of whose inputs we apply a constant (0 or 1). If, for example, $\varphi_1 \equiv 1$, we have $\varphi_{out} \equiv 1$, and the element $E^{\vee}$ can be eliminated from the circuit, and we can apply φ_1 to the inputs of the other elements to which φ_{out} is applied. If, on the other hand, $\varphi_1 \equiv 0$, we have $\varphi_{out} \equiv \varphi_2$, and we can likewise remove $E^{\vee}$ from the circuit and apply φ_2 to the inputs of the other elements to which φ_{out} is applied.

This property can be formulated in a similar way for a conjunctor and an inverter.

Property 2. Suppose that together with the variables $x_1, \ldots, x_n$ we have constants 0 and 1 that can be applied to the inputs of the circuit. Hence if any circuit element realizes a function which is identically zero or unity, it is possible to remove this element from the circuit in such a way that the function realized by the circuit does not change.

This property is evident.

Now let us introduce a concept that will be often used below.

Suppose that the variables $x_1, x_2, \ldots, x_n$ are applied at the inputs of a circuit S that realizes a Boolean function. We shall say that a variable x_i is obstructed by the variables x_j, $x_k, \ldots, x_l$ if the application of certain constants at the circuit inputs corresponding to the variables $x_j, x_k, \ldots, x_l$ will result in an output function of the circuit that does not depend on x_i. In abbreviated form the obstruction of the variable x_i by the variables $x_j, x_k, \ldots, x_l$ (with respect to a circuit S) will be denoted by $S(x_j, x_k, \ldots, x_l \rightarrow x_i)$.

† It is assumed throughout that the configuration of the elements in the circuit can be changed (after certain elements have been removed).

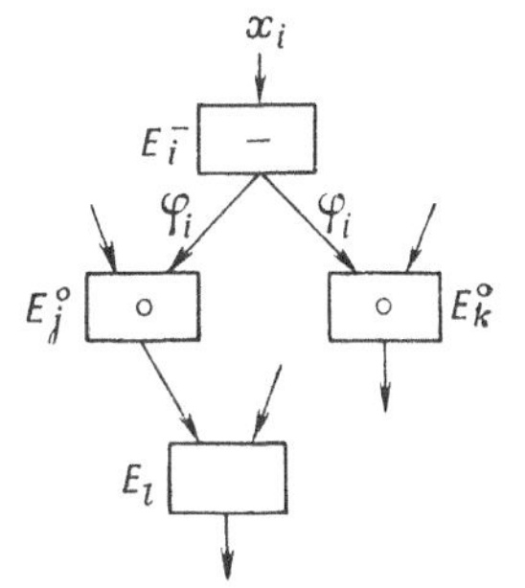

Fig. 4

For example, for the circuit of Fig. 2 we have $S(x_2, x_3 \to x_1)$, $S(x_1 \to x_2)$, and $S(x_1 \to x_3)$, since $\varphi_{out}(x_1, 1, 0) \equiv 0$, $\varphi_{out}(1, x_2, x_3) \equiv x_3$, and $\varphi_{out}(0, x_2, x_3) \equiv \bar{x}_2$. On the other hand, for the circuits of Fig. 3 none of the variables is obstructed.

Below we shall also use the notation

$$\chi(E^\circ) = \begin{cases} 1, & \text{if} \quad E^\circ\text{—disjunctor}, \\ 0, & \text{if} \quad E^\circ\text{—conjunctor}, \end{cases}$$

where $\circ \in \{\vee, \&\}$.

The proof of the lower bounds will be given for the case where at the inputs of the circuit elements we are allowed to apply together with the variables $x_1, \ldots, x_n$ also the constants 0 and 1. It is evident that these bounds will be valid also for the case where only the variables $x_1, \ldots, x_n$ can be applied at the circuit inputs (it is precisely this case that is referred to in Theorems 1-3).

§ 2. Minimal Realization of Sum of n Variables Modulo-2

Theorem 1. If $f_n(x_1, \ldots, x_n) = x_1 \oplus x_2 \oplus \ldots \oplus x_n$, a $f_n'(x_1, \ldots, x_n) = \overline{x_1 \oplus x_2 \oplus \ldots \oplus x_n}$, $n \geqslant 2$, then $L(f_n) = L(f_n') = 4(n-1)$.

Proof of Theorem 1. Upper bound. Since

$$f_n(x_1, \ldots, x_n) = f_{n-1}(x_1, \ldots, x_{n-1}) \oplus x_n,$$
$$f_n'(x_1, \ldots, x_n) = \overline{f_{n-1}(x_1, \ldots, x_{n-1}) \oplus x_n}$$

and (as is easy to see by "trial") $L(f_2) = L(f_2')$, $= 4$, the upper bound can be easily proved by induction on n (the minimal circuits realizing f_2 and f_2' are shown in Fig. 3).

The lower bound is likewise easy to prove by induction on n, by using the following lemma.

Lemma 1. Let S_n^{min} be a minimal circuit realizing one of the functions f_n, f_n', $n \geq 3$. If it is allowed to apply at the inputs of the circuit S_n^{min} the constants 0 and 1, there will always exist four or more elements that can be removed from S_n^{min}, thus yielding a new circuit S_{n-1} which realizes any of the functions f_{n-1} and f_{n-1}'.†

Proof of Lemma 1. Suppose we have a circuit S_n^{min}. Depending on the possible form of this circuit let us consider all the possible cases listed in Tables 1 and 2 (Table 2 lists all the subcases of Case 2.3 of Table 1).

1.1. To this case there corresponds Fig. 4.‡

Since the circuit under consideration is minimal, the output of at least one of the elements E_j° or E_k° (say, the output of the element E_j°) must be applied to the input of some element E_l (and not to the output of the circuit).

Suppose that the constant $\bar{\chi}(E_j^\circ)$ has been applied to the input of the circuit S_n^{min} that corresponds to the variable x_i. Then the circuit S_n^{min} will realize one of the functions f_{n-1} and f_{n-1}',

† The notations used in Theorem 1 and Lemma 1 will be retained in this section throughout.

‡ The constraints corresponding to this case are not formulated in the text, since these constraints are listed in the tables.

TABLE 1. Splitting of Proof of Lemma 1 into Cases

0) General case (the circuit realizing a linear function is minimal; it is allowed to apply the constants 0 and 1 at the circuit inputs)

- 1) At least one variable x_i is applied directly at the input of an inverter (for example, of the inverter E_i^-)
  - 1.1) There exist at least two elements at whose inputs we apply the output of the element E_i
  - 1.2) The output of the element E_i is applied to the input of only one element
- 2) None of the variables $x_1, \ldots, x_n$ is applied directly to the input of an inverter
  - 2.1) There exists at least one variable which is applied to the inputs of more than two elements
  - 2.2) There exists at least one variable which is applied to the input of only one element
  - 2.3) Each variable is applied to the inputs of precisely two elements

constants will be applied at the inputs of the elements E_i^-, E_j°, E_k°, and E_l, and these elements can be removed from the circuit S_n^{min} (Property 1) in such a way that the conditions of Lemma 1 are satisfied.

1.2. Let us note right away that the element E_j° (Fig. 5) cannot be an inverter, since the circuit under consideration is minimal. Furthermore, under our assumptions we cannot confine ourselves to applying x_i only to the input of the element E_i^-. Indeed, if x_i is applied only to the input of E_i^-, it is possible to find constants $\sigma_1, \ldots, \sigma_{i-1}, \sigma_{i-1}, \ldots, \sigma_n$, such that when these constants are applied at the circuit inputs corresponding to the variables $x_1, \ldots, x_{i-1}$, $x_{i+1}, \ldots, x_n$, we shall have the relation $\varphi_1 \equiv \chi(E_j^\circ)$ (this can be always done, since φ_1 does not depend by assumption on x_i and φ_1 cannot be identically equal to a constant, since the circuit under consideration is minimal). In this case $\varphi_j \equiv \chi(E_j^\circ)$ and the output function f_{out} of this circuit will satisfy the relation

$$f_{out}(\sigma_1, \ldots, \sigma_{i-1}, 0, \sigma_{i+1}, \ldots, \sigma_n) = f_{out}(\sigma_1, \ldots, \sigma_{i-1}, 1, \sigma_{i+1}, \ldots, \sigma_n).$$

But this relation signifies that the variable x_i is obstructed by the variables $x_1, x_2, \ldots, x_{i-1}$, $x_{i+1}, \ldots, x_n$, which cannot be the case for a circuit S_n^{min} that realizes one of the functions f_n and f_n'.†

Hence, together with E_i^- there exists an element E_k° at whose input we apply the variable x_i (Fig. 6).

Since neither the output of the element E_j° nor the output of the element E_k° can be the output of the circuit, and we cannot simultaneously apply the output of E_j° to the input of E_k° and the output of E_k° to the input of E_j°, there must exist an element E_l to one of whose inputs we apply either the output of E_j° or the output of E_k°. Let us assume that at the input of E_l we apply the output of E_k° (as shown in Fig. 6).

Let us write $x_i \equiv \chi(E_k^\circ)$ (if the output of E_j° is applied to the input of E_l, we shall write $x_i \equiv \overline{\chi}(E_j^\circ)$ and reason in the same way as below). In this case the circuit will realize one of the functions f_{n-1} and f_{n-1}' of the variables $x_1, \ldots, x_{i-1}, x_{i+1}, \ldots, x_n$, constants will be applied at the inputs of the elements E_i^-, E_j°, E_k°, and E_l, and it is possible to remove these elements from the circuit (Property 1) without changing the output function. Hence Lemma 1 will hold also in this case.

† Below we shall not explain in similar cases in detail the obstruction of variables.

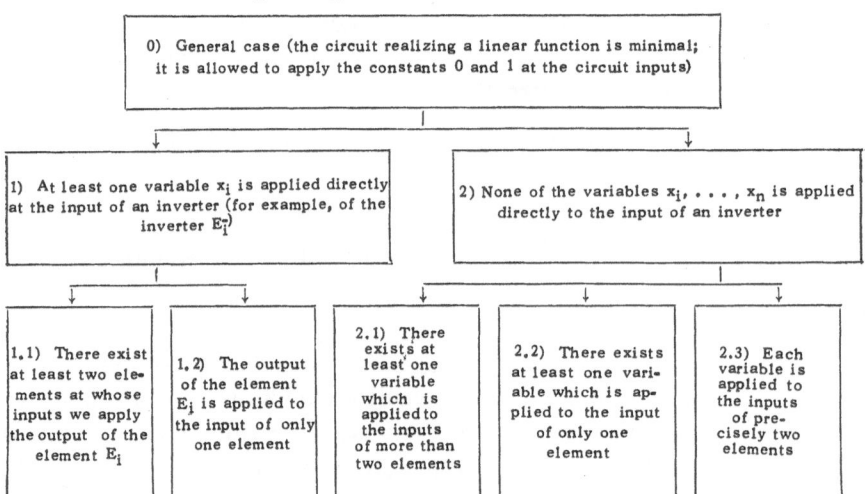

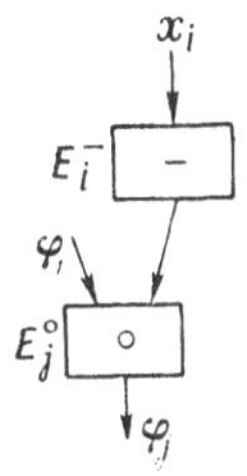
Fig. 5

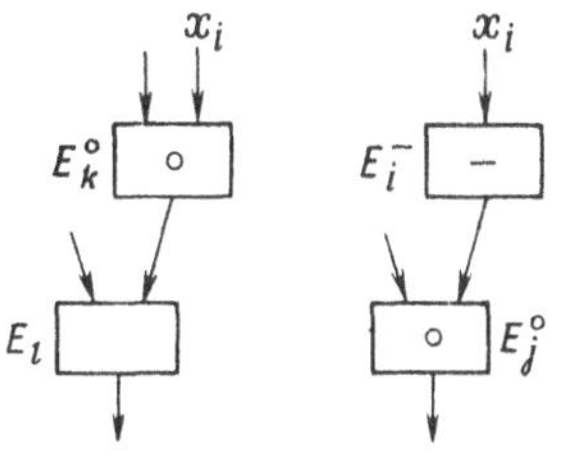
Fig. 6

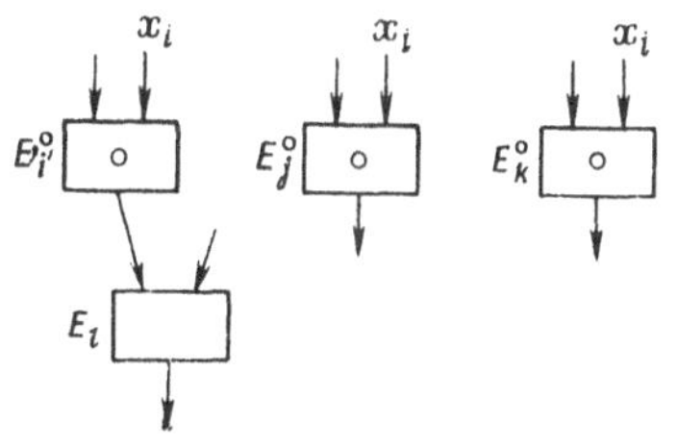
Fig. 7

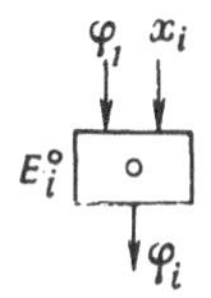
Fig. 8

2.1. In this case it is possible to single out at least three elements E_i°, E_j°, E_k° (Fig. 7) at whose inputs we apply a variable x_i.

Since none of these elements can be the output element of the circuit, and the circuit under consideration is minimal and has no cycles, there must exist an element E_l to one of whose inputs we apply the output of at least one of the elements E_i°, E_j°, E_k° (for example, the output of E_i°, see Fig. 7).

Let us write $x_i \equiv \chi(E_i^\circ)$. In this case we apply constants to the inputs of the elements E_i°, E_j°, E_k°, E_l, and these elements can be removed from the circuit (Property 1) in such a way that the conditions of Lemma 1 hold.

2.2. The Case 2.2 cannot occur.

Indeed, if x_i is applied only to an input of E_i° (Fig. 8), we have $S_n^{min}(x_1, \ldots, x_{i-1}, x_{i+1}, \ldots, x_n \rightarrow x_i)$, which is impossible.

It remains to consider Case 2.3. It is evident that in this case there must exist an element E_i° to both of whose inputs we apply variables belonging to $x_1, \ldots, x_n$. Without loss of generality it can be assumed that the variables x_1 and x_2 are applied to the inputs of E_i°. The variable x_1 is applied also to the input of an element E_j°. Let us consider all possible types of circuits S_n^{min} corresponding to this case (see Table 2).

2.3. 1.1. Let us write $x_1 \equiv \chi(E_j^\circ)$ (Fig. 9). It is then possible to remove (Property 1) the elements E_i° and E_j°, as well as an element E_l to whose input we apply the output of E_j° (such an element E_l always exists, since the circuit under consideration is minimal and the output of E_j° cannot be the output of the circuit, since otherwise we obtain for $x_1 \equiv \chi(E_j^\circ)$ the relation $f_{out}(\chi(E_j^\circ), x_2, \ldots, x_n) \equiv \chi(E_j^\circ))$, and an element E_k, since for $x_1 \equiv \chi(E_j^\circ)$ the value of φ_j does not depend on φ_k, and the output of E_k is not applied to inputs of elements other than E_j°. Hence Lemma 1 holds in this case.

2.3.12. Let us note the following necessary property of the circuit considered in this case (Fig. 10): For $x_2 \equiv \chi(E_i^\circ)$ we must have $\varphi_k \equiv \overline{\chi}(E_j^\circ)$.

Indeed, if this property does not hold, it is possible, for $x_2 \equiv \chi(E_i^\circ)$ and by applying some constants $\sigma_3, \ldots, \sigma_n$ to the circuit inputs corresponding to the variables $x_3, \ldots, x_n$, to obtain the relations

$$\varphi_k \equiv \chi(E_j^\circ), \quad \varphi_j \equiv \chi(E_j^\circ), \quad \varphi_i \equiv \chi(E_i^\circ).$$

But in this case $S_n^{min}(x_2, \ldots, x_n \rightarrow x_1)$, i.e., the circuit under consideration does not realize any of the functions f_n and $f_n^?$.

Thus for the case under consideration the above-formulated property must hold, and for $x_2 \equiv \chi(E_i^\circ)$ constants are applied to the inputs of the elements E_i°, E_j°, and E_l ($\varphi_k \equiv \overline{\chi}(E_j^\circ)$!).

TABLE 2. Splitting of Case 2.3

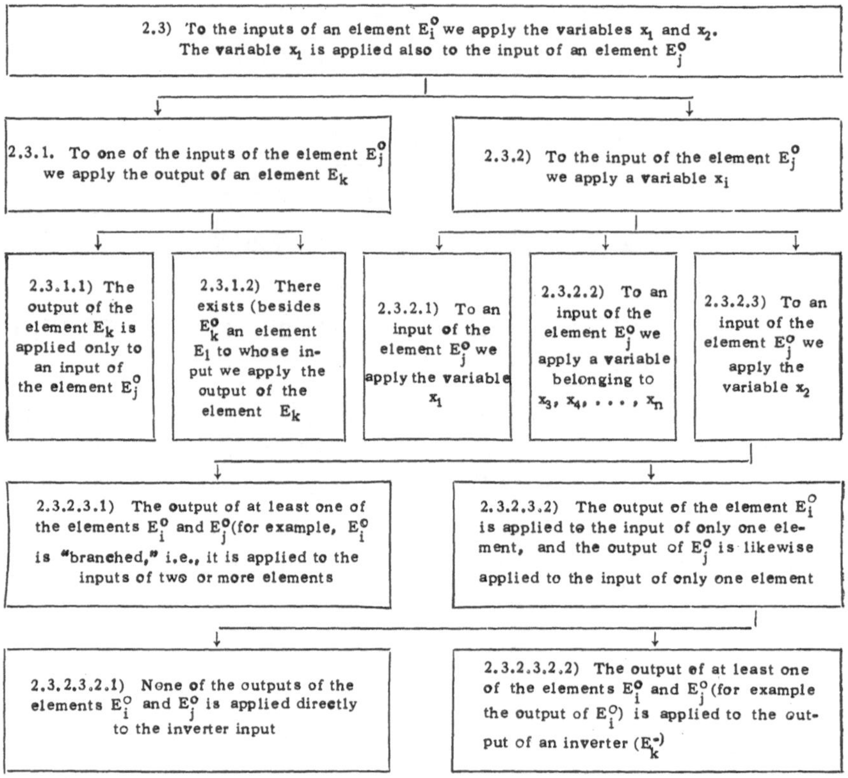

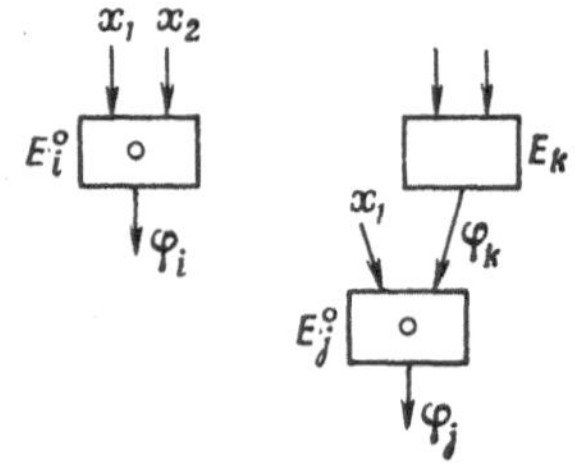

Fig. 9

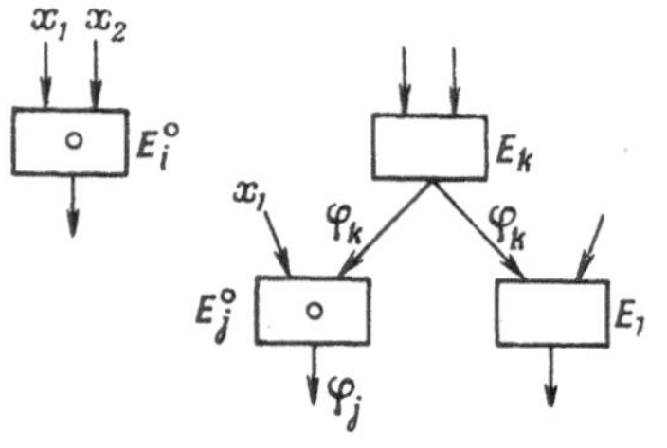

Fig. 10

Thus by satisfying the conditions of Lemma 1, it is possible to remove the elements E_i°, E_j°, E_l (Property 1) and the elements E_k (Property 2).

2.3. 2.1. This case cannot occur, since the circuit S_n^{min} is minimal.

2.3. 2.2. This case (Fig. 11) is likewise impossible, since for it we have S_n^{min} $(x_2, x_i \to x_1)$.

2.3. 2.3. 1. In this case (Fig. 12), for $x_1 \equiv \chi_0(E_i^\circ)$, it is possible to remove four elements (E_i°, E_j°, E_k, and E_l), and satisfy the conditions of Lemma 1.

2.2. 2.3. 2.1. This cannot occur. In fact, the outputs of the elements E_i° and E_j° cannot be applied to the inputs of one and the same element, since otherwise the circuit is not minimal. Hence the output of one of the elements E_i° and E_j° (for example, E_i°) must be applied to the input of an element E_k° (E_k° is not an inverter!), to whose other input we apply one of the variables $x_3, \ldots, x_n$, or a function φ_1 (as is shown in Fig. 13) that does not depend on x_1 and x_2 (x_1 and x_2 are applied only to the inputs of the elements E_i° and E_j°).

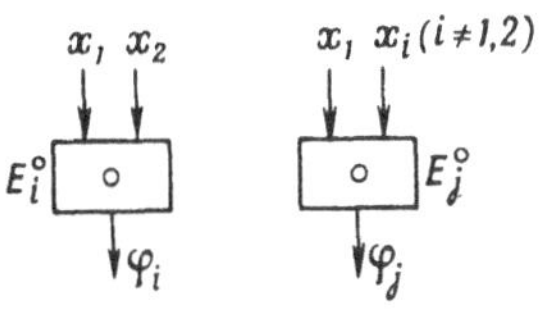

Fig. 11

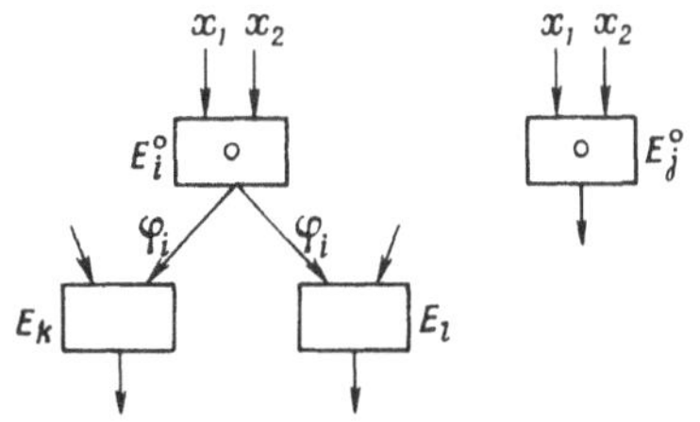

Fig. 12

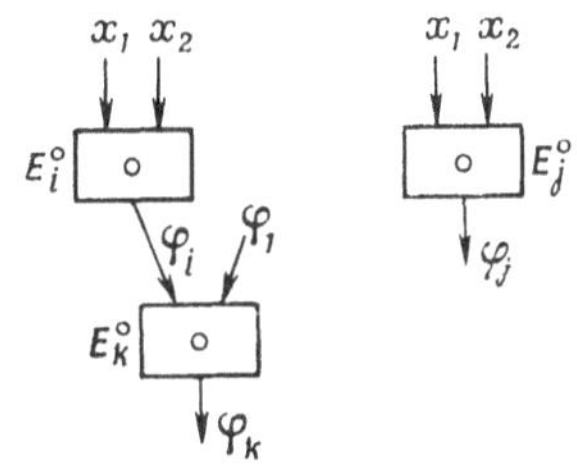

Fig. 13

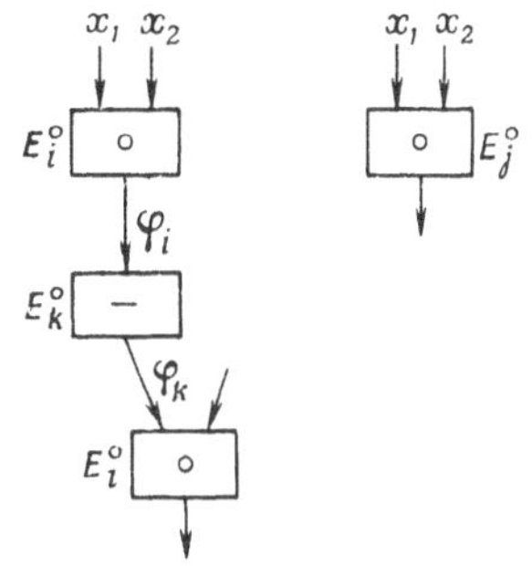

Fig. 14

Let us write $x_2 \equiv \chi(E_j^\circ)$ and apply to the circuit inputs corresponding to the variables $x_3, \ldots, x_k$ constants $\sigma_3, \ldots, \sigma_n$ such that $\varphi_1 = \chi(E_k^\circ)$. Then the values of φ_j and φ_k will not depend on x_1 (the output of E_i° is not "branched") and we finally obtain $S_n^{\min}(x_2, \ldots, x_n \to x_1)$, which cannot be the case.

2.3. 2.3. 2.2. Suppose, for example, that the output of the element E_i° is applied to the input of the inverter E_k^- (Fig. 14). In this case the output of E_k^- (it cannot be the output of the circuit, since the output function of the circuit cannot be identically equal to a constant when a constant is applied to the circuit input that corresponds to the variable x_1) is applied to the input of an element E_l°. Let us write $x_1 \equiv \chi(E_i^\circ)$.

In this case $\varphi_i \equiv \chi(E_i^\circ)$, $\varphi_k \equiv \overline{\chi}(E_i^\circ)$, and constant will be applied to the inputs of the elements E_i°, E_j°, E_k^-, E_l°. Hence follows (by Property 1) the validity of Lemma 1 for this last case.

Thus we have completed the proof of Lemma 1.

Remark. In a similar way it is possible to show that $7(n-1)$ elements are needed for a minimal realization of f_n (and also of f_n') in a base consisting of a disjunctor and negation or of a conjunctor and negation.

§ 3. Minimal Realization of Comparison Operator

The comparison operator F_n is a (2n, 1)-operator (see [4]). It compares two n-digit binary numbers

$$F_n(\tilde{x}, \tilde{y}) = \begin{cases} 1, & \text{if} \quad |\tilde{x}| < |\tilde{y}|, \dagger \\ 0, & \text{if} \quad |\tilde{x}| \geqslant |\tilde{y}|. \end{cases}$$

† If $\tilde{\sigma} = (\sigma_0, \sigma_1, \ldots, \sigma_{k-1})$, then $|\tilde{\sigma}| = \sum_{i=0}^{k-1} \sigma_i 2^i$.

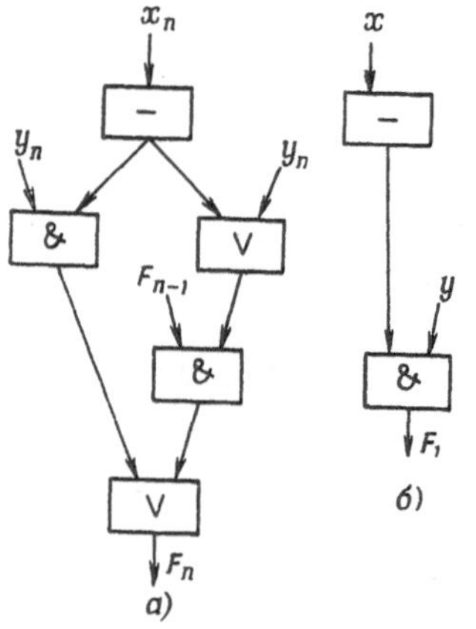

Fig. 15

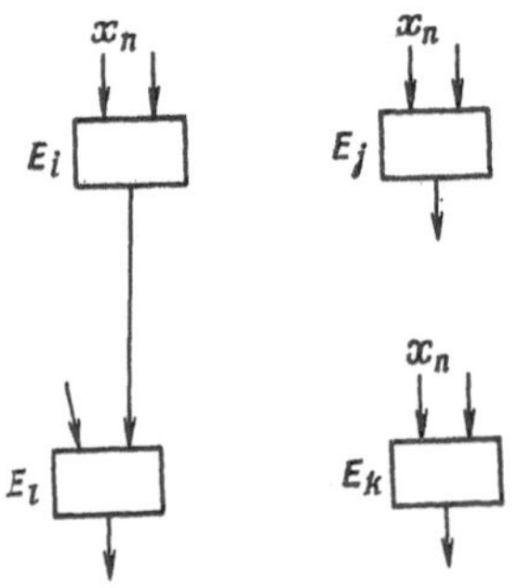

Fig. 16

In the language of formulas, the comparison operator can be defined inductively as follows:

$$F_n(x_1, \ldots, x_n, y_1, \ldots, y_n) = \bar{x}_n y_n \vee (\bar{x}_n \vee y_n) F_{n-1}(x_1, \ldots, x_{n-1}, y_1, \ldots, y_{n-1});$$
$$F_1(x, y) = \bar{x} y.$$

Theorem 2. For any $n \geq 1$ we have

$$L(F_n) = 5n - 3.$$

Proof of Theorem 2. The upper bound follows directly from the inductive definition of the comparison operator (Fig. 15).

The lower bound is easy to obtain by induction on n with the use of the following lemma.

Lemma 2. Let S_n^{min} be a minimal circuit realizing the function F_n, $n \geq 2$. If it is allowed to apply constants 0 and 1 to the inputs of the circuit S_n^{min}, there will always exist five or more elements that can be removed from the circuit S_n^{min}, and from the remaining elements we can construct a new circuit S_{n-1} that realizes F_{n-1}.

Proof of Lemma 2. Suppose we have a circuit S_n^{min}. Let us consider all the possible cases depending on the type of this circuit (Table 3).

1. Suppose that the variable x_n is applied at the inputs of some elements E_i, E_j, and E_k (Fig. 16) (if the variable y_n is applied at the inputs of three elements, the proof will be similar).

Since none of the elements of E_i. E_j, and E_k can be an output element of the circuit, and the circuit itself is minimal and does not contain cycles, there exists an element E_l to whose input we apply the output of at least one of the elements E_i, E_j, and E_k (for example, E_i). If E_i is a conjunctor or disjunctor, we shall write $x_n \equiv \chi(E_i^\circ)$, and if E_i is an inverter, we shall write $x_n \equiv 0$.

By virtue of Property 1 it is possible to remove the elements E_i, E_j, E_k, and E_l, thus obtaining a circuit that realizes the operator $F_n(x_1, \ldots, x_{n-1}, \text{const}, y_1, \ldots, y_{n-1}, y_n)$. Since in this circuit the variable y_n must be applied to the input of at least one element of E_m, it is possible to remove for $y_n \equiv x_n \equiv$ const this element of E_m, thus obtaining a circuit S_{n-1} that realizes F_{n-1}.

TABLE 3. Splitting of Proof of Lemma 2 into Cases

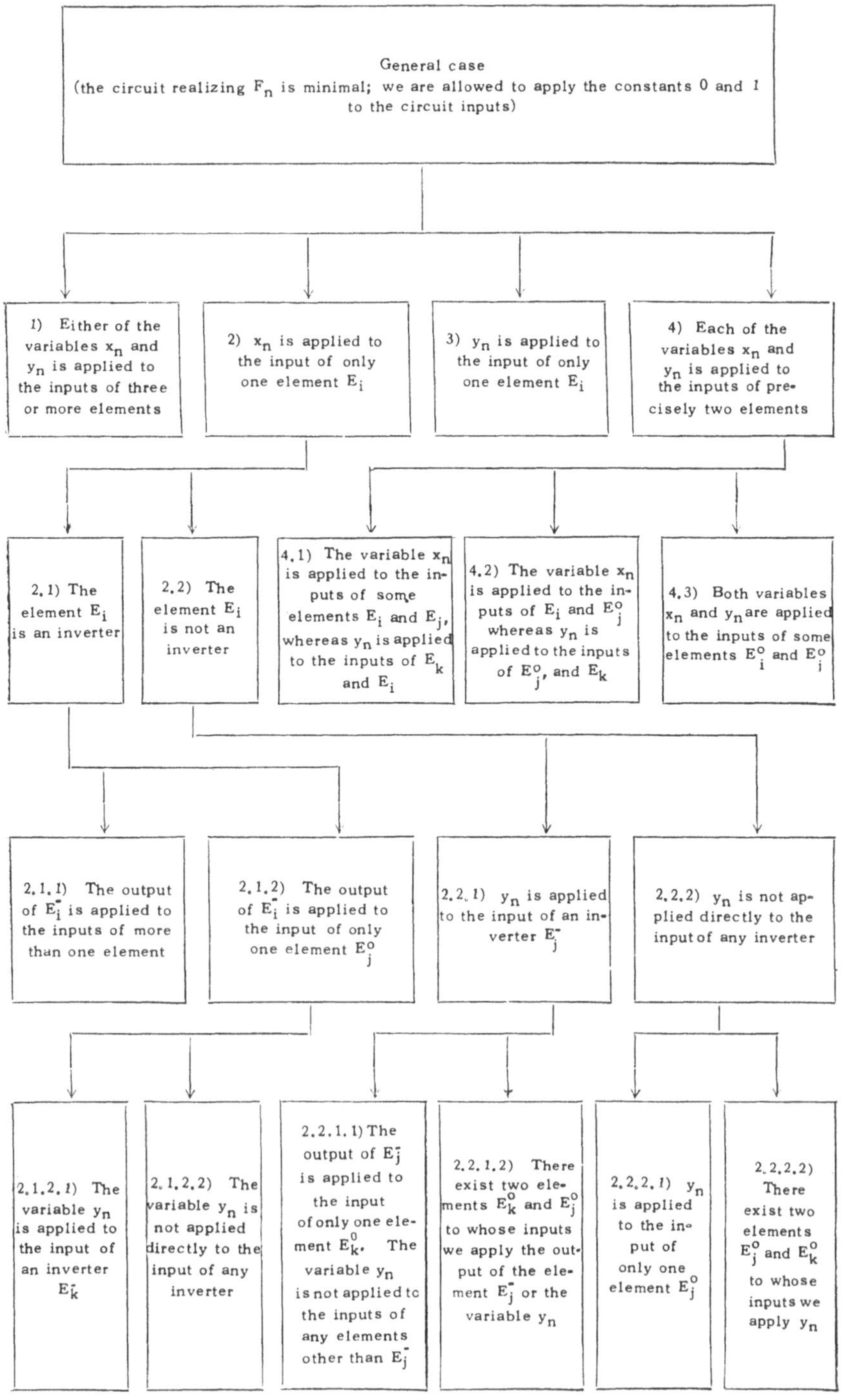
General case (the circuit realizing F_n is minimal; we are allowed to apply the constants 0 and 1 to the circuit inputs)
1) Either of the variables x_n and y_n is applied to the inputs of three or more elements
2) x_n is applied to the input of only one element E_i
3) y_n is applied to the input of only one element E_i
4) Each of the variables x_n and y_n is applied to the inputs of precisely two elements
2.1) The element E_i is an inverter
2.2) The element E_i is not an inverter
4.1) The variable x_n is applied to the inputs of some elements E_i and E_j, whereas y_n is applied to the inputs of E_k and E_l
4.2) The variable x_n is applied to the inputs of E_i and E_j^o whereas y_n is applied to the inputs of E_j^o, and E_k
4.3) Both variables x_n and y_n are applied to the inputs of some elements E_i^o and E_j^o
2.1.1) The output of E_i^- is applied to the inputs of more than one element
2.1.2) The output of E_i^- is applied to the input of only one element E_j^o
2.2.1) y_n is applied to the input of an inverter E_j^-
2.2.2) y_n is not applied directly to the input of any inverter
2.1.2.1) The variable y_n is applied to the input of an inverter E_k^-
2.1.2.2) The variable y_n is not applied directly to the input of any inverter
2.2.1.1) The output of E_j^- is applied to the input of only one element E_k^0. The variable y_n is not applied to the inputs of any elements other than E_j^-
2.2.1.2) There exist two elements E_k^o and E_j^o to whose inputs we apply the output of the element E_j^- or the variable y_n
2.2.2.1) y_n is applied to the input of only one element E_j^o
2.2.2.2) There exist two elements E_j^o and E_k^o to whose inputs we apply y_n

Hence Lemma 2 is valid for Case 1.

2.1.1. Suppose that the output of the element E_i^- is applied to the inputs of the elements E_j° and E_k° (Fig. 17). Since the circuit is minimal and without cycles, and none of the elements E_j° and E_k° can be an output element of the circuit, it follows that there must exist an element E_l to whose input we apply the output of one of the elements E_j° and E_k° (for example, the output of the element E_j°). Let us apply to the circuit input corresponding to the variable x_n the constant

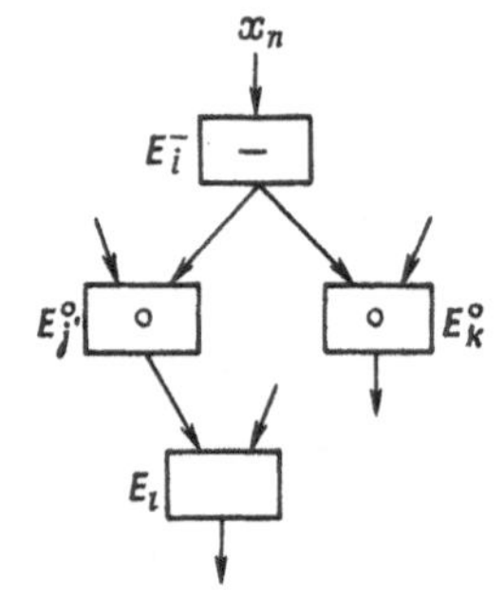

Fig. 17

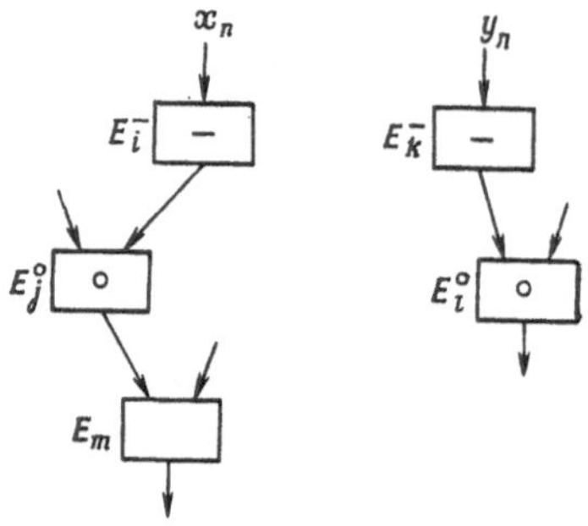

Fig. 18

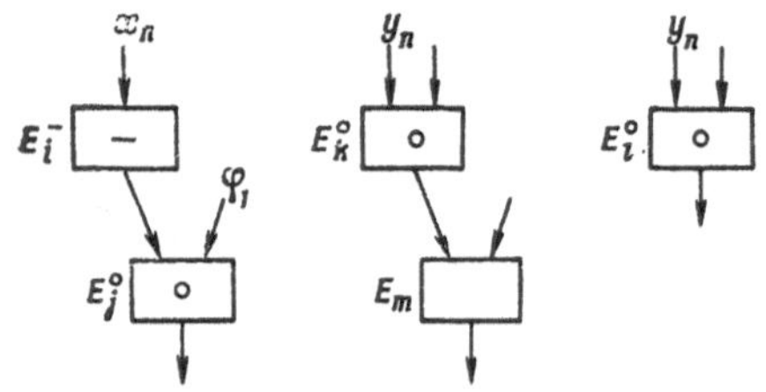

Fig. 19

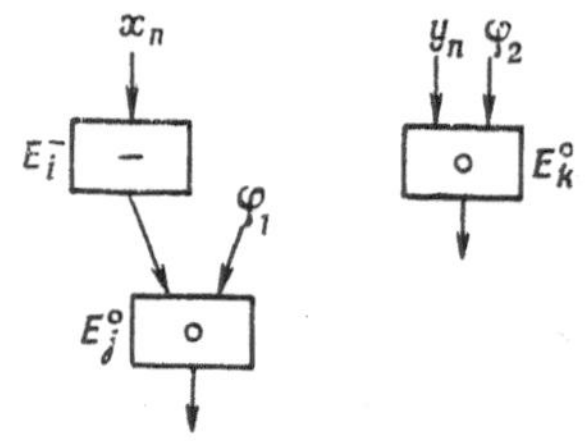

Fig. 20

$\overline{\chi}(E_j^\circ)$. We then remove the elements E_i^-, E_j°, E_k°, and E_l (Property 1), and construct from the remaining elements a circuit S_n' that realizes a function F_n' of $2n-1$ variables:

$$F_n' = y_n\chi(E_j^\circ) \vee (\chi(E_j^\circ) \vee y_n)F_{n-1}.$$

In S_n' there must exist an element E_m to whose input we apply y_n. Let us write $y_n \equiv \overline{\chi}(E_j^\circ)$. By virtue of Property 1 the element E_m can be removed, thus yielding a circuit S_{n-1} that realizes F_{n-1}.

Hence Lemma 2 is valid for the case just considered.

2.1. 2.1. In this case there must exist an element E_l° (Fig. 18), other than E_j°, to whose input we apply the output of E_k^- (otherwise the output function of the circuit will not depend on x_n for some $y_n \equiv$ const). Since the circuit is minimal and without cycles, and none of the elements E_j° and E_l° can be the output of the circuit, there must exist an element E_m ($m \neq j, l$) to whose input we apply the output of one of the elements E_j° and E_l° (for example, the output E_j°). Let us apply the constant $\overline{\chi}(E_j^\circ)$ to the circuit inputs corresponding to the variables x_n and y_n. In this case constants will be also applied to the inputs of the elements E_i^-, E_j°, E_k^-, E_l°, and E_m, and these elements can be removed (according to Property 1) in such a way that the thus-obtained circuit S_{n-1} will realize F_{n-1}.

Hence Lemma 2 is valid for this case.

2.1. 2.2. In this case there must exist two elements E_k and E_l (other than E_j°) to whose inputs we apply the variable y_n. Indeed, otherwise we would have to assume that y_n is applied to the input of E_j°, or that only one element of E_k° exists to whose input we apply y_n. It is not allowed to apply y_n to the input of E_j°, since otherwise we obtain $S_n^{\min}$ ($y_n \to x_n$), which is impossible (for this same reason we cannot apply to the input of E_j° any of the variables $x_1, \ldots, x_{n-1}, y_1, \ldots, y_{n-1}$, but only a function φ_1, see Fig. 19).

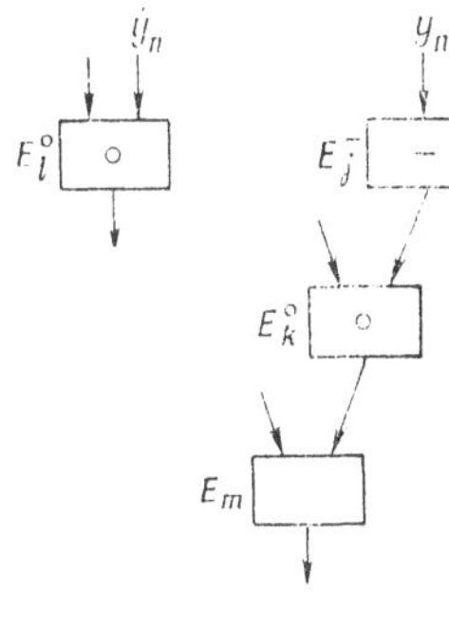

Fig. 21

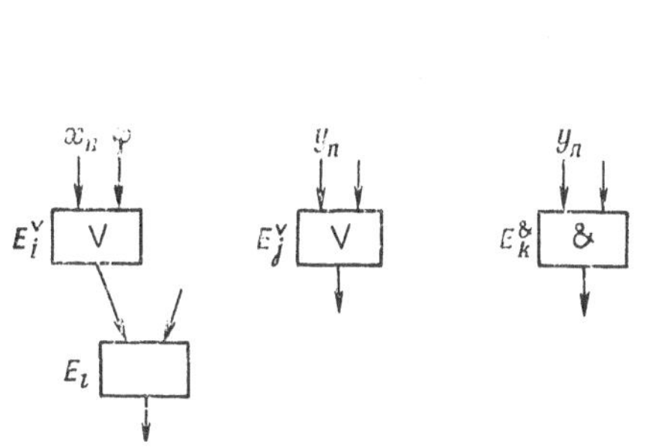

Fig. 22

It is also easy to see that we cannot confine ourselves to applying y_n to the input of only one element of E_k°. In fact, we shall select from the circuit S_n^{min} a subcircuit $S_{\varphi 1}$ that has one input and realizes the function φ_1. This subcircuit must contain the element of E_k° to one of whose inputs we apply y_n (otherwise we have S_n^{min} $(x_1, \ldots, x_{n-1}, y_1, \ldots, y_{n-1} \rightarrow x_n$, which is impossible). Furthermore, if y_n is applied only to an input of E_k°, it is necessary to apply to the second input of this element a function φ_2 that does not depend on x_n and y_n, which is likewise impossible, since it yields S_n^{min} $(x_1, \ldots, x_{n-1}, y_1, \ldots, y_{n-1} \rightarrow y_n)$.

Thus we must assume that the variable y_n is applied to the inputs of the elements E_k° and E_l° (Fig. 20).

Since the circuit has no cycles and none of the outputs of the elements E_k° and E_l° can be the output of the entire circuit, and it is impossible to apply the output of E_k° only to an input of E_l°, or (conversely), the output of E_l° only to an input of E_k° (by virtue of the minimality of the circuit), it follows that there always exists an element E_m ($m \neq j, k, l$) to whose input we apply the output of at least one of the elements E_k° and E_l°. Suppose, for example, that we apply the output of E_k° to an input of E_m (as is shown in Fig. 20). If we now apply the constant $\chi(E_k^\circ)$ to the circuit inputs corresponding to the variables x_n and y_n, it is easy to verify the validity of Lemma 2 for the case under consideration.

2.2. 1.1. This case cannot occur.

Indeed, since cycles are not allowed, it follows that under the limitations mentioned above it is necessary to apply to the input of E_i° or to the input of E_k° a function φ_1 that does not depend on x_n and y_n. But this leads to the obstruction of at least one of the variables x_n and y_n by the variables $x_1, \ldots, x_{n-1}$ and $y_1, \ldots, y_{n-1}$, which is not allowed.

2.2. 1.2. By virtue of the minimality of the circuit S_n^{min}, the output of the element E_j^- must be applied to the input of at least one element E_k°. Moreover, in view of the limitations assumed for this case, there exists an element E_l° to one of whose inputs we apply either the variable $\dot{y}_n$, or the output of the element E_j^- (Fig. 21).†

Since the circuit under consideration is minimal and the output of the element E_k° cannot be the output of the circuit, it is easy to show that there must exist an element E_m (other than E_l°) to whose input we apply the output of the element E_k°.

Let us apply the constant $\overline{\chi}(E_k^\circ)$ to the circuit input corresponding to the variable y_n. By virtue of Property 1 we shall remove the elements E_j^-, E_k°, E_l°, and E_m, thus obtaining a circuit $S_n^!$ that realizes

$$F_n(x_1, \ldots, x_{n-1}, x_n, y_1, \ldots, y_{n-1}, \overline{\chi}(E_k^\circ)).$$

† $\dot{y}$ signifies either y or $\overline{y}$.

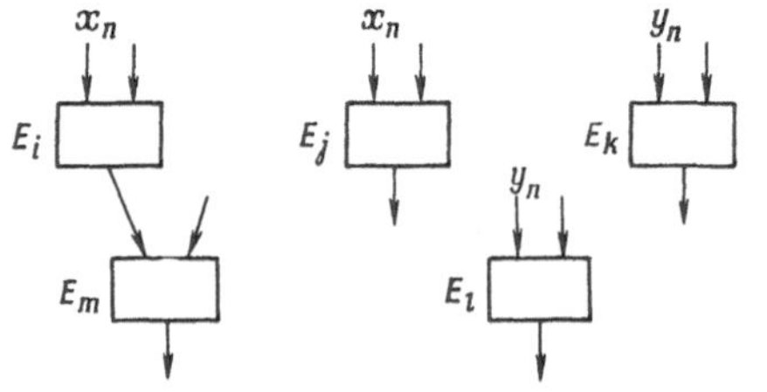

Fig. 23

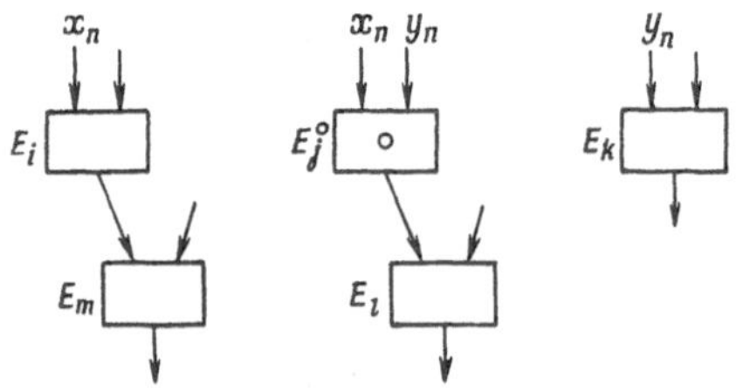

Fig. 24

In the newly obtained circuit S_n' there exists at least one element to whose input we apply x_n and that can be removed by virtue of Property 1 for $x_n \equiv \overline{\chi}(E_k^\circ)$, thus yielding a circuit S_{n-1} which realizes F_{n-1}. Hence Lemma 2 is valid in the case under consideration.

2.2. 2.1. This case cannot occur (see Case 2.2. 1.1).

2.2. 2.2. If E_j° and E_k° are identical elements, the proof of the lemma will be almost evident (it suffices to write $x_n \equiv y_n \equiv \chi(E_j^\circ) \equiv \chi(E_k^\circ)$ and take into account that this circuit is minimal). We shall therefore assume below that E_j° is a disjunctor and E_k° a conjunctor, and, moreover, that y_n is applied only to the inputs of E_j° and E_k° (if y_n is applied to the inputs of three or more elements, see Case 1). Suppose that E_i° is a disjunctor (the case that E_i° is a conjunctor can be analyzed in the same way).

Hence follows that:

1) It is impossible to apply the output of $E_i^\vee$ (Fig. 22) to the inputs of both elements $E_j^\vee$ and $E_k^\&$, since the subcircuit realizing φ must contain at least one of the elements $E_j^\vee$ and $E_k^\&$, since otherwise φ will not depend on y_n, which is impossible (see Case 2.2. 1.1).

2) The output of the element $E_i^\vee$ cannot be the circuit output.

3) We cannot confine ourselves to applying the output of $E_i^\vee$ to the input of only one of the elements $E_j^\vee$ and $E_k^\&$ (if, for example, the output of $E_i^\vee$ is applied only to the input of $E_k^\&$, we obtain $S_n^{\min}(y_n \to x_n)$ (when the identical zero is applied to the circuit input corresponding to the variable y_n).

It follows from 1-3 above that there must exist an element E_l to whose input we apply the output of the element $E_i^\vee$. It is also easy to see that:

4) The output of $E_j^\vee$ cannot be applied to the input of $E_i^\vee$ (see 3)).

5) The output of $E_j^\vee$ cannot be the output of the circuit.

6) We cannot confine ourselves to applying the output of the element $E_j^\vee$ only to the input of the element $E_k^\&$ (if the output of $E_j^\vee$ is applied only to the input of $E_k^\&$, the circuit will not be minimal, which contradicts our original assumption).

It follows from 4-6 that the output of $E_j^\vee$ must be applied to the input of an element, other than $E_i^\vee$ and $E_k^\&$. Here we can have two cases:

a) The output of $E_j^\vee$ is applied to the input of an element E_m, other than E_l;

b) the output of $E_j^\vee$ is applied to the input of E_l.

In Case (a) Lemma 2 is valid, since by applying the identical unity to circuit inputs corresponding to the variables x_n and y_n, it is possible by virtue of Property 1 to remove five elements ($E_i^\vee$, $E_j^\vee$, $E_k^\&$, E_l, E_m) and obtain a new circuit S_{n-1} that realizes F_{n-1}.

In Case (b) the output of the element E_l° (a disjunctor or conjunctor) cannot be the circuit output, since otherwise when we apply the identical unity to the circuit inputs corresponding to the variables x_n and y_n, the output function will not depend on $x_1, \ldots, x_{n-1}$ and $y_1, \ldots, y_{n-1}$, which is impossible. Furthermore, it is likewise not allowed to confine oneself to applying the output of E_l^0 only to the input of E_k^0, the output of $E_j^\vee$ only to the input of E_l°, and the output of $E_i^\vee$ only to the input of E_l° (the application of the output of $E_i^\vee$ only to the inputs of E_l° and E_j and of the output of $E_j^\vee$ only to the input of E_l° cannot take place by virtue of the minimality of the circuit), since otherwise we obtain S_n^{min} $(y_n \to x_n)$.

Hence in the Case (b) there must exist an element E_m to whose input we apply the output of one of the elements $E_i^\vee$, $E_j^\vee$ and E_l°. But in this case, when we apply the identical unity to the circuit inputs corresponding to the variables x_n and y_n, we can remove by virtue of Property 1 the elements $E_i^\vee$, $E_j^\vee$, $E_k^\&$, E_l°, and E_m (to the inputs of the element E_l° we apply "ones"; therefore the output of this element will likewise have a "one"), and obtain a new circuit that realizes F_{n-1}. Thus we have completed the proof of Lemma 2 for Case 2.2. 2.2.

3. The proof of Lemma 2 for the Case 3 is carried out in the same way as for Case 2.

4.1. Since the circuit under consideration is minimal and without cycles, and none of the elements E_i, E_j, E_k, and E_l (Fig. 23) can be an output element of the circuit, there must exist an element E_m $(m \neq i, j, k, l)$ to whose input we apply the output of at least one of the elements E_i, E_j, E_k, and E_l (for example, the output of E_i). For convincing ourselves of the validity of Lemma 2, it suffices to consider the case that a constant $\chi(E_i^\circ)$ is applied to the circuit inputs corresponding to the variables x_n and y_n if E_i is a conjunctor or disjunctor, and zero is applied if E_i is an inverter.

4.2. Let us assume that E_j° is a disjunctor (Fig. 24; if E_j° is a conjunctor, the proof will be similar). If E_i or E_k is an inverter, it is easy to prove Lemma 2 by writing $x_n \equiv y_n \equiv 1$ or $x_n \equiv y_n \equiv 0$. Therefore we shall assume below that none of the elements E_i and E_k is an inverter.

The output of the element E_j° must be applied to the input of an element E_l $(l \neq i, k)$, since otherwise the circuit is not minimal. Next, there must exist an element E_m $(m \neq i, k, l)$ to whose input we apply the output of one of the elements E_i and E_k (for example, E_i). Indeed, if this assumption does not hold, then (by virtue of the minimality of the circuit) the output of E_i must be applied only to the input of E_k° (or conversely, the output of E_k only to the input of E_i°) whereas the output of E_k° (or E_i°) must be applied only to the input of E_l° (Fig. 25). But in this case the element E_l° must be a conjunctor (otherwise we have S_n^{min} $(y_n \to x_n)$). Thus the element E_k° must be a disjunctor (otherwise we likewise have S_n^{min} $(y_n \to x_n)$). But such a case is also impossible, since we likewise obtain S_n^{min} $(y_n \to x_n)$. Hence we must assume the presence of the above-mentioned fifth element E_m (see Fig. 24). To the circuit inputs corresponding to the variables x_n and y_n we shall apply the constant χ (E_i°) if E_i is a disjunctor or conjunctor, and the identical zero if E_i is an inverter. Then we can remove the elements E_i, E_j°, E_k, E_l, and E_m (Property 1) and satisfy the conditions of Lemma 2.

4.3. By virtue of the minimality of the circuit under consideration there exist two elements E_k and E_l to whose inputs we apply the outputs of the elements E_i° and E_j° (Fig. 26). Hence we can have two cases.

a) There exists a fifth element E_m to whose input we apply the output of one of the elements E_k and E_l (for example, E_k). In this case we shall apply to the circuit inputs corresponding to the variables x_n and y_n the constant $\chi(E_k^\circ)$ if E_k is a disjunctor or conjunctor, and zero if E_k is an inverter. By virtue of Property 1 we can then remove five elements E_i°, E_j°, E_k, E_l, and E_m, and satisfy the conditions of Lemma 2.

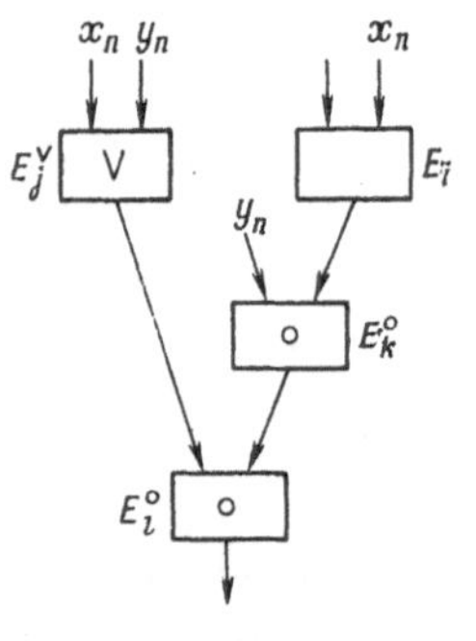

Fig. 25

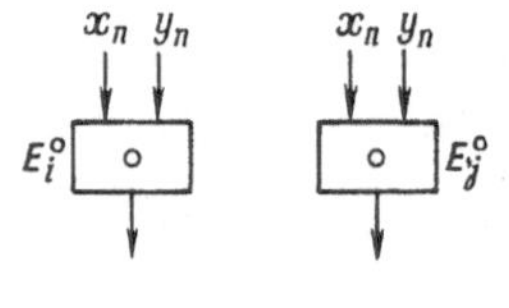

Fig. 26

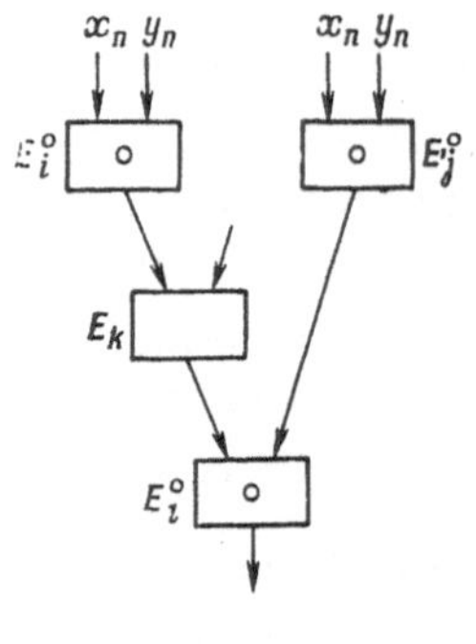

Fig. 27

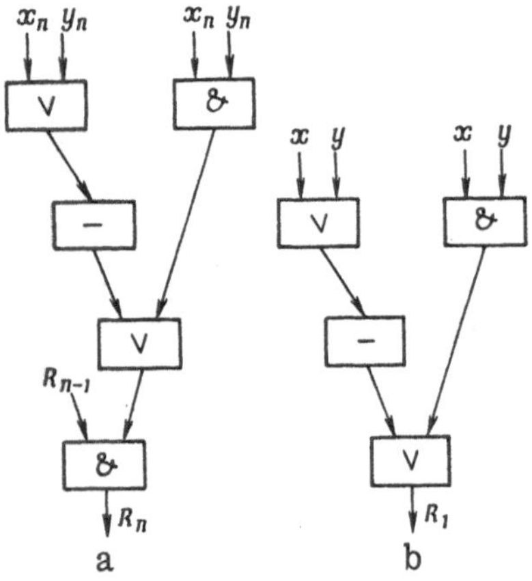

Fig. 28

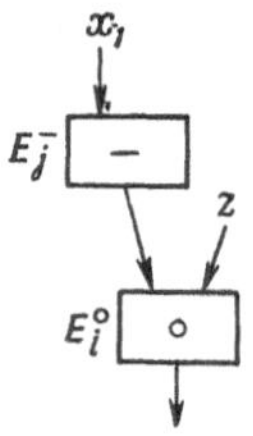

Fig. 29

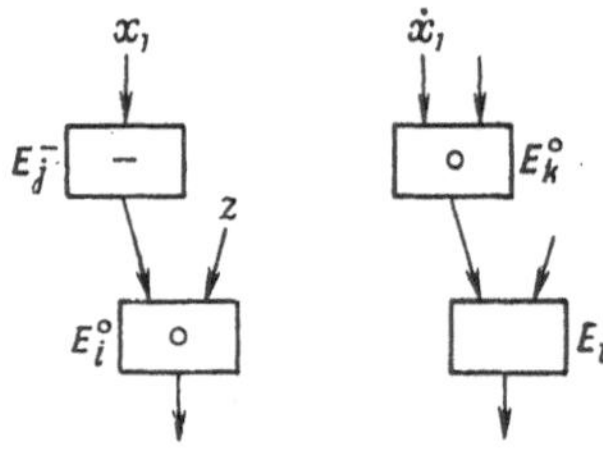

Fig. 30

b) The output of one of the elements E_k and E_l is applied only to the input of the other element, for example, as shown in Fig. 27.

But in this case the output of the element E_l cannot be the output of the circuit, since otherwise for $x_n \equiv y_n \equiv \chi(E_l^\circ)$ the output function will not depend on $x_1, \ldots, x_{n-1}$, and $y_1, \ldots, y_{n-1}$, which is impossible.

Hence there exists a fifth element E_m ($m \neq i, j, k, l$) to whose input we apply the output of E_l°. But in this case, when we apply to the circuit inputs corresponding to the variables x_n and y_n one of the constants ($\chi(E_k^\circ)$ or 0), it is possible by virtue of Property 1 to remove five elements E_i°, E_j°, E_k, E_l, and E_m, thus obtaining a circuit that realizes F_{n-1}. This completes the examination of Case 4.3.

We have also completed the proof of Lemma 2.

§ 4. Minimal Realization of Coincidence Operator

A coincidence operator R_n is a (2n, 1)-operator such that

$$R_n(\tilde{x}, \tilde{y}) = \begin{cases} 1, & \text{if } \tilde{x} = \tilde{y}, \\ 0, & \text{if } \tilde{x} \neq \tilde{y}. \end{cases}$$

This operator can evidently be defined inductively as follows:

$$R_n(x_1, \ldots, x_n, y_1, \ldots, y_n) = \overline{((x_n \vee y_n)} \vee x_n y_n) R_{n-1},$$
$$R_1(x, y) = \overline{(x \vee y)} \vee xy.$$

Theorem 3. For any $n \geq 1$ we have

$$L(R_n) = 5n - 1.$$

Proof of Theorem 3. The upper bound follows directly from the inductive definition of the coincidence operator (Fig. 28).

The lower bound is a simple consequence of Theorem 1 and of the next lemma.

Lemma 3. Let S_n^{min} be a minimal circuit that realizes R_n, $n \geq 2$. If it is allowed to apply the constants 0 and 1 to the inputs of the circuit S_n^{min}, there will always exist five or more elements that can be removed from the circuit S_n^{min}, and from the remaining elements we can construct a new circuit S_{n-1} that realizes R_{n-1}.

Proof of Lemma 3. Suppose we have a minimal circuit S_n^{min} that realizes R_n. In this circuit there evidently exists at least one two-input element E_i° such that to each of its inputs we apply either a variable belonging to $x_1, \ldots, x_n$ and $y_1, \ldots, y_n$, or the output of an inverter to whose input we apply one of the above-mentioned variables. Then the proof of Lemma 3 is continued separately for each of the cases listed in Table 4 (any of the given circuits can be referred to one of the cases listed in Table 4).

1.1. This case cannot occur.

Indeed, if a variable (for example, x_1) is applied only to an input of E_j^- (Fig. 29), whereas the output of E_j^- is applied only to an input of E_i°, with the second input of E_i° being under the action of $z\,(z \in \{x_2, \ldots, x_n, \bar{x}_2, \ldots, \bar{x}_n, y_1, \ldots, y_n, \bar{y}_1, \ldots, \bar{y}_n\})$ we obtain $S_n^{min}\,(z \to x_1)$, which is not permissible.

1.2. Since the circuit under consideration is minimal and the output of E_k° cannot be the circuit output (Fig. 30; without loss of generality it can be assumed that x_1 is applied to the input of E_j^-), it follows that there exists an element E_l to whose input we apply the output of E_k°.

To the circuit input corresponding to the variable x_1 we shall apply a constant 0 or 1 such that $\dot{x}_1 \equiv \chi(E_k^\circ)$. By virtue of Property 1 we shall remove four elements E_i°, E_j^-, E_k°, E_l and construct a new circuit S_n' that realizes R_n (const, $x_2, \ldots, x_n, y_1, y_2, \ldots, y_n$). In this circuit S_n' there must exist an element E_m to whose input we apply the variable y_1. To the input of the circuit S_n' corresponding to the variable y_1 we shall apply the same constant as was applied before to the input of the circuit S_n^{min} corresponding to x_1. Then (by virtue of Property 1) we can remove at least one more element (E_m) and construct a circuit S_{n-1} that realizes R_{n-1}.

2.1. This case cannot occur (see Case 1.1).

2.2. Since the given circuit is minimal and without cycles, and none of the elements E_i°, E_j, and E_k (Fig. 31) can be an output element of the circuit, there exists an element E_l (other than E_i°, E_j, and E_k) to whose input we apply the output of one of the elements E_i°, E_j, and E_k (for example, of E_i°). But in this case, by applying the constant $\chi(E_i^\circ)$ to the circuit inputs corresponding to the variables x_1 and y_1, it is possible to remove no less than five elements and construct a new circuit S_{n-1} that realizes R_{n-1} (in the same way as was done, for example, for Case 1.2).

TABLE 4. Splitting of Proof of Lemma 3 into Cases

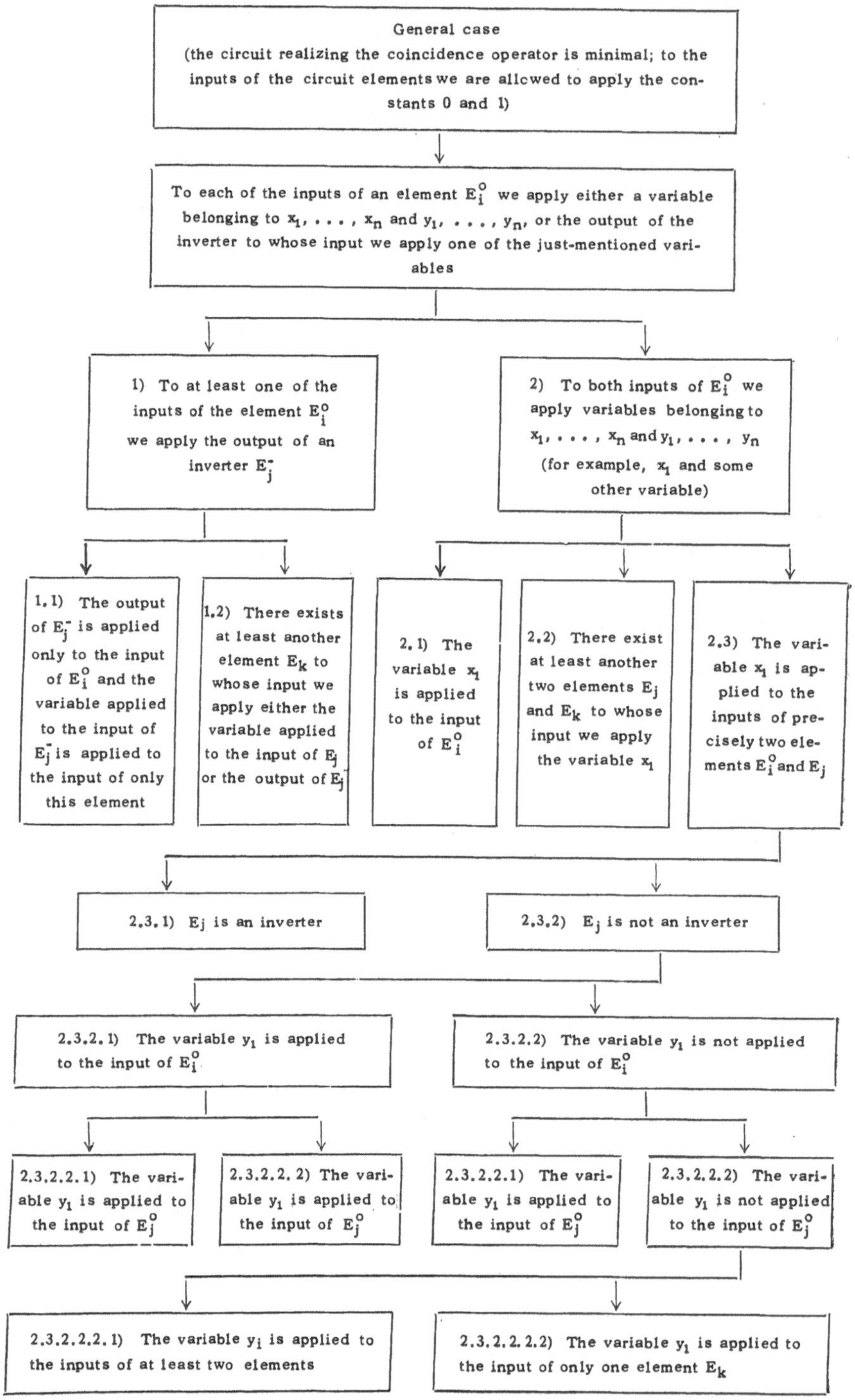

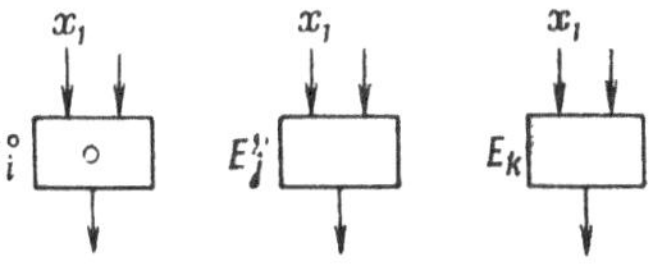

Fig. 31

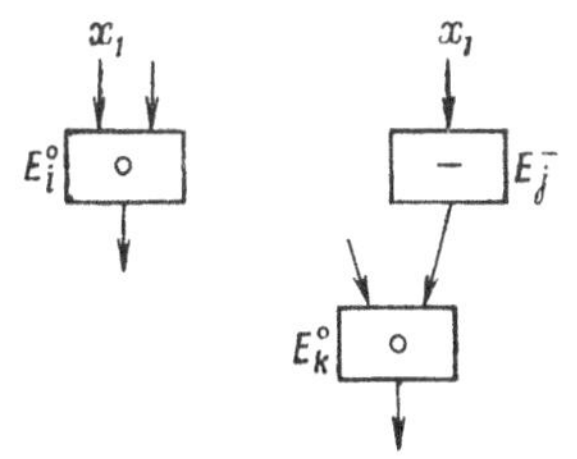

Fig. 32

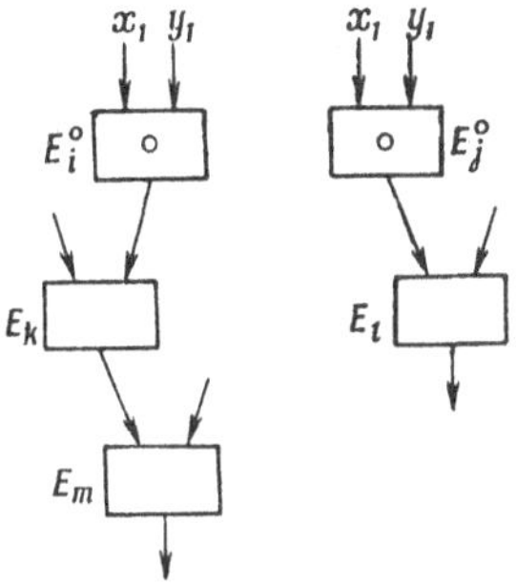

Fig. 33

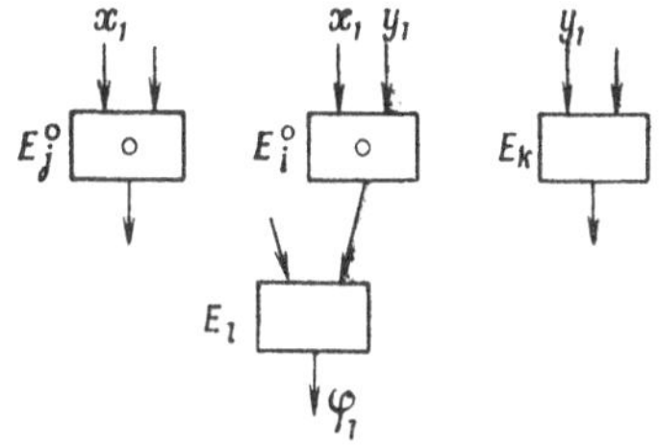

Fig. 34

2.3.1. Suppose that the output of E_j^- is applied to the input of an element E_k° (Fig. 32). By virtue of the minimality of the given circuit there exists an element E_l ($l \neq i$) to whose input we apply the output of the element E_k°. Then the proof can be continued in a similar way as in Case 1.2.

2.3. 2.1.1. By virtue of the minimality of the given circuit there evidently exist elements E_k, E_l, and E_m such that the output of E_i° is applied to the input of E_k, the output of E_j° to the input of E_l, and the output of one of the elements E_k and E_l to the input of E_m (for example, E_k, as shown in Fig. 33).

To the circuit outputs corresponding to the variables x_1 and y_1 let us apply the constant $\chi(E_k^\circ)$ if E_k is a disjunctor or conjunctor, and the identical zero if E_k is an inverter. In this case constants will be applied to the inputs of five elements (E_i°, E_j°, E_k, E_l, E_m), and these elements can be removed and a new circuit constructed that realizes R_{n-1}.

2.3. 2.1.2. In this case there exists an element E_k (other than E_i° and E_j°) to whose input we apply the variable y_1 (otherwise we would have $S_n^{\min}(x_1 \to y_1)$, which is impossible). Furthermore, by virtue of minimality of the circuit $S_n^{\min}$ there exists an element E_l other than E_i°, E_j°, E_k to whose input we apply the output of the element E_i° (Fig. 34).

None of the elements E_j°, E_k, E_l can be the output element of the circuit, since otherwise when a constant (0 or 1) is applied to the circuit inputs corresponding to the variables x_1 and y_1, the output function will not depend on the other variables ($n \geq 2$), which is impossible [for example, for $x_1 \equiv y_1 \equiv \chi(E_l)$, if E_l is a conjunctor or disjunctor, or for $x_1 \equiv y_1 \equiv 0$, if E_l is an inverter, we obtain $\varphi_l \equiv$ const]; since the given circuit is minimal and without cycles, there evidently exists an element E_m (other than E_i°, E_j°, E_k, E_l) to whose input we apply the output of at least one of the elements E_j°, E_k, E_k. But in this case, as is easy to see, when a constant (0 or 1) is applied to the circuit inputs corresponding to the variables x_1 and y_1, constants will be also applied to at least one input of each of the elements E_i°, E_j°, E_k, E_l, E_m, and these elements can be removed from the circuit (Property 1) and a circuit constructed that realizes R_{n-1}.

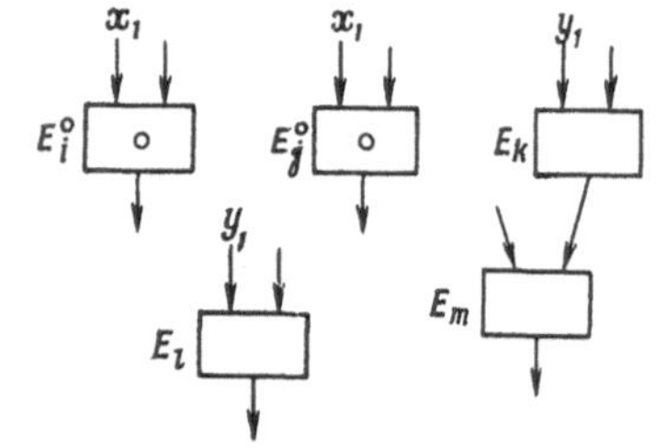

Fig. 35

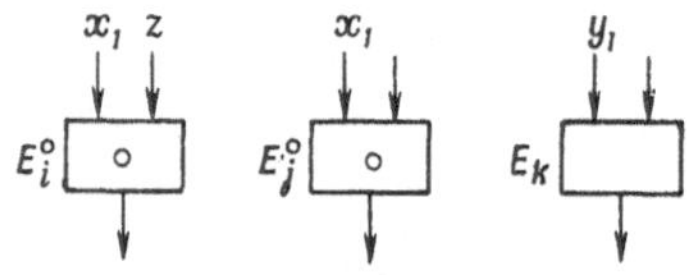

Fig. 36

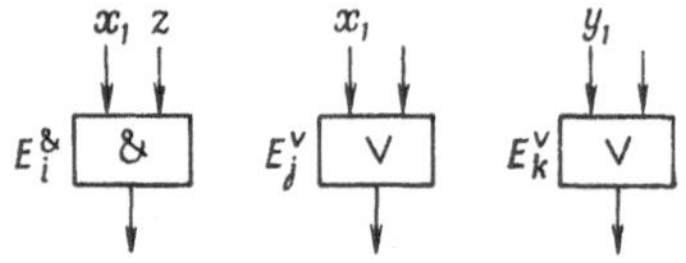

Fig. 37

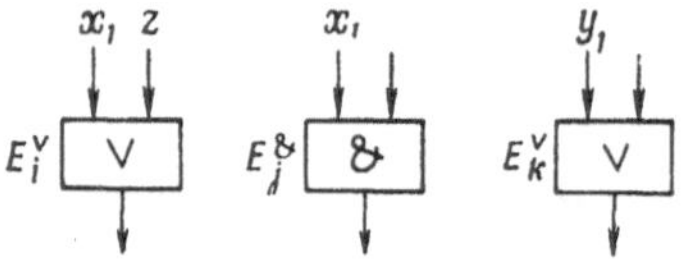

Fig. 38

2.3. 2.2.1. In this case Lemma 3 is proved in the same way as in Case 2.3. 2.1.2.

2.3. 2.2.2.1. In this case the circuit has two elements E_k and E_l (Fig. 35) to whose inputs we apply the variable y_1. Next, it is easy to show that there exists an element E_m (other than E_i°, E_j°, E_k, E_l) to whose input we apply the output of one of the elements E_i°, E_j°, E_k, E_l (for example, E_k). But in this case it is easy to convince oneself of the validity of Lemma 3 for this case [when we apply $\chi(E_k^\circ)$ if E_k is a disjunctor or conjunctor, and the identical zero if E_k is an inverter, to the circuit inputs corresponding to the variables x_1 and y_1, it is possible to remove five elements and obtain a circuit that realizes R_{n-1}].

2.3. 2.2.2.2. Fig. 36 ($z \in \{x_2, \ldots, x_n, y_2, \ldots, y_n\}$).

Let us assume that E_i° and E_j° are the same elements. To the circuit input corresponding to x_1 we shall apply the constant $\chi(E_i^\circ) \equiv \chi(E_j^\circ)$. In this case it is possible to remove no less than four elements (if the circuit has at least two elements E_p and E_q to whose inputs we apply the outputs of E_i° and E_j°, it is possible to remove E_i°, E_j°, E_p, E_q, whereas if the outputs of the elements E_i° and E_j are applied to the inputs of only one element E_r, it is possible to remove E_i°, E_j°, E_r and yet another element E_s to whose input we apply the output of E_r; such an element E_s evidently exists under our assumptions), and obtain a circuit that realizes $R_n(\chi(E_i^\circ), x_2, \ldots, x_n, y_1, \ldots, y_n)$. Furthermore, when $\chi(E_j^\circ)$ is applied to the input of the newly obtained circuit corresponding to the variable y_1, it is possible to remove yet another element and obtain a circuit S_{n-1} that realizes R_{n-1}. Thus Lemma 3 is valid under our assumptions.

It remains to consider the case that the elements E_i° and E_j° are distinct. Let us assume that E_k is a two-input element, for example, a disjunctor (if E_k is an inverter, the reasoning will be the same). Since E_i° and E_j° are distinct elements, one of them (for example, E_j°) will be a disjunctor (Fig. 37).

Under the above assumptions the output of $E_k^\vee$ cannot be applied only to the input of $E_j^\vee$, whereas the output of $E_j^\vee$ cannot be applied at all to the input of $E_k^\vee$ (otherwise we have in both cases $S_n^{\min}(x_1 \to y_1)$). Therefore only the following two cases are possible:

a) There exist two elements E_l and E_m (other than $E_i^\&$, $E_j^\vee$, $E_k^\vee$) to whose inputs we apply the outputs of the elements $E_j^\vee$ and $E_k^\vee$.

b) The outputs of the elements $E_j^\vee$ and $E_k^\vee$ are applied to the inputs of only one element E_l°.

If the identical unity is applied to the circuit inputs corresponding to the variables x_1 and y_1, it is possible (by virtue of Property 1) to remove five elements in Case (a) (the elements $E_i^{\&}$, $E_j^{\vee}$, $E_k^{\vee}$, E_l, E_m), as well as in Case (b) (the elements $E_i^{\&}$, $E_j^{\vee}$, $E_k^{\vee}$, E_l and yet another element at least, to whose input we apply the output of the element E_l) and obtain a circuit that realizes R_{n-1}.

If the element E_i is a disjunctor (Fig. 38), the reasoning will be as follows. The output of $E_k^{\vee}$ cannot be applied only to the input of $E_j^{\&}$ (otherwise we have $S_n^{\min}$ $(x_1 \to y_1)$); hence there exists an element E_l to whose input we apply the output of $E_k^{\vee}$. Furthermore, the output of $E_i^{\vee}$ cannot be applied to the input of $E_k^{\vee}$ (since $S_n^{\min}$ $(x_1 \to y_1)$) and by virtue of the minimality of the circuit we cannot confine ourselves to applying the output of E_i only to the input of $E_j^{\&}$. Hence we can have two cases:

c) The output of $E_i^{\vee}$ is applied to the input of an element E_m other than $E_j^{\&}$, $E_k^{\vee}$, E_l;

d) the output of $E_i^{\vee}$ is applied to the input of E_l. Now it is easy to see that Lemma 3 is valid for Case (c) as well as Case (d).

Thus we have completed the proof of Lemma 3.

Literature Cited

1. O. B. Lupanov, "Synthesis of certain classes of control systems," Problemy Kibernetiki, Vol. 10, Fizmatgiz, Moscow (1963), pp. 63-97.
2. O. B. Lupanov, "A method of synthesis of control systems – the principle of local coding," Problemy Kibernetiki, Vol. 14, Nauka, Moscow (1965), pp. 31-110.
3. E. P. Soprunenko, "Minimal realization of functions by circuits using functional elements," Vol. 15, Problemy Kibernetiki, Nauka, Moscow (1965), pp. 117-134.
4. S. V. Yablonskii, "Functional constructions in k-valued logic," Proc. Steklov Mathematical Institute of the Academy of Sciences of the USSR, 51:5-142 (1958).
4. C. Cardot, "Some results concerning the use of Boolean algebra in the synthesis of relay contact circuits," Annales des Télécommunications, 7(2):75-84 (1952).

FULL TEST FOR NONREPETITIVE SWITCHING CIRCUITS†

Kh. A. Madatyan

Moscow

One of the main subjects of cybernetics is investigation of the reliability of control systems [11]. To test the performance of a control system and to locate failures occurring in it is one of the most difficult and important problems. First investigations in this direction were made by I. A. Chegis and S. V. Yablonskii [9] who proposed an algorithm for testing electric circuits which with a relatively small sequence of instructions (tests) makes it possible to determine not only whether the system operates properly but also to find the nature and location of failures in the system to within electrically distinguishable components.

However, as noted by the authors, the large amount of work involved makes it impracticable even in the most simple cases to use the general algorithm for designing minimal tests and for evaluating their complexity. It is thus of considerable importance to consider the design of minimal tests for specific classes of circuits. Early results in this field can be found in [9] (Chapter 2). The tests discussed there were designed for circuits implementing symmetrical and linear functions, for comparators, and for a binary adder.

Further results were obtained by V. V. Glagolev [2] (tests for block switching circuit), I. V. Kogan [4] (diagnostic tests for nonrepretitive switching circuits), and V. V. Vaksov [1] (single diagnostic tests for nonrepetitive circuits). These works were concerned mainly with the design of minimal tests and with evaluating their complexity.

A comprehensive discussion of full diagnostic tests must also consider the construction of algorithms for carrying out the test procedure.

In this article we consider a class of nonrepetitive circuits. A full diagnostic test is designed for such circuits and its length is evaluated; also, a minimal test is constructed for nonrepetitive Π circuits. A simple algorithm for carrying out the test is formulated for this class of circuits.

Let us now define the basic concepts dealt with in this article.

Let the switching circuit $\mathfrak{A}$ implement the logical algebra function $f(x_1, x_2, \ldots, x_n)$. Assume that the only failures that can occur in the circuit $\mathfrak{A}$ are either short- or open-circuit failures. Thus (if a failure takes place), the circuit $\mathfrak{A}$ turns into the circuit $\mathfrak{A}'$. Let us denote by $f'(x_1, x_2, \ldots, x_n)$ the conduction of the circuit $\mathfrak{A}'$. The function $f'(x_1, x_2, \ldots, x_n)$ is called

†Original article submitted February 15, 1968; revision submitted February 14, 1969.

the failure function. The set of all failure functions is partitioned into the classes F_0, F_1, ..., F_m such that $f(x_1, x_2, ..., x\ )$ belongs to F_0, and the functions f' and f'' belong to one and the same class if and only if

$$f'(x_1, x_2, \ldots, x_n) \equiv f''(x_1, x_2, \ldots, x_n).$$

To distinguish between the above-indicated classes one must carry out a certain experiment that is characterized by the set of sequences applied to the circuit inputs. A general definition of a test is given in [9]. Certain specific types of tests are defined below.

Definition 1. A totality of sequences is called a check test (T_c) with respect to a given list of failures if from the conduction of the circuit with these sequences it is possible to determine whether the circuit is in good working order. Obviously, T_c allows the class F_0 to be distinguished from $\bigcup_{i=1}^{m} F_i$.

Definition 2. A totality of sequences is called a diagnostic test (T_d) with respect to a given list of failures if from the conduction of the circuit with these sequences one can determine the nature of the failure.

Obviously, the test T_d makes it possible to distinguish between the classes F_0, F_1, ..., F_m.

Definition 3. The number of sequences comprising a test is called the length of the test.

From the point of view of the possible failures we shall consider single and full tests.

A single test is a test capable of detecting failures when it is known a priori that the failure can occur in any but only one contact.

A full test is a test capable of detecting failures when the possible failures are either short or open-circuits in any contact (or several contacts simultaneously).

§ 1. Full Test for Switching Circuits

Let $t_{f(x_1, x_2, \ldots, x_n)}$ be the shortest test applicable to all circuit realizations of the function $f(x_1, x_2, ..., x_n)$. Further, let $t(n) = \max_f t_{f(x_1, x_2, \ldots, x_n)}$, where the maximum is taken over all logical algebra functions depending on n arguments. In [9], S. V. Yablonskii asked what is the asymptotics of the function t(n)? We will show that

$$t(n) = 2^n.$$

Definition 4. A test that detects a short- (open-) circuit only in any contact (or contacts) is called a full short- (open-) circuit test.

Let W_f be the set of vertices $(\alpha_1, \alpha_2, ..., \alpha_n)$ of an n-dimensional unit cube such that $f(\alpha_1, \alpha_2, ..., \alpha_n) = 1$. Let $H_{k_0}(f)$ $(B_{k_0}(f))$ be a subset of the set W_f $(W_{\bar{f}})$ whose elements do not overlap with the intervals of an abridged disjunctive normal form of a dimension not less than k_0, and let t_f^o (t_f^s) be the smallest full open- (short-) circuit test.

Lemma 1. $t_f^o \geqslant \frac{|H_{k_0}(f)|}{2^{k_0}}$ $\left(t_f^s \geqslant \frac{|B_{k_0}(f)|}{2^{k_0}}\right)$.†

† $|H_{k_0}(f)|$ is the number of elements in the set $H_{k_0}(f)$.

Proof. To any chain s of the circuit $\mathfrak{A}$ corresponds a conjunction of variables, and to each conjunction corresponds a certain interval in the n-dimensional unit cube.

Let an arbitrary circuit $\mathfrak{A}$ realize the function f, and let $s_1, s_2, \ldots, s_{l_1}$ be chains in the circuit $\mathfrak{A}$, covering the points of the set $H_{k_0}(f)$. Clearly, the chain s_i ($i = 1, 2, \ldots, l_1$) realizes a k-dimensional interval ($k \leq k_0$). In the circuit $\mathfrak{A}$ let us now open all contacts not belonging to the chain s_i. The resulting defective circuit realizes a function equal to unity only with those sequences which realize the chain s_i and, consequently, is distinct from $f \equiv 0$ only with these sequences. Let us do the same for all i ($i = 1, 2, \ldots, l_1$). As a result we obtain a part of the failure function table. The length of the test for such failure functions is not less than $\frac{|H_{k_0}(f)|}{2^{k_0}}$ and thus certainly $t_f^{\text{p}} \geqslant \frac{|H_{k_0}(f)|}{2^{k_0}}$.

To any dead-end cut [7] in the circuit $\mathfrak{A}$ corresponds a disjunction of variables and to each disjunction corresponds a certain interval in the n-dimensional unit cube.

Let $\omega_1, \omega_2, \ldots, \omega_{l_2}$ be the set of dead-end cuts in the circuit $\mathfrak{A}$, that cover the points of the set $B_{k_0}(f)$.

Let us close all contacts not belonging to the cut ω_i. The resulting defective circuit realizes a function equal to zero only with sequences that realize the cut ω_i ($i = 1, 2, \ldots, l_2$). Reasoning similarly we have

$$t_f^{\text{s}} \geqslant \frac{|B_{k_0}(f)|}{2^{k_0}}.$$

This proves the lemma.

Theorem 1. $t(n) = 2^n$, and the fraction of almost all functions of the algebra of logic is

$$t(f) \geqslant (1-\delta)\frac{2^n}{\log_2 n \log_2 \log_2 n} \qquad (0<\delta<1).$$

Proof. For a parity counter $H_0(f) = B_0(f) = 2^{n-1}$, and according to Lemma 1, $t_f \geq t_f^{\text{o}} + t_f^{\text{s}} = 2^n$.

It has been shown in [3] that for any $\varepsilon > 0$ and assuming $K_0 = [\log_2((1+\varepsilon)\log_2 n \cdot \log_2\log_2 n)]$, we have for almost all functions of the algebra of logic

$$|H_{k_0}(f)| \sim 2^{n-1}.$$

In virtue of duality, the same basic premises show that for almost all logic algebra functions, $|B_{k_0}(f)| \sim 2^{n-1}$. Hence we can conclude that for almost all functions

$$|H_{k_0}(f)| + |B_{k_0}(f)| = 2^n(1-o(1)).$$

Since $k_0 = [\log_2((1+\varepsilon)\log_2 n \cdot \log_2\log_2 n)]$, and considering Lemma 1, we obtain the second statement of the theorem, i.e., that for almost all function of the algebra of logic

$$t(f) \geqslant \frac{2^n(1-o(1))}{(1+\varepsilon)\log_2 n \log_2\log_2 n} = (1-\delta)\frac{2^n}{\log_2 n \log_2\log_2 n},$$

where $\delta = \frac{\varepsilon + o(1)}{1+\varepsilon}$.

§ 2. Full Test for Nonrepetitive Switching Circuits

Let $T_c^1(n)$, $T_c(n)$, $T_d^1(n)$, and $T_d(n)$ denote the Shannon functions of respectively a single check test, a full check test, a single diagnostic test, and a full diagnostic test for nonrepetitive switching circuits.

I. V. Kogan [4] proved that $T_c(f) \leq n + 1$, where $T_c(f_1) = n + 1$ for $f_1 = x_1 \cdot x_2 \ldots x_n$.

V. V. Vaksov has shown that $T_d^1(n) = n + 1$. It is easy to observe that $T_c^1(n) = n + 1$.

The number of edges in a dead-end cut will be called the width of the cut [7]. The width of the circuit $\mathfrak{A}$ is said to be the maximum number of edges in the dead-end cut of this circuit. The length of the longest chain of a circuit is called the length of this circuit.

Let the nonrepetitive circuit have a length l and a width b.

In a certain sense we shall improve on the results obtained by I. V. Kogan; thus:

Theorem 2. $T_c(f) \leqslant b + l$.

First, let us formulate certain well-known facts and prove one lemma.

Consider a two-terminal net with terminals A and B. Let each vertex of the net be assigned a certain number. A chain between the terminals is said to be monotonic if the numbers of the vertices of this chain increase in moving from terminal A to terminal B.

Lemma 2. *The vertices of a strongly connected net can be numerated with the numbers 0, 1, 2, ..., l (where l is the length of the net) so that a monotonic chain passes through every edge.*

A proof of this lemma is given in [7].

A segment of any arbitrary chain δ from the vertex c to vertex b is denoted $\delta(c, d)$.

Let c and d be two adjacent vertices of contact of two monotonic chains δ_1 and δ_2, and γ be an edge through which only one of these chains passes in the interval (c, d) (Fig. 1). Denote $\delta_1(c, d)$ by $(\delta_1, \gamma, \delta_2)$ and $\delta_2(c, d)$ by $(\delta_2, \gamma, \delta_1)$.

Lemma 3. *Let the vertices of a strongly connected net be numerated so that a monotonic chain passes through every edge, and let the width of the net be b. One can then indicate b monotonic chains such that at least one such chain passes through every edge of the net.*

Proof. Let S(b) be a two-terminal strongly connected net of width b. Let us draw b monotonic chains $\sigma_1, \sigma_2, \ldots, \sigma_b$ which altogether cover a maximum number of edges of the net S(b). We shall show that these chains cover all edges of the net S(b). Let us assume that the opposite is true, i.e., no matter how we choose the chains $\sigma_1, \ldots, \sigma_b$ there exists one edge α such that none of the chosen chains passes through it. We will show that the width of such a net is greater than b. Through α let us draw a monotonic chain σ.

The following statement is then true:

For any monotonic chain δ that does not pass through α there exists in the segment (δ, α, σ) at least one edge α^δ through which passes at most one chain of the set $\{\sigma_1, \sigma_2, \ldots, \sigma_b\}$. Let us assume the contrary, i.e., that there exists a monotonic chain δ' in which at least two chains from the set $\{\sigma_1, \sigma_2, \ldots, \sigma_b\}$ pass through every edge of the segment $(\delta_1, \alpha, \sigma)$.

Two cases are possible.

1. There exists a chain σ_i in which the segment $(\sigma_i, \alpha, \sigma)$ coincides with the segment $(\delta', \alpha, \sigma)$.

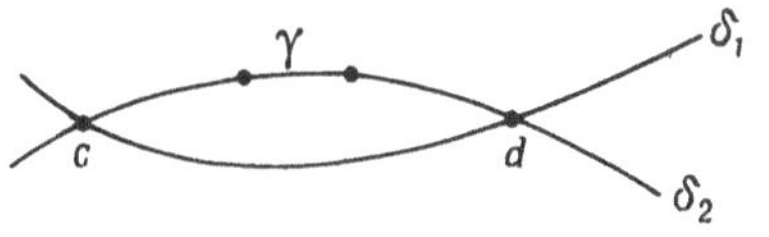

Fig. 1

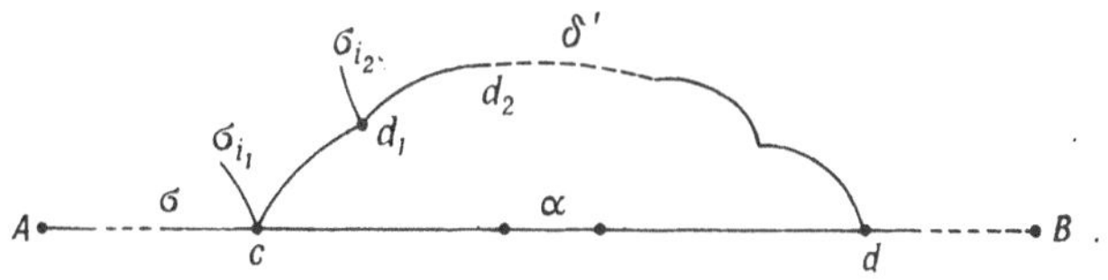

Fig. 2

2. There is no chain σ_i such that σ_i covers the entire segment $(\delta', \alpha, \sigma)$.

If the second case is true, then on the segment $(\delta', \alpha, \sigma)$ let us select the vertices d_1 and d_2 (Fig. 2) so that the chain σ_{i_1} passes through (c, d_1) and the chain σ_{i_2} passes through (d_1, d_1). In place of the chains σ_{i_1} and σ_{i_2} let us take the chains σ'_{i_1} and σ'_{i_2}, where

$$\sigma'_{i_1}\colon \sigma_{i_1}(A, d_1) - \sigma_{i_2}(d_1, B),$$
$$\sigma'_{i_2}\colon \sigma_{i_2}(A, d_1) - \sigma_{i_1}(d_1, B)$$

and σ'_{i_1} covers the segment $\delta'(c, d_2)$, greater than the segment covered by the chain σ_{i_1}. Repeating this argument, we arrive at the first case, i.e., the chain σ_i covers the segment $(\delta', \alpha, \sigma)$. Instead of the chain σ let us take the chain

$$\sigma'_i\colon \sigma_i(A, c) - \sigma(c, d) - \sigma_i(d, B).$$

The chains $\sigma_1, \sigma_2, \ldots, \sigma'_i, \ldots, \sigma_b$ cover then also the edge α, which is impossible.

The statement is thus true.

Obviously, the set of edges $\{\alpha^\delta\} \cup \alpha$ forms a cut of which not a single one of the edges $\alpha, \alpha^{\sigma_1}, \alpha^{\sigma_2}, \ldots, \alpha^{\sigma_3}$ can be rejected. Hence follows that the width of the net exceeds b, which is impossible.

This proves the lemma.

In the nonrepetitive circuit let us isolate a certain dead-end cut (chain). To all variables in the cut (chain) let us assign the value 0 (1), the value 1 (0) being assigned to all other variables. Such a sequence of values of the variables is denoted by ω(s). In the following no distinction will be made between the cut (chain) and the sequence which realizes this cut (chain). The set of sequences ω is denoted by Ω, and the set of sequences s by S.

A chain is said to be a *unit chain with respect to a certain cut* if it crosses it exactly once.

Theorem 3. *The subsets of the set $\Omega \cup S$ form a full check test for the circuit $\mathfrak{A}$, if through every contact of the circuit passes a cut and a unit chain with respect to this cut.*

A proof of this theorem is given [4].

Now let us prove Theorem 2.

Take $l + 1$ parallel planes $L_1, L_2, \ldots, L_{l+1}$ arranged in order of their numeration. On the plane L_i let us place all vertices of the net to which the number $i - 1$ $(i = 1, 2, \ldots, l + 1)$ has been assigned. From Lemma 2 follows that two ends of one edge cannot lie in one and the same plane. Consequently, any net of length l can be geometrically realized so that any monotonic chain crosses the plane L_i $(i = 1, 2, \ldots, l + 1)$ exactly once.

The set of all edges that cross the portion of space between the planes L_i and L_{i+1} is denoted by ω_i. Evidently, the sets of edges ω_i form dead-end cuts. Let us choose all dead-end cuts

Fig. 3

$\omega_1, \omega_2, \ldots, \omega_l$. According to Lemma 3, in a net of width b we can select b monotonic chains so that at least one of these nets passes through every edge of the net. These chains are unit chains with respect to all chosen cuts. According to Theorem 3, the selected dead-end cuts and the chains form a full check test for the given circuit.

Consequently, a minimal full check test does not exceed the sum of the length and width of the given circuit.

This proves Theorem 2.

This theorem actually improves the results obtained by I. V. Kogan. If the width and length of a net are of the order $\sqrt{n}$, the check test for this net is less than the $c\sqrt{n}$ and not n + 1, where c is some constant.

In a nonrepetitive circuit let us take the dead-end cut ω and a unit chain s with respect to this cut. Let us denote by $\omega_s(s_\omega)$ a sequence coinciding with the sequence $\omega(s)$ except that variable through which the chain s (cut ω) intersects the cut ω (chain s).

Example. In the circuit shown in Fig. 3 let us take the cut ω = (001101) and s = (101000); then ω_s = (101101) and s_ω = (001000).

The net of sequences $\omega_s(s_\omega)$ will be denoted by $\Omega_S(S_\omega)$.

Lemma 4. The set of sequences $\Omega \cup \Omega_s$ forms a full diagnostic test for nonrepetitive circuits.

Proof. Let $\mathfrak{A}'$ and $\mathfrak{A}''$ be distinct defective states of the circuit $\mathfrak{A}$, that realize the functions f' and f'' respectively.

Two cases are possible:

1. There is a variable x_i such that one of the functions (either f' or f'') depends (essentially) on the variable x_i while the other is independent of x_i.

2. The functions f' and f'' depend on the same variables.

In the first case let x_i be a variable such that f' depends essentially on x_i and f'' is independent of x_i.

In virtue of the fact that f' depends essentially on x_i, we can make in the circuit $\mathfrak{A}'$ a cut ω' which passes through the contact x_i ($f'(\tilde{\omega}')) = 0$.† Let us consider the cut ω in circuit $\mathfrak{A}$, corresponding to the cut ω' in the circuit $\mathfrak{A}$. Thus, $f'(\tilde{\omega}) = 0$ since both $\tilde{\omega}$ and $\tilde{\omega}'$ differ only in those variables on which f' does not depend essentially. If $f''(\tilde{\omega}) = 1$, the lemma is true since $\tilde{\omega} \in \Omega$. If, however, $f''(\tilde{\omega}) = 0$, let us consider the sequence $\tilde{\tilde{\omega}}$, which coincides with the sequence $\tilde{\omega}$ except in the variable x_i; then, $f'(\tilde{\tilde{\omega}}) = 1$, and $f''(\tilde{\tilde{\omega}}) = 0$ and $\tilde{\tilde{\omega}} \in \Omega_S$.

In the second case, as f' is distinct from f'' we can find a cut ω' such that $f'(\tilde{\omega}) \neq f''(\tilde{\omega}')$, and, consequently, $f'(\tilde{\omega}) \neq f''(\tilde{\omega})$ is satisfied also on the cut ω of the circuit $\mathfrak{A}$, corresponding to the cut ω'.

This proves the lemma.

The following lemmas are proved in exactly the same way.

Lemma 5. The set of sequences $S \cup S_\Omega$ forms a full diagnostic test for nonrepetitive circuits.

† $\tilde{\omega}'$ is a sequence that realizes the cut ω'.

Fig. 4

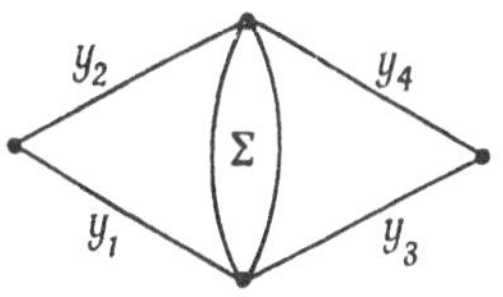

Fig. 5

Lemma 6. The set of sequences $\Omega \cup S$ forms a full diagnostic test for nonrepetitive circuits.

Note. As will become evident below, tests constructed according to Lemmas 4, 5, and 6 are minimal tests for some circuits.

Let A(n) be the maximum number of chains in a two-terminal net with n edges.

Lemma 7. $A(n) = 3^{\frac{n}{3}}$.

Proof. For every two-terminal net (A, B) with n edges one can find a two-terminal $\Pi(\Sigma_1, \Sigma_2, \ldots, \Sigma_l)$ net [8] with n edges (Fig. 4) in which the number of chains is not less than the number of chains in the given net.

In fact, let us denote the edges incidental to one of the terminals, say A, by the numbers 1, 2, 3, ..., k.

Let a_i (i = 1, 2, ..., k) be the number of chains passing through the i-th edge.

Let also $a = \max a_i$ and let a be reached at the j-th edge. Let us disconnect the i-th edge (i ≠ j) from its corresponding internal vertex and attach it to the internal vertex c to which the edge j is incidental. If the net resulting from this operation has a branch, we identify the internal vertices of the edges i and j. Let us carry out this operation for all i (i ≠ j). Obviously, the number of chains in the new net is not less than the number of chains in the original net. Let us now take the two-terminal subnet (C, B) and apply to it the same operation. Let us continue this process till the result is a $\Pi(\Sigma_1, \Sigma_2, \ldots, \Sigma_l)$ net. Since A(n) is attained on $\Pi(\Sigma_1, \Sigma_2, \ldots, \Sigma_l)$ nets, it suffices if we consider only these nets.

Let σ_i be the number of edges in each subnet Σ_i of the $\Pi(\Sigma_1, \Sigma_2, \ldots, \Sigma_l)$ net. Then, the number of chains in this net, A(Π), is equal to $\sigma_1 \cdot \sigma_2 \cdot \sigma_3 \ldots \sigma_l$, where $\sigma_1 + \sigma_2 + \ldots + \sigma_l = n$. It is clear that $\max A(\Pi) = 3^{n/3}$.

This proves the lemma.

Lemma 8. $T_d(n) \leqslant n 3^{\frac{n}{3}}$.

The proof of this lemma follows immediately from Lemmas 5 and 7 as to every chain s correspond not more than n − 1 sequences of the type S_Ω.

Let B(r) be the number of dead-end cuts in a two-terminal net with r vertices.

Lemma 9. $B(r) \leqslant 2^{r-2}$.

Proof. Every dead-end cut partitions the internal vertices of a net into two parts. Thus, $B(r) \leqslant C^0_{r-2} + C^1_{r-2} + \ldots$

Theorem 4. For any ε > 0 and sufficiently large n, for almost all nonrepetitive circuits that realize the function $f(x_1, x_2, \ldots, x_n)$

$$T_d(f) \leqslant 2^{\frac{2n}{\ln n}(1+\varepsilon)}.$$

TABLE 1

	$y_1\, y_2\, y_3\, y_4\, \tilde{x}$	F_2^0	F_2^i $i=1, 2, \ldots, s$	F_3^0	F_3^i $i=1, 2, \ldots, s$	F_1^0	F_1^i $i=1, 2, \ldots, s$	F_4^0	F_4^i $i=1, 2, \ldots, s$
	$1\,0\,0\,1\, \tilde{\alpha}_i$	1		1		0		0	
A_1	. . . . .	.	f_i	.	f_i	.	0	.	0
	$1\,0\,0\,1\, \tilde{\alpha}_i$	1		1		0		0	
	$1\,1\,0\,1\, \tilde{\alpha}_i$	1		1		0		1	
A_2	. . . . .	.	f_i	.	1	.	0	.	1
	$1\,1\,0\,1\, \tilde{\alpha}_i$	1		1		0		1	
	$1\,0\,1\,1\, \tilde{\alpha}_i$	1		1		1		0	
A_3	. . . . .	.	1	.	f_i	.	1	.	0
	$1\,0\,1\,1\, \tilde{\alpha}_i$	1		1		1		0	
	$0\,1\,1\,0\, \tilde{\alpha}_i$	0		0		1		1	
A_4	. . . . .	.	0	.	0	.	f_i	.	f_i
	$0\,1\,1\,0\, \tilde{\alpha}_i$	0		0		1		1	
	$0\,1\,1\,1\, \tilde{\alpha}_i$	0		1		1		1	
A_5	. . . . .	.	0	.	1	.	f_i	.	1
	$0\,1\,1\,1\, \tilde{\alpha}_i$	0		1		1		1	
	$1\,1\,1\,0\, \tilde{\alpha}_i$	1		0		1		1	
A_6	. . . . .	.	1	.	0	.	1	.	f_i
	$1\,1\,1\,0\, \tilde{\alpha}_i$	1		0		1		1	

Proof. Let r_n be the number of vertices in a two-terminal net with n edges. In [6] it is shown that for almost all nets $r \leq (2n/\ln n)(1+\varepsilon)$, where $\varepsilon \to 0$ when $n \to \infty$. According to Lemma 9, the number of dead-end cuts in such nets $9\, B(r_n) \leqslant 2^{r_n-2}$ and according to Lemma 4, $T_d(f) \leqslant 2^{r_n-2}(n+1) = 2^{\frac{2n}{\ln n}(1+\varepsilon)}$, where $\varepsilon \to 0$ when $n \to \infty$.

The theorem is thus proved.

Let the nonrepetitive circuit $\mathfrak{A}$ realize the function $f(x_1, x_2, \ldots, x_n)$ and let $T_d^0(\mathfrak{A}) = t$. Let us construct a new nonrepetitive circuit Δ (Fig. 5) that realizes the function $F(y_1, y_2, y_3, y_4, x_1, x_2, \ldots, x_n)$.

Lemma 10. $T_d^0(\Delta) \geqslant 2t+2$.

Proof. Let us compile a fractional table of failure functions for the circuit Δ and show that for these failures the length of the test is not less than $2t + 2$.

Let $\mathfrak{A}_0, \mathfrak{A}_1, \ldots, \mathfrak{A}_s$ be all the possible defective states of the circuit $\mathfrak{A}$ (it is assumed that the circuit contacts can only open as a result of action of the failure source) that realize the functions $f_0, f_1, \ldots, f_s$ $(f_0 = f)$ respectively.

Consider the following defective states Δ_j^i ($j = 1, 2, 3, 4$; $i = 1, 2, \ldots, s$) of the circuit Δ. The defective states Δ_j^i occur when the contact y_j is open and the circuit $\mathfrak{A}$ is in the defective state $\mathfrak{A}_i$.

The failure

Δ_1^i has the conduction $F_1^i = y_2 y_3 f_i \vee y_2 y_4$,

Δ_2^i has the conduction $F_2^i = y_1 y_4 f_i \vee y_1 y_3$,

Δ_3^i has the conduction $F_3^i = y_1y_4f_i \vee y_2y_4$,

Δ_4^i has the conduction $F_4^i = y_2y_3f_i \vee y_1y_3$,

where i = 0, 1, 2, ..., s.

Let us compile a table of failure functions for these defective states (Table 1). In the table we shall consider only those sequences in which the defective states Δ_j^i (j = 1, 2, 3, 4; i = 0, 1, 2, ..., s) are distinct.

Sequences of this kind are:

	y_1	y_2	y_3	y_4	$\tilde{x}$
A_1	1	0	0	1	$\tilde{\alpha}_i$
A_2	1	1	0	1	$\tilde{\alpha}_i$
A_3	1	0	1	1	$\tilde{\alpha}_i$
A_4	0	1	1	0	$\tilde{\alpha}_i$
A_5	0	1	1	1	$\tilde{\alpha}_i$
A_6	1	1	1	0	$\tilde{\alpha}_i$,

where $\tilde{\alpha}_i$ (i = 1, 2, ..., l) are all the possible sequences in which $f(\tilde{\alpha}_i) = 1$. The minimal set of sequences that distinguishes between all defective states Δ_2^i (i = 0, 1, 2, ..., s) and between the states Δ_3^i is denoted by $T_{2,3}$, and those concerning Δ_1^i and Δ_4^i by $T_{1,4}$. As seen in Table 1,

$$T_{2,3} \subset A_1 \cup A_2 \cup A_3 \qquad (|T_{2,3}| \geqslant t),$$
$$T_{1,4} \subset A_4 \cup A_5 \cup A_6 \qquad (|T_{1,4}| \geqslant t),$$

so that

$$T_{2,3} \cap T_{1,4} = 0. \tag{1}$$

If in the set of sequences $T_{2,3}$ there exists at least one sequence $\tilde{\alpha}$ of the set A_2 or A_3, $|T_{2,3}| \geq t + 1$. In fact, let us assume that $|T_{2,3}| < t + 1$ and $\tilde{\alpha} \in T_{2,3} \cap (A_2 \cup A_3)$ (for example, $\tilde{\alpha} \in A_2$); then there are two defective states Δ_2' and Δ_2'' which differ only in the sequence $\tilde{\alpha}$ of the set $T_{2,3}$. In this case the defective states Δ_3' and Δ_3'' do not differ in the sequences of the set $T_{2,3}$ which is impossible.

Similarly, if in the set of sequences $T_{1,4}$ there is at least one sequence of the set A_5 or A_6, then $|T_{1,4}| \geq t + 1$.

The following cases are possible:

I.	$T_{2,3} \subset A_1$,	$T_{1,4} \subset A_4$,	then	$T_d^0(\Delta) \geqslant 2t$,
II.	$T_{2,3} \subset A_1$,	$T_{1,4} \cap (A_5 \cup A_6) \neq 0$,	then	$T_d^0(\Delta) \geqslant 2t + 1$,
III.	$T_{2,3} \cap (A_2 \cup A_3) \neq 0$,	$T_{1,4} \subset A_4$,	then	$T_d^0(\Delta) \geqslant 2t + 1$,
IV.	$T_{2,3} \cap (A_2 \cup A_3) \neq 0$,	$T_{1,4} \cap (A_5 \cup A_6) \neq 0$,	then	$T_d^0(\Delta) \geqslant 2t + 2$.

It is sufficient to consider the cases I and II, since II and III are not essentially different, and the lemma is true for case IV.

I. The defective states Δ_2^0 and Δ_3^0 differ only in sequences belonging to the set $A_5 \cup A_6$, a Δ_1^0 while Δ_1^0 and Δ_4^0 differ in sequences of $A_2 \cup A_3$, that do not belong to $T_{2,3} \cup T_{1,4}$. Thus, in order to distinguish between the defective states Δ_j^0 (j = 1, 2, 3, 4) we need at least two more sequences. Thus, $T_d^0(\Delta) \geq 2t + 2$.

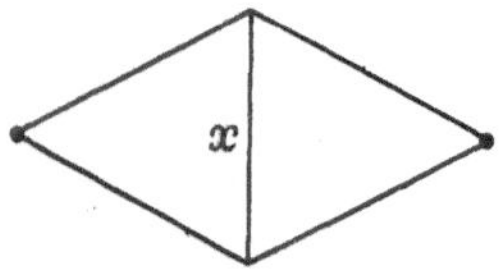

Fig. 6

II. The defective states Δ_1^0 and Δ_4^0 differ only in sequences of the set $A_2 \cup A_3$, that are not included in the test $T_{2,3} \cup T_{1,4}$. Thus, $T_d^0(\Delta) \geq 2t + 2$. The lemma is proved completely.

In the bridge circuit M_1 (Fig. 6) let us replace the contact x by the subcircuit M_1; we obtain the circuit M_2, etc. The circuit M_k results from replacing the contact x in the circuit M_1 by the circuit M_{k-1}.

Lemma 11. $T_d(M_k) = 3 \cdot 2^{\frac{n+3}{4}} - 4$, where $n = 4k + 1$.

Proof. According to Lemma 10, $T_d^0(M_k) \geq 2T_d^0(M_{k-1}) + 2$, whence $T_d^0(M_k) \geq 3 \cdot 2^k - 2$. In virtue of duality: $T_d^s(M_k) \geq 3 \cdot 2^k - 2$. Whence

$$T_d(M_k) \geqslant T_d^0(M_k) + T_d^s(M_k) \geqslant 3 \cdot 2^{k+1} - 4. \tag{2}$$

In the circuit M_k $|\Omega| = |S| = 3 \cdot 2^k - 2$, so that from Lemma 6 follows that

$$T_d(M_k) \leqslant 3 \cdot 2^{k+1} - 4. \tag{3}$$

From (2) and (3) we have $T_d(M_k) = 3 \cdot 2^{k+1} - 4 = 3 \cdot 2^{\frac{n+3}{4}} - 4$.

Remark. The minimal test of circuit M_k coincides with the test designed in accordance with Lemma 6 (see note on p. 111). From Lemmas 7 and 11 follows immediately

Theorem 5. $3 \cdot 2^{\frac{n+3}{4}} - 4 \leqslant T_d(n) \leqslant n3^{\frac{n}{3}}$.

§ 3. Full Test for Nonrepetitive Π Circuits

Let $T_d^{\Pi}(n)$ be the Shannon function of a full diagnostic test for nonrepetitive Π circuits.

Theorem 6. $T_d^{\Pi}(n) = \frac{3}{2} n$.

To prove this theorem we must first consider several lemmas.

Let $\mathfrak{A}_1$ and $\mathfrak{A}_2$ be two nonrepetitive circuits which realize the functions $f_1(x_1, x_2, \ldots, x_n)$ and $f_2(y_1, y_2, \ldots, y_m)$, and let the sets of sequences t_1 and t_2 be their full diagnostic tests respectively. Joining these circuits in series (parallel) we obtain the circuit $\mathfrak{A}$, which realizes the function

$$f = f_1(x_1, \ldots, x_n) \cdot f_2(y_1, \ldots, y_m) \quad [f = f_1(x_1, \ldots, x_n) \vee f_2(y_1, \ldots, y_m)].$$

Lemma 12. $T_d^{\Pi}(f) \leqslant |t_1| + |t_2|$.

Proof. Consider the case when $\mathfrak{A}_1$ and $\mathfrak{A}_2$ are connected in series. To every sequence from the set t_1 we add m ones on the right and to every sequence from t_2 we add n ones on the left. The result is a set of sequences $t = t_1' \cup t_2'$ of length n + m. Clearly, $|t| \leq |t_1| + |t_2|$. We shall now prove that the set of sequences t is a full diagnostic test for the circuit $\mathfrak{A}$. To do this we take two different failures of the circuit

$$f' = f_1' \cdot f_2'; \quad f'' = f_1'' \cdot f_2''$$

and show that they are distinct in the sequences t. Since $f' \not\equiv f''$, either $f_1' \not\equiv f_1''$ or $f_2' \not\equiv f_2''$. Assume that f_1' differs from f_1'' in the sequence $\tilde{\alpha}$ from t_1, i.e., $f_1'(\tilde{\alpha}) \neq f_1''(\tilde{\alpha})$; then $f'(\tilde{\alpha}') \neq f''(\tilde{\alpha}')$ (where $\tilde{\alpha}'$ is a sequence obtained from $\tilde{\alpha}$ by the above method) since f_2' and f_2'' cannot be identically equal to zero.

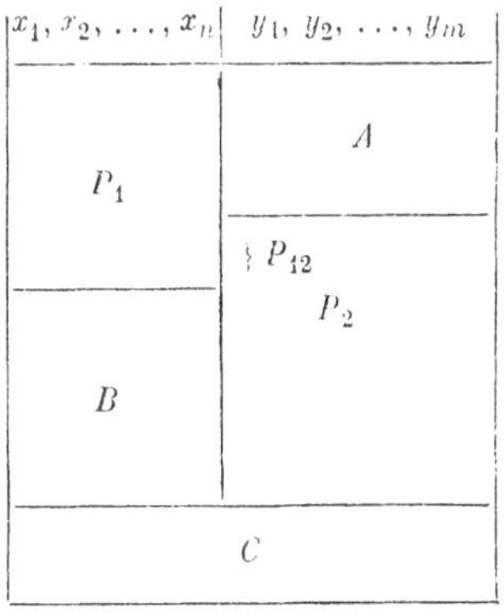

Fig. 7

Consider now the case when $\mathfrak{A}_1$ and $\mathfrak{A}_2$ are connected in parallel. To every sequence belonging to the set t_1 we add m zeros on the right and to every sequence from t_2, n zeros on the left.

The proof that the set of sequences $t = t_1' \cup t_2'$ is a full diagnostic test for $\mathfrak{A}$ is similar to that above.

This proves the lemma.

Corollary. The connection of a single contact in series (parallel) to a nonrepetitive circuit increases the length of its full diagnostic test by not more than unity.

Lemma 13. $T_d^{\Pi}(n) \leqslant \frac{3}{2}n, \quad n > 1.$

Proof. The lemma is proved by applying the method of mathematical induction to the number of contacts. The statement is clearly true for n = 2.

Let us assume that the statement is true for all $k \leq n$ and prove that it is also true for n + 1.

A Π circuit with n + 1 edges is a parallel or series connection of two Π circuits with l and s edges: $l + s = n + 1$. According to Lemma 12,

$$T_d^{\Pi}(n+1) \leqslant \frac{3}{2} l + \frac{3}{2} s = \frac{3}{2}(n+1).$$

This proves the lemma.

Let $T_d^{s,\Pi}(f)$ ($T_d^{o,\Pi}(f)$) be a full short- (open-) circuit test of a nonrepetitive Π circuit Σ that realizes the function f. Let $f = f_1(x_1, x_2, \ldots, x_n) \vee f_2(y_1, y_2, \ldots, y_m)$,

$$T_d^{s,\Pi}(f_1) = |t_1|; \quad T_d^{s,\Pi}(f_2) = |t_2|.$$

If $f_1 = \bigvee_{i=1}^{n} x_i$, then $T_d^{s,\Pi}(f) = |t_2|$, since there exists only one failure function for f_1 which is identically equal to 1.

Lemma 14. If $f = f_1 \vee f_2$, where $f_1 \neq \bigvee_{i=1}^{n} x_i$, $f_2 \neq \bigvee_{j=1}^{m} y_i$, then

$$T_d^{s,\Pi}(f) \geqslant |t_1| + |t_2|.$$

Proof. Assume that the set of sequences t is a test† for the function f. This set should contain tests for both f_1 and f_2.

Let us take the set of sequences $P = P_1 \cup P_2$ from t, where the set P_1 forms a dead-end test minimal with respect to t for the function f_1 and P_2 is a similar test for the function f_2. Assume that the sets of sequences P_1 and P_2 intersect on the sequences forming the set P_{12}. Let us first arrange the set of sequences P_1 so that all sequences belonging to P_{12} are at the end, next the remaining sequences of P_2, and at last the set of sequences C, where C is formed of all sequences that to not belong to the set P (Fig. 7). We will prove that the number of rows

† A short-circuit test is meant here.

in the matrix C is not less than $|P_{12}|$. Without loss of generality we can presuppose that the number of $|A'|$ of nonzero rows of the matrix A is not less than the number of nonzero rows of the matrix B. Let us show that the matrices B and C should contain $|P_{12}| + |A'|$ nonzero rows. Let us take the first sequence $\tilde{\alpha}$ of the set P_1 to which a nonzero sequence from the matrix A has been added. Since the set P_1 forms a dead-end test for the function f_1, there are two failure functions f_1' and f_1'' that differ in the sequence $\tilde{\alpha} \in P_1$ only. Let in this sequence the variable $y_i = 1$. Let us take in the circuit $\mathfrak{A}_2$, a chain that passes through the contact y_i and close all contacts (except y_i) along this chain. The result is a failure function f_2' of the circuit $\mathfrak{A}_2$, $f_2' \not\equiv 1$. The failure functions $f' = f_1' \vee f_2'$ and $f'' = f_1'' \vee f_2'$ do not differ in the sequences P_1, so that the matrix B or C should have a sequence β ($\beta \neq 0$) such that $f'(\beta) \neq f''(\beta)$. The same procedure is applied also to any two failure functions of the circuit Σ that differ in the sequence α only.

An analogous reasoning is carried out for any arbitrary sequences $P_{12} \cup A'$ of the type α; as a result we obtain sequences of the type β in the matrices B and C. Let us show that $s \geq |P_{12}| + |A'|$. Assume that $s < |P_{12}| + |A'|$; from the set of sequences P_1 let us eliminate the sequences $P_{12} \cup A'$, considered above and replace them with sequences of the type β. The result is a test for the circuit $\mathfrak{A}_1$, whose length is less than $|P_1|$; this is impossible as the set of sequences P_1 is a minimal test with respect to t for the circuit $\mathfrak{A}_1$. However, since the number of nonzero rows in the matrix B is less than $|A'|$, the number of rows in matrix C is not less than $|P_{12}|$.

This proves the lemma.

Obviously, if $f = f_1 \vee f_2$, then $T_d^{o,\Pi}(f_1) \geq T_d^{o,\Pi}(f_1) + T_d^{o,\Pi}(f_2)$ and $T_d^{\Pi}(f) \geq T_d^{o,\Pi}(f_1) + T_d^{s,\Pi}(f_1) + T_d^{o,\Pi}(f_2) + T_d^{s,\Pi}(f_2)$. In virtue of duality from $f = f_1 \cdot f_2$ follows that $T_d^{\Pi}(f) \geq T_d^{o,\Pi}(f_1) + T_d^{s,\Pi}(f_1) + T_d^{o,\Pi}(f_2) + T_d^{s,\Pi}(f_2)$.

Thus, a full test constructed in accordance with Lemma 12 is minimal provided t_1 and t_2 are minimal tests for f_1 and f_2 respectively. Hence and from Lemma 13 follows the proof of Theorem 6.

Remark. From Lemmas 12 and 14 follows that there exists a sufficiently simple method for constructing minimal full diagnostic tests for nonrepetitive Π circuits. This method does not presuppose the construction of a failure-function table. The question which now arises is how to apply this test to the localization of the failure, i.e., how to find which particular contacts are defective to within electrically distinguishable components.

Each sequence of a minimal test checks the failure of contact x_i if the value of the correct function differs from the value of the incorrect function generated by the failure of the contact x_i. To every sequence corresponds a set of single failures which it checks. In front of each sequence let us write out the contacts checked out by the given sequence.

Example. Let us design a minimal test for the circuit shown in Fig. 8a. For this purpose let us first consider its individual subcircuits.

The subcircuits shown in Fig. 8, b, c, and d have the following sequences as full tests:

x_1	x_2	x_3	x_4	x_5	x_6
1	0	1	0	1	0
0	1	0	1	0	1
0	0	0	0	1	1

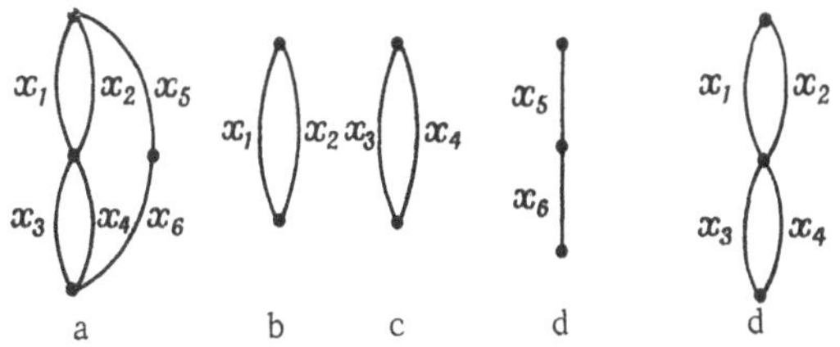

Fig. 8

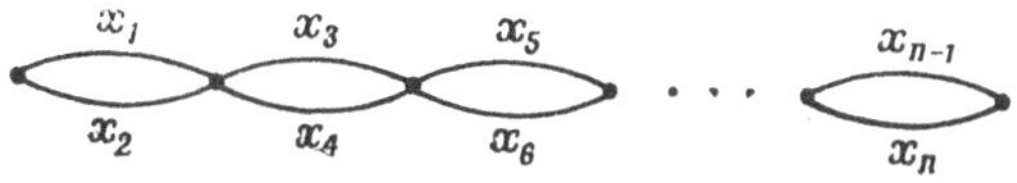

Fig. 9

Subcircuits b) and c) are connected in series. Consequently, in accordance with Lemma 12 the following set of sequences is a full test for the subcircuit e) in Fig. 8:

x_1	x_2	x_3	x_4
1	0	1	1
0	1	1	1
0	0	1	1
1	1	1	0
1	1	0	1
1	1	0	0

Now, taking into account that the subcircuits d) and e) are connected in parallel, the minimal full test of the complete circuit is

	x_1	x_2	x_3	x_4	x_5	x_6
$(x_1)^o$	1	0	1	1	0	0
$(x_2)^o$	0	1	1	1	0	0
$(x_1)^s\,(x_2)^s$	0	0	1	1	0	0
$(x_3)^o$	1	1	1	0	0	0
$(x_4)^o$	1	1	0	1	0	0
$(x_3)^s\,(x_4)^s$	1	1	0	0	0	0
$(x_6)^s$	0	0	0	0	1	0
$(x_5)^s$	0	0	0	0	0	1
$(x_5)^o\,(x_6)^o$	0	0	0	0	1	1

The failure checked out by the given sequence is shown at the left. The opening and closing of contact is denoted by $(x)^o$ and $(x)^s$ respectively.

Remark. A minimal test for Π (Σ) circuits designed in accordance with Lemma 12 coincides with the test constructed according to Lemma 4 (5) (see note on p. 111).

Let the nonrepetitive Π circuit that realizes the function $f(x_1, x_2, \ldots, x_n)$ have a length l and width b. Let $T_c^{1,\Pi}(f)$ and $T_c^{\Pi}(f)$ be respectively a single and a full check test for the nonrepetitive circuit which realizes the function f.

Theorem 7. $T_c^{1,\Pi}(f) = T_c^{\Pi}(f) = b + l$.

TABLE 2

No. of variables \ Length of test	2	3	4	5	6	7	8	9	10	11	12
$n=1$	1										
$n=2$		2									
$n=3$			4								
$n=4$				8	2						
$n=5$					16	8					
$n=6$						32	30	4			
$n=7$							64	92	24		
$n=8$								128	318	68	10

Proof. The upper bound follows from Theorem 2. Considering that $T_c^{1,\Pi}(f) \leq T_c^{\Pi}(f)$, it is sufficient to prove that

$$T_c^{1,\Pi}(f) \geqslant b+l. \tag{4}$$

Since the width of the circuit is b, there exists a dead-end cut with b contacts. An open-circuit in each of the contacts entering into this dead-end cut can be detected only if the unit chain passes through one of contacts in the cut. Thus, in order to detect the opening of all contacts we must have at least b sequences. Hence

$$T_c^{1,0,\Pi}(f) \geqslant b. \tag{5}$$

In virtue of duality:

$$T_c^{1,s,\Pi}(f) \geqslant l. \tag{6}$$

From (5) and (6) follows (4). The theorem is thus proved.

Corollary. $b+l \leqslant T_d^{1,\Pi}(f) \leqslant n+1$.

From Lemmas 13 and 14 follows that:

$$n+1 \leqslant T_d^{\Pi}(f) \leqslant \left[\frac{3}{2}n\right].$$

The length of the full diagnostic test for the circuit in Fig. 9 is exactly 3n/2.

The subset of Π nets whose minimal diagnostic test has a length n + k is denoted by $s_{n,k}$.

It can be easily observed that for any k (k = 1, 2, ..., [n/2]) the subset $s_{n,k}$ is nonempty.

Let $\Pi_{n,k}$ be the number of $s_{n,k}$ nets.

The subset $s_{n,1}$ results from a parallel or series connection of one edge to nets with n − 1 edges which, in turn, are obtained by parallel or series connection of one edge to nets with n − 2 edges, etc., $\Pi_{n,1} = 2^{n-1}$. In fact, $\Pi_{n,1} = 2\Pi_{n-1,1} = 2^2\Pi_{n-2,1} = \ldots = 2^{n-1}\Pi_{1,1} = 2^{n-1}$. From Lemmas 12 and 14 follows that for such nets

$$T_d^{\Pi}(f) = n+1, \tag{7}$$

and from Theorem 7 follows that

$$T_c^{1,\Pi}(f) = n + 1, \tag{8}$$

since the sum of the length and width of such circuits is n + 1. From (7) and (8) follows:

Theorem 8. If $f \in s_{n,1}$, then $T_c^{1,\Pi}(f) = T_c^{\Pi}(f) = T_d^{1,\Pi}(f) = T_d^{\Pi}(f) = n + 1$.

Table 2 indicates the length of the full diagnostic test and the number of circuits having a test of this length for small n. Let $H(f)$ be the number of distinct failure functions of a nonrepetitive Π circuit $\mathfrak{A}$, that realizes the function $f(x_1, x_2, \ldots, x_n)$.

Theorem 9. $H(f) = 2^n$.

The proof is obtained by mathematical induction applied to the number of contacts.

The statement is obviously true for n = 1.

Assume that the statement is true for all k < n and prove that it is also true for n.

The Π circuit $\mathfrak{A}$ with n contacts is a parallel or series connection of two Π circuits ($\mathfrak{A}_1$ and $\mathfrak{A}_2$) with k_1 and k_2 contacts, where $k_1 + k_2 = n$. Let $f = f_1 \cdot f_2$; then to each defective state of the circuit $\mathfrak{A}_1$, whose conduction is not identically equal to zero correspond 2^{k_2} defective states of the circuit $\mathfrak{A}_2$ with a conduction not identically equal to zero (including also the circuit $\mathfrak{A}_2$). To the correct state of $\mathfrak{A}_1$ correspond 2^{k_2} failure functions of the circuit $\mathfrak{A}_2$.

Consequently,

$$H(f) = (2^{k_1} - 1) \cdot 2^{k_2} + 2^{k_2} = 2^n.$$

The theorem is thus proved.

In conclusion the author expresses his gratitude to S. V. Yablonskii under whose guidance this work was conducted.

Literature Cited

1. V. V. Vaksov, "On tests for nonrepetitive switching circuits," Avtomat. i Telemekh., 26(3):521-524 (1965).
2. V. V. Glagolev, "Design of tests for block circuits," Dokl. Akad. Nauk SSSR, 144:6 (1962).
3. V. V. Glagolev, "Some bounds for disjunctive normal forms of functions of the algebra of logic," in: Systems Theory Research, Vol. 19, Consultants Bureau, New York (1970), p. 74.
4. I. V. Kogan, "On tests for nonrepetitive switching circuits," Problemy Kibernetiki, Vol. 12, Fizmatgiz, Moscow (1964), pp. 39-44.
5. A. V. Kuznetsov, "On nonrepetitive switching circuits and nonrepetitive superposition of logical-algebra functions," Trudy MIAN SSSR, Vol. 60, Moscow (1958), pp. 186-225.
6. O. B. Lupanov, "On asymptotic estimates of graphs and nets with edges," in: Problemy Kibernetiki, No. 4, Fizmatgiz, Moscow (1960), pp. 5-21.
7. Kh. A. Madatyan, "Synthesis of switching circuit of limited width," Problemy Kibernetiki, Vol. 14, Nauka, Moscow (1965), pp. 301-307.
8. B. A. Trakhtenbrot, "On the theory of nonrepetitive switching circuit," Trudy MIAN SSSR, Vol. 60, Moscow (1958), pp. 226-269.
9. I. A. Chegis and S. V. Yablonskii, "Logical methods for testing electrical circuits," Trudy MIAN SSSR, Vol. 60, Moscow (1958), pp. 270-362.
10. S. V. Yablonskii, "Functional designs in k-valued logic," Trudy MIAN SSSR, Vol. 60, Moscow (1958), pp. 5-142.

11. S. V. Yablonskii, "Basic concepts of cybernetics," Problemy Kibernetiki, Vol. 2, Fizmatgiz, Moscow (1959), pp. 7-38.
12. J. Riordan and C. E. Shannon, "The number of two-terminal series-parallel networks," J. Math. and Phys., 21(2):83-83 (1942).

ON FINITE MODEL SCHEMES HAVING DISCRETE FUNCTIONING†

Yu. A. Vinogradov

Moscow

Schemes with discrete functioning are ordinarily presented in the form of graphs whose directed edges are lines for transmitting information while the nodes are functional elements. Functions of memoryless schemes can be obtained as superpositions of functions of operating (functioning) elements.

Combined with the fact that such schemes are often very complex is the fact that calculation of their functioning is negotiated by a comparatively simple mathematical model of the functional elements. Serving as such models are the usual models of the algebra of logic. Obviously, faithful representation of the functioning of the scheme as a whole can be obtained only with well fitting functions of the elements and their models. The necessary good fit is ordinarily obtained by constructing special elements whose characteristics admit a good binary interpretation. However, there are cases of certain unnatural matchings of characteristics of elements to binary models in which such matchings are found not to conduce to the objective. In this case the binary model must be provided with certain restrictions within which the model has meaning. The elaboration of these, of course, is not characteristically a binary affair.

There is in principle another way of approximating functions of elements and their models, namely, the construction of more refined discrete models for the scheme elements. Such models may be functions of k-valued logic.

The possibility in principle of using many-valued logic in synthesis of discretely functioning electrical schemes was proved by S. V. Yablonskii [4], and in 1956-57 a group of diploma students at Gorky University built such a model for various electronic computer schemes [1, 2, 3]. The present work is a continuation of this and deals with certain general questions of construction of finite models of schemes realized by functional elements.

§ 1. Finite-Valued Models

It is clear that a specific functional element f with n inputs and one output can be posed in the form of a function $f(x_1, x_2, \ldots, x_n) = x_{n+1}$ expressing the dependence of the signal at the element's output on its inputs. Such functions of elements we call natural and shall, in the sequel, study not only elements but also their natural functions. We shall construct finite models

† Original article submitted September 10, 1968.

TABLE 1

X	$F_1(X)$	$F_2(X)$	$F_3(X)$
0	4	4	1
1	4	0	2
2	3	1	3
3	2	2	4
4	1	3	0

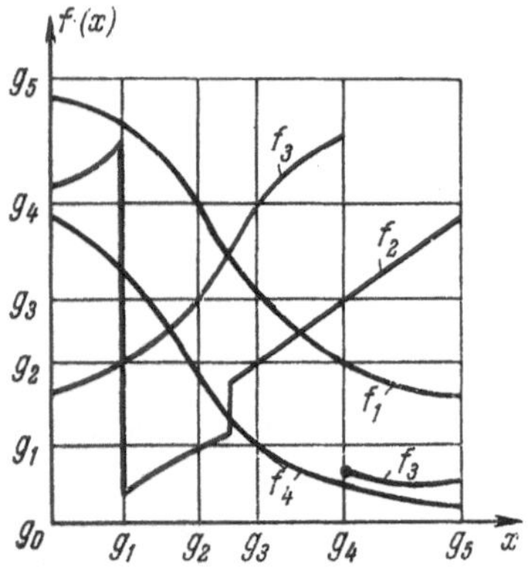

Fig. 1

for schemes whose elements are realizations of natural functions defined on the segment $A = [a, b]$ for each variable (in space A^n), taking values, also, from A.

Consider a partition H of the set A into points of division $a = g_0, g_1, \ldots, g_{k-1}, g_k = b$, $k > 1$, and intervals (g_0, g_1), $(g_1, g_2), \ldots, (g_{k-1}, g_k)$. Suppose the division points are arranged in increasing order. Denote the set of points $\{g_i\}$ by I; the set of intervals (g_i, g_{i+1}) by G. We shall use the terminology, H-partition, I-points, and G-intervals.

Definition. The function $f(x_1, x_2, \ldots, x_n) = x_{n+1}$ has k-valued model (k-valued description) if there exists an H-partition such that

$$x_{n+1} = f(x_1, x_2, \ldots, x_n) \in (g_l, g_{l+1}) \in G, \text{ if } x_i \in (g_{s_i}, g_{s_i+1}) \in G. \tag{1}$$

Substituting for the variable x_i respectively the discrete variable X_i, taking the value l if $x_i \in (g_l, g_{l+1})$, we transform a function $f(x_1, x_2, \ldots, x_n)$ defined on A and taking values from A into a finitely many-valued function $F(X_1, X_2, \ldots, X_n)$ defined on the set $\{0, 1, \ldots, k-1\}$ and taking values from this same set. We call the function F an I-model of the function f. In this work, we confine our attention to models of continuous single-valued functions of one variable.

§ 2. k-Valued Models of Functions of One Variable

The continuous function $f(x)$ has a finite-valued model if there exists an H-partition such that, for any $(g_i, g_{i+1}) \in G$ there exists a $(g_l, g_{l+1}) \in G$ and

$$f[(g_i, g_{i+1})] \subseteq (g_l, g_{l+1}). \tag{1'}$$

Condition (1') is fulfilled if none of the sets $f[(g_i, g_{i+1})]$, $i = 0, 1, \ldots, k-1$ include values from I.

Example 1. Consider some functions (not confined to continuous and single valued, with the aim of demonstrating the modelling possibility) prescribed by the graphs in Fig. 1, along with their models under the given H-partition. Number the G-intervals. Condition (1') is fulfilled for functions f_1, f_2, and f_3. These functions will correspond respectively to the model functions from P_5 (Table 1). As for the function f_4, we cannot, for the given H-partition, place it in correspondence with a k-valued model function, since Condition (1') is not fulfilled for f_4. Actually $f_4[(g_1, g_2)]$ includes the I-points g_2 and g_3.

We note that the H-partition considered in the example is suitable for some functions, and the grating formed by the prescribed I-points is common for these functions. The existence of a common grating for the system of elements forming a single scheme is necessary for the construction of a finite model of that scheme.

It is evident from the example just introduced that taking the definition of a finite model permits construction of models of functions of various forms. However, by no means can all functions have finite models. Below we consider necessary and sufficient conditions for the existence of finite models as well as methods of obtaining such models.

Lemma 1. If $f(x)$ has an I-model and $g \in I$, then $f^{-1}(g) \in I$.

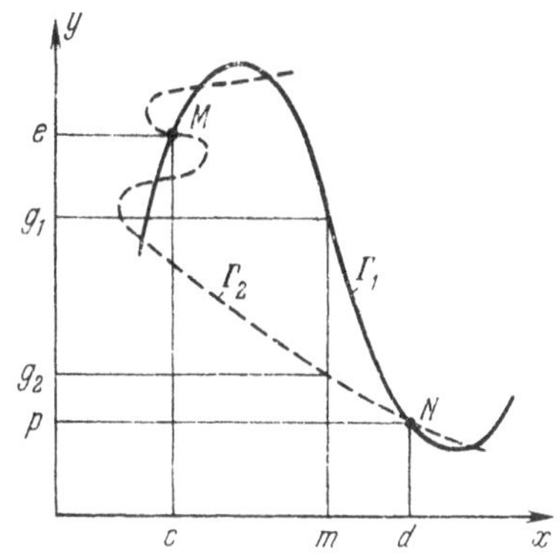

Fig. 2

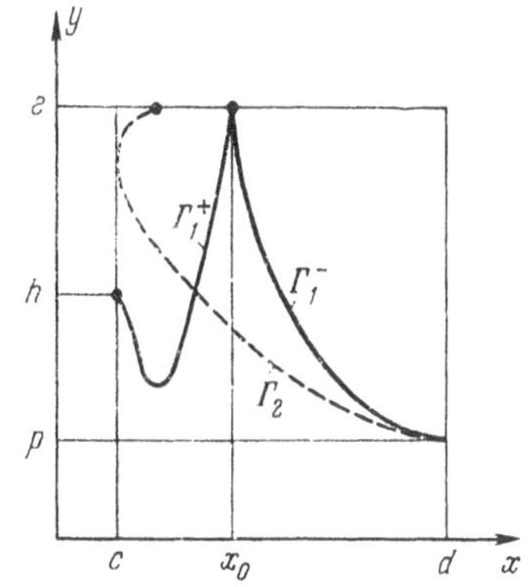

Fig. 3

<u>Proof.</u> In the contrary case, we can find $x \bar{\in} I$, such that $f(x) = g$, which contradicts Condition (1').

<u>Corollary.</u> If $g \in I$, then $f^{-1}(g)$ is finite.

Introduce the notation

$$f_i^{-1}(g) = \underbrace{f^{-1}(f^{-1}(\ldots(f^{-1}}_{i \text{ times}}(g))\ldots))$$

and call the sum $\bigcup_{i=1}^{\infty} f_i^{-1}(g) = f^*(g)$ the *track* of g. We remark that the tracks of two different points either do not intersect or else one contains the other. The set consisting of a, g, and b we denote by E_g.

<u>Theorem 1.</u> The continuous function $f(x)$ has a k-valued model if there exists $g \in (a, b)$ such that $f^*(E_g)$ is finite.

<u>Proof.</u> <u>Necessity.</u> Let $f(x)$ have an I-model for a finite set I. Then there exists $g \in I \cap (a, b)$, and tracks $f^*(a)$, $f^*(g)$, and $f^*(b)$ are finite (Corollary of Lemma 1).

<u>Sufficiency.</u> Let $f^*(E_g)$ be finite. Then, in $[a, b]$ there is a finite nonempty set of points $E_g \cup f^*(E_g) = \dot{I}$. We show that $\dot{I}$ can be a set in which an I-model of the given function is constructed. The condition of existence of a k-valued model for $f(x)$ fails to be fulfilled only if there exists a value $g_s \in \dot{I}$ whose pre-image does not belong to the set $\dot{I}$; this is impossible, since $\dot{I}$ contains all its pre-images.

However, this theorem does not give an effective method of finding a set of I-points. More effective methods will be considered below.

We prove two preliminary lemmas.

Denote by Γ_1 and Γ_2 the graphs of the functions $y = f_1(x)$ and $x = f_2(y)$ defined on the segments [c, d] and [p, e], respectively.

<u>Lemma 2.</u> If

$$f_1([c, d]) \supseteq [p, e],$$
$$f_2([p, e]) \supseteq [c, d],$$

and in the closed rectangle $[c, d] \times [p, e]$ there are precisely two (distinct) points of intersection of the graphs Γ_1 and Γ_2: M(c, e) and N(d, p), then for any point $g_1 \in (p, e)$ there can be found a point $g_2 \in (p, e)$, $g_2 \in f_2^{-1}(f_1^{-1}(g_1))$, such that

$$g_1 \neq g_2.$$

Proof. It follows from continuity of $f_1(x)$ that for any $g_1 \in (p, e)$ there exists $m \in (c, d)$, such that $m \in f_1^{-1}(g_1)$. It is easy to see that $g_1 \neq g_2$ since otherwise the point (m, g_1) would be a point of intersection of the graphs Γ_1 and Γ_2. (Shown in Fig. 2 is one of four varieties of possible mutual placement of the graphs Γ_1 and Γ_2 and the points M and N.)

Corollary. Since g_1 from (p, e) generates g_2 on (p, g_1) (or on (g_1, e)) it follows that unlimited repetition of this procedure would lead to the formation of an infinite set of points $g_1, g_2, \ldots, g_s, \ldots$, located on the interval (p, e).

Lemma 3. If functions $f_1(x)$ and $f_2(y)$ are defined respectively on L and Q, where L and Q (or Q and L) are either

1) semiintervals or

2) semiinterval and interval or

3) segments, and if

$$f_1(L) = Q, \quad f_2(Q) = L,$$

then the graphs Γ_1 and Γ_2 have a common point.

Proof. We prove only Case 1). Let L = [c, d), Q = (p, e], $f_1(c) = h$, and let x_0 be a point close to c at which $f_1(x_0) = e$. We denote by Γ_1^+ the part of the graph Γ_1 included between the points (c, h) and (x_0, e), while by Γ_1^- we denote the remaining part of the graph Γ_1 along with the point (x_0, e) adjoined to it (Fig. 3).

Consider the values of the function $f_2(e)$:

1) $f_2(e) = x_0$,

2) $f_2(e) < x_0$, and

3) $f_2(e) > x_0$.

In Case 1), Γ_1 and Γ_2 have a common point (x_0, e). In Case 2) the point $(f_2(e), e)$ is separated from all points of the rectangle $[x_0, d] \times (p, e)$ by the arc Γ_1^+, and Γ_2 must intersect the arc Γ_1^+. In Case 3) the same point is separated by the arc Γ_1^- from all points of the rectangle $[c, x_0] \times (p, e]$, and Γ_2 must intersect the arc Γ_1^-.

Remark 1. It is easy to see that Lemma 3 is also true in case

$$f_1([c, d)) = (p', e] \supseteq (p, e], \quad f_2((p, e]) = [c, d') \supseteq [c, d).$$

Remark 2. If L and Q (or Q and L) are intervals, then the lemma is false; if they are segment and interval or segment and semiinterval, then its hypothesis is inconsistent.

Denote by Γ_f the graph of the function $f(x)$. Plotting for each point $B(x_i, y_i) \in \Gamma_f$ the corresponding point $B'(y_i, x_i)$ we get $\Gamma_{f^{-1}}$, the graph of the inverse function $f^{-1}(x)$. Let $\mathfrak{R} = \Gamma_f \cap \Gamma_{f^{-1}}$. We note that the set of coordinates of points from $\mathfrak{R}$ coincides with the set of solutions of the equation $f(x) = f^{-1}(x)$ since to each point $B(x_i, y_i) \in \mathfrak{R}$ there corresponds the point $B'(y_i, x_i) \in \mathfrak{R}$, and the abscissas of points in $\mathfrak{R}$ are solutions of the equation $f(x) = f^{-1}(x)$. Denote the set of solutions of the equation $f(x) = f^{-1}(x)$ by S. Since $f(x)$ and $f^{-1}(x)$ are continuous functions, it follows that $\mathfrak{R}$ and S are closed linearly ordered sets (finite, countable, or continual).

2.1. Models of Functions for Which $f(A) = A$

Lemma 4. If $f(A) = A$ and $g \bar{\in} S$, then $f^*(g)$ is infinite.

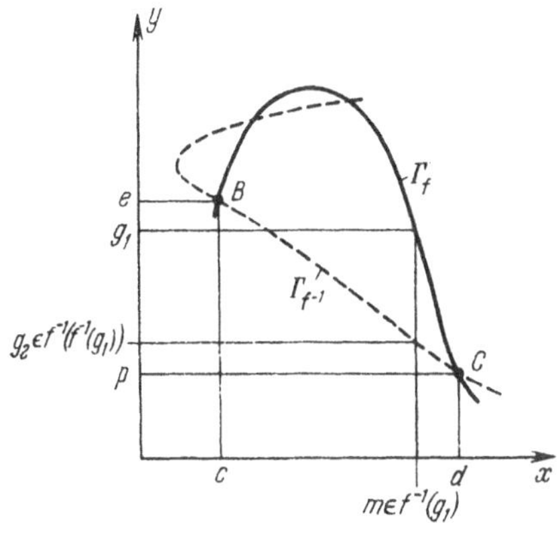

Fig. 4

P r o o f. Suppose $\overline{S} = A \setminus S$. Then $\overline{S}$ is either empty or consists of intervals and (or) semiintervals (abutting on the ends of A).

The points B(c, e) and $C(d, p) \in \mathfrak{R}$ are said to be n e i g h b o r s i n $\mathfrak{R}$, if the interval (B, C) (on the contour $\Gamma_{f^{-1}}$) contains no points of $\mathfrak{R}$ (Fig. 4). Denote a' and b' the least and the greatest values from S. We note that $\mathfrak{R} \subset [a', b'] \times [a', b']$.

I. We show that $f^*(g_1)$ is infinite if g_1 belongs to an interval in $\overline{S}$. Suppose points B(c, e) and C(d, p) are neighbors in $\mathfrak{R}$. Then $(p, e) \in \overline{S}$, and the hypothesis of Lemma 2 is fulfilled for the points B and C and the closed rectangle [c, d] × [p, e]. Here $f_1 = f_2 = f$ and for any $g_1 \in (p, e)$, $f^*(g_1)$ is infinite.

II. We show that $f^*(g)$ is infinite if g belongs to a semiinterval in $\overline{S}$. We consider the following cases.

1. If $x \in (b', b]$, then $f(x) \neq b$. Otherwise, in the square $(b', b] \times (b', b]$ the graph Γ_f would have a point in common with the diagonal $y \equiv x$. Such a point would belong to $\mathfrak{R}$, but by hypothesis, all points from $\mathfrak{R}$ belong to $[a', b'] \times [a', b']$.

Analogously $f(x) \neq a$ if $x \in [a, a')$.

2. If $x_1 \in [a, a')$ and $x_2 \in (b', b]$, then it is impossible that $f(x_1) = b$ and $f(x_2) = a$ can both be fulfilled. Otherwise

$$f((b', b]) = [a, a'') \supseteq [a, a'),$$

$$f([a, a')) = (b'', b'] \supseteq (b', b]$$

and the rectangle $[a, a') \times (b', b]$ contains a point of $\mathfrak{R}$ (Lemma 3), but by hypothesis all points of $\mathfrak{R}$ belong to $[a', b'] \times [a', b']$.

Thus we have either 3) $f([a', b']) = A$ or 4) $f(x_1) = b$, $x_1 \in [a, a')$ and $f(x_2) = a$, $x_2 \in [a', b']$, or 5) $f(x_1) = b$, $x_1 \in [a', b']$ and $f(x_2) = a$, $x_2 \in (b', b]$. We consider these cases.

3) $f([a', b']) = A$. Suppose $g \in [a, a') \cup (b', b]$. Then there is a point P(m, g), $m \in f^{-1}(g) \cap [a', b']$, such that $P \notin \mathfrak{R}$, since P is not contained in $[a', b'] \times [a', b']$. But then m belongs to an interval from $\overline{S}$ and the track $f^*(g)$ is infinite since $f^*(m)$ is infinite.

4) $f(x_1) = b$, $x_1 \in [a, a')$ and $f(x_2) = a$, $x_2 \in [a', b']$. Suppose $g \in [a, a')$. But then, as shown in 3, the track $f^*(g)$ is infinite. Suppose $g \in (b', b]$. But since $f([a, a')) \supseteq (b', b]$, there exists $m \in f^{-1}(g) \cap [a, a')$, and the track $f^*(g)$ is infinite by reason of the fact that $f^*(m)$ is infinite.

5) $f(x_1) = b$, $x_1 \in [a', b']$ and $f(x_2) = a$, $x_2 \in (b', b]$. The proof of infinitude of the track $f^*(g)$ for $g \in [a, a') \cup (b', b]$ is analogous to the proof in Point 4.

Accordingly, the track $f^*(g)$ is infinite if $g \in \overline{S}$, as was to be shown.

Let $f(x)$ be an arbitrary continuous function. Construct for it a set $\mathfrak{J}$ consisting of those and only of those g for which the following conditions† are fulfilled: 1) g belongs to S, 2) $\overline{\overline{f^{-1}(g)}} = 1$ and 3) $\overline{\overline{f^{-1}(f^{-1}(g))}} = 1$.

† $\overline{\overline{f^{-1}(g)}}$ is the cardinality of the set $f^{-1}(g)$.

TABLE 2

$\overline{\overline{\mathfrak{I}}}$	$\exists g,\ g \in \mathfrak{I}\cdot(a, b): f^{-1}(g) = g$	$\forall g,\ g \in \mathfrak{I}\cdot(a, b): f^{-1}(g) \neq g$
$< \aleph_0$	Arbitrary finite, k-valuedness $2 \leqslant k \leqslant \overline{\overline{\mathfrak{I}}}-1$	Arbitrary odd k-valuedness $3 \leqslant k \leqslant \overline{\overline{\mathfrak{I}}}-1$
$\aleph_0$	Arbitrary finite and denumerable valuedness	Arbitrary odd and denumerable valuedness
c	Arbitrary finite, denumerable, and continual valuedness	Arbitrary odd, denumerable, and continual valuedness

Theorem 2. If the continuous function $f(x)$ defined on A takes all values in A, then the track $f^*(g)$, $g \in A$, is finite if and only if $g \in \mathfrak{I}$.

Proof. Necessity of the condition $g \in S$ follows from Lemma 4. We prove necessity of the conditions $\overline{\overline{f^{-1}(g)}} = 1$ and $\overline{\overline{f^{-1}(f^{-1}(g))}} = 1$.

a) $\overline{\overline{f^{-1}(g)}} > 0$, since $f(A) = A$.

b) We show that $\overline{\overline{f^{-1}(g)}} < 2$. Let $f^{-1}(g) = \{g_m, g_n, \ldots\}$. Then there exist points $M(g_m, g)$ and $N(g_n, g) \in \Gamma_f$. But the graph $\Gamma_{f^{-1}}$ has a unique point with ordinate g, and either M or $N \overline{\in} \mathfrak{R}$. Then either g_m or $g_n \overline{\in} S$ and either $f^*(g_m)$ or $f^*(g_n)$ is infinite. Consequently the track $f^*(g)$ too is infinite since $f^*(g) \supseteq f^*(g_m) \cup f^*(g_n)$.

Necessity of $\overline{\overline{f^{-1}(f^{-1}(g))}} = 1$ is proved analogously, putting $f^{-1}(g) = g'$.

Sufficiency. Let $g \in S$, $f^{-1}(g) = g_\alpha$, and $f^{-1}(g_\alpha) = g_\beta$. Then there exist points $B(g_\alpha, g)$, $C(g_\beta, g_\alpha)$ and $B'(g, g_\alpha)$, $C'(g_\alpha, g_\beta) \in \mathfrak{R}$. But since $\overline{\overline{f^{-1}(g_\alpha)}} = 1$, it follows that $g_\beta = g$ (C coincides with B' and C' coincides with B). Consequently, $f^*(g) = \{g_\alpha, g\}$, as was to be proved.

Corollary 1. A continuous function $f(x)$ defined in $A = [a, b]$ and taking all values of A has a finite model if and only if there exists a $g \in (a, b)$ such that $\{a, g, b\} \in \mathfrak{I}$. In this case $f(x)$ will have a model with $I = \{a, g, f^{-1}(g), b\}$.

Corollary 2. If $f(x)$ has a finite model, then at least one of its modelled functions is contained either in P_2 or in P_3.

Note. A k-valued model of the function $f(x)$ defined in A and taking all values in A can only be a strictly monotone k-valued function, i.e., either $F_1(X) = X$ or $F_2(X) = k - 1 - X$, where F and X take values from $\{0, 1, \ldots, k-1\}$.

Thus for the function $f(x)$ taking all values from A, the set of I-points is formed only of values lying in the set $\mathfrak{I}$. It is obvious that the number of finite-valued models of a given function depends on the cardinality of $\mathfrak{I}$ and is equal to the number of admissible subsets of $\mathfrak{I}$ (an admissible subset is a subset including in itself a and b and along with each g the preimage $f^{-1}(g)$). The set $\mathfrak{I}$ can be finite, infinite, or continual.

The valuedness of the model is tied to the cardinality of $\mathfrak{I}$. The possible cases are exhibited in Table 2.

Example 2. We construct a finite model for the function $f(x)$ given by the graph Γ_f of Fig. 5.

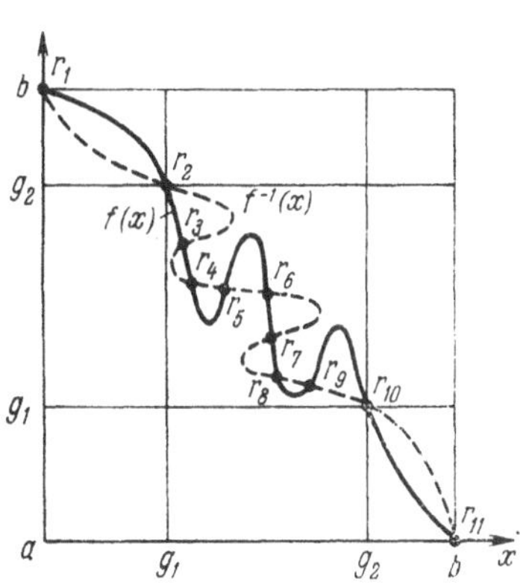

Fig. 5

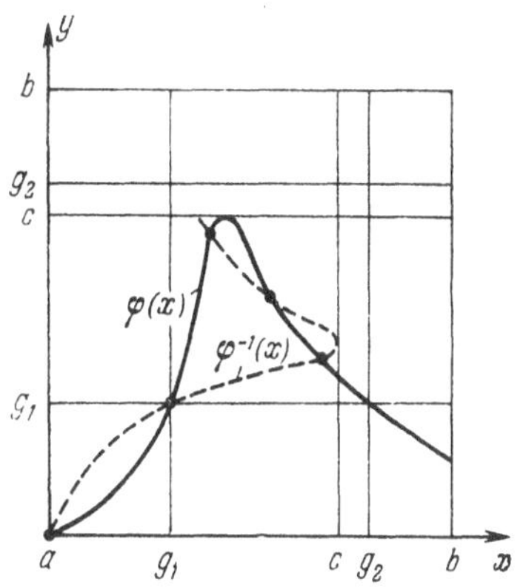

Fig. 6

The mirror image of the graph Γ_f relative to the diagonal y = x gives the graph of the inverse function Γ_{f-1} and the intersection $\Gamma_f \cap \Gamma_{f-1}$ in the set $\mathfrak{R} = \{r_1, r_2, \ldots, r_{11}\}$. The projection of the set $\mathfrak{R}$ on the x axis forms the set S and the values a, g_1, g_2, and b from S form the set   Putting $I = \mathfrak{J}$ and numbering the G-intervals we get the model of largest valuedness $F(X) = 2 - X$, $X \in \{0, 1, 2,\}$. Since for any g from $\mathfrak{J} \cap (a, b)$, $f^{-1}(g) \neq g$, and $\bar{\bar{\mathfrak{J}}} = 4$, it follows that this model will be the only possible model of the function $f(x)$.

2.2. Models of Functions for Which $f(A) \subset A$. We consider conditions under which continuous functions not taking all values from A have finite models.

For such functions the sequence A, $f(A)$, $f(f(A))$, ... forms a system of segments, nested in one another. Their intersection is either a segment (and we denote it by [u, v]) or a point. The segment [u, v] is either the intersection of a finite system of nested segments or else such a finite system does not exist.

Consider semiintervals from $A \setminus [u, v]$. Denote $A \setminus f(A), f(A) \setminus f(f(A)), \ldots$ by A_1, A_2, ... We note that if $g \in A_i$, then $f^{-1}(g) \in A_{i-1}$, and the complete preimage of each $g \in A_1$ is empty. We select from each A_i a subset A'_i such that $g \in A'_i$ if its complete preimage $f^{-1}(g)$ is finite and is contained in A'_{i-1} if A'_{i-1} is not empty.

Any values lying in the sum $\bigcup A'_i = \mathfrak{A}'$, have a finite track since $f_i^{-1}(g) = \varnothing$ for any $g \in A_i$.

Consider the segment [u, v]. Here $f([u, v]) = [u, v]$. We define the set $\mathfrak{J}_{uv}$ for the function $f(x)$ and the segment [u, v] in the same way that in 2.1 the set $\mathfrak{J}$ was defined for the function $f(x)$ and the segment $[a, b]$.

It is readily seen that a finite track has only those values $g \in [u, v]$, for which $f^{-1}(g) \in \mathfrak{J}_{uv} \cup \mathfrak{A}'$ Since to serve as an I-point in the interval (a, b) there can be taken any value from the whole nonempty set $\mathfrak{A}'$, it follows from Theorem 1 that a necessary and sufficient condition for the existence of a finite model of a function not taking all values from A is finiteness of the tracks $f^*(a)$ and $f^*(b)$. We consider the following cases.

1. $a, b \bar{\in} f(A)$. Here $f^*(a)$ and $f^*(b)$ are empty and such functions always have finite-valued models.

2. $b \bar{\in} f(A)$. Here $f^*(b) = \varnothing$, u = a, and the track $f^*(a)$ is finite when and only when

a) $f^{-1}(a) = \{a, C_a\}$ or

b) $f^{-1}(a) = \{v, C_a\}$, and $f^{-1}(v) = \{a, C_a\}$, where C_a and $C_v \in \mathfrak{A}'$.

3. $a \bar{\in} f(A)$. The condition is analogous to Conditions a) and b) in Point 2.

Example 3. Construct a finite-valued model for the function $\varphi(x)$, given by the graph in Fig. 6.

Here $b \bar{\in} \varphi(A)$, $\varphi^{-1}(a) = a$, and hence the function $\varphi(x)$ has a model (Case 2 of the ones considered) in which $\mathfrak{A}' = A_1 = (c, b]$, $[u, v] = [a, c]$, $\mathfrak{F}_{ac} = \{a, g_1\}$, $\varphi^{-1}(g_1) = \{g_1, g_2\}$, $g_2 \in \mathfrak{A}'$. The I-points of the model are chosen from the set $\{a, g_1, (c, b]\}$. The set $I \cap (c, b)$ is arbitrarily formed by excluding the values g_2 which necessarily fall in I if $g_1 \in I$.

Taking for instance $I = \{a, g_1, g_2, b\}$ and numbering the G intervals, we get the model

$$\Phi(X) = \begin{cases} X, & X \neq 2, \\ 0, & X = 2. \end{cases}$$

Literature Cited

1. G. A. Andreev, "Application of k-valued logic to schematic synthesis of mathematical machines" (diploma thesis), RFF GGU (1956).
2. T. I. Kir'yanova, "Synthesis of electronic computers and control schemes by means of valued logic" (diploma thesis), RFF GGU (1956).
3. N. A. Loginova, "Synthesis of digital computers and control schemes by means of many-valued logic" (diploma thesis), RFF GGU (1957).
4. S. V. Yablonskii, Proceedings of the All-Union Mathematical Society, Vol. 3, Akad. Nauk SSSR (1956), pp. 425-431.

ON A CERTAIN GENERALIZATION OF FINITE AUTOMATA, WHICH FORMS A HIERARCHY ANALOGOUS TO THE GRZEGORCZYK CLASSIFICATION OF PRIMITIVELY RECURSIVE FUNCTIONS†

V. A. Kozmidiadi

Moscow

This paper considers a certain generalization of the notion of a finite automaton. A sequence of expanding classes of n-automata (n = 0, 1, 2, ...) is formed. Each of the classes is formed by closure via a composition in the class of primitive n-automata. Under these conditions a primitive n-automaton operates similarly to a conventional finite automaton: it has an initial state and is stipulated by a certain function of transitions that determine the new state as a function of the previous state and the next input level. However, the states of the automaton are words in the input alphabet; the output word is formed as a sequence of states that are passed through by the automaton due to the action of the input word. The function of transitions for a primitive automaton of the (n + 1)-st rank is stipulated by means of an automaton of the n-th rank.

Chapter I gives the definition of an automaton of the n-th rank (n = 0, 1, 2, ...) and presents a number of examples of such automata.

Chapter II proves combination theorems for n-automata: extension, branching, union, repetition, and configuration. Moreover, a theorem of rank elevation is proved which states that any transformation which is performable on an n-automaton may also be performed on an (n + 1)-automaton.

In Chapter III an example is constructed of the word numbering carried out in the alphabet A by 1-automata. On the basis of this numbering it is proved that any primitively recursive function is calculable on the appropriate n-automaton; on the other hand, any n-automaton is equivalent in a definite sense to a certain primitively recursive function.

Chapter IV compares classes of n-automata and classes of primitively recursive functions from the A. Grzegorczyk classification. It is established that the class of functions calculated on n-automata ($n \geq 1$) coincides with the (n + 2) class of Grzegorczyk.

† Original article submitted October 26, 1967.

CHAPTER I

The Notion of an n-Automaton

In the present chapter a basic notion of the paper is defined — the notion of an n-automaton for n = 0, 1, 2, ... Any n-automaton is fully equivalent, relative to the alphabet A (A is the alphabet of an n-automaton), to a certain normal algorithm† [5]. The reverse, however, is not true. Moreover, n-automata in a certain sense occupy the same position among all dictionary calculable functions as primitively recursive functions do among all partially recursive functions.

In the present chapter we shall consider a series of examples of n-automata besides defining an n-automaton. Many of these examples will be used in the later exposition.

§ 1. Definition of an n-Automaton

1. The definition of an n-automaton is given inductively (by induction with respect to n).

A primitive 0-automaton $\mathfrak{A}^{(0)}$ in the alphabet A is stipulated as follows. The word S_0 in the alphabet $A \setminus \{*\}$ and the finite collection of substitution formulas of the form $\xi S \to T$, where $\xi \in A \cup \{*\}$, S, $T \Omega A \setminus \{*\}$. are indicated. The word S_0 is called the initial state of the primitive 0-automaton $\mathfrak{A}^{(0)}$. The primitive 0-automaton $\mathfrak{A}^{(0)}$ prescribes sequential transformation of the word P in the alphabet A according to the following rule.

Assume $P \,\bar{\underline{\circ}}\, \xi_1\xi_2 \ldots \xi_k$ $(k \geqslant 0, \xi_1, \ldots, \xi_k$ are letters of the alphabet A). Let us form the word P*. Let us assume, moreover, that $\xi_{k+1} \,\bar{\underline{\circ}}\, *$. Let us form the word $\xi_1 S_0$. In the collection of substitution formulas let us find the formula $\xi_1 S_0 \to S_1$ (if, of course, such a formula exists there); then we use the word $\xi_2 S_1$ to find the formula $\xi_2 S_1 \to S_2$, etc. A string of words $S_1, S_2, \ldots, S_{k+1}$ is formed. If such a string cannot be formed (this is the case when for a certain i $(0 \leq i \leq k)$ the collection of substitution formulas does not contain formulas of the type $\xi_{i+1} S_i \to S_{i+1}$), we assume that the result of the transformation of the original word by the primitive 0-automaton $\mathfrak{A}^{(0)}$ is not defined. However, if such a string may be formed, then we assume that the result of applying the primitive 0-automaton $\mathfrak{A}^{(0)}$ to the word P is the word $S_1 S_2 \ldots S_{k+1}$; in other words, the primitive 0-automaton $\mathfrak{A}^{(0)}$ transforms (manipulates) the original word P into the word $S_1 S_2 \ldots S_{k+1}$. Any word S in the alphabet $A \setminus \{*\}$ is called a state of $\mathfrak{A}^{(0)}$.

Assume that $\mathfrak{A}^{(0)}$ is a primitive 0-automaton in the alphabet A; then $\mathfrak{A}^{(0)}$ is called intrinsic if $* \notin A$, and nonintrinsic if $* \in A$.

Assuming that the notion of composition of algorithms is known, we give the following definition of the notion of a 0-automaton.

A primitive 0-automaton in the alphabet A is a 0-automaton in the alphabet A.

Assume $\mathfrak{A}^{(0)}$ is a primitive 0-automaton in the alphabet A, while $\mathfrak{B}^{(0)}$ is a 0-automaton in the same alphabet. Then the algorithm that is a composition of $\mathfrak{A}^{(0)}$ and $\mathfrak{B}^{(0)}$, (i.e., the algorithm prescribing first the application of the primitive 0-automaton $\mathfrak{A}^{(0)}$, to arbitrary initial data, and then the application of the 0-automaton $\mathfrak{B}^{(0)}$, to the result of its work) is a 0-automaton in the alphabet A. This 0-automaton is intrinsic if the 0-automata $\mathfrak{A}^{(0)}$, $\mathfrak{B}^{(0)}$ are intrinsic, and it is nonintrinsic if $\mathfrak{A}^{(0)}$ or $\mathfrak{B}^{(0)}$ are nonintrinsic.

† In this paper we shall strive to adhere to the notation in [5]. Moreover, the plan of the paper itself in many ways coincides with the plan of the monograph [5]: a number of terms, etc., are taken from there.

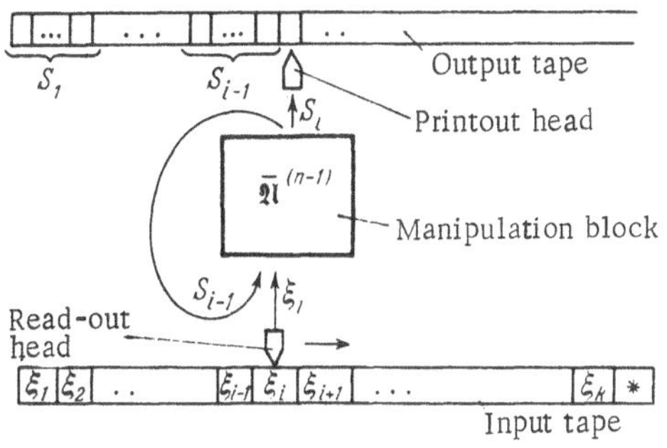

Fig. 1

A **primitive (n + 1)-automaton** $\mathfrak{A}^{(n+1)}$ is stipulated in the alphabet A as follows. The word S_0 is indicated in the alphabet A along with the nonintrinsic n-automaton $\overline{\mathfrak{A}^{(n)}}$ in the alphabet $A \cup \{*\}$. The word S_0 is called **the initial state of the primitive (n + 1)-automaton** $\mathfrak{A}^{(n+1)}$.

$\mathfrak{A}^{(n+1)}$ prescribes sequential transformation of the word P in the alphabet A thus:

Assume $P \,\bar{\underline{\circ}}\, \xi_1\xi_2 \ldots \xi_k$ ($k \geqslant 0$, $\xi_1, \xi_2, \ldots, \xi_k$ are letters of the alphabet A). Let us form the word $P*$. Let us place $\xi_{k+1} \,\bar{\underline{\circ}}\, *$. We construct the word $\xi_1 S_0$ and apply the n-automaton $\overline{\mathfrak{A}^{(n)}}$ to it; as a result we obtain the word S_1; then we apply $\overline{\mathfrak{A}^{(n)}}$ to the word $\xi_2 S_1$, etc.

A string of words S_1, S_2, ... is formed. If such a string cannot be formed (this is the case when for a certain i ($0 \leq i \leq k$) the results of transforming the word $\xi_{i+1}S$ by the n-automaton $\overline{\mathfrak{A}^{(n)}}$ is not defined), we assume that the result of transforming the original word P by the primitive (n + 1)-automaton $\mathfrak{A}^{(n+1)}$ is not defined.

If such a string may be formed, then we assume that the result of applying the primitive (n + 1)-automaton $\mathfrak{A}^{(n+1)}$ to the word P is the word $S_1S_2 \ldots S_{k+1}$; in other words, $\mathfrak{A}^{(n+1)}$ transforms (manipulates) the word P into the word $S_1S_2 \ldots S_{k+1}$.

Any word S in the alphabet $A \setminus \{*\}$ is called a **state of the automaton** $\mathfrak{A}^{(n+1)}$.

The primitive (n + 1)-automaton $\mathfrak{A}^{(n+1)}$ is called **intrinsic** if $* \notin A$, and **nonintrinsic** if $* \in A$.

Let us continue the definition of an (n + 1)-automaton. A primitive (n + 1)-automaton in the alphabet A is an **(n + 1)-automaton** in the alphabet A.

Assume $\mathfrak{A}^{(n+1)}$ is a primitive (n + 1)-automaton in the alphabet A, and that $\mathfrak{B}^{(k)}$ ($k \leqslant n+1$) is a k-automaton in the same alphabet. Then the algorithm which is a composition of $\mathfrak{A}^{(n+1)}$ and $\mathfrak{B}^{(k)}$, is an (n + 1)-automaton in the alphabet A; this (n + 1)-automaton is **intrinsic** if the (n + 1)- and k-automata $\mathfrak{A}^{(n+1)}$ and $\mathfrak{B}^{(k)}$ are intrinsic, and it is **nonintrinsic** if $\mathfrak{A}^{(n+1)}$ or $\mathfrak{B}^{(k)}$ are nonintrinsic.

2. The primitive n-automaton $\mathfrak{A}^{(n)}$ functions in a certain sense in almost the same way as a conventional finite automaton (see [4]). In order to understand the functioning of an n-automaton more easily, it is convenient to represent it as a machine consisting of a readout head and a manipulation block $\overline{\mathfrak{A}^{(n-1)}}$ (Fig. 1); we shall not enter into the arrangement of the manipulation block at present. The original word P is fed letter-by-letter into the $\overline{\mathfrak{A}^{(n-1)}}$ block which forms a new state from the preceding state and the next letter.

3. Taking account of the external similarity between the functioning of finite automata and n-automata, it would be desirable to provide the possibility of discussing primitive n-automata in the same terms in which conventional finite automata are discussed. We shall sometimes use such a terminology. Let us consider the primitive n-automaton $\mathfrak{A}^{(n)}$ in the alphabet A and the original word P:

$$P \,\bar{\underline{\circ}}\, \xi_1\xi_2 \ldots \xi_i\xi_{i+1} \ldots \xi_k,$$

where $k \geq 0$; $\xi_1, \xi_2, \ldots, \xi_k$ are letters of the alphabet A. Assume

$$S_1, S_2, \ldots, S_i, S_{i+1}, \ldots, S_k$$

is the string of words described in subsection 1. In this case we say that upon receiving the next, (i + 1)-st, letter of the original word the primitive n-automaton $\mathfrak{A}^{(n)}$ is in the state S_i; due to the effect of the input letter ξ_{i+1} it transits from the state S_i to the state S_{i+1}. Hereafter " $\mathfrak{A}^{(n)}: S \vdash_{\xi} T$ " means that the primitive n-automaton $\mathfrak{A}^{(n)}$ transits from the state S to the state T due to the action of the letter ξ.

" $\mathfrak{A}^{(n)}: S \vdash_{\xi} T\,(T)$ " means that the n-automaton $\mathfrak{A}^{(n)}$ transits from the state S to the state T and prints out the word T due to the action of the letter ξ.

Instead of

$$\mathfrak{A}^{(n)}: S_1 \vdash_{\xi_1} S_2;\ \mathfrak{A}^{(n)}: S_2 \vdash_{\xi_2} S_3;\ \ldots;\ \mathfrak{A}^{(n)}: S_{k-1} \vdash_{\xi_{k-1}} S_k$$

and instead of

$$\mathfrak{A}^{(n)}: S_1 \vdash_{\xi_1} S_2\,(S_2);\ \mathfrak{A}^{(n)}: S_2 \vdash_{\xi_2} S_3\,(S_3);\ \ldots;\ \mathfrak{A}^{(n)}: S_{k-1} \vdash_{\xi_{k-1}} S_k\,(S_k)$$

we shall usually write the abbreviated relation

$$\mathfrak{A}^{(n)}: S_1 \vdash_{\xi_1} S_2 \vdash_{\xi_2} S_3 \vdash \ldots \vdash S_{k-1} \vdash_{\xi_{k-1}} S_k$$

and, correspondingly,

$$\mathfrak{A}^{(n)}: S_1 \vdash_{\xi_1} S_2\,(S_2) \vdash_{\xi_2} S_3\,(S_3) \vdash \ldots \vdash S_{k-1}\,(S_{k-1}) \vdash_{\xi_{k-1}} S_k\,(S_k)$$

Assume P is a word in the alphabet A, the condition $P \,\overline{\underline{\circ}}\, \xi_1\xi_2 \ldots \xi_k,\ k > 0$ being valid. Assume that upon receiving the first letter ξ_1 of the word P the n-automaton $\mathfrak{A}^{(n)}$ is in the state S_i, and assume

$$\mathfrak{A}^{(n)}: S_i \vdash_{\xi_1} S_{i+1}\,(S_{i+1}) \vdash_{\xi_2} S_{i+2}\,(S_{i+2}) \vdash_{\xi_3} \ldots \vdash_{\xi_{k-1}} S_{i+k-1}\,(S_{i+k-1}) \vdash_{\xi_k} S_{i+k}\,(S_{i+k}).$$

In this case we shall say that the n-automaton $\mathfrak{A}^{(n)}$ transits from the state S_i to the state S_{i+k} due to the action of the word P and shall denote this event thus:

$$\mathfrak{A}^{(n)}: S_i \vDash_P S_{i+k},$$

or we shall say that the n-automaton $\mathfrak{A}^{(n)}$ transits from the state S_i to the state S_{i+k} due to the action of the word P; this is denoted by

$$\mathfrak{A}^{(n)}: S_i \vDash_P S_{i+k}\,(S);$$

by definition we place

$$\mathfrak{A}^{(n)}: S_i \vDash_\Lambda S_i$$

and

$$\mathfrak{A}^{(n)}: S_i \vDash_\Lambda S_i\,(\Lambda).$$

4. Let us introduce certain additional simplifications of the terminology, and let us likewise give certain definitions.

Assume there is an n-automaton $\mathfrak{A}^{(n)}$. We shall call the number n the r a n k of $\mathfrak{A}^{(n)}$.

We shall often speak loosely, saying "n-automaton" or simply "automaton" instead of "primitive n-automaton."

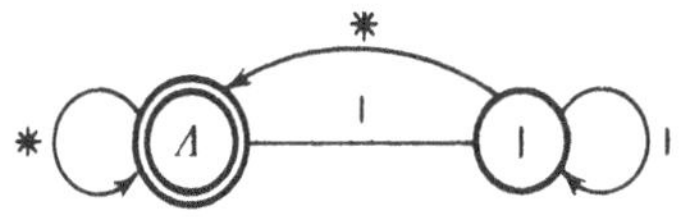

Fig. 2

Hereafter we shall denote automata and collections of substitution formulas by capital Gothic letters (sometimes with indices). Under these conditions a superscript in parentheses will always denote the rank of the corresponding automaton. It is allowable, in denoting a certain n-automaton, to drop the rank in its designation if for some reason we do not wish to indicate it.

We shall likewise call 0-automata finite automata, while n-automata having $n > 0$ shall be called infinite automata.

5. For the case in which the n-automaton $\mathfrak{A}^{(n)}$ transforms a certain word P into a certain word Q we shall say that the result of the transformation of the original word by the n-automaton $\mathfrak{A}^{(n)}$ is defined, while the n-automaton $\mathfrak{A}^{(n)}$ is applicable to the word P.

If the n-automaton $\mathfrak{A}^{(n)}$ is applicable to the word P, then it manipulates it into a certain completely defined word. We shall denote this word by $\mathfrak{A}^{(n)}(P)$.

Hereafter we shall sometimes make use of the symbol !. By placing this symbol in front of a certain expression, we shall thereby assert that this expression has meaning.

6. An n-automaton over the alphabet A is called an n-automaton in any expansion of the alphabet A (i.e., in an alphabet B which is such that $A \subseteq B$). By analogy with the theory of algorithms [5], the notion of equivalence and complete equivalence of automata relative to the alphabet A is introduced.

Later on we shall sometimes make use of the symbol of conditional equality $\simeq$. By placing this symbol between two expressions we thereby state that these expressions denote the same thing if just one of them has meaning. Here we shall write the additional conditions imposed on the composite part of the expression considered in parentheses if necessary.

7. Let us consider a certain primitive automaton $\mathfrak{A}$ in the alphabet A. According to the definition given in subsection 1, any word in the alphabet $A \setminus \{*\}$ is a state of the automaton $\mathfrak{A}$.

The state S of the automaton $\mathfrak{A}$ is called accessible if there exists a word P in the alphabet A such that

$$\mathfrak{A}\colon S_0 \models {}_P S,$$

where S_0 is the initial state of the automaton $\mathfrak{A}$. From the definition it is evident that the initial state of any automaton is accessible, since

$$\mathfrak{A}\colon S_0 \models {}_\Lambda S_0.$$

It is not difficult to see that the number of states for which such a word P exists (i.e., accessible states) is finite for a primitive 0-automaton. Actually, the initial state S_0 and, perhaps, certain of those states which are encountered in the right sides of the substitution formulas from the collection $\overline{\mathfrak{A}}$, are accessible for a 0-automaton. However, since the collection of substitution formulas of a primitive 0-automaton is finite, it follows that the number of accessible states is also finite. This fact explains why primitive 0-automata are also called finite automata.

8. We shall use the convenient representation of primitive 0-automata in the form of graphs, as is the usual practice.

The vertices of the graph shall be depicted by circles. Within the circle we shall write the state which denotes the given vertex. The initial state is depicted by a circle drawn with a double line.

9. The number of accessible states of a primitive n-automaton for n > 0 may be infinite. It is precisely for this reason that we have called n-automata infinite for n > 0.

The fact that a primitive 1-automaton having an infinite number of states is possible shall be shown in the next example. We shall soon be provided with the possibility of proving the same thing for automata of rank higher than 1 as well.

Let us consider the 1-automaton $\mathfrak{A}^{(1)}$ in the one-letter alphabet $A = \{ | \}$ having the initial state Λ. Under these conditions $\overline{\mathfrak{A}^{(0)}}$ is a 0-automaton in the alphabet $\{|, *\}$, which has the graph shown in Fig. 2. In other words, $\overline{\mathfrak{A}^{(0)}}$ is a 0-automaton in the alphabet $\{ |, * \}$, which has the initial state Λ and the following collection of substitution formulas:

$$\overline{\overline{\mathfrak{A}}}: \begin{cases} | \rightarrow | \\ * \rightarrow \Lambda \\ || \rightarrow | \\ *| \rightarrow \Lambda \end{cases}$$

Let us prove that this 1-automaton has infinitely many accessible states. For this purpose we shall show that

$$\mathfrak{A}^{(1)}: \Lambda \models_P P \qquad (P\Omega A). \tag{1}$$

The latter statement derives (by induction from the length of the word P) from the relationships

$$\mathfrak{A}^{(1)}: \Lambda \models_\Lambda \Lambda \tag{2}$$

and

$$\mathfrak{A}^{(1)}: P \models_| |P \qquad (P\Omega A). \tag{3}$$

The first of these relationships derives directly from the definition of a transition from state to state due to the action of an empty word. In order to substantiate the second, it is necessary to consider the functioning of the automaton $\overline{\mathfrak{A}}^{(0)}$. For the latter the following relationship holds:

$$\overline{\mathfrak{A}}^{(0)}: \Lambda \models_{|P*} \Lambda(|P) \qquad (P\Omega A), \tag{4}$$

$$\overline{\mathfrak{A}}^{(0)}: \Lambda \models_{*P*} \Lambda(P) \qquad (P\Omega A). \tag{5}$$

From this (3) and (1) derive.

Thus, due to the action of various words P_1 and P_2 ($P_1, P_2 \Omega A$) the automaton $\mathfrak{A}^{(1)}$ enters various states P_1 and P_2. Since there is an infinitely large number of words in the alphabet $A = \{ | \}$, there are also infinitely many accessible states of the automaton $\mathfrak{A}^{(1)}$.

§ 2. Examples of n-Automata

1. In this section we shall consider a number of examples of n-automata. Basically these will be n-automata which carry out the same transformations as do normal algorithms given in §4 of Chapter I in [5]. Almost all of these transformations may be realized by means of 0- and 1-automata.

In view of the unwieldiness of the formulations, we shall not present the constructions of the corresponding automata; it is assumed that where necessary the reader can carry them out himself. As an example, a description of one 1-automaton is given at the end of the section.

2. A 0-automaton $\mathfrak{A}^{(0)}_{\mathrm{A}, A}$ may be constructed in the alphabet A, which is applicable to any word in A and which attaches the word AΩA to this word on the left; i.e., it functions thus:

$$\mathfrak{A}^{(0)}_{\mathrm{A}, A}(P) \mathrel{\underline{\overline{\circ}}} AP \qquad (P\Omega\mathrm{A}).$$

3. An empty n-automaton (n = 0, 1, 2, ...) may be constructed in the alphabet A (i.e., an automaton which is not applicable to any word in A).

4. One may construct† a 0-automaton $\mathfrak{B}^{(0)}_{\mathrm{A}, A}$ in the alphabet A, which attaches the word A at the right to any word in A; i.e., it functions thus:

$$\mathfrak{B}^{(0)}_{\mathrm{A}, A}(P) \mathrel{\underline{\overline{\circ}}} PA \qquad (P\Omega\mathrm{A}).$$

5. One may construct a 0-automaton $\mathfrak{C}^{(0)}_{\mathrm{A}, \alpha}$ in the alphabet A, which manipulates any word P in the alphabet A into the word obtained from P by discarding all α $(\alpha \in \mathrm{A})$.

6. One may construct a cancelling 0-automaton $\mathfrak{C}^{(0)}_{\mathrm{A}}$ in the alphabet A, i.e., an automaton which operates thus:

$$\mathfrak{C}^{(0)}_{\mathrm{A}}(P) \mathrel{\underline{\overline{\circ}}} \Lambda \qquad (P\Omega\mathrm{A}).$$

7. A 0-automaton $\mathfrak{C}^{A}_{\mathrm{A}}$ in the alphabet A may be constructed which operates thus:

$$\mathfrak{C}^{A}_{\mathrm{A}}(P) \mathrel{\underline{\overline{\circ}}} A \qquad (P\Omega\mathrm{A},\ A\Omega\mathrm{A}).$$

8. A 0-automaton $\mathfrak{D}_{\mathrm{A}}$ in the alphabet $\mathrm{A} \cup \{|\}$ may be constructed which manipulates any word P in A into the length of this word:

$$\mathfrak{D}_{\mathrm{A}}(P) \mathrel{\underline{\overline{\circ}}} [P^{\partial} \qquad (P\Omega\mathrm{A}).$$

9. Assume that the letter α does not belong to the alphabet A. Truncating 0-automata $\mathfrak{I}_{\mathrm{A}, \alpha}$ and $\mathfrak{G}_{\mathrm{A}, \alpha}$ may be constructed in the alphabet $\mathrm{A} \cup \{\alpha\}$, i.e., automata which operate thus:

$$\mathfrak{I}_{\mathrm{A}, \alpha}(P\alpha Q) \mathrel{\underline{\overline{\circ}}} P \qquad (P,\ Q\Omega\mathrm{A}),$$
$$\mathfrak{G}_{\mathrm{A}, \alpha}(P\alpha Q) \mathrel{\underline{\overline{\circ}}} Q \qquad (P,\ Q\Omega\mathrm{A}).$$

10. Assume that the letters α, β do not belomg to the alphabet A. An excising 0-automaton $\mathfrak{K}_{\mathrm{A}, \alpha, \beta}$ may be constructed in the alphabet $\mathrm{A} \cup \{\alpha, \beta\}$, i.e., an automaton that operatues thus:

$$\mathfrak{K}_{\mathrm{A}, \alpha, \beta}(P\alpha Q\beta R) \mathrel{\underline{\overline{\circ}}} Q \qquad (P,\ Q,\ R\Omega\mathrm{A}).$$

11. A 0-automaton $\mathfrak{A}_{\mathrm{A}, A, B, C}$ may be constructed over the alphabet A, for which

$$\mathfrak{A}_{\mathrm{A}, A, B, C}(A) \mathrel{\underline{\overline{\circ}}} B \qquad (A\Omega\mathrm{A}),$$
$$\mathfrak{A}_{\mathrm{A}, A, B, C}(P) \mathrel{\underline{\overline{\circ}}} C \qquad (P\Omega\mathrm{A},\ P \not\equiv A).$$

† The role of $*$, which terminates the manipulated word, is evident in this example; if we were to agree to apply the word P, rather than the word $P*$, to the automaton input in calculating $\mathfrak{A}(P)$ it would be impossible to construct the indicated automaton.

12. A 1-automaton $\mathfrak{H}_A^{(1)}$ may be constructed which manipulates any word in A into the inversion† of this word:

$$\mathfrak{H}_A^{(1)}(P) \underline{\underline{\circ}} [P^{\cup} \qquad (P\Omega A).$$

13. A doubling 1-automaton $\mathfrak{F}_A$ may be constructed in the alphabet A, which operates thus:

$$\mathfrak{F}_A(P) \underline{\underline{\circ}} PP \qquad (P\Omega A).$$

14. Let us juxtapose each letter ξ of the alphabet Б with a specific word A_ξ in the alphabet A. Then, replacing each letter ξ by the word A_ξ in an arbitrary word P in the alphabet B, we obtain a certain word in the alphabet A – the result of replacing the letters ξ by the word A_ξ ($\xi \in$ B) in the word P.

If B = $\{\alpha_1, \alpha_2, ..., \alpha_n\}$, then the result of replacing the letters ξ by the word A_ξ ($\xi \in$ B) shall be denoted by the symbol

$$S_{B_1, B_2, \ldots, B_n}^{\alpha_1, \alpha_2, \ldots, \alpha_n} P \,|,$$

where $B_i \underline{\underline{\circ}} A_{\alpha_i}$ $(0 < i \leq n)$.

One may construct a 0-automaton $\mathfrak{K}_{A, B_1, \ldots, B_n}^{Б, \alpha_1, \ldots, \alpha_n}$ over Б which is such that

$$\mathfrak{K}(P) \underline{\underline{\circ}} S_{B_1, \ldots, B_n}^{\alpha_1, \ldots, \alpha_n} P \,|\, **).$$

The discarding of certain letters from a word is a particular case of the substitution of words for letters; this operation has already been partially considered in subsection **5**.

The discarding of certain letters from a word is a particular case of replacement of letters by words for which A = Б, $A_\xi \underline{\underline{\circ}} \Lambda$ for certain ξ and $A_\xi \underline{\underline{\circ}} \xi$ for the remaining ξ.

We shall apply the following terminology. Assume that the alphabet A is an expansion of the alphabet Б, i.e., Б $\subseteq$ A. The result of discarding the letters of the alphabet A$\setminus$Б from the word P in A shall be called the projection of the word P onto the alphabet Б and shall be denoted by $[P^{Б}$.

Assume A $= \{\alpha_1, \ldots, \alpha_n\}$, Б $= \{\alpha_1, \ldots, \alpha_k\}$, the condition $k \leq n$ being valid. Then, obviously,

$$[P^{Б} \underline{\underline{\circ}} S_{\alpha_1, \ldots, \alpha_k, \Lambda, \ldots, \Lambda}^{\alpha_1, \ldots, \alpha_k, \alpha_{k+1}, \ldots, \alpha_n} P \,|.$$

15. Assume A and Б are alphabets without common letters, the following equalities being valid:

$$A = \{\alpha_1, \alpha_2, \ldots, \alpha_m\}, \qquad Б = \{\beta_1, \beta_2, \ldots, \beta_n\}.$$

† The inversion of the word P is called the word Λ if $P \underline{\underline{\circ}} \Lambda$; if $P \not\underline{\underline{\circ}} \Lambda$ and $P \underline{\underline{\circ}} \xi_1\xi_2 \ldots \xi_k$ $(k > 0)$, then the inversion of the word is called the word $\xi_k \ldots \xi_2\xi_1$. The inversion of the word P is denoted by $[P^{\cup}$.

‡ Here and later on we shall drop the indices of $\mathfrak{K}$. for purposes of brevity.

For each word P in the union of these alphabets — $A \cup Б$ — one may construct both the projection onto the alphabet A — $[P^A$ and the projection onto the alphabets $Б — [P^Б$.

One may construct a 1-automaton $\mathfrak{L}_{A,Б}$, which manipulates each word P in $A \cup Б$ into the word $[P^A [P^Б$.

16. Let us now consider a series of automata which transform natural numbers or systems of natural numbers into natural numbers. A natural number is a word in the alphabet $Ч = \{|\}$. The system of natural numbers is a word in the alphabet $C = \{|, \alpha\}$.

17. One may construct a 0-automaton $\mathfrak{A}_1$, which manipulates any natural number N into the remainder from the division of this number by 5.

18. One may now construct a 0-automaton $\mathfrak{A}_2$, which manipulates any number N into the integer part obtained by the division of this number N by 5 (this integer part is denoted by [N/5]).

19. One may construct a 0-automaton $\mathfrak{A}_3$, which manipulates each natural number N into a pair of natural numbers $K\alpha L$, where $L = N - 5[N/5]$ (i.e., the remainder of the division of N by 5), while $K = [N/5]$.

20. A 1-automaton $\mathfrak{A}_4$, may be constructed which manipulates a pair of natural numbers $M\alpha N$ into the product $M \cdot N$ of these numbers.

21. One may construct a 1-automaton $\mathfrak{A}_5$, which manipulates each natural number N into its square; i.e., the automaton is such that

$$\mathfrak{A}_5(N) \underset{\circ}{=} N \cdot N.$$

22. One may construct a 1-automaton $\mathfrak{A}_{6,N}$ which manipulates the natural number M into the number N^M (N is a natural number, $N > 0$).

23. One may construct a 1-automaton $\mathfrak{A}_7$, that manipulates any pair of natural numbers $M\alpha N$ into a number equal to the modulus of their difference; i.e., this is an automaton for which

$$\mathfrak{A}_7(M\alpha N) \underset{\circ}{=} |M - N|.$$

24. Let us now consider several automata which perform more special functions.

One may construct a 1-automaton $\mathfrak{A}_8$, which manipulates each natural number N into a number equal to the distance to the largest complete square that does not exceed N.†

25. One may construct a 1-automaton $\mathfrak{A}_9$, that manipulates a natural number to zero if it is the complete square of a certain natural number, while it manipulates it to 1 if this number is not a complete square.‡

26. One may construct a 1-automaton $\mathfrak{A}_{10}$, which manipulates a pair of natural numbers $M\alpha N$ into their arithmetic (or allowed) difference.

† The corresponding primitively recursive function is denoted in [7] by quadres (n) and expressed as:

$$\text{quadres}(n) = n \dot{-} [\sqrt{n}]^2.$$

‡ The corresponding primitively recursive function is denoted in [7] by quad (n) and may be defined thus:

$$\text{quad}(n) = \text{sg}(\text{quadres}(n)).$$

27. One may construct a 1-automaton $\mathfrak{A}_{11}$, which manipulates a pair of natural numbers $M\alpha N$ into the remainder from the division of M by N (usually the number obtained is denoted thus: rm(M, N)).

28. One may construct a 1-automaton $\mathfrak{A}_{12}$, which manipulates the pair of natural numbers $M\alpha N$ into the integer part of the quotient from the division of M by N. The number obtained is usually denoted thus: [M/N].

29. As an example, let us consider the construction of the 1-automaton $\mathfrak{A}_8$ from **24.**

$\mathfrak{A}_8$ is constructed thus: $\mathfrak{A}_8 = \mathfrak{A}_{8,2} \circ \mathfrak{A}_{8,1}$.

The 1-automaton is $\mathfrak{A}_{8,1}$. The initial state is Λ.

The 0-automaton is $\overline{\mathfrak{A}}_{8,1}$. We assume that $\overline{\mathfrak{A}}_{8,1} = \overline{\mathfrak{A}}_{8,1,2} \circ \overline{\mathfrak{A}}_{8,1,1}$.

Now we stipulate the 0-automaton $\overline{\mathfrak{A}}_{8,1,1}$. The initial state is $\beta\beta$.

$$\overline{\overline{\mathfrak{A}}}_{8,1,1}: \begin{cases} |\beta\beta \to \Lambda \\ *\beta\beta \to \alpha\alpha \\ | \to \alpha \\ * \to |||0 \\ |\alpha\alpha \to \alpha\alpha \\ 0\alpha\alpha \to \alpha\alpha\alpha \\ *\alpha\alpha \to \Lambda \\ |\alpha \to | \\ 0\alpha \to ||| \\ |\alpha\alpha\alpha \to \gamma \\ *\alpha\alpha\alpha \to \Lambda \\ || \to | \\ 0| \to 0| \\ *| \to \Lambda \\ |||| \to \beta \\ |\gamma \to \gamma \\ *\gamma \to \Lambda \\ |0| \to | \\ *0| \to \Lambda \\ |\beta \to \beta \\ *\beta \to 0 \end{cases}$$

The graph of this automaton is depicted in Fig. 3. As the 0-automaton $\overline{\mathfrak{A}}_{8,1,2}$ we choose the 0-automaton

$$\mathfrak{R}^{\{|,0,\alpha,\beta,\gamma\},|,0,\alpha,\beta,\gamma}_{\{|,0,\gamma\},|,0,\Lambda,|,\gamma}.$$

Let us consider the operation of the 1-automaton $\mathfrak{A}_{8,1}$ on the number N. Assume N = 0. Let us calculate $\mathfrak{A}_{8,1}(0)$:

$$\mathfrak{A}_{8,1}: \Lambda \vdash_* \Lambda(\Lambda),$$

and since

$$\overline{\mathfrak{A}}_{8,1,1}(*) \mathrel{\underset{\circ}{=}} \alpha\alpha,$$

it follows also that

$$\overline{\mathfrak{A}}_{8,1,2}(\alpha\alpha) \mathrel{\underset{\circ}{=}} \Lambda.$$

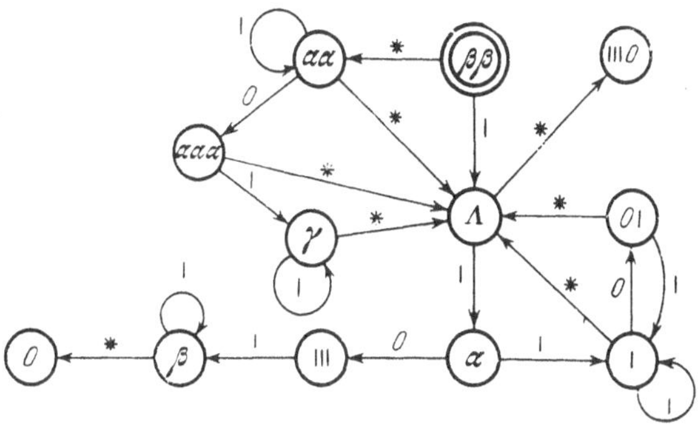

Fig. 3

Thus,

$$\mathfrak{A}_{8,1}(0) \stackrel{\circ}{=} \Lambda.$$

Then let us consider the following relationships:

$$\mathfrak{A}_{8,1}\colon \Lambda \vdash_{|} |||0 \vdash_{*} \Lambda,$$

since

$$\overline{\mathfrak{A}}_{8,1,1}\colon \beta\beta \vdash_{|} \Lambda(\Lambda) \vdash_{*} |||0\,(|||0)$$

and

$$\overline{\mathfrak{A}}_{8,1,2}(|||0) \stackrel{\circ}{=} |||0.$$

From here on we reason by induction. However, we shall first introduce a certain notation.

Assume N is a natural number. $\underline{N}$ shall denote the nearest natural number which is not larger than N and is a complete square.†

$\overline{N}$ shall denote the nearest natural number which is not smaller than N and is a complete square.‡

The inductive assumption resides in the following: if $N \geq 1$, then

$$\mathfrak{A}_{8,1}\colon \Lambda \vDash_{N} (([\sqrt{\underline{N}}]+1)^2 \dot{-} N)\,0\,(N \dot{-} \underline{N}) \vdash_{*} \gamma^{N \dot{-} \underline{N}}.$$

Let us determine the state in which the automaton $\mathfrak{A}_{8,1}$ arrives due to the action of the number N + 1.

† In the notation of [7] we have

$$\underline{N} = [\sqrt{N}]^2.$$

‡ The corresponding primitively recursive function has the following definition: $\overline{N} = ([\sqrt{N}] + \mathrm{sg}\,(N - [\sqrt{N}]^2))^2$, or

$$\overline{N} = \begin{cases} N, & \text{if } N = \underline{N}, \\ ([\sqrt{\underline{N}}]+1)^2, & \text{if } N \neq \underline{N}. \end{cases}$$

For this purpose it is necessary to calculate:

$$\overline{\mathfrak{A}}_{8,1,2}\ (\overline{\mathfrak{A}}_{8,1,1}((1+(([\sqrt{N}]+1)^2 \dot{-} N))\,0\,(N \dot{-} \underline{N}))).$$

Let us begin by considering the case in which $N + 1 \neq \bar{N}$. In this case, as is easily seen,

$$([\sqrt{N}]+1)^2 \dot{-} N > 1.$$

Then the word $1+(([\sqrt{N}]+1)^2 \dot{-} N)$ begins with the word |||. Carrying out the calculations, we obtain

$$\overline{\mathfrak{A}}_{8,1,1}\colon \beta\beta \vdash_{|} \Lambda(\Lambda) \vdash_{|} \alpha(\alpha) \vdash_{|} |(|) \models_{(([\sqrt{N}]+1)^2 \dot{-} N) \dot{-} 2} |((((([\sqrt{N}]+1)^2 \dot{-} N) \dot{-} 2)) \vdash_0 0|(0|) \models$$
$$\models_{N \dot{-} \underline{N}} |(N \dot{-} \underline{N}) \vdash_* \Lambda(\Lambda).$$

Thus,

$$\overline{\mathfrak{A}}_{8,1,1}\ (((1+([\sqrt{N}]+1)^2) \dot{-} N)\,0\,(N \dot{-} N)) \ \overline{\circ}\ \alpha(((([\sqrt{N}]+1)^2 \dot{-} N) \dot{-} 1)\,0\,((N \dot{-} \underline{N})+1). \quad (1)$$

Since we are considering the case $N + 1 \neq \bar{N}$, it follows that $\underline{N+1} = \underline{N}$.

Therefore from (1) we have

$$\overline{\mathfrak{A}}_{8,1,1}\ (((1+([\sqrt{N}]+1)^2) \dot{-} N)\,0\,(N \dot{-} \underline{N}))\ \overline{\circ}\ \alpha(([\sqrt{N+1}]+1)^2 \dot{-} (N+1))\,0\,((N+1) \dot{-} \underline{N+1})).$$

Now applying the 0-automaton $\overline{\mathfrak{A}}_{8,1,2}$, we obtain

$$\mathfrak{A}_{8,1}\colon (([\sqrt{N}]+1)^2 \dot{-} N)\,0\,(N \dot{-} \underline{N}) \vdash_{|} (([\sqrt{N+1}]+1)^2 \dot{-} (N+1))\,0\,((N+1) \dot{-} \underline{N+1}).$$

Moreover, the calculations yield

$$\overline{\mathfrak{A}}_{8,1,1}\colon \beta\beta \vdash_* \alpha\alpha(\alpha\alpha) \models_{([\sqrt{N+1}]+1)^2 \dot{-} (N+1)} \alpha\alpha((\alpha\alpha)^{([\sqrt{N+1}]+1)^2 \dot{-} (N+1)}) \vdash$$
$$\vdash_0 \alpha\alpha\alpha(\alpha\alpha\alpha) \models_{(N+1) \dot{-} \underline{N+1}} \gamma(\gamma^{(N+1) \dot{-} \underline{N+1}}) \vdash_* \Lambda(\Lambda).$$

Thus,

$$\overline{\mathfrak{A}}_{8,1,1}(*(([\sqrt{N+1}]+1)^2 \dot{-} (N+1))\,0\,((N+1) \dot{-} \underline{N+1}))\ \overline{\circ}\ \alpha\alpha(\alpha\alpha)^{([\sqrt{N+1}]+1)^2 \dot{-} (N+1)}\alpha\alpha\alpha\gamma^{(N+1) \dot{-} \underline{N+1}}.$$

Now applying the automaton $\overline{\mathfrak{A}}_{8,1,2}$, we obtain

$$\mathfrak{A}_{8,1}\colon (([\sqrt{N+1}]+1)^2 \dot{-} (N+1))\,0\,((N+1) \dot{-} \underline{N+1}) \vdash_* \gamma^{(N+1) \dot{-} \underline{N+1}}.$$

This completes the induction step for the case in which $N + 1 \neq \bar{N}$.

Let us now consider the case $N + 1 = \bar{N}$. In this case, as can easily be seen,

$$([\sqrt{N}]+1)^2 \dot{-} N = 1. \quad (2)$$

Then the word $(1+([\sqrt{N}]+1)^2) \dot{-} N$ begins with the word ||. Carrying out the calculations, we obtain

$$\overline{\mathfrak{A}}_{8,1,1}\colon \beta\beta \vdash_{|} \Lambda(\Lambda) \vdash_{|} \alpha(\alpha) \vdash_0 |||(|||) \models_{N \dot{-} \underline{N}} \beta(\beta^{N \dot{-} \underline{N}}) \vdash_* 0(0).$$

Thus,

$$\overline{\mathfrak{A}}_{8,1,1}\ (((1+([\sqrt{\underline{N}}]+1)^2) \dot{-} N)\,0\,(N \dot{-} \underline{N})) \,\underline{\underline{\circ}}\, \alpha\,|||\,\beta^{N \dot{-} \underline{N}}0.$$

It can easily be seen that by virtue of (2) we obtain the following result by applying the automaton $\overline{\mathfrak{A}}_{8,1,2}$:

$$\mathfrak{A}_{8,1}\colon (([\sqrt{\underline{N}}]+1)^2 \dot{-} N)\,0\,(N \dot{-} \underline{N}) \vdash_{|} (([\sqrt{\underline{N+1}}]+1)^2 \dot{-} (N+1))\,0.$$

Further,

$$\overline{\mathfrak{A}}_{8,1,1}\colon \beta\beta \vdash_{*} \alpha\alpha\,(\alpha\alpha) \models_{K} \alpha\alpha\,((\alpha\alpha)^{K}) \vdash_{0} \alpha\alpha\alpha\,(\alpha\alpha\alpha) \vdash_{*} \Lambda\,(\Lambda),$$

where $K=([\sqrt{\underline{N+1}}]+1)^2 \dot{-} (\underline{N}+1)$.

Thus, applying the automaton $\overline{\mathfrak{A}}_{8,1,2}$, we obtain

$$\mathfrak{A}_{8,1}\colon (([\sqrt{\underline{N+1}}]+1)^2 \dot{-} (N+1))\,0 \vdash_{*} \Lambda\,(\Lambda).$$

The inductive proof has been completed. Let us now choose the automaton $\mathfrak{R}^{\{0,|,\gamma\},0,|,\gamma}_{\{|\},\Lambda,\Lambda,|}$ as the automaton $\mathfrak{A}_{8,2}$; we obtain the required equality:

$$\mathfrak{A}_{8}(N) \,\underline{\underline{\circ}}\, N \dot{-} \underline{N}.$$

CHAPTER II

The Construction of n-Automata

In this chapter we consider a number of constructions that allow new n-automata to be constructed on the basis of given m-automata (m ≤ n).

§ 1. The Extension of n-Automata

1. Assume $\mathfrak{A}$ is an m-automaton in the alphabet A, while Б is an expansion of the alphabet A. We shall say that an n-automaton $\mathfrak{B}$ in the alphabet Б is an *extension* of the m-automaton $\mathfrak{A}$ on the alphabet Б, if

$$\mathfrak{A}(P) \simeq \mathfrak{B}(P) \qquad (P\Omega\text{A}).$$

Just as in the theory of normal algorithms, we shall consider certain special forms of extensions of m-automata.

2. Assume $\mathfrak{A}$ is an m-automaton in the alphabet A while Б is an expansion of the alphabet A. Let us stipulate the m-automaton $\mathfrak{B}$ in Б, having taken the stipulation of the m-automaton $\mathfrak{A}$, as its stipulation; this, of course, is allowable, since each word in A is also a word in Б at the same time.

Expansion of the alphabet A in no way affects the operation of the automaton on a word in A. Therefore, the constructed automaton $\mathfrak{B}$ is the expansion of the automaton $\mathfrak{A}$ to the alphabet Б.

3. The extension of the m-automaton $\mathfrak{A}$ to the alphabet Б, which was described above, shall be called the *direct extension* of this m-automaton to the alphabet Б.

It is obvious that if $\mathfrak{B}$ is the direct extension of the automaton $\mathfrak{A}$ to the alphabet Б, then $\mathfrak{B}$ may not be applied to a word P for which $P\Omega Б$, but $P\Omega A$ is invalid. Thus, the automaton $\mathfrak{b}$ in the alphabet Б is such that it is the extension of the automaton $\mathfrak{A}$ to the alphabet Б, while at the same time it is applicable only to words in the alphabet A (and thereby only to those words to which $\mathfrak{A}$ is applicable).

§ 2. The Theorem of Rank Elevation

1. In this section we shall show that, roughly speaking, all transformations of words in the alphabet A which can be performed by means of automata of rank n can also be performed by means of automata of rank m (n ≤ m), i.e., we shall show that in a certain sense m-automata are no weaker than n-automata.

1.1. The Theorem of Rank Elevation. *For each automaton $\mathfrak{A}^{(n)}$ in the alphabet A one may construct an automaton $\mathfrak{B}^{(n+1)}$ in the same alphabet, which is such that*

$$\mathfrak{A}^{(n)}(P) \simeq \mathfrak{B}^{(n+1)}(P) \qquad (P\Omega A).$$

2. Proof. For each n ≥ 0 an identical n-automaton $\mathfrak{A}_A^{(n)}$ will be constructed in the alphabet A, for which

$$\mathfrak{A}_A^{(n)}(P) \mathrel{\underset{\circ}{=}} P \qquad (P\Omega A).$$

Further, the automaton $\mathfrak{B}^{(n+1)}$ that we are required to construct in the alphabet A for the proof of Theorem 1.1 shall be defined thus:

$$\mathfrak{B}^{(n+1)} = \mathfrak{A}_A^{(n+1)} \circ \mathfrak{A}^{(n)}.$$

Thus, the proof has been reduced to the construction of the automaton $\mathfrak{A}_A^{(n)}$ (for n = 0, 1, 2, ...). Assume that the initial state of the automaton $\mathfrak{A}_A^{(n)}$ is Λ. We shall construct the automaton $\mathfrak{A}_A^{(n)}$ in such a way that it operates on the word $P \mathrel{\underset{\circ}{=}} \xi_1\xi_2 \ldots \xi_k$ $(k \geqslant 0)$ in the following manner:

$$\mathfrak{A}_A^{(n)}\colon \Lambda \vdash_{\xi_1} \xi_1(\xi_1) \vdash_{\xi_2} \xi_2(\xi_2) \vdash_{\xi_3} \cdots \vdash_{\xi_k} \xi_k(\xi_k) \vdash_{*} \Lambda(\Lambda).$$

Thus, the following relationships must hold:

$$\overline{\mathfrak{A}}_A(\xi\eta) \mathrel{\underset{\circ}{=}} \xi \qquad (\xi,\ \eta \in A), \tag{1}$$

$$\overline{\mathfrak{A}}_A(\xi) \mathrel{\underset{\circ}{=}} \xi \qquad (\xi \in A), \tag{2}$$

$$\overline{\mathfrak{A}}_A(*\xi) \mathrel{\underset{\circ}{=}} \Lambda \qquad (\xi \in A), \tag{3}$$

$$\overline{\mathfrak{A}}_A(*) \mathrel{\underset{\circ}{=}} \Lambda. \tag{4}$$

Let us denote the automaton satisfying the conditions (1)-(4) by $\mathfrak{B}$. It is obvious that if we construct the (n − 1)-automaton $\mathfrak{B}^{(n-1)}$ (n ≥ 1), then one can determine the $\mathfrak{A}_A^{(n)}$ by choosing $\overline{\mathfrak{A}}_A$ to be the automaton $\mathfrak{B}^{(n-1)}$; the initial state of the automaton $\mathfrak{A}_A^{(n)}$ is Λ.

We inductively construct the automata $\mathfrak{A}_A^{(n)}$ and $\mathfrak{B}^{(n)}$.

For n = 0 the automaton $\mathfrak{A}_A^{(0)}$ was constructed in [Chap. I, §2, subsection 2]; this is the automaton $\mathfrak{A}_{A,A}^{(0)}$. The automaton $\mathfrak{B}^{(0)}$ is stipulated thus: the initial state is $\alpha\alpha$. (It is assumed that $\alpha \in A$.)

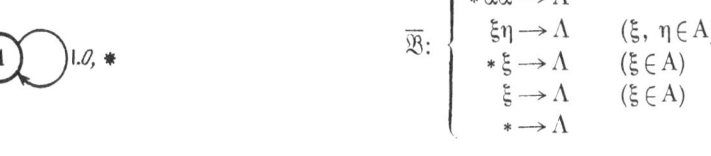

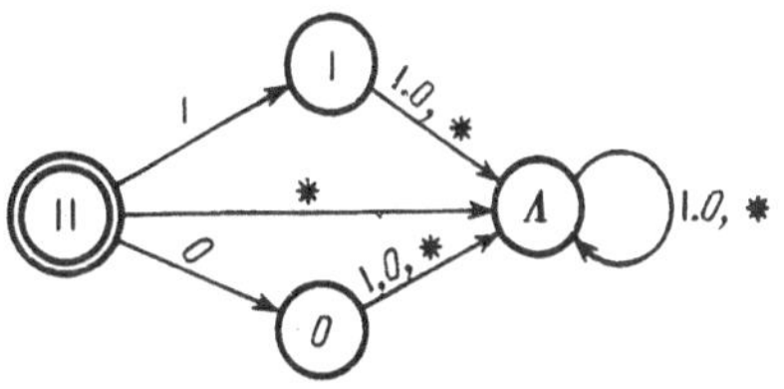

Fig. 4

$$\overline{\mathfrak{B}}:\left\{\begin{array}{ll} \xi\alpha\alpha \to \xi & (\xi \in A)^{\dagger} \\ *\alpha\alpha \to \Lambda & \\ \xi\eta \to \Lambda & (\xi,\ \eta \in A). \\ *\xi \to \Lambda & (\xi \in A) \\ \xi \to \Lambda & (\xi \in A) \\ * \to \Lambda & \end{array}\right.$$

The graph of this automaton for A = {1, 0} is displayed in Fig. 4. As α we have taken |.

We further assume that the automata $\mathfrak{A}_A^{(n)}$ and $\mathfrak{B}^{(n)}$ have already been constructed. Let us construct the automata $\mathfrak{A}_A^{(n+1)}$ and $\mathfrak{B}^{(n+1)}$. As has already been indicated above, the automaton $\mathfrak{A}_A^{(n+1)}$ may be constructed thus: its initial state is Λ. As $\overline{\mathfrak{A}}_A$ we choose the automaton $\mathfrak{B}^{(n)}$.

If we now construct $\mathfrak{A}_A^{(n+1)}$, it follows that $\mathfrak{B}^{(n+1)}$ may be constructed thus:

$$\mathfrak{B}^{(n+1)} = \mathfrak{A}_A^{(n+1)} \circ \mathfrak{B}^{(0)}.$$

Since $\mathfrak{A}_A^{(n+1)}$ is an identical automaton, the same relationships (1)-(4) are fulfilled for the automaton $\mathfrak{B}^{(n+1)}$ as are fulfilled for the automaton $\mathfrak{B}^{(0)}$. Thus, the construction of an identical n-automaton $\mathfrak{A}_A^{(n)}$ has been completed for any n (and thereby the proof of Theorem 1.1 has been completed).

§ 3. The Branching of Automata

1. Sometimes it is necessary to construct instructions in the following form: "the automaton $\mathfrak{A}$ or the automaton $\mathfrak{B}$ is to be applied to the original word, depending on whether or not the original word begins with a given letter."

Thus, a certain new algorithm is stipulated which is a combination of the automata $\mathfrak{A}$ and $\mathfrak{B}$. The question naturally arises as to whether the calculations instructed to be performed by the algorithm described may be performed on some automaton. The answer to this question is given by the following theorem.

1.1. The Branching Theorem. *Assume $\mathfrak{A}$ and $\mathfrak{B}$ are automata having the ranks n_a and n_b, respectively, in the alphabet A. Assume α is a certain letter of the alphabet A. Then one can construct an automaton $\mathfrak{C}$ of rank $n = \max(n_a, n_b)$ over the alphabet A such that*

$$\mathfrak{C}(\Lambda) \simeq \mathfrak{A}(\Lambda), \tag{1}$$

$$\mathfrak{C}(\xi P) \simeq \left\{\begin{array}{lll} \mathfrak{A}(\xi P), & \text{if} \quad \xi \,\bar{\bar{\circ}}\, \alpha & (P\Omega A), \\ \mathfrak{B}(\xi P), & \text{if} \quad \xi \not\equiv \alpha & (P\Omega A). \end{array}\right. \tag{2}$$

Proof. The proof is carried out by induction with respect to n. The basis is n = 0. Let us define the automaton $\mathfrak{C}$ as follows:

$$\mathfrak{C} = \mathfrak{C}_3 \circ \mathfrak{C}_2 \circ \mathfrak{C}_1. \tag{3}$$

† Following [5], we use abridged notation for the substitution formulas. The letters ξ, η are arbitrary letters of the alphabet A, as is stated in the condition written to the right of the formula. These letters take the values of any letters of the alphabet A. Thus, for example, the first row replaces only that number of substitution formulas which is equal to the number of letters in the alphabet A.

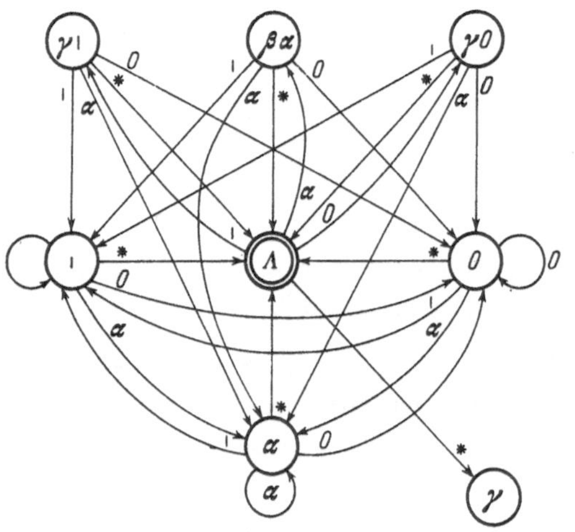

Fig. 5

Under these conditions the automaton $\mathfrak{C}_1$ operates thus in the alphabet $A \cup \{\beta, \gamma\}$ $(\beta \notin A,\ \gamma \notin A)$: its initial state is Λ.

$$\overline{\mathfrak{C}}_1:\ \left\{\begin{array}{ll} * \to \gamma & \\ \xi \to \gamma\xi & (\xi \in A \setminus \{\alpha\}) \\ \alpha \to \beta\alpha & \\ \xi\beta\alpha \to \xi & (\xi \in A) \\ \xi\gamma\eta \to \xi & (\xi \in A,\ \eta \in A \setminus \{\alpha\}) \\ *\beta\alpha \to \Lambda & \\ *\gamma\xi \to \Lambda & (\xi \in A \setminus \{\alpha\}) \\ \xi\eta \to \xi & (\xi,\ \eta \in A) \\ *\xi \to \Lambda & (\xi \in A) \end{array}\right.$$

The graph of this automaton (for the alphabet $A = \{0,\ |,\ \alpha\}$) is depicted in Fig. 5. For this automaton the following holds:

$$\mathfrak{C}_1(\Lambda) \eqcirc \gamma,$$

$$\mathfrak{C}_1(\xi P) \eqcirc \left\{\begin{array}{lll} \beta\xi P, & \text{if } \xi \eqcirc \alpha & (P\Omega A), \\ \gamma\xi P, & \text{if } \xi \not\eqcirc \alpha & (P\Omega A). \end{array}\right.$$

Let us construct the automaton $\mathfrak{C}_2$ in the alphabet $A \cup \overline{A} \cup \{\beta, \gamma, \delta\}$ $(\delta \notin A)$. The initial state is Λ.

$$\overline{\mathfrak{C}}_2:\ \left\{\begin{array}{ll} \beta \to \beta & \\ \gamma \to \gamma & \\ \xi\gamma \to Q_\xi & (\xi \in A \cup \{*\}) \\ \xi\beta \to R_\xi & (\xi \in A) \\ \overline{\mathfrak{A}}' & \\ \overline{\mathfrak{B}} & \end{array}\right.$$

Let us clarify how the words Q_ξ and R_ξ are chosen. Assume that the initial state of the automaton $\mathfrak{A}$ is S, while that of the automaton $\mathfrak{B}$ is T; in $\overline{\mathfrak{B}}$ (the collection of substitution formulas of the automaton $\mathfrak{B}$) we find the formula of the form $\xi T \to A$, where $\xi \in A \cup \{*\}$. If there is no such formula, the substitution formula for the corresponding ξ is also absent in $\overline{\mathfrak{C}}_2$. If there is such a formula, we assume that $Q_\xi \eqcirc A$. The words R_ξ are defined somewhat differently from the procedure described above. In $\overline{\mathfrak{A}}$ we seek the formula of the form $\xi S \to B$ $(\xi \in A)$. If $B \eqcirc \Lambda$, we assume that $R_\xi \eqcirc \delta$; if $B \not\eqcirc \Lambda$, then we assume that $R_\xi \eqcirc \overline{B}$, where B is the twin† of the word B.

Let us now clarify what $\overline{\mathfrak{A}}'$ is. $\overline{\mathfrak{A}}'$ is obtained from the table $\overline{\mathfrak{A}}$ as follows. Each formula of the form $\xi P \to Q$ from the table $\overline{\mathfrak{A}}$ is replaced by the formula $\xi P' \to Q'$, where the operation " ' " is defined as follows:

† The twin of a word is defined exactly as in [5]. Let us give the necessary definitions. Each letter ξ of the alphabet A is juxtaposed with a new letter – the twin of the letter ξ (denoted by $\overline{\xi}$), different letters of A being juxtaposed with different new letters. The twins of letters from A comprise an alphabet of twins $\overline{A}$ containing the same number of letters as the alphabet A but having no letters in common with A. The twin of a word in the alphabet A is defined obviously.

$$A' \mathrel{\underline{\overline{\circ}}} \begin{cases} \delta, & \text{if} \quad A \mathrel{\underline{\overline{\circ}}} \Lambda, \\ \bar{A} \text{ (the twin of the word } A), & \text{if} \quad A \mathrel{\underline{\overline{\circ}}} \Lambda. \end{cases}$$

Finally, $\mathfrak{C}_2$ operates as follows:

$$\mathfrak{C}_2(\mathfrak{C}_1(\Lambda)) \simeq \gamma\mathfrak{A}(\Lambda)$$

$$\mathfrak{C}_2(\mathfrak{C}_1(\xi P)) \simeq \begin{cases} \beta\overline{\mathfrak{A}(\xi P)}, & \text{if} \quad \xi \mathrel{\underline{\overline{\circ}}} \alpha \quad (P\Omega A), \\ \gamma\mathfrak{B}(\xi P), & \text{if} \quad \xi \not\mathrel{\underline{\overline{\circ}}} \alpha \quad (P\Omega A). \end{cases}$$

And, finally, there is the automaton $\mathfrak{C}_3$. As this automaton we shall take

$$\mathfrak{K}_{A,}^{A \cup \bar{A} \cup \{\beta, \gamma, \delta\},} {}^{\alpha_1, \ldots, \alpha_m, \bar{\alpha}_1, \ldots, \bar{\alpha}_m, \beta, \gamma, \delta,}_{\alpha_1, \ldots, \alpha_m, \alpha_1, \ldots, \alpha_m, \Lambda, \Lambda, \Lambda}$$

(let us assume that $A = \{\alpha_1, \ldots, \alpha_m\}$).

This ends consideration of the basis.

The Induction Step. Let us assume that the theorem is valid for automata having a rank not exceeding n. Let us show that it is also valid for automata of rank (n + 1).

We shall construct the automaton $\mathfrak{C}$ in the form

$$\mathfrak{C} = \mathfrak{D}_3 \circ \mathfrak{D}_2 \circ \mathfrak{D}_1.$$

As the automaton $\mathfrak{D}_1$ let us take the 0-automaton $\mathfrak{C}_1$ from the basis. Then let us define the (n + 1)-automaton $\mathfrak{D}_2$ over the alphabet $A \cup \{\beta, \gamma\}$. The initial state is Λ. Based on the inductive assumption, let us construct the n-automaton $\overline{\mathfrak{D}}_2$:

$$\overline{\mathfrak{D}}_2(\xi P) \simeq \begin{cases} \beta S, & \text{if} \quad \xi \mathrel{\underline{\overline{\circ}}} \beta, \\ \gamma T, & \text{if} \quad \xi \mathrel{\underline{\overline{\circ}}} \gamma, \\ \mathfrak{E}(\xi P), & \text{if} \quad \xi \not\mathrel{\underline{\overline{\circ}}} \beta \text{ and } \xi \not\mathrel{\underline{\overline{\circ}}} \gamma; \end{cases}$$

here S, T are the initial states of the automata $\mathfrak{A}$ and $\mathfrak{B}$, respectively.

The n-automaton $\mathfrak{E}$ is defined as:

$$\mathfrak{E} = \mathfrak{E}_2 \circ \mathfrak{E}_1.$$

Under these conditions the 0-automaton $\mathfrak{E}_1$ operates thus:

$$\mathfrak{E}_1(\xi\eta Q) \mathrel{\underline{\overline{\circ}}} \eta\xi Q \qquad (\xi\eta Q \Omega A \cup \{\beta, \gamma, *\}).$$

We omit the details of the construction of this automaton.

Once again, we define the n-automaton $\mathfrak{E}_2$: on the basis of the inductive proposition:

$$\mathfrak{E}_2(\xi P) \simeq \begin{cases} \beta\mathfrak{K}^{A,\, \alpha_1, \ldots, \alpha_m}_{\bar{A},\, \bar{\alpha}_1, \ldots, \bar{\alpha}_m} (\overline{\mathfrak{K}} (\mathfrak{K}_{A,}^{A \cup \bar{A} \cup \{\beta\},} {}^{\alpha_1, \ldots, \alpha_m, \bar{\alpha}_1, \ldots, \bar{\alpha}_m, \beta}_{\alpha_1, \ldots, \alpha_m, \alpha_1, \ldots, \alpha_m, \Lambda} (\xi P))), & \text{if} \quad \xi \mathrel{\underline{\overline{\circ}}} \beta, \\ \gamma\overline{\mathfrak{B}}(\mathfrak{K}_{A,}^{A \cup \{\gamma\},} {}^{\alpha_1, \ldots, \alpha_m, \gamma}_{\alpha_1, \ldots, \alpha_m, \Lambda} (\xi P)), & \text{if} \quad \xi \mathrel{\underline{\overline{\circ}}} \gamma. \end{cases}$$

Let us finally construct the last automaton $\mathfrak{D}_3$, in the capacity of which we choose the 0-automaton

$$\mathfrak{K}_{A,}^{A \cup \bar{A} \cup \{\beta, \gamma\},} {}^{\alpha_1, \ldots, \alpha_m, \bar{\alpha}_1, \ldots, \bar{\alpha}_m, \beta, \gamma}_{\alpha_1, \ldots, \alpha_m, \alpha_1, \ldots, \alpha_m, \Lambda, \Lambda}.$$

As is easily seen, the (n + 1)-automaton $\mathfrak{C}$ constructed in this way satisfies the conditions of the theorem.

This completes the proof of the branching theorem.

§ 4. A Union of n-Automata

1. For many constructions it is necessary jointly to consider the results of the operation of two or several automata on the same initial data. In such cases the construction of a system of all of these results is useful, since this system may itself serve as the initial condition for the operation of some other automaton.

In general, if the automata $\mathfrak{A}_1, \mathfrak{A}_2, \ldots, \mathfrak{A}_k$ are given in the alphabet A which does not contain the letter α, one may consider the algorithm $\mathfrak{A}$, satisfying the condition

$$\mathfrak{A}(P) \simeq \mathfrak{A}_1(P)\,\alpha\,\mathfrak{A}_2(P)\,\alpha \ldots \alpha\,\mathfrak{A}_k(P).$$

We shall call this algorithm the "union of the given automata $\mathfrak{A}_1, \mathfrak{A}_2, \ldots, \mathfrak{A}_k$».

It is natural to ask whether a union of automata may be calculated on a certain automaton. The answer to this question is given by the following theorem which constitutes the basic content of the given section.

1.1. The Union Theorem. Assume $\mathfrak{A}_1, \mathfrak{A}_2, \ldots, \mathfrak{A}_k$ $(k \geq 1)$ are automata in the alphabet A which have the ranks $n_1, n_2, \ldots, n_k$, respectively. Then an automaton $\mathfrak{A}$ over $A \cup \{\alpha\}$ of rank n may be constructed, where $n = \max(n_1, n_2, \ldots, n_k, 1)$, such that

$$\mathfrak{A}(P) \simeq \mathfrak{A}_1(P)\,\alpha\,\mathfrak{A}_2(P)\,\alpha \ldots \alpha\,\mathfrak{A}_k(P) \qquad (\alpha \notin A,\ P\Omega A).$$

2. We begin with the proof of the following fact.

2.1. Regardless of the automata $\mathfrak{A}$ and $\mathfrak{B}$ in the alphabet A which have the ranks n_1 and n_2, respectively, one can construct an automaton $\mathfrak{C}$ over A of rank n, where $n = \max(n_1, n_2, 1)$, such that

$$\mathfrak{C}(P) \simeq \mathfrak{A}(P)\,\mathfrak{B}(P) \qquad (P\Omega A).$$

In the beginning let us construct a 1-automaton which operates as:

$$\mathfrak{C}_1(P) \mathrel{\overline{\underline{\circ}}} P\alpha\bar{P},$$

where $\bar{P}$ is the twin of the word P, while $\alpha \notin A \cup \bar{A}$. It is not difficult to construct such an automaton. Then we define the automaton $\mathfrak{C}$ thus:

$$\mathfrak{C} = \mathfrak{C}_3 \circ \mathfrak{C}_2 \circ \mathfrak{C}_1.$$

$\mathfrak{C}_2$ is stipulated as follows. The initial state is S_1, where S_1 is the initial state of the automaton $\mathfrak{A}$.

$$\bar{\mathfrak{C}}_2(\xi P) \simeq \begin{cases} \bar{\mathfrak{A}}(\xi P), & \text{if } \xi \in A, \\ \bar{S}_1\bar{\mathfrak{A}}(*P), & \text{if } \xi \mathrel{\overline{\underline{\circ}}} \alpha, \\ \bar{\mathfrak{B}}(\mathfrak{R}^{A\cup\bar{A},\ \alpha_1, \ldots, \alpha_m,\ \bar{\alpha}_1, \ldots, \bar{\alpha}_m}_{A,\ \Lambda, \ldots, \Lambda,\ \alpha_1, \ldots, \alpha_m}(\xi P)), & \text{if } \xi \in \bar{A} \cup \{*\}. \end{cases}$$

Here $\bar{S}_1$ is the twin of the initial state of the automaton $\mathfrak{A}$. In constructing $\bar{\mathfrak{C}}_2$, the branching theorem was used. As $\mathfrak{C}_3$ we choose the automaton

$$\mathfrak{K}_{A,\quad \alpha_1, \ldots, \alpha_m, \alpha_1, \ldots, \alpha_m}^{A \cup \bar{A}, \alpha_1, \ldots, \alpha_m, \bar{\alpha}_1, \ldots, \bar{\alpha}_m}.$$

Now let us prove the following statement.

2.2. Regardless of the automata $\mathfrak{A}$ and $\mathfrak{B}$ in the alphabet A having the rank n_1 and n_2, respectively, one can construct an automaton $\mathfrak{C}$ of rank n over the alphabet $A \cup \{\alpha\}$, such that

$$\mathfrak{C}(P) \simeq \mathfrak{A}(P)\alpha\mathfrak{B}(P).$$

Under these conditions $n = \max(n_1, n_2, 1)$.

This statement can easily be obtained from the previous one, 2.1, in the following way:

$$\mathfrak{C}(P) \simeq \mathfrak{A}(P)\mathfrak{A}^{(0)}_{A\cup\{\alpha\},\alpha}(\mathfrak{B}(P)),$$

where $\mathfrak{A}^{(0)}_{A\cup\{\alpha\},\alpha}$ is a 0-automaton that attaches the letter α to any word in the alphabet $A \cup \{\alpha\}$ at the left.

From (2.2) the union theorem can easily be obtained by induction with respect to k.

§ 5. Repetition of an n-Automaton

In constructing n-automata it is sometimes necessary to use the following construction: it is required to manipulate the system of words in the alphabet A:

$$P_1\alpha P_2\alpha \ldots \alpha P_r \qquad (\alpha \notin A)$$

into the word $\mathfrak{A}(P_1)\alpha\mathfrak{A}(P_2)\alpha \ldots \alpha\mathfrak{A}(P_r)$, $\mathfrak{A}$ being a certain n-automaton in the alphabet A. The natural question arises as to whether the operation described above can be carried out on some automaton. The answer to this question is given by the following theorem.

1.1. The Repetition Theorem. *Assume $\mathfrak{A}$ is an n-automaton ($n \geq 0$) in the alphabet A, while α is a letter that does not belong to A. Then one can construct an n-automaton $\mathfrak{B}$ over the alphabet $A \cup \{\alpha\}$ such that*

$$\mathfrak{B}(P_1\alpha P_2\alpha \ldots \alpha P_r) \simeq \mathfrak{A}(P_1)\alpha\mathfrak{A}(P_2)\alpha \ldots \alpha\mathfrak{A}(P_r) \qquad (r \geqslant 1;\ P_i \Omega A,\ 0 < i \leqslant r). \tag{1}$$

Proof. Assume

$$\mathfrak{A} = \mathfrak{A}_k \circ \mathfrak{A}_{k-1} \circ \ldots \circ \mathfrak{A}_1 \qquad (k \geqslant 1),$$

where $\mathfrak{A}_i$ $(0 < i \leq k)$ are primitive automata in the alphabet A which all have (according to the theorem of rank elevation) the rank n. Assume that the initial state of $\mathfrak{A}_i$ is S_0. Let us define the n-automaton $\mathfrak{C}_i$ in the alphabet $A \cup \bar{A} \cup \{\alpha, \beta, \gamma\}$ $(\beta \notin A \cup \bar{A},\ \gamma \notin A \cup \bar{A})$. Its initial state is S_0. $\bar{\mathfrak{C}}_i$ is defined as:

$$\bar{\mathfrak{C}}_i(\xi P) \simeq \begin{cases} \bar{\mathfrak{A}}_i(*P), & \text{if} \quad \xi \underset{\circ}{=} \alpha \quad \text{or} \quad \xi \underset{\circ}{=} *, \\ \gamma, & \text{if} \quad \xi \underset{\circ}{=} \beta, \\ \mathfrak{D}_i(\xi P), & \text{if} \quad \xi \in A. \end{cases}$$

Let us now show how to construct the (n − 1)-automaton $\mathfrak{D}_i$ (or the collection of substitution formulas, if n = 0):

$$\mathfrak{D}_i = \mathfrak{D}_{i,2} \circ \mathfrak{D}_{i,1}.$$

$\mathfrak{D}_{i,1}$ is a 0-automaton such that

$$\mathfrak{D}_{i,1}(\xi P) \underset{\circ}{=} P\xi.$$

The (n − 1)-automaton $\mathfrak{D}_{i,2}$ is constructed as follows:

$$\mathfrak{D}_{i,2}(\eta R) \simeq \begin{cases} \overline{\mathfrak{A}}_i(\mathfrak{D}_{i,3}(\eta R)), & \text{if} \quad \eta \underset{\circ}{=} \gamma, \\ \overline{\mathfrak{A}}_i(\mathfrak{D}_{i,4}(\eta R)), & \text{if} \quad \eta \in A; \end{cases}$$

under these conditions

$$\mathfrak{D}_{i,3} = \mathfrak{R}^{A \cup \{\gamma\},\, \alpha_1, \ldots, \alpha_m,\, \gamma}_{A,\, \alpha_1, \ldots, \alpha_m,\, \Lambda} \circ \mathfrak{B}_{A \cup \{\gamma\},\, S_0}$$

(see [1, §2.14] and [1, §2.4]), while the 0-automaton $\mathfrak{D}_{i,4}$ is such that

$$\mathfrak{D}_{i,4}(P\xi) \underset{\circ}{=} \xi P.$$

As is easily seen, the following relationship holds:

$$\mathfrak{C}_i(Q_1\alpha\beta Q_2\alpha\beta \ldots \alpha\beta Q_r) \simeq \mathfrak{A}_i(Q_1)\gamma\mathfrak{A}_i(Q_2)\gamma \ldots \gamma\mathfrak{A}_i(Q_r).$$

Let us now introduce the 0-automata $\mathfrak{B}_1$ and $\mathfrak{B}_2$. $\mathfrak{B}_1$ is the automaton

$$\mathfrak{R}^{A \cup \{\alpha, \gamma\},\, \alpha_1, \ldots, \alpha_m,\, \alpha,\, \gamma}_{A \cup \{\alpha, \beta\},\, \alpha_1, \ldots, \alpha_m,\, \alpha\beta,\, \alpha\beta},$$

while $\mathfrak{B}_2$ is the automaton

$$\mathfrak{R}^{A \cup \{\gamma\},\, \alpha_1, \ldots, \alpha_m,\, \gamma}_{A \cup \{\alpha\},\, \alpha_1, \ldots, \alpha_m,\, \alpha}.$$

Now we may define the n-automaton $\mathfrak{B}$ thus:

$$\mathfrak{B} = \mathfrak{B}_2 \circ \mathfrak{C}_k \circ \mathfrak{B}_1 \circ \mathfrak{C}_{k-1} \circ \mathfrak{B}_1 \circ \ldots \circ \mathfrak{B}_1 \circ \mathfrak{C}_2 \circ \mathfrak{B}_1 \circ \mathfrak{C}_1 \circ \mathfrak{B}_2.$$

It can easily be checked that (1) holds for it.

§6. The Continuation of n-Automata

1. Sometimes the necessity arises of "supplementing" a function stipulated by a certain n-automaton in the alphabet A in some fashion so that as a result a function is obtained that is defined on all words in the alphabet A. The question arises as to whether the function obtained in this manner can be calculated on an automaton. In order to answer this question, we shall formulate the following theorem.

1.1. The Continuation Theorem. *Assume $\mathfrak{A}$ is an n-automaton (n ≥ 0) in the alphabet A. An n-automaton $\mathfrak{B}$ may be constructed such that in the same alphabet A it is applicable to all words in A, and if $!\mathfrak{A}(P)$, then*

$$\mathfrak{A}(P) \underset{\circ}{=} \mathfrak{B}(P).$$

We omit the proof of this theorem.

CHAPTER III

Automata and Primitively Recursive Functions

In this chapter we consider the problem of the equivalence of automata and primitively recursive functions. In order to compare these notions it is necessary either to find a method which allows primitively recursive functions to be used for manipulating the word in any alphabet into a word in the same alphabet, as automata do, or to consider only those automata which manipulate systems of natural numbers into numbers. In this chapter we consider both approaches.

Throughout this chapter the word "function" denotes a partial arithmetic function (i.e., a function whose argument values are natural numbers and which takes natural values).

§ 1. m-Automaton Functions

1. The function $f(x_1, x_2, \ldots, x_n)$ is called an m-automaton function (an m-function, an automaton function) if there exists an m-automaton $\mathfrak{F}$ over the alphabet C such that

$$f(x_1, x_2, \ldots, x_n) \simeq \mathfrak{F}(x_1 \alpha x_2 \alpha \ldots \alpha x_n).^{\dagger} \tag{1}$$

The function $f(x_1, x_2, \ldots, x_n)$ is called fully m-automaton (fully automaton) if there exists an m-automaton $\mathfrak{F}$ over the alphabet C which is applicable to any word of the form $x_1 \alpha x_2 \alpha \ldots \alpha x_n$, such that (1) holds. The class of completely m-automaton functions shall be denoted by A_f^m.

2. We shall assume that the notion of a primitively recursive function is known.

The rank of primitively recursive function is called the length of its shortest primitively recursive description.

3. Let us take up the proof of the fact that any primitively recursive function is an automaton function. The proof is based on several lemmas.

3.1. Assume $g(x_1, \ldots, x_m)$, $h_1(x_1, \ldots, x_k)$, ..., $h_m(x_1, \ldots, x_k)$ ($m \geq 1$, $k \geq 1$) are respectively n_0-, n_1-, ..., n_m-automaton functions. Then the function

$$f(x_1, \ldots, x_k) \simeq g(h_1(x_1, \ldots, x_k), \ldots, h_m(x_1, \ldots, x_k))$$

is a $\max(n_0, n_1, \ldots, n_m, 1)$-automaton function. (The lemma on substitution.)

Assume $\mathfrak{G}, \mathfrak{H}_1, \ldots, \mathfrak{H}_m$ are automata such that

$$\begin{gathered}\mathfrak{G}(x_1 \alpha x_2 \alpha \ldots \alpha x_m) \simeq g(x_1, \ldots, x_m), \\ \mathfrak{H}_1(x_1 \alpha x_2 \alpha \ldots \alpha x_k) \simeq h_1(x_1, x_2, \ldots, x_k), \\ \ldots\ldots\ldots\ldots\ldots\ldots\ldots\ldots \\ \mathfrak{H}_m(x_1 \alpha x_2 \alpha \ldots \alpha x_k) \simeq h_m(x_1, x_2, \ldots, x_k).\end{gathered}$$

On the basis of the union theorem one can construct an automaton $\mathfrak{H}$ having the rank $\max(n_1, n_2, \ldots, n_m, 1)$ such that

$$\mathfrak{H}(x_1 \alpha x_2 \alpha \ldots \alpha x_k) \simeq \mathfrak{H}_1(x_1 \alpha x_2 \alpha \ldots \alpha x_k) \alpha \mathfrak{H}_2(x_2 \alpha x_2 \alpha \ldots \alpha x_k) \alpha \ldots \alpha \mathfrak{H}_m(x_1 \alpha \ldots \alpha x_k).$$

† We identify natural numbers and words in the alphabet $\text{Ч} = \{|\}$.

Now it is obvious that the automaton $\mathfrak{F}$ having the rank $\max(n_0, n_1, \ldots, n_m, 1)$, which is defined as

$$\mathfrak{F} = \mathfrak{G} \circ \mathfrak{H},$$

satisfies the condition

$$\mathfrak{F}(x_1\alpha \ldots \alpha x_k) \simeq g(h_1(x_1, \ldots, x_k), \ldots, h_m(x_1, \ldots, x_k)),$$

whence the lemma derives.

3.2. Assume the functions $g(x_1, \ldots, x_k)$ and $h(x_1, \ldots, x_k, x_{k+1}, x_{k+2})$ are respectively n_1- and n_2-automaton functions. Then the function $f(x_1, \ldots, x_k, x_{k+1})$, which is defined by the system

$$\begin{cases} f(x_1, \ldots, x_k, 0) \simeq g(x_1, \ldots, x_k), \\ f(x_1, \ldots, x_k, y+1) \simeq h(f(x_1, \ldots, x_k, y), x_1, \ldots, x_k, y) \end{cases}$$

is a $\max(n_1, n_2 + 1, 2)$-automaton function.

Assume $\mathfrak{G}$ and $\mathfrak{H}$ are automata such that

$$\mathfrak{G}(x_1\alpha \ldots \alpha x_k) \simeq g(x_1, \ldots, x_k),$$
$$\mathfrak{H}(x_1\alpha \ldots \alpha x_k \alpha x_{k+1} \alpha x_{k+2}) \simeq h(x_1, \ldots, x_k, x_{k+1}, x_{k+2}).$$

Let us construct the automaton $\mathfrak{F}$ as follows:

$$\mathfrak{F} = \mathfrak{F}_4 \circ \mathfrak{F}_3 \circ \mathfrak{F}_2 \circ \mathfrak{F}_1.$$

Here $\mathfrak{F}_1$ is an automaton constructed by means of the union theorem and having the rank $\max(n_1, 1)$. Under these conditions

$$\mathfrak{F}_1(x_1\alpha \ldots \alpha x_{k+1}) \simeq \mathfrak{G}(\mathfrak{A}(x_1\alpha \ldots \alpha x_{k+1}))\alpha x_1 \alpha \ldots \alpha x_{k+1},$$

$\mathfrak{A}$ being a 0-automaton that operates thus:

$$\mathfrak{A}(x_1\alpha \ldots \alpha x_k \alpha x_{k+1}) \mathrel{\underset{\circ}{=}} x_1\alpha \ldots \alpha x_k;$$

one can also construct a 0-automaton $\mathfrak{F}_2$, such that

$$\mathfrak{F}_2(\mathfrak{F}_1(x_1\alpha \ldots \alpha x_k \alpha x_{k+1})) \simeq \mathfrak{G}(x_1\alpha \ldots \alpha x_k)\alpha x_1 \alpha \ldots \alpha x_k \alpha\beta x_{k+1}.$$

Let us now define the $(\max(n_2, 1) + 1)$-automaton $\mathfrak{F}_3$ thus. The initial state is Λ. The $\max(n_2, 1)$-automaton $\overline{\mathfrak{F}}_3$ is stipulated as:

$$\overline{\mathfrak{F}}_3(\xi\eta P) \simeq \begin{cases} \eta P\xi, & \text{if } \xi \not\mathrel{\underset{\circ}{=}} \beta, \\ \mathfrak{B}(\xi\eta P), & \text{if } \xi \not\mathrel{\underset{\circ}{=}} *, \ \eta \mathrel{\underset{\circ}{=}} \beta, \\ \mathfrak{C}(\xi\eta P), & \text{if } \xi \mathrel{\underset{\circ}{=}} *, \\ \xi\eta P, & \text{if } \xi \mathrel{\underset{\circ}{=}} \beta \end{cases}$$

$(P\Omega\{|, \alpha, \beta\})$.

Let us describe the automata $\mathfrak{B}$ and $\mathfrak{C}$. The automaton $\mathfrak{B}$ is described by:

$$\mathfrak{B} = \mathfrak{B}_3 \circ \mathfrak{B}_2 \circ \mathfrak{B}_1.$$

The 0-automaton $\mathfrak{B}_1$ operates thus:

$$\mathfrak{B}_1(|\beta P) \mathrel{\underset{\circ}{=}} P \qquad (P\Omega\{|, \alpha\}) \qquad \text{(see [I. § 2.9])}.$$

The automaton $\mathfrak{B}_2$ is defined by means of the union theorem. Its rank is $\max(n_2, 1)$

$$\mathfrak{B}_2(x_0\alpha x_1\alpha \ldots \alpha x_k\alpha y) \simeq \mathfrak{H}(x_0\alpha x_1\alpha \ldots \alpha x_k\alpha y)\,\alpha x_0\alpha x_1\alpha \ldots \alpha x_k\alpha y.$$

The 0-automaton $\mathfrak{B}_3$ operates as follows:

$$\mathfrak{B}_3(x\alpha x_0\alpha \ldots \alpha x_k\alpha y) \mathrel{\underset{\circ}{=}} \beta x\alpha x_0\alpha \ldots \alpha x_k\alpha (y+1).$$

Thus,

$$\mathfrak{B}(|\beta x_0\alpha x_1\alpha \ldots \alpha x_k\alpha y) \simeq \beta\mathfrak{H}(x_0\alpha x_1\alpha \ldots \alpha x_k\alpha y)\,\alpha x_1\alpha x_2\alpha \ldots \alpha x_k\alpha (y+1).$$

Now we stipulate the 0-automaton $\mathfrak{C}$:

$$\mathfrak{C}(*\beta x_0\alpha x_1\alpha \ldots \alpha x_k\alpha y) \mathrel{\underset{\circ}{=}} \gamma^{x_0}.$$

It can easily be shown that

$$\mathfrak{K}^{\{|,\ \alpha,\ \beta,\ \gamma\},\ |,\ \alpha,\ \beta,\ \gamma}_{\{|\},\qquad\quad \Lambda,\ \Lambda,\ \Lambda,\ |}(\mathfrak{F}_3(\mathfrak{F}_2(\mathfrak{F}_1(x_1\alpha x_2\alpha \ldots \alpha x_k\alpha y)))) \simeq f(x_1, x_2, \ldots, x_k, y).$$

It is obvious that the rank of an automaton $\mathfrak{F}$ constructed in this way is equal to

$$\max(\max(n_1, 1),\ \max(n_2, 1)+1) = \max(n_1, n_2+1, 2).$$

Since the original functions are automaton functions, we arrive at the following statement.

3.3. The Theorem of Primitively Recursive Functions. Any primitively recursive function is completely an automaton function.

§ 2. Numeration of the Words in the Alphabet A

1. The correspondence Γ between the words in the alphabet A (as usual, it is assumed that $* \notin A$) and natural numbers is called the m-numeration of the words in the alphabet A, provided that there exists n-, n_0-, n_1-automata $\mathfrak{G}$, $\mathfrak{G}_0$, and $\mathfrak{G}_1$, respectively ($m = \max(n, n_0, n_1)$), such that:

a) $\mathfrak{G}_0$ is applicable to all natural numbers and manipulates into an empty word those and only those numbers x for which $!\mathfrak{G}_1(x)$ and $\mathfrak{G}_1(x)\Omega A$.

b) $\mathfrak{G}$ is applicable to each word P in the alphabet A and manipulates such a word P into a natural number; the corresponding natural number $\mathfrak{G}(P)$ is such that $\mathfrak{G}_0(\mathfrak{G}(P)) \mathrel{\underset{\circ}{=}} \Lambda$, while $\mathfrak{G}_1(\mathfrak{G}(P)) \mathrel{\underset{\circ}{=}} P$.

If $\mathfrak{G}_0(n) \mathrel{\underset{\circ}{=}} \Lambda$ and $\mathfrak{G}_1(n) \mathrel{\underset{\circ}{=}} P$, then the natural number n is called the number P in the m-numeration Γ.

It is obvious that if $P\Omega A$, $Q\Omega A$, where $P \not\mathrel{\underset{\circ}{=}} Q$, it follows that if n_1 and n_2 are respectively the numbers of the words P and Q, $n_1 \neq n_2$ (i.e., the numbers of different words in the alphabet A are different). However, from the definitions it does not follow that each word in the alphabet has only one number. On the contrary, examples can be constructed of m-numeration such that at least certain words have more than one number.

For any word P the natural number $\mathfrak{G}(P)$ is called the *principal number of the word P*. Thus, each word has a unique principal number. The m-numeration of the words in the alphabet A is called *complete* if any natural number is the number of a certain word in the alphabet A.

Assume there exists an alphabet A such that $* \notin A$, and also a certain complete m-numeration Γ of the words in the alphabet A.

We shall say that *the automaton $\mathfrak{A}$ in the alphabet A is a primitively recursive automaton* if there exists a primitively recursive function f for which the following holds:

a) if $\mathfrak{A}(P) \overline{\overline{\circ}} Q$, $(P\Omega A)$, $\mathfrak{G}_1(n_1) \overline{\overline{\circ}} P$ (i.e., n_1 is the number of the word P), then $\mathfrak{G}_1(f(n_1)) \overline{\overline{\circ}} Q$ (i.e., $f(n_1)$ is the number of the word Q).

b) if $f(n_1) = n_2$, then $\mathfrak{A}(\mathfrak{G}_1(n_1)) \overline{\overline{\circ}} \mathfrak{G}_1(n_2)$.

2. The problem of this section is to construct an example of the 1-numeration of words in the alphabet A.

Assume $A = \{\alpha_1, \ldots, \alpha\}$ $(k > 0)$. Let us assume, furthermore, that the letter 0 does not belong to the alphabet A.

We shall assume that the letters $0, \alpha_1, \ldots, \alpha_k$ are digits in a $(k+1)$-ary numbering system; under these conditions 0 denotes 0, α_1 denotes the natural number 1, α_2 denotes 2, ..., α_k denotes k.

Each word in the alphabet A is then the notation for a certain natural number in the $(k+1)$ numbering system. Let us juxtapose a certain natural number with each word P in A; if $P \overline{\overline{\circ}} \Lambda$, then we juxtapose the number 0 with P; if $P \overline{\overline{\circ}} \Lambda$ and $P \overline{\overline{\circ}} \alpha_{i_0}\alpha_{i_1} \ldots \alpha_{i_s}$ $(s \geqslant 0)$, we juxtapose the number

$$i_0 + i_1 \cdot k + \ldots + i_s \cdot k^s$$

with it.

Each natural number, of course, has its notation in the $(k+1)$-ary numbering system. This notation R is a word in the alphabet $A \cup \{0\}$. The word $[R^A$ is already a word in the alphabet A. Thus, each natural number corresponds to a completely specified word in A.

Let us now construct the n-automata $\mathfrak{G}$ and $\mathfrak{G}_1$. Under these conditions $\mathfrak{G}$ must manipulate Λ into the number 0, and the nonempty word $P\Omega A$ into a natural number whose notation (in the $(k+1)$-ary system) is the word P. $\mathfrak{G}_1$ must manipulate any natural number x into the word $[R^A$, where R is the notation of the number x.

Let us begin with the construction of the automaton $\mathfrak{G}$. We split up this construction into a series of stages. First, one may construct the 1-automaton $\mathfrak{A}$ in the alphabet A such that

$$\mathfrak{A}(\Lambda) \overline{\overline{\circ}} \Lambda$$

$$\mathfrak{A}(\xi_1\xi_2 \ldots \xi_{l-1}\xi_l) \overline{\overline{\circ}} (l \dot{-} 1)\, \xi_1\, (l \dot{-} 2)\, \xi_2 \ldots |\, \xi_{l-1}\xi_l \qquad (l > 0;\ \xi_i \in A,\ 0 < i \leqslant l)$$

(we assume that $| \notin \mathrm{A}$). Then we take the 0-automaton

$$\mathfrak{K} = \mathfrak{R}\,{}^{\mathrm{A} \cup \{|\},\ \alpha_1, \ldots, \alpha_k, |}_{\{|,\ \alpha,\ \beta\},\ \alpha|\beta, \ldots, \alpha|^k\beta, |};$$

then the following holds for $l > 0$:

$$\mathfrak{K}(\mathfrak{A}(\alpha_{i_1}\alpha_{i_2}\ldots\alpha_{i_l})) \mathrel{\underset{\circ}{=}} (l \dot{-} 1)\,\alpha i_1\beta\,(l \dot{-} 2)\,\alpha i_2\beta \ldots |\,\alpha i_{l-1}\beta\alpha i_l\beta;$$

for $l = 0$:

$$\mathfrak{K}(\mathfrak{A}(\Lambda)) \mathrel{\underset{\circ}{=}} \Lambda.$$

For any $k \geq 0$

$$j_k(x, y) = (k+1)^x \cdot y, \tag{1}$$

one may construct the automaton $\mathfrak{P}_k$, which calculates the function (1); i.e., one can construct an automaton such that the following relationship holds:

$$\mathfrak{P}_k(x\alpha y) \mathrel{\underset{\circ}{=}} (k+1)^x \cdot y.$$

It is not difficult to see that the automaton $\mathfrak{P}_k$ has the rank 1.

Now let us use the repetition theorem [II. §5.1.1]. However, we are not able to apply it immediately.

The cause of this lies in the fact that the word $\mathfrak{K}(\mathfrak{A}(P))$ ends with a separating letter β, and for the repetition theorem this should not be done. It is easy to construct the 0-automaton $\mathfrak{C}$ in the alphabet $\mathrm{A} \cup \{|,\ \alpha,\ \beta\}$, which is such that

$$\mathfrak{C}(P\beta) \mathrel{\underset{\circ}{=}} P, \quad \text{where} \quad P\Omega\mathrm{A} \cup \{|,\ \alpha,\ \beta\}.$$

Now we do have the right to use the repetition theorem; using the 1-automaton $\mathfrak{P}_k$, we construct the automaton $\mathfrak{B}$ thus:

$$\mathfrak{B}((l-1)\,\alpha i_1\beta\,(l-2)\,\alpha i_2\beta \ldots |\,\alpha i_{l-1}\beta\alpha i_l\beta) \mathrel{\underset{\circ}{=}} i_1(k+1)^{l-1}\beta i_2(k+1)^{l-2}\beta \ldots i_{l-1}(k+1)\,\beta i_l.$$

Now we apply the 0-automaton $\mathfrak{C}^{(0)}_{\{|,\ \beta\},\ \beta}$, to the results and discard all entries of the letter β from the word to which the automaton is applied (see [I. §2.5]). As a result we obtain (for $l > 0$):

$$\mathfrak{C}^{(0)}_{\{|,\ \beta\},\ \beta}(\mathfrak{B}(\mathfrak{C}(\mathfrak{K}(\mathfrak{A}(\alpha_{i_1}\alpha_{i_2}\ldots\alpha_{i_l}))))) \mathrel{\underset{\circ}{=}} \sum_{j=1}^{l} i_j(k+1)^{l-j}.$$

Let us assume that

$$\mathfrak{G} = \mathfrak{C}^{(0)}_{\{|,\ \beta\},\ \beta} \circ \mathfrak{B} \circ \mathfrak{C} \circ \mathfrak{K} \circ \mathfrak{A}.$$

It is not difficult to check the fact that for $l = 0$,

$$\mathfrak{G}(\Lambda) \mathrel{\underset{\circ}{=}} \Lambda.$$

Having traced the construction of the automaton $\mathfrak{G}$, one can verify the fact that the rank of $\mathfrak{G}$ is equal to 1.

Let us now go over to the construction of the automaton $\mathfrak{G}_1$.

One may construct a 1-automaton $\mathfrak{A}$ in the alphabet $\{|, \alpha, \beta\}$, which is such that

$$\mathfrak{A}(\Lambda) \simeq \Lambda,$$
$$\mathfrak{A}(x) \simeq x\alpha x\beta(x \dot{-} 1)\alpha x\beta \ldots \beta 1\alpha x\beta\alpha x\beta \qquad (x > 0).$$

Let us now determine the two-placed partially recursive function $f_k(x, y)$ $(k > 0)$, which causes a natural number z $(0 \le z \le k)$ equal to the digits having the weight $(k + 1)^x$ in the notation of the number y in the (k + 1)-ary numbering system to correspond to each pair of natural numbers x and y.

The function $f_k(x, y)$ may be stipulated thus:

$$f_k(x, y) = [y/(k+1)^x] \dot{-} (k+1)[y/(k+1)^{x+1}]. \tag{2}$$

Assume $\mathfrak{B}$ is an automaton which calculates the function $f_k(x, y)$. Let us estimate the rank of the automaton $\mathfrak{B}$, considering the fact that the function $f_k(x, y)$ calculated by the automaton is stipulated by the relationship (2). The functions $(k + 1)^x$, $[x/y]$, xy, $x \dot{-} y$, $x + 1$, can be calculated on 1-automata (see, respectively, [I. §2.22], [I. §2.28], [I. §2.20], [I. §2.26]).

From this it is easy to obtain the fact (using the theorem from Chapter II) that the rank of the automaton $\mathfrak{B}$ is equal to 1.

Making use of the repetition theorem, we construct an automaton $\mathfrak{C}$, such that

$$\mathfrak{C}(\mathfrak{A}(x)) \simeq n_1\beta n_2\beta n_3\beta \ldots \beta n_m\beta$$

holds. Then we construct the 0-automaton $\mathfrak{D}$ in such a way that

$$\mathfrak{D}(\mathfrak{C}(\mathfrak{A}(x))) \simeq i_1\alpha_{i_1}i_2\alpha_{i_2} \ldots i_m\alpha_{i_m}.$$

Finally, we obtain

$$\mathfrak{K}^{A \cup \{|\},\ \alpha_1,\ \alpha_2, \ldots, \alpha_k,\ |}_{A,\qquad \alpha_1,\ \alpha_2, \ldots, \alpha_k,\ \Lambda}(\mathfrak{D}(\mathfrak{C}(\mathfrak{A}(x)))) \simeq \alpha_{i_1}\alpha_{i_2} \ldots \alpha_{i_m}.$$

Thus, we have constructed the automaton $\mathfrak{G}_1$. Its rank, as is easily verified, is equal to 1.

So, the construction of an example of 1-numbering Γ of words in the alphabet A has been completed.

§ 3. Primitively Recursive Automata

1. We shall say that an n-automaton $\mathfrak{A}$ is a **primitively recursive automaton** if the function F defined by the relation (1) is primitively recursive:

$$F(x) \simeq \mathfrak{G}(\mathfrak{A}(\mathfrak{G}_1(x))). \tag{1}$$

Our problem now is to prove the following:

1.1. Theorem. *Any everywhere-determinate n-automaton $\mathfrak{A}$ in the alphabet A is a primitively recursive automaton.*

2. In the beginning we shall deal with the construction of certain primitively recursive functions.

Let us consider the alphabet $A = \{\alpha_1, \ldots, \alpha_k\}$ and the numeration Γ_A of the words in the alphabet A, which was constructed in the preceding section. Each word P in A corresponds to at least one natural number – the number of the word – in this numeration. Let us consider the alphabet $Б = \{\alpha_1, \ldots, \alpha_k, \ldots, \alpha_{k+l}\}$ $(l \geq 0)$ which is an expansion of the alphabet A; for words in this alphabet we construct the numeration $\Gamma_Б$. in accordance with §2. Each word P in A receives at least one number in the numeration $\Gamma_Б$ (as a word in Б). We shall call this number *the quasinumber of the word P*.

Let us construct a primitively recursive function f such that for every number of any word in the alphabet A the function gives some quasinumber of this word.

It is stipulated as follows:

$$f(x) = \sum_{i=0}^{x} (k + l + 1)^i f_k(i, x). \tag{1a}$$

The function f_k appearing here was constructed in [§2.2]. If x is the principal number of the word P, then we shall call the number $f(x)$ *the principal quasinumber of the word P*. It is likewise easy to construct the primitively recursive function g that uses each quasinumber of any word P in the alphabet A to produce the number of this word. This function is defined thus:

$$g(x) = \sum_{i=0}^{x} (k + 1)^i f_{k+l}(i, x). \tag{2}$$

It is evident that if x is the principal quasinumber of P, then g(x) is its principal number.

Assume P and Q are numbers in the alphabet Б, while x and y are their numbers in the numeration $\Gamma_Б$. Let us construct a primitively recursive function h such that h(x, y) is the number of the word PQ in the numeration $\Gamma_Б$. Assume r(y) is the number of the highest nonzero digit of the notation of the number y in the $(k + l + 1)$-ary numbering system.

The function r(y) may be defined in a primitively recursive manner thus:

$$r(y) = \begin{cases} 0, & \text{if } y = 0, \\ (\mu x_{x \leq y}((k + l + 1)^x > y)) \dot{-} 1. \end{cases} \tag{3}$$

Now, of course, one can define the function h(x, y) thus:

$$h(x, y) = x \cdot (k + l + 1)^{r(y)+1} + y. \tag{4}$$

Under these conditions it is clear from the construction of the functions h and r that if x and y are the principal numbers of the words P and Q, respectively, then h(x, y) is the principal number of the word PQ.

Finally, let us construct the last primitively recursive function a that we need, which uses the number (in the numeration Γ_A) of the word P (PΩA) to produce the principal number of this word.

For this purpose we begin by constructing the auxiliary function b which for $y > 0$ produces the number of the x-th nonzero digit in the $(k + 1)$-ary notation of the number y from x.

The function b may be stipulated as follows:

If y = 0, then

$$b(x, y) = 0 \qquad \text{for all} \qquad x \geqslant 0. \tag{5}$$

If y > 0, then

$$b(0, y) = \mu i_{0 \leq i \leq y}(f_k(i, y) > 0), \tag{6}$$

$$b(x + 1, y) = [\mu i_{b(x, y) < i \leq y+2}(f_k(i, y) > 0)] \,\mathrm{sg}\,[y \dot{-} (k + 1)^{b(x, y)+1}] + (y + 1)\, \overline{\mathrm{sg}}\,[y \dot{-} (k + 1)^{b(x, y)+1}].$$

It is obvious that

$$b(x, y) \leqslant y+1. \tag{7}$$

Now one can easily go over to the function a:

$$a(x) = \sum_{i=0}^{x} (k+1)^i \cdot f_k(b(i, x), x). \tag{8}$$

In order to prove the Theorem 1.1 we shall consider the alphabet Б to be $A \cup \{*\}$. Thus, $l = 1$.

3. Let us now go over to the proof of Theorem 1.1. We shall conduct this proof by induction with respect to n – the rank of the automaton $\mathfrak{A}$.

<u>Basis: n = 0.</u> Assume $\mathfrak{A}$ is a primitive 0-automaton having the initial state S_0. The substitution formulas from $\overline{\mathfrak{A}}$ have the form $\xi S \to T$, where $\xi \in A \cup \{*\}$, $S \Omega A$, $T \Omega A$. We shall juxtapose each such formula with a pair of natural numbers (s, t) such that s is the principal quasi number of the word ξS, while t is the principal quasi number of the word T. As a result, we obtain a finite collection of pairs $(s_0, t_0), \ldots, (s_p, t_p)$. This collection is such that if $s_i = s_j$, then $t_i = t_j$. This provides the possibility of defining the primitively recursive function φ_0 in the following manner:

$$\varphi_0(x) = \sum_{i=0}^{p} t_i \cdot \overline{\mathrm{sg}}\, |x - s_i|.$$

In other words,

$$\varphi_0(x) = \begin{cases} t_0, & \text{if } x = s_0, \\ t_1, & \text{if } x = s_1 \\ \ldots\ldots\ldots\ldots\ldots & \\ t_p, & \text{if } x = s_p, \\ 0, & \text{if } x \neq s_i \quad (0 \leqslant i \leqslant p). \end{cases}$$

Assume P is a word in the alphabet A having the number x. Then its principal number is equal to $a(x)$ (see [2.(5)-2.(6)]), and the principal quasi number is $f(a(x))$ (see [2.(1)]), and the principal quasi number of the word $P*$ is equal to $h(f(a(x)), y_0)$ (see [2.(4)]); here y_0 is the principal number (in the numeration $\Gamma_{A \cup \{*\}}$) of the word $*$. Let us now construct the auxiliary function $\chi_0(x, y)$ which uses the pair of natural numbers (x, y) to produce the principal quasi number of the state into which the automaton $\mathfrak{A}$ transfers due to the action of the word having the quasi number $[y/(k+2)^{r(y) \dot{-} x}]$. If the word having the principal quasi number y is $P*$, where $P * \,\overline{\underline{\circ}}\, \xi_{r(y)}\xi_{r(y)\dot{-}1} \ldots \xi_1\xi_0 \; (\xi_0 \,\overline{\underline{\circ}}\, *)$, then the word having the quasi number $[y/(k+2)^{r(y) \dot{-} x}]$ is the word

$$\xi_{r(y)}\xi_{r(y)\dot{-}1} \ldots \xi_{r(y)\dot{-}x} \qquad (0 \leqslant x \leqslant r(y)).$$

Let us define the functions $\chi_0(x, y)$ by means of the recursion

$$\left.\begin{array}{l} \chi_0(0, y) = \begin{cases} n_0, & \text{if } f_{k+1}(r(y), y) = 0, \\ \varphi_0(h(f_{k+1}(r(y), y), n_0)), & \text{if } f_{k+1}(r(y), y) > 0; \end{cases} \\ \quad \text{here } n_0 \text{ is the principal quasinumber of the initial} \\ \quad \text{state } S_0 \text{ of the automaton } \mathfrak{A}, \\ \chi_0(x+1, y) = \begin{cases} \chi_0(x, y), & \text{if } x+1 > r(y) \text{ or if } x+1 \leqslant \\ & \leqslant r(y) \text{ and } f_{k+1}(r(y) \dot{-} (x+1), y) = 0, \\ \varphi_0(h(f_{k+1}(r(y) - (x+1), y), \chi_0(x, y))), & \\ \quad \text{if } x+1 \leqslant r(y) \text{ and } f_{k+1}(r(y) \dot{-} (x+1), y) > 0. \end{cases} \end{array}\right\} \tag{1}$$

And, finally, let us define the function $\omega_0(x, y)$ which uses the pair of natural numbers (x, y) to produce the principal quasi number of the words that the automaton $\mathfrak{A}$ prints out by manipulating the word having the number $[y/(k+2)^{r(y)\dot{-}x}]$. This function is defined thus:

$$\left.\begin{array}{l}\omega_0(0, y)=\begin{cases}0, & \text{if } f_{k+1}(r(y), y)=0,\\ \chi_0(0, y), & \text{if } f_{k+1}(r(y), y)>0,\end{cases}\\ \omega_0(x+1, y)=\begin{cases}\omega_0(x, y), & \text{if } x+1>r(y) \text{ or if}\\ & x+1\leqslant r(y) \text{ and } f_{k+1}(r(y)\dot{-}(x+1), y)=0,\\ h(\omega_0(x, y), \chi_0(x+1, y)), & \text{if}\\ & x+1\leqslant r(y) \text{ and } f_{k+1}(r(y)\dot{-}(x+1), y)>0.\end{cases}\end{array}\right\} \tag{2}$$

It is obvious that the functions χ_0 and ω_0 are primitively recursive.

For a primitive 0-automaton $\mathfrak{A}$ the function F_0 may be defined thus:

$$F_0(x)=g(\omega_0(h(f(a(x)), y_0), h(f(a(x)), y_0))) \tag{3}$$

(here the function g was constructed in [2.(2)], h was constructed in [2.(4)], f was constructed in [2.(1)], a was constructed in (2.(8)]; from this it follows that the function F_0 is primitively recursive.

Assume $\mathfrak{A}$ is a 0-automaton in the alphabet A such that

$$\mathfrak{A}=\mathfrak{A}_l\circ\mathfrak{A}_{l-1}\circ\ldots\circ\mathfrak{A}_1, \qquad (l\geqslant 1),$$

the automata $\mathfrak{A}_1, \ldots, \mathfrak{A}_l$ being primitive 0-automata in the same alphabet. Assume, moreover, that the functions $F_{0,1}, \ldots, F_{0,l}$ have been constructed for them according to the discussion above. Then the function

$$F_0(x)=F_{0,l}(F_{0,l-1}(\ldots(F_{0,1}(x))\ldots))$$

obviously proves Theorem 1.1 for all 0-automata.

The Induction Step. According to the induction proposition each everywhere-determinate automaton having a rank not exceeding n in the alphabet A is primitively recursive.

In exactly the same way as in the basis we shall begin by considering only primitive (n + 1)-automata in the alphabet A. For them we define the functions χ_{n+1} and ω_{n+1} in a manner completely analogous to the constructions in the basis. The sole difference will be the fact that in the basis it was necessary to construct the function φ_0 on the foundation of the collection of substitution formulas. Now, however, the function φ_{n+1} may be defined thus:

$$\varphi_{n+1}(x)=f(F_n(g(x)));$$

under these conditions the function F_n is constructed for $\overline{\mathfrak{A}}$, having the rank n on the basis of the induction proposition.

In exactly the same way as in the basis one may carry out the transition from primitive (n + 1)-automata to all (n + 1)-automata.

This completes the proof of Theorem 1.1.

§ 4. Completely-Automaton Functions

1. In this section we consider the second approach which was mentioned in the introduction to the present chapter: namely, we shall consider only those automata which manipulate systems of natural numbers into natural numbers. The functions stipulated by such automata shall be called n-automaton functions in accordance with §1, subsection 1 of this chapter.

Our problem is to prove the following:

1.1. Theorem. Any completely n-automaton function is primitively recursive.

Assume $\mathfrak{F}$ is an automaton in the alphabet A, where $\{|, \alpha\} \subseteq A$. Based on the continuation theorem it may be assumed that $\mathfrak{F}$ is applicable to any word in the alphabet A. Let us introduce the numeration of words in the alphabet A (see [§2]). Assume x is the number of the word $x_1 \alpha x_2 \alpha \ldots \alpha x_m$. As was proved in [§3.1.1.], the function F, which is defined as

$$F(x) = \mathfrak{G}(\mathfrak{F}(\mathfrak{G}_1(x))), \tag{1}$$

is a primitively recursive function.

We shall assume that $A = \{\alpha_1, \alpha_2, \ldots, \alpha_k\}$, $(k \geq 2)$; assume that the principal number of the word | is a, while the principal number of the word α is b. We shall construct the primitively recursive functions θ_m $(m \geq 1)$ which use the m-tuple of natural numbers $x_1, \ldots, x_m$ to produce the principal number (in the numeration Γ of the words in the alphabet A) of the word $x_1 \alpha x_2 \alpha \ldots \alpha x_m$.

The one-place function θ_1 is defined as

$$\begin{cases} \theta_1(0) = 0, \\ \theta_1(x+1) = (k+1)\,\theta_1(x) + a. \end{cases} \tag{2}$$

Assume that the function $\theta_m(x_1, \ldots, x_m)$ is defined. Let us define $\theta_{m+1}(x_1, \ldots, x_{m+1})$:

$$\begin{cases} \theta_{m+1}(x_1, \ldots, x_m, 0) = (k+1)\,\theta_m(x_1, \ldots, x_m) + b, \\ \theta_{m+1}(x_1, \ldots, x_m, x_{m+1}+1) = (k+1)\,\theta_{m+1}(x_1, \ldots, x_m, x_{m+1}) + a. \end{cases} \tag{3}$$

After this we construct a primitively recursive function $\varkappa$ which uses the number (in the numeration Γ of the words in the alphabet A) of a natural number to produce the number itself. This function is constructed thus:

$$\varkappa(x) = \sum_{i=0}^{x} \overline{\mathrm{sg}}\,|f_k(i, x) - a|, \tag{4}$$

where f_k is the function constructed in [§2.2]. And, finally, we define the new primitively recursive function:

$$\varkappa(F(\theta_m(x_1, x_2, \ldots, x_m)));$$

for this function the obvious equation

$$f(x_1, x_2, \ldots, x_m) = \varkappa(F(\theta_m(x_1, x_2, \ldots, x_m))),$$

holds, from which Theorem 1.1 derives.

CHAPTER IV

n-Automata, Türing Machines, and the A. Grzegorczyk Classes

A. Grzegorczyk [9] considered the classification of the class $\mathfrak{R}$ of primitive recursive functions, having defined the class $\mathcal{E}^n$ of primitively recursive functions which are such that $\mathfrak{R} = \bigcup_{n=0}^{\infty} \mathcal{E}^n$, $\mathcal{E}^n \subset \mathcal{E}^{n+1}$, but $\mathcal{E}^n \neq \mathcal{E}^{n+1}$ $(n \geq 0)$.

In this chapter we introduce for consideration the classes $\mathcal{T}^n_A$ of Türing machines whose length of calculations in operating on words of length x $(x \geq 0)$ is majorized by a certain function $f(x)$ from $\mathcal{E}^n$, and $\mathcal{U}^n_A$ of Türing machines whose operating zone for operation on words of length x is majorized by a certain function $g(x)$ from $\mathcal{E}^n$.

On the other hand, we have classes of n-automata available. The purpose of this chapter is to establish the following facts:

$$A^n_A = \mathcal{T}^{n+2}_A = \mathcal{U}^{n+2}_A \qquad (n \geqslant 1),$$
$$A^n_\Phi = \mathcal{T}^{n+2}_\Phi = \mathcal{U}^{n+2}_\Phi = \mathcal{E}^{n+2} \qquad (n \geqslant 1).$$

The subscript "f" shows that a class of arithmetic functions is being considered.

§ 1. Determination of the Grzegorczyk Classes and Türing Machines with a Constrained Operating Time and a Constrained Operating Zone

1. Let us present the definition of the classes $\mathcal{E}^n$ $(n \geq 0)$ given by Grzegorczyk [9].

Let us construct the sequence of functions $f_n(x, y)$ for $n \geq 0$:

$$f_0(x, y) = y + 1, \quad f_1(x, y) = x + y, \quad f_2(x, y) = (x + 1)(y + 1);$$

for $n \geq 2$:

$$\begin{cases} f_{n+1}(0, y) = f_n(y + 1, y + 1), \\ f_{n+1}(x + 1, y) = f_{n+1}(y, f_{n+1}(x, y)). \end{cases}$$

The class $\mathcal{E}^n$ $(n \geq 0)$ is the least class of functions which contains the functions $S(x) = x + 1$, $U_1(x, y) = x$, $U_2(x, y) = y$, $f_n(x, y)$ and is closed relative to the operation with superposition and bounded recursion.

Following Grzegorczyk, we shall say that the class $\mathcal{X}$ is closed relative to the superposition operation if it is closed relative to the following three operations:

a) The operation of substitution of functions. If

$$f(x_1, \ldots, x_{k-1}, x_k, x_{k+1}, \ldots, x_l) \in \mathcal{X} \qquad (1 \leqslant k \leqslant l),$$
$$g(y_1, \ldots, y_m) \in \mathcal{X} \qquad (m \geqslant 1),$$

then

$$f(x_1, \ldots, x_{k-1}, g(y_1, \ldots, y_m), x_{k+1}, \ldots, x_l) \in \mathcal{X}.$$

b) The operation of identification of variables. If

$$f(x_1, \ldots, x_{j-1}, x_j, x_{j+1}, \ldots, x_{k-1}, x_k, x_{k+1}, \ldots, x_l) \in \mathcal{X}, \qquad 1 \leqslant j < k \leqslant l,$$

then

$$f(x_1, \ldots, x_{j-1}, y, x_{j+1}, \ldots, x_{k-1}, y, x_{k+1}, \ldots, x_l) \in \mathcal{X}.$$

c) The operation of substitution of a constant. If

$$f(x_1, \ldots, x_{k-1}, x_k, x_{k+1}, \ldots, x_l) \in \mathcal{X} \qquad (1 \leqslant k \leqslant l),$$

then

$$f(x_1, \ldots, x_{k-1}, 0, x_{k+1}, \ldots, x_l) \in \mathcal{X}.$$

The class $\mathcal{X}$ is called closed relative to the operation of bounded recursion if the function f is defined as

$$\begin{cases} f(x_1, \ldots, x_l, 0) = g(x_1, \ldots, x_l), \\ f(x_1, \ldots, x_l, x+1) = h(x_1, \ldots, x_l, x, f(x_1, \ldots, x_l, x)), \end{cases}$$

where

$$f(x_1, \ldots, x_l, x) \leqslant j(x_1, \ldots, x_l, x), \qquad (g \in \mathcal{X}, h \in \mathcal{X}, j \in \mathcal{X}),$$

belongs to the class $\mathcal{X}$.

2. Now let us go over to the definition of certain classes of Türing machines.

We shall consider Türing machines which operate on a two-way type and have an external output at A, including the empty letter λ (machines in the alphabet A).

As usual, a machine over the alphabet A shall be called a machine in any extension of the alphabet A.

The machine may be in any of a finite list $Q = \{q_0, \ldots, q_k\}$ $(k \geq 1)$ of states. In this list the initial state q_0 and the concluding state q_k are isolated.

A situation may be described thus. Assume the machine is in state q and accepts the letter ξ with its head. Assume $P_L \equiv \xi_{r_1}\xi_{r_1-1} \ldots \xi_1$ is a word printed on the tape and situated to the left of the head, r_1 being the minimal number such that all cells are empty to the left of the cell containing the letter ξ_{r_1}. Analogously, we define $P_R \equiv \eta_1\eta_2 \ldots \eta_{r_2}$ as a word printed on the tape and situated to the right of the heads. Then the situation of the Türing machine may be stipulated by the word $S \equiv P_L q \xi P_R$; here P_L, P_R may be empty words. Under these conditions it is assumed that $Q \cap A = \varnothing$. The operating zone of the given situation is called the minimal zone containing the scanned cell plus all cells in which nonempty letters are printed. The length of the operating zone of the given situation is called the number $[P_L \xi P_R^{\partial}$, if $\xi P_R \not\equiv \Lambda$ and the number $[P_L^{\partial} + 1$, if $\xi P_R \equiv \Lambda$. We shall denote the length of the operating zone of the situation S thus: $[S^3$.

In order to describe an elementary action we shall make use of expressions of one of the following three types:

$$\begin{aligned} &\xi q_i \to \eta L q_j, \\ &\xi q_i \to \eta C q_j \qquad (\xi \in A, \ \eta \in A, \ q_i \in Q, \ q_j \in Q), \\ &\xi q_i \to \eta R q_j; \end{aligned}$$

L, C, R indicate the fact that the readout head is moved to the left, remains in place, or is moved to the right.

Assume the machine $\mathfrak{T}$ operates on the word P. It begins to operate from the initial situation $S_0 \equiv q_0 P$ and successively passes through the situations $S_1, S_2, \ldots$.

This sequence of situations is called the *process of manipulating the word P*; the manipulation process either terminates (at the very first appearance of the situation containing q_k) or continues without limit. If the process of the manipulation of the word P ends, then we shall characterize it by the following quantities:

$[\mathfrak{T}(P)^{\theta}$ is the duration of the process (i.e., the number of its situations);

$[\mathfrak{T}(P)^{3}$ is the length of its operating zone. The operating zone of the process is the minimal zone encompassing the operating zones of all of its situations.

We shall consider only those Türing machines over alphabet A for which the process of manipulating any word PΩA ends, a word in A being obtained as the result. Let us define the classes $\mathcal{T}^n_A$, $n \geq 0$ of machines over the alphabet A such that for any machine $\mathfrak{T}$ from the class $\mathcal{T}^n_A$ one can find a one-place function $f \in \mathcal{E}^n$, which is such that

$$[\mathfrak{T}(P)^{\theta} \leqslant f([P^{\partial}) \qquad (P\Omega A).$$

Analogously, we define the classes $\mathcal{U}^n_A$ $(n \geq 0)$ of Türing machines over the alphabet A such that for any machine $\mathfrak{T}$ from the class $\mathcal{U}^n_A$ one can find a one-place function $f \in \mathcal{E}^n$, which is such that

$$[\mathfrak{T}(P)^{3} \leqslant f([P^{\partial}) \qquad (P\Omega A).$$

By analogy with the definition of the classes A^n_Φ of fully n-automaton functions one can introduce the classes of $\mathcal{T}^n_\Phi$ and $\mathcal{U}^n_\Phi$ of functions that are calculable by means of Türing machines from $\mathcal{T}^n_c$ (from $\mathcal{U}^n_c$), correspondingly). The class of n-automata over the alphabet A shall be denoted by A^n_A.

§ 2. Basic Theorems

1. Let us begin by dealing with the relationships between $\mathcal{T}^n_A$ and $\mathcal{U}^n_A$. Let us prove the following:

1.1. Theorem. $\mathcal{T}^n_A = \mathcal{U}^n_A \quad (n \geqslant 3)$.

We shall show first that

$$\mathcal{T}^n_A \subseteq \mathcal{U}^n_A. \tag{1}$$

Assume that the Türing machine $\mathfrak{T}$ over the alphabet A is such that $\mathfrak{T} \in \mathcal{T}^n_A$. This means that one can find a function $f(x) \in \mathcal{E}^n$, such that

$$[\mathfrak{T}(P)^{\theta} \leqslant f([P^{\partial}) \qquad (P\Omega A). \tag{2}$$

But since in the process of operating on the word P the machine $\mathfrak{T}$ cannot scan more than $f([P^{\partial})$ cells in the course of making no more than $f([P^{\partial})$ steps, it follows that

$$[\mathfrak{T}(P)^{3} \leqslant f([P^{\partial}) + [P^{\partial} \qquad (P\Omega A). \tag{3}$$

Further, the function x + y belongs to $\mathcal{E}^n$ $(n \geq 1)$. From this it follows that if $f(x) \in \mathcal{E}^n$, then $g(x) = f(x) + x \in \mathcal{E}^n$ $(n \geq 1)$ also. Thus, the function $g(x) \in \mathcal{E}^n$ such that

$$[\mathfrak{T}(P)^3 \leqslant g([P^\partial) \qquad (P\Omega A) \tag{4}$$

has been constructed. The relation (1) derives from this.

Let us go over to the proof of the opposite

$$\mathcal{U}_A^n \subseteq \mathcal{T}_A^n \qquad (n \geqslant 3). \tag{5}$$

Assume that (4) holds. This means that the Türing machine completes its operation on the word P without departing from a zone of length $g([P^\partial)$. Let us calculate the maximum number of situations through which the machine might pass during the process of manipulating the word P. These situations can be described by words of the form

$$S \mathrel{\underline{\overline{\circ}}} P_L q \xi P_R.$$

The inequality

$$[S^3 \leqslant g([P^\partial)$$

holds. It can easily be seen that the number of different situations for a length $g([P^\partial)$ of the operating band does not exceed

$$(k+1) \cdot g([P^\partial) \cdot r^{g([P^\partial)},$$

where (k + 1) is the number of plates of the Türing machine; r is the number of different letters of the alphabet Б (A ⊆ Б). From this it follows that

$$[\mathfrak{T}(P)^e \leqslant f([P^\partial) \qquad (P\Omega A),$$

where

$$f(x) = (k+1) \cdot g(x) \cdot r^{g(x)}.$$

But from [9] it is well known that $h_1(x, y) = xy$ and $h_2(x, y) = x^y$ belong to $\mathcal{E}^n$ ($n \geq 3$). Therefore, if $g(x) \in \mathcal{E}^n$, then $f(x) \in \mathcal{E}^n$ ($n \geq 3$) also. And this means that (5) holds. Thus, the proof of Theorem 1.1. has been completed.

2. Now we shall consider one auxiliary (but important) lemma:

<u>2.1. Lemma.</u> *Assume the function $f(x)$ is a completely m-automaton function ($m \geq 1$). Assume, moreover, that $\mathfrak{T}$ is a Türing machine over the alphabet A and is such that $\mathfrak{T}$ is applicable to any word PΩA and*

$$[\mathfrak{T}(P)^e \leqslant f([P^\partial).$$

Then one can construct an m-automaton $\mathfrak{A}$ over the alphabet A, which is such that

$$\mathfrak{A}(P) \mathrel{\underline{\overline{\circ}}} \mathfrak{T}(P) \qquad (P\Omega A). \tag{1}$$

We shall assume that $\mathfrak{T}$ is a Türing machine in the alphabet Б (A ⊆ Б) which has the separate states Q; here $Б \cap Q = \varnothing$, and the letters α and | are such that $\alpha, | \notin Б \cup Q$. Further, we shall assume that $Б = \{\beta_1, \ldots, \beta_r\}$ ($r \geq 1$), while Б is the alphabet of twins.

We shall construct the automaton $\mathfrak{A}$ in the form of the composition

$$\mathfrak{A} = \mathfrak{A}_7 \circ \mathfrak{A}_6 \circ \mathfrak{A}_5 \circ \mathfrak{A}_4 \circ \mathfrak{A}_3 \circ \mathfrak{A}_2 \circ \mathfrak{A}_1.$$

The automaton $\mathfrak{A}_1$ performs the following transformation of the word P:

$$\mathfrak{A}_1(P) \mathrel{\overline{\underline{\circ}}} [P^{\cup} q_0 \alpha \mathfrak{F}([P^{\partial});$$

here $\mathfrak{F}$ is an m-automaton ($m \geq 1$) corresponding to the function f. The automaton $\mathfrak{A}_1$ may be constructed thus:

$$\mathfrak{A}_1(P) \mathrel{\overline{\underline{\circ}}} \mathfrak{B}_{A \cup \{q_0\}, q_0}(\mathfrak{H}_A^{(1)}(P)) \alpha \mathfrak{F}(\mathfrak{D}_A(P)).$$

The description of the 1-automaton $\mathfrak{H}_A^{(1)}$ may be found in [I. §2.12], that of the 0-automaton $\mathfrak{D}_A$ may be found in [I. § 2.8], and that of the 0-automaton $\mathfrak{B}_{A \cup \{q_0\}, q_0}$ may be found in [I.§2.4].

Further, the automaton $\mathfrak{A}_1$ is constructed from these automata on the basis of the union theorem [II. §4.1.1]. From this theorem it follows that the rank of the automaton $\mathfrak{A}_1$ is equal to max(m, 1); since $m \geq 1$, the rank of $\mathfrak{A}_1$ is equal to m.

Let us now consider the 0-automaton $\mathfrak{E}_T$, which performs the elementary action of a Türing machine $\mathfrak{T}$. Assume $\mathfrak{T}$ is a Türing machine in the alphabet Б, which has the set of states Q. Assume that the machine $\mathfrak{T}$, being in the active situation S, performs one elementary action and transits to the situation R. The 0-automaton $\mathfrak{E}_T$ in the alphabet $\text{Б} \cup Q$ operates as follows:

$$\mathfrak{E}_T(S) \mathrel{\overline{\underline{\circ}}} R.$$

If S is a passive situation (i.e., a situation containing a concluding state), then

$$\mathfrak{E}_T(S) \mathrel{\overline{\underline{\circ}}} S.$$

The construction of the automaton $\mathfrak{E}_T$ presents no difficulty, and we shall not dwell on it.

Now let us construct the 1-automaton $\mathfrak{A}_2$. Its initial state is Λ. Making use of the branching theorem [II. §2.1.1], we construct $\overline{\mathfrak{A}}_2$ thus:

$$\overline{\mathfrak{A}}_2(\xi R) \mathrel{\overline{\underline{\circ}}} \begin{cases} \mathfrak{A}^{(0)}_{A \cup \{q_0\}, \Lambda}(\xi R), & \text{if} \quad \xi \in A \cup \{q_0\}, \\ \mathfrak{C}^{(0)}_{A \cup \{q_0, \alpha\}, \alpha}(\xi R), & \text{if} \quad \xi \mathrel{\overline{\underline{\circ}}} \alpha, \\ \mathfrak{E}_T(\mathfrak{C}^{(0)}_{\text{Б} \cup Q \cup \{|\}, |}(\xi R)), & \text{if} \quad \xi \mathrel{\overline{\underline{\circ}}} |, \\ \mathfrak{K}_{\overline{\text{Б}}}^{\text{Б} \cup Q \cup \{*\},\ \beta_1, \ldots, \beta_r, q_0, q_1, \ldots, q_k, *}{}_{\bar{\beta}_1, \ldots, \bar{\beta}_r, \Lambda, \Lambda, \ldots, \Lambda, \Lambda}(\xi R), & \text{if} \quad \xi \mathrel{\overline{\underline{\circ}}} *. \end{cases}$$

Here R is a word in the alphabet $\text{Б} \cup Q$. The automaton $\mathfrak{A}_2$ is applied to words of the form $[P^{\cup} q_0 \alpha \mathfrak{F}([P^{\partial})$. Assume

$$P \mathrel{\overline{\underline{\circ}}} \xi_1 \xi_2 \ldots \xi_{l-1} \xi_l \qquad (l \geqslant 0).$$

Then the automaton $\mathfrak{A}_2$ manipulates the word

$$\xi_l \xi_{l-1} \ldots \xi_2 \xi_1 q_0 \alpha \mathfrak{F}(l)$$

thus:

$$\mathfrak{A}_2\colon \Lambda \vdash_{\xi_l} \xi_l \vdash_{\xi_{l-1}} \xi_{l-1} \xi_l \vdash \ldots \vdash_{\xi_2} \xi_2 \ldots \xi_{l-1} \xi_l \vdash_{\xi_1}$$
$$\vdash_{\xi_1} \xi_1 \xi_2 \ldots \xi_{l-1} \xi_l \vdash_{q_0} q_0 P \vdash_{\alpha} q_0 P \vdash_{|} S_1 \vdash_{|} S_2 \vdash \ldots \vdash_{|} S_{\mathfrak{F}(l)} \vdash_{*} \bar{R}_1 \mathfrak{T}\overline{(P)}\bar{R}_2 \quad (\bar{R}_1, \bar{R}_2 \Omega \{\bar{\lambda}\}).$$

Here q_0P, S_1, S_2, ..., S_q is the process of manipulating the word P by the Türing machine $\mathfrak{T}\,(q \leqslant \mathfrak{F}\,(l))$. The inequality

$$q \leqslant \mathfrak{F}\,(l)$$

derives from the fact that from the condition of the theorem,

$$[\mathfrak{T}\,(P)^{\theta} \leqslant \mathfrak{F}\,(l).$$

S_q is a passive situation; $S_{q+i} \mathrel{\overline{\underline{\circ}}} S_q \;(0 \leqslant i \leqslant \mathfrak{F}\,(l) - q)$. Now the choice of the automata $\mathfrak{A}_3 - \mathfrak{A}_5$ is obvious. As the automaton $\mathfrak{A}_3$ let us choose the 0-automaton from [I. §2.14]

$$\mathfrak{K}_{\text{Б},\qquad\quad \Lambda,\;\ldots,\;\Lambda,\;\beta_1,\;\ldots,\;\beta_r,\;\Lambda,\;\Lambda,\;\ldots,\;\Lambda}^{\text{Б}\cup\bar{\text{Б}}\cup Q,\;\beta_1,\;\ldots,\;\beta_r,\;\bar{\beta}_1,\;\ldots,\;\bar{\beta}_r,\;q_0,\;q_1,\;\ldots,\;q_k}.$$

We obtain

$$\mathfrak{A}_3\,(\mathfrak{A}_2\,(\mathfrak{A}_1\,(P))) \mathrel{\overline{\underline{\circ}}} R_1\mathfrak{T}\,(P)\,R_2 \qquad (P\Omega\text{A},\; R_1\Omega\,\{\lambda\},\; R_2\Omega\,\{\lambda\}).$$

Now it remains only for the words R_1, R_2 to be "removed." For this purpose let us use the 0-automaton $\mathfrak{B}$ in the alphabet Б, which can be described as follows: its initial state is Λ,

$$\bar{\mathfrak{B}}: \begin{cases} \lambda \to \Lambda & \\ \xi \to \xi & (\xi \in \text{Б} \setminus \{\lambda\}) \\ \xi\eta \to \xi & (\xi,\, \eta \in \text{Б}) \\ * \to \Lambda & \\ *\xi \to \Lambda & (\xi \in \text{Б}). \end{cases}$$

It is clear that if $\mathfrak{A}_4$ is chosen to be the automaton $\mathfrak{B}$, then

$$\mathfrak{A}_4\,(R_1\mathfrak{T}\,(P)\,R_2) \mathrel{\overline{\underline{\circ}}} \mathfrak{T}\,(P)\,R_2.$$

Then we also choose $\mathfrak{A}_5$ and $\mathfrak{A}_7$ to be the 1-automaton $\mathfrak{H}_{\text{Б}}^{(1)}$ (from [I. §2.12]). As the automaton $\mathfrak{A}_6$ we again use $\mathfrak{F}$. Then we obtain

$$\mathfrak{A}_7\,(\mathfrak{A}_6\,(\mathfrak{A}_5\,(\mathfrak{A}_4\,(R_1\mathfrak{T}\,(P)\,R_2)))) \mathrel{\overline{\underline{\circ}}} \mathfrak{H}_{\text{Б}}^{(1)}\,(\mathfrak{B}\,(\mathfrak{H}_{\text{Б}}^{(1)}\,(\mathfrak{B}\,(R_1\mathfrak{T}\,(P)\,R_2)))) \mathrel{\overline{\underline{\circ}}} \mathfrak{T}\,(P).$$

Thus, an m-automaton $\mathfrak{A}$, has been constructed for which (1) is fulfilled.

3. Our next problem is to show that the following theorem holds:

3.1. Theorem. $A_{\text{A}}^{n} \subseteq \mathcal{U}_{\text{A}}^{n+2}, \quad A_{\text{A}}^{n} \subseteq \mathcal{T}_{\text{A}}^{n+2} \qquad (n \geqslant 1).$

We need to prove that for each n-automaton $\mathfrak{A}$ over the alphabet A one can construct a Türing machine $\mathfrak{T}$ over the alphabet A, which is such that

$$\mathfrak{A}\,(P) \mathrel{\overline{\underline{\circ}}} \mathfrak{T}\,(P) \qquad (P\Omega\text{A}).$$

Under these conditions there exists a function $f\,(x) \in \mathscr{E}^{n+2}$, which is such that

$$[\mathfrak{T}\,(P)^{3} \leqslant f\,([P^{\partial}).$$

We shall assume that $\lambda \notin \text{A}$, and also that if $\mathfrak{A}$ is an automaton in the alphabet Б, it follows that $\lambda \notin \text{Б}$.

We shall show that furthermore the function f may be made nondecreasing.

We shall prove this theorem by induction with respect to n.

The Basis, $n = 0$. To begin with, let us consider the case in which $\mathfrak{A}$ is a primitive 0-automaton. Let us describe how the corresponding Türing machine operates. At first the machine determines whether the word that it is manipulating is empty. If it scans an empty cell in the initial situation, then the original word is empty. It writes $*$ into the scanned cell. If it scans a nonempty cell in the initial situation, then it moves its head (without altering the word written on the tape) until it finds the first empty cell, and then writes $*$ into this cell. Then the machine returns to the cell into which it wrote the first letter $*$ of the word P. After this, cyclic operation of the machine begins. The machine scans the next letter of the original word (at the beginning this letter was the first one). Moreover, it remembers the state in which the automaton is. The machine erases the letter in the scanned cell, remembers the state into which the automaton $\mathfrak{A}$ transits, and, if the word which the automaton $\mathfrak{A}$ is about to print is not empty, it moves the head to the right until it encounters the first empty cell. Beginning with this cell, the machine writes into the subsequent cells that word which the automaton $\mathfrak{A}$ prints. After this the machine moves the head in such a way that it scans the first nonempty cell to the left. If the word that the automaton $\mathfrak{A}$ must print is empty, then the head simply moves one step to the right. The operation is repeated from the beginning of the cycle. The machine goes out of this cycle for the case in which it reveals $*$ in the scanned cell. It remembers this fact, and, in performing its operation according to the cycle for the last time, it returns the head to the first nonempty cell to the left and stops.

Let us note that the machine constructed is such that its head moves only one cell to the left of the first letter of the original word during the process of its operation. We shall call this property of the machine property A.

It is obvious that for the case in which $\mathfrak{A}$ is a nonprimitive automaton and the relation

$$\mathfrak{A} = \mathfrak{A}_k \circ \mathfrak{A}_{k-1} \circ \ldots \circ \mathfrak{A}_1$$

holds, where $\mathfrak{A}_i$ $(1 \leq i \leq k)$ are already primitive 0-automata, the machine does not stop after it has operated for $\mathfrak{A}_1$ but continues to operate for the automaton $\mathfrak{A}_2$ and so forth. The machine constructed in this way also has the property A.

It is not difficult to see that for the case of a primitive 0-automaton

$$[\mathfrak{T}(P)^3 \leqslant C_1 \cdot [P^{\partial} + C_2.$$

Simple operations show that this same inequality also holds for the case in which $\mathfrak{A}$ is a nonprimitive automaton. The function in the right side of the inequality is nondecreasing.

On the other hand, the functions xy and x + y belong to the class $\mathcal{E}^2$. Therefore the theorem for the case $n = 0$ has been proved.

The Induction Step. Let us assume that the theorem holds for n-automata. Let us show that in this case it also holds for $(n + 1)$-automata. To begin with, we consider the case in which $\mathfrak{A}$ is a primitive $(n + 1)$-automaton $(n \geq 0)$. Once again let us describe the operation of the machine $\mathfrak{T}$ on the word $P\Omega A$.

The initial stage of operation of the machine $\mathfrak{T}$ is exactly the same as it is in the basis: namely, it attaches $*$ to the word P at the right. In the cells which follow the one occupied by $*$, the machine writes in the word $\lambda\alpha S_0$ (S_0 is the initial state of the automaton $\mathfrak{A}$; we assume that $\alpha \notin B$). After this that portion of the operation of this machine begins which will be repeated many times. The head returns to the beginning of the word, the machine remembers the first letter on the left, erases it, and writes this letter into the cell in which the letter α was written previously. After this the control is transferred to that portion of the machine which simulates the operation of the automaton $\overline{\mathfrak{A}}$. According to the inductive assumption, this portion of the machine satisfies the property A. When this operation ends, the following word will be written on tape:

$$P_1 * \lambda^r S_1. \tag{1}$$

Now the machine proceeds thus: first it obtains the word

$$P_1 * \alpha\lambda^{r-1} S_1 \tag{2}$$

on the tape. After this it shifts the word S_1 until the empty cells within the word vanish.

As a result it obtains

$$P_1 * \alpha S_1. \tag{3}$$

Then it attaches the word $\lambda\alpha S_1$ to the word obtained on the tape; as a result the word

$$P_1 * \alpha S_1 \lambda\alpha S_1 \tag{4}$$

is formed on the tape. Finally, the machine deletes the first entry of the letter α and again shifts the word:

$$P_1 * S_1 \lambda\alpha S_1. \tag{5}$$

For all of these operations the machine, as can easily be seen, is not required to depart beyond the zone required to write the longest of the words given in (1)-(5).

After this a word of the same type as that with which the description of the cyclic part of the operation began turns out to be on the tape. The cycle is repeated until the machine erases the letter * which follows the original word P. After this the machine performs the next cycle; however, at the end of it, it merely shifts the newly obtained word while "removing" the empty cells; now it stops.

Let us estimate the operating zone required by the machine $\mathfrak{T}$ in order to perform the operation described.

Since the automaton $\bar{\mathfrak{A}}$ has the rank n, it follows that according to the induction proposition there exist a Turing machine $\bar{\mathfrak{T}}$ and a nondecreasing function $\bar{f}(x) \in \mathscr{E}^n$ such that

$$\bar{\mathfrak{T}}(P) \,\underline{\overline{\circ}}\, \bar{\mathfrak{A}}(P) \qquad (P\Omega\text{Б} \cup \{*\}),$$
$$[\mathfrak{T}(P)^3 \leqslant \bar{f}([P^\partial).$$

Let us define the primitively recursive function g thus:

$$g(0) = s_0 + 2,$$
$$g(x+1) = \bar{f}(g(x)+1).$$

This function is nondecreasing, since the function $\bar{f}(x)$ does not decrease. This function obviously belongs to the class $\mathscr{E}^{n+1}$, since $\bar{f}(x) \in \mathscr{E}^n$ and since one recursion applied to the functions $\mathscr{E}^n$ may yield only a function from $\mathscr{E}^{n+1}$.

Before the machine $\mathfrak{T}$ began to operate on the word P, an operating zone of length $[P^\partial$ was used to write the original word. Before the beginning of cyclic operation a zone of length

$$[P^\partial + s_0 + 3$$

was used, where s_0 is the length of the initial state.

Assume that the function h(x) yields the length of the operating zone used after the x-th cycle has been carried out $(0 \leqslant x \leqslant [P^{\partial}+1)$. In all, the machine carries out $[P^{\partial}+1$ cycles. Before the beginning of cyclic operation (i.e., for x = 0),

$$h(0) = [P^{\partial} + s_0 + 3 = [P^{\partial} + 1 + g(0).$$

Further,

$$h(x+1) \leqslant h(x) + 2g(x+1) + 2.$$

Since the function g(x) is nondecreasing, it follows that

$$h(x) \leqslant [P^{\partial} + 1 + 2\sum_{i=0}^{x}(g(i)+2) = [P^{\partial} + 1 + 4(x+1) + 2\sum_{i=0}^{x} g(i).$$

Since the summation operation does not extend beyond the class $\mathscr{E}^n$ (n ≥ 2), while $g(x) \in \mathscr{E}^{n+1}$, it follows that the function appearing in the right side of the inequality belongs to $\mathscr{E}^{n+1}$. Finally, in order to calculate $\mathfrak{A}(P)$ the Türing machine $\mathfrak{T}$ uses an operating zone which is bounded as follows:

$$[\mathfrak{T}(P)^3 \leqslant C_1 \sum_{i=0}^{[P^{\partial}+1} g(i) + C_2 [P^{\partial} + C_3,$$

where C_1, C_2, C_3 are certain constants determined by the automaton $\mathfrak{A}$. Once again we have constructed a nondecreasing function. Using a method analogous to the one used in the basis, one can go over from primitive (n + 1)-automata to nonprimitive ones. This ends the proof of Theorem 3.1.

4. Let us now prove a series of auxiliary statements.

4.1. Lemma on Superposition. The class of arithmetic functions $\mathcal{U}_{\Phi}^n$ is closed relative to the superposition operation.

In accordance with the definitions of the superposition operation (§1, subsection 1) the proof breaks down into three points.

a. Assume that the functions $f(x_1, \ldots, x_{k-1}, x_k, x_{k+1}, \ldots, x_l)$ $(1 \leq k \leq l)$ and $g(y_1, \ldots, y_m)$ belong to the class $\mathcal{U}_{\Phi}^n$ and are calculated on the machines $\mathfrak{F}$ and $\mathfrak{G}$, respectively.

Let us show how to construct a Turing machine $\mathfrak{H}$ from $\mathcal{U}_{\Phi}^n$, which calculates the function

$$h(x_1, \ldots, x_{k-1}, y_1, \ldots, y_m, x_{k+1}, \ldots, x_l) = f(x_1, \ldots, x_{k-1}, g(y_1, \ldots, y_m), x_{k+1}, \ldots, x_l).$$

Before the beginning of operation the word

$$x_1\alpha \ldots \alpha x_{k-1}\alpha y_1\alpha \ldots \alpha y_m\alpha x_{k+1}\alpha \ldots \alpha x_l$$

is written on the tape of the machine. In the beginning the machine $\mathfrak{H}$ operates in almost the same way as the machine $\mathfrak{G}$ in manipulating the word $y_1\alpha \ldots \alpha y_m$. The differences reside in the following: for the operation of $\mathfrak{G}$ empty cells are situated to the left and right of $y_1\alpha \ldots \alpha y_m$; for the machine $\mathfrak{H}$ this is not so. Therefore, the machine $\mathfrak{H}$ starts by noting the boundaries of the word $y_1\alpha \ldots \alpha y_m$ (i.e., it writes the word $x_1\alpha \ldots \alpha x_{k-1}\beta y_1\alpha \ldots \alpha y_m\gamma x_{k+1}\alpha \ldots \alpha x_l$ on the tape (we shall assume that β and γ do not belong to the alphabets of the machines $\mathfrak{F}$ and $\mathfrak{G}$).

In the process of operation the machine $\mathfrak{H}$, in simulating the activity of $\mathfrak{G}$, constantly monitors it, and in the case in which $\mathfrak{G}$ attempts to write some letter in a cell containing β, it moves the word ending in the letter β one cell to the left, thereby freeing one cell for the operation of $\mathfrak{G}$. Analogously, the machine moves the word beginning with the letter γ one cell to the right when this is necessary. It is obvious that the operating zone required for these calculations does not exceed the following magnitude:

$$z+\bar{g}([y_1\alpha\ldots\alpha y_m^\vartheta),$$

where g is a function belonging to $\mathscr{E}^n$ such that

$$[\mathfrak{G}(y_1\alpha\ldots\alpha y_m)^\vartheta\leqslant\bar{g}(\sum_{i=1}^{m}y_i+m\dot{-}1),$$

while $z=[x_1\alpha\ldots\alpha x_{k-1}\alpha y_1\alpha\ldots\alpha y_m\alpha x_{k+1}\alpha\ldots\alpha x_l^\vartheta$. Since $[y_1\alpha\ldots\alpha y_m^\vartheta<z$, the zone may be bounded by the function of the following type:

$$\bar{h}_1(z)=z+\bar{g}(z).$$

It is clear that the function $\bar{h}_1$ belongs to the class $\mathscr{E}^n$.

Then the machine $\mathfrak{H}$, after the calculation of $g(y_1, \ldots, y_m)$ has been completed, shifts the word that has been obtained so as to achieve a situation in which the word

$$x_1\alpha\ldots\alpha x_{k-1}\alpha g(y_1,\ldots,y_m)\alpha x_{k+1}\alpha\ldots\alpha x_l$$

is formed on the tape. Now the machine $\mathfrak{H}$ operates on this word in exactly the same way as the machine $\mathfrak{F}$; as a result, $h(x_1, \ldots, x_{k-1}, y_1, \ldots, y_m, x_{k+1}, \ldots, x_l)$ is calculated. The zone which would be required for this calculation does not exceed

$$\bar{h}(z)=\bar{f}(\bar{h}_1(z));$$

here $\bar{f}$ is a function from $\mathscr{E}^n$ such that

$$[\mathfrak{F}(x_1\alpha\ldots\alpha x_l)^\vartheta\leqslant\bar{f}(\sum_{i=1}^{l}x_i+l\dot{-}1).$$

It can be understood that $\bar{h}$ belongs to the class $\mathscr{E}^n$, which proves point a) of the lemma.

We shall not consider points b) and c) by virtue of their obvious nature.

4.2. Lemma on Bounded Recursion. The class of arithmetic functions $\mathscr{U}_\Phi^n$ is closed relative to the operation of bounded recursion.

Assume the function $f(x_1, \ldots, x_l, x)$ is obtained by the operation of bounded recursion from the functions $g(x_1, \ldots, x_l)$, $h(x_1, \ldots, x_l, x, y)$, $j(x_1, \ldots, x_l, x)$ in accordance with [§1.1]. Assume that the functions g, h, j belong to the class $\mathscr{U}_\Phi^n$, the functions $\bar{g}, \bar{h}, \bar{j}$ from $\mathscr{E}^n$ bounding the zones required for calculating the functions g, h, j on the machines $\mathfrak{G}$, $\mathfrak{H}$, $\mathfrak{J}$, respectively.

As the function $j(x_1, \ldots, x_l, x)$ one may always choose a nondecreasing function. Actually, from the function $j(x_1, \ldots, x_l, x)$ one can construct a function of the following type:

$$j'(x_1,\ldots,x_l,x)=\sum_{i_1=0}^{x_1}\sum_{i_2=0}^{x_2}\cdots\sum_{i_l=0}^{x_l}\sum_{i=0}^{x}j(i_1,i_2,\ldots,i_l,i).$$

Since the summation operation does not extend beyond the limits of the class $\mathscr{E}^n$ ($n \geq 2$) (see [9]), the function $j'\in\mathscr{E}^n$. It is obvious that the function j' is nondecreasing and

$$j(x_1,\ldots,x_l,x)\leqslant j'(x_1,\ldots,x_l,x).$$

Therefore, further on we shall assume in proving this lemma that the function j is nondecreasing.

Let us construct the machine $\mathfrak{F}$ from $\mathcal{U}^n_\Phi$, which calculates the function f. Before the beginning of operation, the word

$$x_1\alpha \dots \alpha x_l\alpha x.$$

is written on the tape of the machine. Assume z is the length of this word:

$$z = \sum_{i=1}^{l} x_i + x + l.$$

First we double the word written on the tape; i.e., we obtain the word

$$x_1\alpha \dots \alpha x_l\alpha x\beta x_1\alpha \dots \alpha x_l\alpha x.$$

Under these conditions β is a letter that does not belong to the alphabets of the machines $\mathfrak{G}$, $\mathfrak{H}$, $\mathfrak{I}$. For this purpose we require an operating zone having a length 2z + 2. After this we calculate $j(x_1, \dots, x_l, x)$ in such a way as to obtain the word

$$x_1\alpha \dots \alpha x_l\alpha x\beta j(x_1, \dots, x_l, x)$$

on the tape. For this we require a zone no larger than

$$2z + 1 + \bar{j}(z).$$

Then we construct the operation of the machine $\mathfrak{F}$ in such a way as to obtain the word

$$x_1\alpha \dots \alpha x_l\alpha x\beta 0\alpha 0\alpha P_{0,0}\beta 0\alpha \mid \alpha P_{0,1}\beta \dots \beta 0\alpha j(x_1, \dots, x_l, x)\,\alpha P_{0,\,j(x_1,\dots,x_l,x)}\beta \mid \alpha 0\alpha P_{1,0}\beta \dots$$
$$\dots \beta x\alpha 0\alpha P_{x,0}\beta \dots \beta x\alpha j(x_1, \dots, x_l, x)\,\alpha P_{x,\,j(x_1,\dots,x_l,x)},$$
$$P_{i,k} \eqcirc x_1\alpha \dots \alpha x_l\alpha i\alpha k$$
$$(0 \leqslant i \leqslant x,\ 0 \leqslant k \leqslant j(x_1, \dots, x_l, x))$$

on the tape. Obviously, in order to construct this word we require a zone which does not exceed

$$[(x+1)\,j(x_1, \dots, x_l, x) + 1]\cdot(z + 2j(x_1, \dots, x_l, x) + x + 4) \leqslant [(z+1)\cdot i(z) + 1]\,[2z + 2i(z) + 4],$$

where $i(z) = j(z, \dots, z, z)$.

Since the function j is nondecreasing, while $z \geq x_i$, $(1 \leq i \leq l)$, $z \geq x$, it follows that

$$j(x_1, \dots, x_l, x) \leqslant i(z).$$

After this the machine $\mathfrak{H}$ applies the machine $\mathfrak{F}$ to each word $P_{i,k}$. As a result the word

$$x_1\alpha \dots \alpha x_l\alpha x\beta 0\alpha 0\alpha h(x_1, \dots, x_l, 0, 0)\,\beta$$
$$0\alpha \mid \alpha h(x_1, \dots, x_l, 0, 1)\,\beta \dots \beta 0\alpha j(x_1, \dots, x_l, x)$$
$$\alpha h(x_1, \dots, x_l, 0, j(x_1, \dots, x_l, x))\,\beta \dots \beta x\alpha 0\alpha h(x_1, \dots, x_l, x, 0)\,\beta \dots$$
$$\dots \beta x\alpha j(x_1, \dots, x_l, x)\,\alpha h(x_1, \dots, x_l, x, j(x_1, \dots, x_l, x))$$

is obtained. For this calculation we require a zone no larger than

$$z + 1 + (x+1)\,j(x_1, \dots, x_l, x)\cdot[\bar{h}(2z + j(x_1, \dots, x_l, x)) + 2z + 3].$$

This zone may be bounded as follows:

$$(z+1)\cdot i(z)\cdot[\bar{h}(2z + i(z)) + 2z + 3] + z + 1.$$

Now the machine $\mathfrak{F}$ transforms the word by applying the machine $\mathfrak{G}$ to the piece $x_1\alpha \ldots \alpha x_l$. As a result the word

$$g(x_1, \ldots, x_l)\,\alpha x\beta 0\alpha 0\alpha h(x_1, \ldots, x_l, 0, 0)\,\beta$$
$$0\alpha \,|\, \alpha h(x_1, \ldots, x_l, 0, 1)\,\beta \ldots \beta x\alpha j(x_1, \ldots, x_l, x)\,\alpha$$
$$h(x_1, \ldots, x_l, x, j(x_1, \ldots, x_l, x))$$

turns out to be written on the tape. The zone required for the calculations indicated is bounded as follows:

$$\bar{g}(z) + (z+1)\cdot i(z)\,[\bar{h}(2z + i(z)) + 2z + 3] + z + 1.$$

For the subsequent calculations we already require no expansion of the zone. The machine $\mathfrak{F}$ operates further in the following way. It clarifies whether or not it is true that $x = 0$. If $x = 0$, then the calculated value is equal to $g(x_1, \ldots, x_l)$. The machine leaves this value on the tape, while it erases everything else.

Let us now consider the case in which $x > 0$. Let us show how the machine, knowing $f(x_1, \ldots, x_l, k)$ and k, calculates $f(x_1, \ldots, x_l, k+1)$. For $k = 0$, $f(x_1, \ldots, x_l, 0)$; as we have already said – this is $g(x_1, \ldots, x_l)$. This number written on the tape is tagged in some way by the machine so that the machine can find it. Moreover, it also tags the number x, the tag denoting that the case $k = 0$ is being considered. Then the machine seeks an entry of the word $\beta P\alpha Q\alpha R\beta$ $(P, Q, R\Omega Ч)$, such that $P = 0$, $Q = g(x_1, \ldots, x_l)$. Such an entry can certainly be found, since

$$f(x_1, \ldots, x_l, x) \leqslant j(x_1, \ldots, x_l, x),$$

while the word written on the tape is such that for any $P \leq x$, $Q \leq j(x_1, \ldots, x_l, x)$ an entry of the word $\beta P\alpha Q\alpha R\beta$ can be found in it.

Thus, the necessary entry has been found. Then

$$R = h(x_1, \ldots, x_l, 0, g(x_1, \ldots, x_l)) = f(x_1, \ldots, x_l, 1).$$

The machine tags this word (and erases the tag on the preceding value of the function f). Furthermore, it shifts the tag on the word x, noting that the calculated value corresponds to $k = 1$. Then it checks whether or not k and x coincided. If $k < x$, then the operation described above is repeated. If $k = x$, the machine erases everything on the tape with the exception of the tagged word.

Considering the constraints on the length of the operating zone which we gave during the description of the operation of the machine $\mathfrak{F}$, it can easily be seen that

$$[\mathfrak{F}(x_1\alpha \ldots \alpha x_l\alpha x)^3 \leqslant \bar{f}(z),$$

where $\bar{f}(z) \in \mathscr{E}^n$. Thus, the proof of the lemma has been completed.

5. Let us consider the sequence of functions f_n $(n \geq 0)$ from [§1.1]. The description of the function f_{n+1} $(n \geq 2)$ [§1.1(4)] is not primitively recursive (f_{n+1} is defined by means of so-called recursion with insertions; see [7]). Let us give a primitively recursive description of f_{n+1}. Note initially that the functions f_0, f_1, f_2 are obviously primitively recursive. Let us now construct the auxiliary function g_{n+1} $(n \geq 2)$:

$$\begin{cases} g_{n+1}(0, y) = y, \\ g_{n+1}(x+1, y) = f_n(g_{n+1}(x, y) + 1,\ g_{n+1}(x, y) + 1). \end{cases} \quad (1)$$

It is not difficult to see that the function f_{n+1} ($n \geq 2$) may be expressed in terms of the function g_{n+1} thus:

$$f_{n+1}(x, y) = g_{n+1}(2^x, y).$$

Then we prove primitive recursiveness of the functions g_{n+1} and f_{n+1} by induction with respect to n on the basis of the fact that since f_2 is primitively recursive, g_3 is also primitively recursive.

5.1. One may construct a Türing machine from $\mathcal{T}^3_\Phi$, such that it calculates the function $f_3(x, y)$ (i.e., $f_3(x, y) \in \mathcal{T}^3_\Phi$).

First let us show that the function $g_3(x, y)$ belongs to $\mathcal{T}^3_\Phi$. Actually, this function is defined as follows in accordance with (1):

$$g_3(0, y) = y,$$
$$g_3(x+1, y) = (g_3(x, y) + 2)^2.$$

From Theorem 4.4. of [9] it follows that

$$g_3(x, y) < (y+2)^{2^{2^{2x}}}$$

Thus, obviously,

$$y + 2 < 2^{y+2}.$$

Therefore,

$$g_3(x, y) < 2^{(y+2)\cdot 2^{2^{2x}}}.$$

The function appearing in the right side of the latter inequality belongs to A^1_Φ, since A^1_Φ belong to the functions N^x, $x \cdot y$, $x + y$ and this class is closed relative to substitution in accordance with [III. §1.3.1].

Since, as proved in Theorem 3.1, $A^n_\Phi \subseteq \mathcal{T}^{n+2}_\Phi$, it follows that the function $2^{(y+2)2^{2^{2x}}}$ belongs to $\mathcal{T}^3_\Phi$. Thus, applying the lemma on bounded recursion 4.2, we find that $g_3(x, y) \in \mathcal{T}^3_\Phi$. Further, the function 2^x belongs to A^1_Φ, and therefore it belongs to $\mathcal{T}^3_\Phi$. Applying the lemma on superposition 4.1, we find that $f_3(x, y) \in \mathcal{T}^3_\Phi$.

6. We shall say that the function $f(x_1, \ldots, x_l)$ majorizes the function $g(x_1, \ldots, x_l)$ (or the function g is majorized by the function f) if the function f is nondecreasing and for any collection of values of the variables $x_1, \ldots, x_l$ the inequality

$$g(x_1, \ldots, x_l) \leqslant f(x_1, \ldots, x_l)$$

holds.

First let us prove several lemmas.

<u>6.1. Lemma.</u> Assume that the functions $h_1(x_1, \ldots, x_{k-1}, x_k, x_{k+1}, \ldots, x_l)$ and $h_2(y_1, \ldots, y_m)$ are such that for them $\bar{h}_1$ and $\bar{h}_2$ from A^n_Φ, exist which majorize them. Then for the function h obtained by substituting the function h_2 into the function h_1 in place of the variable x_k, there exists in A^n_Φ a function $\bar{h}$ that majorizes it.

Thus,

$$h(x_1, \ldots, x_{k-1}, y_1, \ldots, y_m, x_{k+1}, \ldots, x_l) = h_1(x_1, \ldots, x_{k-1}, h_2(y_1, \ldots, y_m), x_{k+1}, \ldots, x_l);$$

$$h(x_1, \ldots, x_{k-1}, y_1, \ldots, y_m, x_{k+1}, \ldots, x_l) \leqslant \overline{h}_1(x_1, \ldots, x_{k-1}, h_2(y_1, \ldots, y_m), x_{k+1}, \ldots, x_l) \leqslant$$
$$\leqslant \overline{h}_1(x_1, \ldots, x_{k-1}, \overline{h}_2(y_1, \ldots, y_m), x_{k+1}, \ldots, x_l).$$

Since the class A^n_Φ is closed relative to superposition of functions, it follows that a function $\overline{h}$ has been constructed that belongs to A^n_Φ and majorizes the function h.

6.2. Lemma. Assume that the function h_1 is such that a function $\overline{h}_1$ from A^n_Φ exists for it which majorizes it. Then for the function h, obtained from h_1 by the operation of identification of variables, there exists in A^n_Φ a function $\overline{h}$ that majorizes it.

The function $\overline{h}$ is derived from the function $\overline{h}_1$ by identification of those same variables that were identified in the transition from the function h_1 to the function h.

6.3. Lemma. Assume the function h_1 is such that for it there exists a function $\overline{h}_1$ from A^n_Φ, which majorizes it. Then for the function h obtained from h_1 by the operation of substitution of 0 there exists in A^n_Φ a function $\overline{h}$ that majorizes it.

The function $\overline{h}$ is obtained from the function $\overline{h}_1$ by substituting 0 in place of the very same variable as in the transition from the function h_1 to the function h.

6.4. Lemma. Assume that the functions $h_1(x_1, \ldots, x_l)$, $h_2(x_1, \ldots, x_l, x, y)$, $h_3(x_1, \ldots, x_l, x)$ are such that for them the functions $\overline{h}_1, \overline{h}_2, \overline{h}_3$, respectively, from A^n_Φ, exist which majorize them. Then for the function h obtained from the function h_1, h_2, h_3 by the operation of bounded recursion $(h(x_1, \ldots, x_l, x) \leq h_3(x_1, \ldots, x_l, x))$ there exists in A^n_Φ a function $\overline{h}$ that majorizes it.

It is obvious that h is majorized by the function $\overline{h}_3$.

Let us now prove the following theorem.

6.5. Theorem. a) For each function $g(x_1, \ldots, x_l)$ from $\mathscr{E}^{n+2}$ one can find a function $f(x_1, \ldots, x_l)$ from A^n_Φ $(n \geq 1)$ that majorizes it;

b) $\mathscr{T}^{n+2}_{\mathrm{A}} = A^n_{\mathrm{A}}$;

c) $f_{n+2} \in A^n_\Phi$.

We shall carry out the proof of the theorem by induction with respect to n.

The basis, n = 1. By analogy with the definition from [III. §1.2] one can introduce the concept of the rank r of a function from $\mathscr{E}^3$. Let us prove statement a) of the theorem (for the case n = 1) by induction with respect to r.

The basis, r = 1.

It is required to show that the original function can be majorized by certain functions from A^1_Φ. For the functions

$$f(x) = x + 1, \quad U_1(x, y) = x, \quad U_2(x, y) = y$$

this is obvious:

$$U_1(x, y) \leqslant x + y, \quad U_2(x, y) \leqslant x + y.$$

The function x + y belongs to A_Φ^0 – this is the automaton $\mathfrak{S}^{(0)}_{\{1,\alpha\},\alpha}$ from [II. §3.5].

Let us deal with the function $f_3(x, y)$. This function is defined as follows:

$$\begin{cases} f_3(0, y) = (y+2)^2, \\ f_3(x+1, y) = f_3(x, f_3(x, y)). \end{cases}$$

If we now define the function g_3 according to 5.1 as follows:

$$\begin{cases} g_3(0, y) = y, \\ g_3(x+1, y) = (g_3(x, y)+2)^2, \end{cases}$$

it follows that the equation

$$f_3(x, y) = g_3(2^x, y)$$

holds.

In §5.1 it was proved that the function $2^{(y+2)\cdot 2^{2^{2x}}}$ majorizes $g_3(x, y)$ and belongs to the class A_Φ^1. It is not difficult to see that this function is nondecreasing. Further,

$$f_3(x, y) < 2^{(y+2)\cdot 2^{2^{2^{x+1}}}},$$

where the function appearing in the right side of the inequality belongs to A_Φ^1.

The induction step (in the basis). This step is carried out on the basis of Lemmas 6.1-6.4. Thus, for the basis (n = 1) the statement a) of the theorem has been proved. But then, taking account of Lemma 2.1, it can easily be seen that

$$\mathcal{T}_A^3 \subseteq A_A^1 \text{ and } \mathcal{T}_\Phi^3 \subseteq A_\Phi^1.$$

However, from Theorem 3.1 it follows that

$$\mathcal{T}_A^3 = A_A^1 \text{ and } \mathcal{T}_\Phi^3 = A_\Phi^1.$$

Since $f_3(x, y) \in \mathcal{T}_\Phi^3$ (see 5.1) it follows that $f_3(x, y) \in A_\Phi^1$.

The Induction Step. Let us begin by proving statement c) of the theorem. In accordance with the inductive proposition $f_{n+2}(x, y) \in A_\Phi^n$ ($n \geq 1$). The function $g_{n+3}(x, y)$, which is defined according to [5.(1)], belongs to the class A_Φ^{n+1} (see [III. §1.3.2]). The function $y = 2^x$ belongs to any class A_Φ^i ($i \geq 1$). Since A_Φ^{n+1} is closed relative to the substitution [III. §1.3.1], it follows that $f_{n+3}(x, y) \in A_\Phi^{n+1}$.

Now in order to prove statement a) of the theorem we again carry out reasoning by induction (induction with respect to the rank r).

The basis, r = 1. Only the case of the function $f_{n+3}(x, y)$ is of interest. Thus, as proved above, the function $f_{n+3}(x, y)$ belongs to the class A_Φ^{n+1}. Therefore, the majorizing function $f_{n+3}(x, y)$ may be constructed as follows:

$$f_{n+3}(x, y) \leqslant f_{n+3}(x, y).$$

The Induction Step. The inductive step is conducted in exactly the same way as in the basis. Reasoning as in the basis, we obtain

$$\mathcal{T}_A^{n+3} = A_A^{n+1}.$$

With this the proof of the theorem ends. Let us go over to the next statement.

6.6. $\mathscr{E}^n \subseteq \mathscr{T}^n_\Phi$.

Actually, according to 6.5, $f_n(x, y) \in \mathscr{T}^n_\Phi$, other original functions of the class $\mathscr{E}^n$ obviously likewise belong to $\mathscr{T}^n_\Phi$. Moreover, from Lemmas 6.1-6.4 it follows that $\mathscr{T}^n_\Phi$ is closed relative to the operations of superposition and bounded recursion. From this we obtain 6.6.

7. The next problem of this section is to show that $A^{n+1}_\Phi \subseteq \mathscr{E}^{n+3}$ $(n \geq 0)$.

For this purpose we carry out the arithmetization of the automata from A^{n+1}_A, using $\mathscr{E}^{n+3}$ as the sole resource.

Let us modify the definition of a primitively recursive automaton given in [III. §3.1] as follows.

An n-automaton $\mathfrak{A}$ is an m-recursive automaton if the function $f(x)$ defined by the relationship

$$f(x) \simeq \mathfrak{G}(\mathfrak{A}(\mathfrak{G}_1(x))),$$

is a function from $\mathscr{E}^m$.

Now we have the possibility of formulating such a theorem.

7.1. Theorem. Any everywhere-determinate (n + 1)-automaton in the alphabet A is an (n + 3)-recursive automation, $n \geq 0$.

For purposes of the proof we will have to show that certain functions constructed in [III. §2] and [III. §3] belong to $\mathscr{E}^3$, i.e., they are elementary in the Kalmar sense.

First, the function $f_k(x, y)$ defined in [III. §2.2(2)] is obviously elementary. But then the function $f(x)$ from [III. §3.2(1)] is likewise elementary, since the class $\mathscr{E}^3$ is closed relative to the operation of bounded summation. Further, the function $g(x)$ from [III. §3.2(2)] likewise belongs to $\mathscr{E}^3$. It can easily be shown that the function $r(y)$ defined in [III. §3.2(3)] is likewise elementary. And, finally, the elementarity of the functions $h(x, y)$, $b(x, y)$, and $a(x)$ from [III. §3.2(4)-(8)] is obvious. Let us now show that the functions χ_0 (defined according to [III. §3.3(1)] and χ_1 are elementary, just as are the functions ω_0 (from [III. §3.3(2)]) and ω_1, while the functions χ_i and ω_i belong to the class $\mathscr{E}^{i+2}$ $(i \geq 2)$. Actually, the function χ_0 uses the pair of natural numbers (x, y) to produce the principal quasi number of the state into which the primitive 0-automaton $\mathfrak{A}$ transfers due to the action of the word having the number $[y/(k+2)^{r(y) \dot{-} x}]$. Therefore,

$$\chi_0(x, y) \leqslant C_1,$$

where C_1 is a constant which does not depend on x, y.

Further, $\omega_0(x, y)$ yields the principal quasi number of the word that the automaton $\mathfrak{A}$ prints due to the action of the word having the number $[y/(k+2)^{r(y) \dot{-} x}]$. If the presented word P has the length x, then it is obvious that the automaton prints a word whose length does not exceed

$$C_2([P^\partial + 1) = C_2(x+1).$$

But then

$$\omega_0(x, y) \leqslant (k+2)^{C_2(x+1)+1}.$$

It turns out to be possible to use bounded recursion to define the functions χ_0 and ω_0 defined in [§3.3(1)] and [§3.3(2)] by means of recursions. Therefore, $\chi_0 \in \mathscr{E}^3$ and $\omega_0 \in \mathscr{E}^3$.

Let us go on. The function F_0, as follows from its definition [III. § 3.3(3)] and what has been said above, is likewise elementary. We now go over to the functions χ_1 and ω_1.

Due to the action of the word $\xi_1 \ldots \xi_x$ $(x > 0)$ the primitive automaton $\mathfrak{A}$ transits from the initial state S_0 to the state S_x. Let us estimate $[S_x^\partial$.

For a certain constant C_3 the following inequality is fulfilled for the automaton $\overline{\mathfrak{A}}$:

$$[\overline{\mathfrak{A}}(P)^\partial \leqslant C_3([P^\partial + 1).$$

Let us define the function q(x) thus:

$$\begin{cases} q(0) = [S_0^\partial, \\ q(x+1) = (C_3+2)(q(x)+1). \end{cases}$$

Then

$$[S_x^\partial \leqslant q(x). \tag{1}$$

On the other hand, by induction it can be proved that

$$q(x) = [S_0^\partial \cdot (C_3+2)^x + (C_3+2)\frac{(C_3+2)^x - 1}{C_3+1}. \tag{2}$$

Further, the length of the word that the automaton $\mathfrak{A}$ prints when the words $\xi_1\xi_2 \ldots \xi_x$ $(x > 0)$ are presented to it does not exceed the quantity

$$\sum_{i=1}^{x} q(i) \leqslant \sum_{i=1}^{x} ([S_0^\partial (C_3+2)^i + (C_3+2)^{i+1}) \leqslant x \cdot [S_0^\partial ((C_3+2)^x + (C_3+2)^{x+1}). \tag{3}$$

Let us assume that

$$p(x) = x \cdot [S_0^\partial \cdot ((C_3+2)^x + (C_3+2)^{x+1}). \tag{4}$$

If the length of the word does not exceed z, then the principal quasi number of this word does not exceed $(k+2)^{z+1}$.

Let us now return to the definition of the function χ_1. This function is defined by means of recursion via the elementary functions $f_{k+1}(x, y)$, $r(y)$, $h(x, y)$, etc., and likewise via the functions φ_1. However, the function φ_1 is likewise elementary, since

$$\varphi_1(x) = f(F_0(g(x))),$$

while the elementarity of the functions f, g, and F_0 has been proved.

Further, the function χ_1 may be bounded by an elementary function as follows by virtue of (1)-(2):

$$\chi_1(x, y) \leqslant (k+2)^{q(x)+1}.$$

Thus, χ_1 is elementary.

Now concerning ω_1. According to the definition [III. § 3.3(2)], ω_1 can be defined in terms of elementary functions by means of recursion. On the other hand, ω_1 may be bounded by an elementary function in the following way by virtue of (3)-(4):

$$\omega_1(x, y) \leqslant (k+2)^{p(x)+1}.$$

Therefore, ω_1 is elementary; then F_1 is also elementary.

Reasoning inductively, we then go over to automata having higher ranks. Let us consider the functions χ_{n+1} and ω_{n+1}. The function χ_{n+1} can be defined by recursion in terms of elementary functions and the function φ_{n+1} that belongs, according to the induction assumption, to the class $\mathscr{E}^{n+2}$; then the function χ_{n+1} will belong to the class $\mathscr{E}^{n+3}$. Estimating the function ω_{n+1} as in the basis, one can bound it by means of a function from $\mathscr{E}^{n+3}$. Therefore, ω_{n+1} and thus F_{n+1} belongs to the class $\mathscr{E}^{n+3}$ too.

With this the proof of Theorem 7.1 ends.

Now let us prove the following:

7.2. Theorem. Any completely m-automaton function belongs to the Grzegorczyk class $\mathscr{E}^{n+2}$ ($m \geq 1$).

The proof of this theorem is analogous to the proof of the theorem [III. §4.1.1]. It is merely necessary to show that the functions θ_i ($i \geq 1$) and $\varkappa$ constructed there are elementary (belong to $\mathscr{E}^3$).

Elementarity of the function $\varkappa$ derives directly from its definition [III. §4.1(4)], it being necessary to consider the fact that, as shown above, the functions $f_k(i, x)$ are elementary.

In order to prove elementarity of the function θ_i defined in [III. §4.1(3)], it is necessary to construct the elementary function that majorizes it.

It is not difficult to check the fact that

$$\theta_1(x) = a\left[\frac{(k+1)^x \dot{-} 1}{k}\right];$$

the elementarity of θ_1 derives from this.

In exactly the same way one can immediately write the formula for calculating θ_m. Actually

$$\theta_{m+1}(x_1, x_2, \ldots, x_m, x_{m+1}) = (k+1)^{x_{m+1}}((k+1)\,\theta_m(x_1, \ldots, x_m) + b) + b\cdot\left[\frac{(k+1)^{x_{m+1}} \dot{-} 1}{k}\right].$$

Thus, Theorem 7.2 has been proved.

8. Let us make several remarks. Based on the Grzegorczyk result from [9] to the effect that an $(s+1)$-place function ($s \geq 0$) exists from the class $\mathscr{E}^{n+1}$ ($n \geq 2$), which is universal in the class of s-place functions from $\mathscr{E}^n$, one can prove the following statement.

8.1. Assume A is an alphabet such that $\alpha \in A$, while $\delta \notin A$. One can construct an $(n+1)$-automaton $\mathfrak{B}^{(n+1)}$ in the alphabet $A \cup \{\delta\}$ ($n \geq 0$) which is such that regardless of the nature of the n-automaton $\mathfrak{A}^{(n)}$, one can find a word $\{\mathfrak{A}^{(n)}\}$ $(\{\mathfrak{A}^{(n)}\}\,\Omega\,\{\alpha\})$, in the alphabet A such that the following equation is valid for it:

$$\mathfrak{B}^{(n+1)}(\{\mathfrak{A}^{(n)}\}\,\delta P) \simeq \mathfrak{A}^{(n)}(P)\;(P\Omega A).$$

The automaton $\mathfrak{B}$ is called an $(n+1)$-automaton, which is universal in the class of n-automata.

Using the arithmetization presented above, it is easy to obtain the following statement:

8.2. Regardless of the nature of the n-automaton $\mathfrak{A}$ ($n \geq 1$) in the alphabet A, which is applicable to any word P in A, the function $f(x)$ defined as

$$f(x) = \max_{[P^{\partial}=x} [\mathfrak{A}(P)^{\partial},$$

belongs to the class $\mathscr{E}^{n+2}$.

Then it may be shown that:

8.3. For each $n \geq 0$ one can construct an (n+1)-automaton $\mathfrak{A}^{(n+1)}$ in the alphabet A, which is such that no n-automaton $\mathfrak{B}^{(n)}$ over the alphabet A exists for it such that the relationship

$$\mathfrak{A}^{(n+1)}(P) \simeq \mathfrak{B}^{(n)}(P) \qquad (P\Omega A)$$

holds.

Literature Cited

1. V. M. Glushkov, "Abstract theory of automata," Uspekhi Matem. Nauk, Vol. 16, No. 5 (101), pp. 3-62 (1961).
2. S. C. Kleeny, Introduction to Mathematics [Russian translation], IL, Moscow (1957).
3. V. A. Kozmidiadi, "On sets which are decidable and enumerable by automata," Dokl. Akad. Nauk SSSR, 142(5):1005-1006 (1962).
4. V. A. Kozmidiadi, "On sets which are enumerable and decidable by automata," in: Problems in Logic, Philosophy Institute, Academy of Sciences of the USSR (1963), pp. 102-115.
5. A. A. Markov, Theory of Algorithms, Transactions of the V. A. Steklov Mathematics Institute, Academy of Sciences of the USSR, Vol. 42 (1964).
6. V. A. Trakhtenbrot, Türing Calculations with Logarithmic Delay, Algebra and Logic, 3(4):33-48 (1964).
7. R. Peter, Recursive Functions [Russian translation], IL, Moscow (1954).
8. V. S. Chernyavskii, On a Certain Class of Normal Markov Algorithms, in: Logic Investigations, Philosophy Institute, Academy of Sciences of the USSR (1959), pp. 263-299.
9. A. Grzegorczyk, "Some classes of recursive functions," Rozprawy Matematyczne (Warsaw), Vol. 4 (1953).

GENERAL LINEAR AUTOMATA†

A. A. Muchnik

Moscow

Introduction

Many problems in the theory of control systems reduce to the objects considered in the present paper – general linear automata. The necessity of studying them is manifested most clearly in the problem of reducing finite probabilistic automata [12, 13, 15, 19] which are an important particular case of general linear automata. On the other hand, many properties which apply to finite deterministic automata can be generalized successfully for the case of general linear automata, their proof turning out to be sometimes simpler than well-known proofs of analogous theorems for finite automata.

§ 1. The Definition of General Linear Automata

A general linear automaton $\mathfrak{A}$ over the field P is called the system

$$\mathfrak{A} = (V, M, \Sigma, Z, \varphi, \psi),$$

where V and Z are linear spaces over the field P; M is a certain set of points in V, $M \subseteq V$; $\Sigma = \{\sigma_1, \sigma_2, \ldots, \sigma_k\}$ is an input alphabet; $\varphi_{\sigma_i}(\vec{v}) = \varphi(\vec{v}, \sigma_i)$ is a linear transformation of the space V which maps the set M in itself and depends on the input symbol σ_i, $\vec{z} = \psi_{\sigma_i}(\vec{v}) = \psi(\vec{v}, \sigma_i)$ is a linear operator which maps V in Z. The points (vectors) $\vec{v}$ of the set M are called states of the general linear automaton $\mathfrak{A}$ while the vectors $\vec{z} = \psi(\vec{v}, \sigma_i)$, where $\vec{v} \in M$ are called outputs of $\mathfrak{A}$. The entire system operates in time that is considered to be discrete:

$$t = 0, 1, 2, \ldots$$

If $\mathfrak{A}$ is in the state $\vec{v}(t)$ at time t and a signal $\sigma = \sigma(t)$ is applied at that time, then the output of the general linear automaton $\mathfrak{A}$ at time t is

$$\vec{z}(t) = \psi_{\sigma(t)}[\vec{v}(t)] = \psi[\vec{v}(t), \sigma(t)], \tag{1}$$

while the state of $\mathfrak{A}$ at the next instant in time is

$$\vec{v}(t+1) = \varphi_{\sigma(t)}[\vec{v}(t)] = \varphi[\vec{v}(t), \sigma(t)]. \tag{2}$$

† Original article submitted October 26, 1967.

As we see, this definition is analogous to the definition of a finite deterministic Mealy automaton [4], while general linear automata are deterministic Mealy automata which are finite or infinite (this depends on the power of Σ, M, and V).

Initial general linear automata are called general linear automata having an isolated initial state $\vec{v}_0 \in M$.

The dimensionality $r(\mathfrak{A})$ *of a general linear automaton* $\mathfrak{A}$ *is called* the dimensionality of the minimal hyperplane Γ_M in V, which contains the set M. *The dimensionality of the output of a general linear automaton* $\mathfrak{A}$ is called the dimensionality of the space Z. Further on we shall be interested chiefly in general linear automata having finite-dimensioned Γ_M and Z. We shall call such general linear automata *finite-dimensioned general linear automata*.

Two cases are possible: 1) $\Gamma_M = L(M)$, where L(M) denotes the linear hull of the set M, and 2) $\Gamma_M \subset L(M)$ (the symbol $\subset$ denotes rigorous insertion throughout this paper).

For example, if V is a three-dimensional space, while the set M consists of two points $a = (1, 0, 0)$ and $b = (0, 1, 0)$, then Γ_M is the straight line $x + y = 1$ in the plane Oxy, while L(M) coincides with Oxy. Now assume M contains, besides a and b, an additional point $c = (x_c, y_c, 0)$, which does not lie on the straight line $x + y = 1$. Then $\Gamma_M = L_M = Oxy$.

In the first case $r(\mathfrak{A})$ is equal to the dimensionality of L(M).

In the second case (for finite-dimensioned general linear automata) $r(\mathfrak{A})$ is equal to $r[L(M)] - 1$, where $r[L(M)]$ is equal to the dimensionality of L(M).

Actually, having chosen from the vectors M the basis $\vec{v}_1, \vec{v}_2, \ldots$ for L(M) and having expressed all the coordinates of points M in this basis, we obtain: either (a) a point $x = (x_1, x_2, \ldots)$ can be found in M such that $\Sigma x_i \neq 1$, or (b) $\Sigma x_i = 1$ for all points $(x_1, x_2, \ldots) \in M$.

It is not difficult to see that cases a) and b) coincide with cases 1 and 2 considered above, respectively.

General linear automata for which case 1 [a] holds are called *general-linear automata of the I type*, while general linear automata for which case 2 [b] holds are called *general linear automata of the II type*.

Since the set M goes over into itself for any mapping of φ_{σ_i}, it follows that the functioning of general linear automata depends solely on the set M, and on Σ, Z, and φ, while it is independent of the choice of the linear space V, provided only that the mappings of φ and ψ on the set $M \times \Sigma$ remain unchanged. The least of possible V containing M is obviously L(M), and therefore hereafter we shall not distinguish between $\mathfrak{A}(V, M, \ldots)$ and $\mathfrak{A}(L(M), M, \ldots)$, and V will sometimes be dropped in the notation of a general linear automaton for the sake of brevity.

If $M = L(M)$, then the general linear automaton $\mathfrak{A}$ is called *complete*. If $M = \Gamma_M \subset L(M)$, then the general linear automaton $\mathfrak{A}$ is called *weakly complete*.

The general linear automaton $\mathfrak{A}' = (M', \Sigma, Z, \varphi', \psi')$ is called *a subautomaton of the general linear automaton* $\mathfrak{A}$, if $M' \subseteq M$ and the mappings φ'_{σ_i} and ψ'_{σ_i} coincide with φ_{σ_i} and ψ_{σ_i} of the space L(M') respectively (it is obvious that $L(M') \subseteq L(M)$). In this case the general linear automaton $\mathfrak{A}$ is called a *hyperautomaton or supplement* of $\mathfrak{A}'$. The general linear automation $\overline{\mathfrak{A}} = (\overline{M}, \Sigma, Z, \varphi, \psi)$ with $\overline{M} = L(M)$ is called the *supplement of the general linear automaton* $\mathfrak{A}(M, \ldots)$, while the general linear automaton $\mathfrak{A}(\Gamma_M, \Sigma, Z, \varphi, \psi)$ is called the *weak supplement of the general linear automaton* $\mathfrak{A}(M, \ldots)$, if $\Gamma_M \subset L(M)$.

Since general linear automata are automata (perhaps, infinite ones), one may speak of their automaton isomorphism, homomorphism, etc. Various general linear automata may be isomorphic to one and the same deterministic automaton $\mathfrak{B}(S, \Sigma, U, \varphi, \psi)$; this depends on the choice of the field P and the representations of S and U as sets in linear spaces over P.

The mappings φ_{σ_i} can be generalized naturally for words x:

$$\left.\begin{array}{l}\varphi_\Lambda = e \quad (e \text{ is an identical transformation,} \\ \Lambda \text{ is an empty word}), \\ \varphi_{x\sigma} = \varphi_\sigma \cdot \varphi_x.\end{array}\right\} \tag{3}$$

If systems of coordinate vectors are chosen in the spaces L(M) and Z, then the linear transformations φ_{σ_i} and the linear operators ψ_{σ_i} correspond to the matrices $A(\sigma_i)$ and $B(\sigma_i)$. Equations (1) and (2) then yield

$$\vec{z}(t) = \vec{v}(t)\, B\, [\sigma(t)], \tag{1'}$$

$$\vec{v}(t+1) = \vec{v}(t)\, A\, [\sigma(t)]. \tag{2'}$$

For general linear automata of the II type the sum of the elements of each row of each matrix $A(\sigma)$ is equal to 1.

The system of coordinate vectors (the basis) in L(M) is called proper if it is contained in M. Sometimes it is necessary for us to choose linear dependent systems of coordinate vectors M and Z. In these cases the coordinates of the points M and the matrices $A(\sigma_i)$ and $B(\sigma_i)$ are ambiguously defined if additional constraints are not imposed on them.

Let us note that the term "linear automata over the field P" in the literature is currently assumed to define automata whose states, inputs, and outputs at time t are vectors over the fields P: $\vec{s}(t)$, $\vec{x}(t)$, and $\vec{y}(t)$, respectively, while the transitions and outputs are described by the equations

$$\vec{s}(t+1) = \vec{s}(t)\, A + \vec{x}(t)\, B, \quad \vec{y}(t) = \vec{s}(t)\, C + \vec{x}(t)\, D,$$

where **A, B, C,** and **D** are matrices which do not depend on time.†

Let us show that linear automata are a particular case of general linear automata. As the states of the general linear automaton $\mathfrak{A}$ we shall take the vector $\vec{v} = (\vec{s}, \underbrace{1, 1, \ldots, 1}_{l \text{ ones}})$,

where l is the dimensionality of the input vector $\vec{x}(t)$. The transition matrix $\mathbf{A}_1(\vec{x})$ is assumed to equal

$$\left(\begin{array}{c|c} A & 0 \\ \hline B^*(\vec{x}) & \begin{array}{cccc} 1 & 0 & \ldots & 0 \\ 0 & 1 & \ldots & 0 \\ \cdot & \cdot & \cdot & \cdot \\ 0 & 0 & \ldots & 1 \end{array}\end{array}\right),$$

where the elements $\mathbf{B}^*(\vec{x})$ are $b^*_{ij}(\vec{x}) = x_i b_{ij}$, while the output matrix $\mathbf{B}_1(\vec{x})$ is assumed to equal

† Recently, a paper by D. R. Deuel has appeared [J. Comp. Syst. Sci., 3(1):93-118 (1969)] in which linear automata with time-varying matrices A, B, C, and D are considered.

$$\begin{pmatrix} C \\ D^*(\vec{x}) \end{pmatrix},$$

where $d^*_{ij} = x_i d_{ij}$. It is obvious that

$$\vec{v}(t) A_1(\vec{x}) = \vec{v}(t+1) = (\vec{s}(t+1), \underbrace{1, 1, \ldots, 1}_{l \text{ ones}}), \qquad \vec{y}(t) = \vec{v}(t) B_1(\vec{x}).$$

Thus, the general linear automaton $\mathfrak{A}$ describes the operation of a linear automaton.

Although the dimensionality of the vector $\vec{v}$ is equal to the sum of the dimensionalities of the vectors $\vec{s}$ and $\vec{x}$, the set M of all vectors $\vec{v} = (\vec{s}, 1, 1, \ldots, 1)$ nevertheless lies in the hyperplane defined by the condition that the last l coordinates $\vec{v}$ are equal to 1. Therefore, the dimensionality of this hyperplane is equal to the dimensionality of $\vec{s}$, and the dimensionality of the general linear automaton $\mathfrak{A}$ is the same.

One may define the notion of a Moore general linear automaton by choosing the linear operator ψ to be independent of σ.

The initial general linear automaton $\mathfrak{A}$ maps any string of inputs $\sigma_{i_1}, \sigma_{i_2}, \ldots, \sigma_{i_k}, \ldots$ into a string of outputs

$$\vec{z}(\sigma_{i_1}), \vec{z}(\sigma_{i_1}, \sigma_{i_2}), \ldots, \vec{z}(\sigma_{i_1}, \sigma_{i_2}, \ldots, \sigma_{i_k}), \ldots,$$

this mapping being deterministic [5, 22].

Let us also introduce the linear space of inputs U_Σ having the dimensionality $m = |\Sigma|$, having juxtaposed each input signal σ_i with a basis vector $\vec{\sigma}_i$. The elements of U_Σ will be the vectors $k_1\vec{\sigma}_1 + k_2\vec{\sigma}_2 + \ldots + k_m\vec{\sigma}_m$, where $k_1, k_2, \ldots, k_m$ are arbitrary elements of the field P.

The input $\vec{k} = k_1\vec{\sigma}_1 + k_2\vec{\sigma}_2 + \ldots + k_m\vec{\sigma}_m$ (hereafter we shall write: $(k_1, k_2, \ldots, k_m)$) corresponds to the linear transformations

$$\varphi_{\vec{k}}(\vec{v}) = \sum_{i=1}^{m} k_i \varphi_{\sigma_i}(\vec{v})$$

and the linear operators

$$\psi_{\vec{k}}(\vec{v}) = \sum_{i=1}^{m} k_i \psi_{\sigma_i}(\vec{v}).$$

Instead of the matrices $\mathbf{A}(\sigma_j)$ and $\mathbf{B}(\sigma_j)$ it is necessary to take the vectors $a_i^{jh} = a_{ih}(\sigma_j)$ and $b_l^{jh} = b_{lh}(\sigma_j)$.

Due to the action of the vector $\vec{r} = r_j$, the state $\vec{v} = v_k$ goes over into the state

$$\vec{v}' = v'_i = r_j v_h a_i^{jh} \tag{1''}$$

and yields the output – the vector

$$\vec{z} = z_l = r_j v_h b_l^{jh} \tag{2''}$$

(the multiplication in (1") and (2") is tensor multiplication; i.e., $r_j a_i^{jh} = \sum_j r_j a_i^h(\sigma_j)$; compare with (1') and (2')).

In matrix form the transition and output equations will be the following:

$$\vec{v}(t+1) = \vec{v}(t) \boldsymbol{A}(\vec{r}), \quad \vec{z}(t) = \vec{v}(t) \boldsymbol{B}(\vec{r}),$$

where $\mathbf{A}(\vec{r})$ and $\mathbf{B}(\vec{r})$ are defined as:

$$A(\vec{r}) = \sum_j A(\sigma_j) r_j, \qquad B(\vec{r}) = \sum_j B(\sigma_j) r_j.$$

Now any input string of vectors $\vec{r}, \vec{r}_2, \ldots, \vec{r}_k, \ldots$ is already juxtaposed with the output string $\vec{z}_1, \vec{z}_2, \ldots, \vec{z}_k$ by an initial general linear automaton, this mapping θ being linear and deterministic (i.e., without anticipation [4]).

We use $\theta_{\vec{s}_1, \vec{s}_2, \ldots, \vec{s}_j}$ to denote the mapping of the set of strings $\{\vec{r}_k\}$ into the set of strings $\{\vec{z}_k\}$, which is such that

$$\theta(\vec{s}_1, \vec{s}_2, \ldots, \vec{s}_j, \vec{r}_1, \vec{r}_2, \ldots, \vec{r}_k) = \vec{w}_1, \ldots, \vec{w}_j, \vec{z}_1, \vec{z}_2, \ldots, \vec{z}_k, \ldots$$

The mapping $\theta_{\vec{s}_1 \vec{s}_2 \ldots \vec{s}_j}$ is achieved by a general linear automaton $\mathfrak{A}$ having the initial state $v(\vec{s}_1, \vec{s}_2, \ldots, \vec{s}_j)$, which is that state into which $\vec{v}_0$ transits due to the action of the input $\vec{s}_1, \vec{s}_2, \ldots, \vec{s}_j$. The set of mappings $\theta_{\vec{s}_1 \vec{s}_2 \ldots \vec{s}_j}$ is thus a linear space having a dimensionality no higher than the dimensionality of $\mathfrak{A}$. These considerations are analogous to the investigation of finite-automaton mappings by Raney [22]. Hereafter we, however, will be more interested in mappings connected with input symbols $\{\sigma_i\}$, rather than with their vector generalizations.

It is not difficult to prove the equivalence of Mealy and Moore general linear automata from the point of view of the mappings which they produce. Only the dimensionality L(M) of a general linear Moore automaton may be, in general, higher than the dimensionality $L(M_1)$ of the corresponding general linear Mealy automaton by a factor m, where m is the number of input letters.

§ 2. Examples of General Linear Automata

A physical system considered in discrete time, whose space can be described by n real (or complex) numerical parameters which vary due to the input of external signals, these variations being linear transformations of the parameters, may serve as the interpretation of an n-dimensional general linear automaton. A controllable Markov chain having n states may serve as an example of such a system.

Below these objects – finite probabilistic automata and sources – will be considered in detail, and it will be shown that they are a particular case of a general linear automaton.

Let us give a number of other examples of general linear automata.

1. Let us show that finite deterministic automata having states are (n – 1)-dimensional general linear automata over any field P.

Assume $\mathfrak{B}(S, \Sigma, \Xi, \varphi^*, \psi^*)$ is a finite deterministic Mealy automaton, where $S = \{s_1, s_2, \ldots, s_n\}$ is its set of states, Σ is the input alphabet, $\Xi = (\xi_1, \xi_2, \ldots, \xi_m)$ is the output alphabet, $\varphi^*(s, \sigma)$ and $\psi^*(s, \sigma)$ are the functions of transitions and outputs, respectively [4].

Let us consider the general linear automaton $\mathfrak{A}(V, M, \Sigma, Z, \varphi, \psi)$, where M is the set of basic vectors $P_n = \{\vec{e}_1, \vec{e}_2, \ldots, \vec{e}_n\}$, Z is an m-dimensional space over the field P having the basis $(\vec{f}_1, \vec{f}_2, \ldots, \vec{f}_m)$, and the linear transformations φ_{σ_i} and the operators ψ_{σ_i} are defined as follows:

$$\varphi_{\sigma_i}(\vec{e}_h) = \vec{e}_j,$$

when $\varphi^*(s_h\sigma) = s_j$ ($i = 1, 2, \ldots, k$; $h = 1, \ldots, n$) and

$$\psi_{\sigma_i}(\vec{e}_h) = \vec{f}_l,$$

when $\psi(s_h, \sigma_i) = \xi_l$ ($h = 1, 2, \ldots, n$; $i = 1, 2, \ldots, k$). It is not difficult to verify the fact that $\mathfrak{A}$ is a general linear automaton of the II type.

Juxtaposing the state S_h with the vectors $\vec{e}_h$, and the outputs ξ_l with the vectors $\vec{f}_l$, we thereby establish isomorphism of the automaton $\mathfrak{B}$ and the general linear automaton $\mathfrak{A}$.†

The dimensionality of $\mathfrak{A}$ is equal to $n - 1$, since all of the elements belong to the hyperplane $x_1 + x_2 + \ldots + x_n = 1$.‡

2. In exactly the same way one may construct a general linear automaton having the dimensionality $n - 1$ which is isomorphic to any probabilistic automaton having n states [13, 20]. One need only take the vectors of the probability distribution of the states of the probabilistic automaton as the "states" of this automaton, and the probability distribution vectors of the output symbols of the probabilistic automaton as its "outputs."

After this the probabilistic automaton becomes an automaton (having an infinite number of states and outputs), and one can speak of its isomorphism relative to a general linear automaton (of the II type).

3. Probabilistic sources may be interpreted by means of a general linear automaton. However, instead of considering probabilistic sources, it is better to consider probabilistic automata having a Carlyle output which generalize these sources; such automata are essentially Shannon communication channels having a finite number of states [7, 13, 14]. They are defined by stipulating sets of inputs X, sets of states $S = (s_1, s_2, \ldots, s_n)$, sets of outputs Y, and the conditional probabilities $p(y, s'/s, x)$ that the system will go over from state s to state s' due to the input x (a channel transmitting the signal x) and produce y at the output.

(Finite probabilistic sources correspond to the case in which X contains one symbol.)

Each string of inputs-outputs $(x_1y_1), (x_2y_2) \ldots (x_ky_k)$ and each state s corresponds to a number $p_s(y_1y_2 \ldots y_k/x_1x_2 \ldots x_k)$ – the probability that the automaton in the state s, having accepted the input word $x_1x_2 \ldots x_k$, will deliver the word $y_1y_2 \ldots y_k$.

Let us now construct a general linear automaton $\mathfrak{A}(M, X \times Y, \varphi, \psi)$, where the set of states M is the set of points $\vec{p} = (p_1, p_2, \ldots, p_k)$, $1 \geq p_i \geq 0$; $X \times Y$ is the set of inputs of $\mathfrak{A}$; the set of real numbers is the space of outputs of $\mathfrak{A}$; the mapping $\varphi(x, y)$ of the set M is determined by the matrix $\|a_{ij}(x, y)\|$, $a_{ij}(x, y) = \sum p(y, s_j/s_i, x)$, and the linear functional $\psi(x, y) = B(\vec{p}) = \vec{b} \cdot \vec{p}$ will be defined by the vector $\vec{b}(x, y) = (b_1, b_2, \ldots, b_n)$, where $b_i(x, y) = \sum_{s' \in S} p(y, s'/s_i, x)$. The number p_i ($i = 1, \ldots, n$) is the probability that the probabilistic automaton (starting from a certain state) delivers the word y due to the input of the input word x, and arrives at the state s_i. Therefore, the sum $\sum_{i=1}^{n} p_i$ may be less than unity. A Carlyle probabilistic automaton is a general linear automaton of the I type.

The input–output pair (x, y) of a probabilistic Carlyle automaton corresponds to the input of a general linear automaton $\mathfrak{A}$. From the general linear automaton $\mathfrak{A}$ one may restore the

† Above we noted that general linear automata are automata, and therefore one can speak of isomorphism of $\mathfrak{B}$ and $\mathfrak{A}$.

‡ Such a representation of finite deterministic automata was first given in a paper by M. L. Tsetlin [11].

Carlyle automaton (communication channel). In this sense general linear automata allow Carlyle automata to be interpreted, since the latter are not in general automata even when they have an infinite number of states and outputs, if X is taken as the input alphabet; this is true because the probabilities of the transitions of states depend not only on the input symbols but also on the output symbols, and a transition to probabilitic Moore automata requires a time shift of the outputs. We shall discuss the difference between Carlyle automata and other probabilistic automata in greater detail in the chapter on probabilistic automata.

4. Assume that there is a finite set of identical finite automata $\mathfrak{B}(S, \Sigma, U, \varphi, \psi)$.

At each time identical input signals $\sigma(t)$ are applied to the inputs of all the automata (which may be in different states). The output of such a system is the vector $(t_1, t_2, \ldots, t_m)$, where t_i is the number of automata producing the symbol u_i at the output. Obviously, t_i is an integer nonnegative number $\sum_{i=1}^{n} t_i = d$.

The operation of such a system may be described by means of a general linear automaton $\mathfrak{A}$. Let us assume that the vectors $(a_1, a_2, \ldots, a_n)$ are its states, where a_j is the number of automata $\mathfrak{B}$, that are in the state s_j at the given time.

The function φ_σ shall be defined by means of the transition matrix $\| a_{ij}(\sigma) \|$, where

$$\underset{(i,j=1,\ldots,n)}{a_{ij}(\sigma)} = \begin{cases} 1 & \text{for } \varphi(s_i, \sigma) = s_j, \\ 0 & \text{otherwise;} \end{cases}$$

ψ_b shall be defined in terms of the output matrix $\mathbf{B}(\sigma) = \| b_{ij}(\sigma) \|$,

$$\underset{\substack{(i=1,2,\ldots,n;\\ j=1,2,\ldots,m)}}{b_{ij}(\sigma)} = \begin{cases} 1, & \text{if } \psi(s_i, \sigma) = t_j, \\ 0 & \text{otherwise.} \end{cases}$$

A more general case holds in the consideration of a finite set D of finite automata among which there may be automata of different kinds ($D = D_1 \cup D_2 \cup \ldots \cup D_h$ where D_i is the set of automata of the i-th kind) but with a common input and output alphabet.

Then the coordinates of the vectors of state will be partitioned into eight groups in the corresponding general linear automata; the coordinates of the i-th group give the distribution of the set of automata D_i over the states $s_i = \{q_1^{(i)}, \ldots, q_{n_i}^{(i)}\}$, while the matrices $\mathbf{A}(\sigma)$ will have a diagonally cellular structure in which the i-th cell of $\mathbf{A}(\sigma)$ will contain the matrices $\mathbf{A}_i(\sigma)$ that are transition matrices of the automata of the i-th group.

The matrices $\mathbf{B}(\sigma)$ of the output will be constructed from the matrices $\mathbf{B}_i(\sigma)$ formed in the column

$$\begin{pmatrix} B_1 \\ B_2 \\ \cdots \\ B_h \end{pmatrix}$$

An interesting scheme is obtained if we allow for the possibility of p r o d u c t i o n and "a n n i h i l a t i o n" of automata. Assume that the finite automaton $\mathfrak{A}$ in the state s_i is caused by the action σ to generate $a_{ik}(\sigma)$ automata in the state s_k. The states of the corresponding general linear automata will be vectors whose coordinates a_i constitute the number of automata $\mathfrak{A}$ in the state s_i. The elements of the matrix $\mathbf{A}(\sigma)$ are integer nonnegative numbers.

Let us expand this model by allowing the generation of both automata and "antiautomata." The corresponding matrices $\mathbf{A}(\sigma)$ may have any integer numbers as their elements. Let us define the notation of an antiautomaton $\mathfrak{A}^-(S, \Sigma, U^-, \varphi, \psi)$ for a finite automaton $\mathfrak{A}$. This finite automaton differs from $\mathfrak{A}$ only in that the symbols $u_1^-, u_2^-, \ldots, u_m^-$ serve as its outputs, and zero is obtained for "addition" of the same outputs $u_i + u_i^-$ of the automaton and antiautomaton. Then one may assume that the automaton $\mathfrak{A}$ and the antiautomaton $\mathfrak{A}^-$ "annihilate" if they are in the same state.†

From theorems on experiments with general linear automata there will derive basically the same estimates for "automaton media" formed from finite automata of a single type having n states, both in examples analyzed in this subsection and for finite automata with n states.

5. Let us consider the four-pole in Fig. 1. Alternating currents of specified amplitude and phase are applied to poles 1, 2 at each instant of discrete time. The instants of discrete time may be represented as intervals of a certain length Δ, the gaps between them being of sufficient duration so that the transients in the network can establish a specified state at the next instant. At the initial instant t_0 the pulses $i_1(0)$ and $i_2(0)$ are applied to poles 1, 2, respectively.

The state of the network is characterized by the amplitude and phase of the currents i_1 and i_2 which are applied to the poles 1, 2 (Fig. 2).

Each of the currents i_k (k = 1, 2) may be characterized by a complex number $r_k e^{i\varphi_k}$, where r_k is the amplitude and φ_k is the phase of the current.

The block of the network u_{kl} (k, l = 1, 2) transmits the current from the pole k to the pole l, amplifies or attenuates it, and changes its phase. Consequently, the action of each block may similarly be characterized by a complex number a_{kl}. Thus, the current applied to the pole j' is equal to

$$i_1 a_{1j} + i_2 a_{2j}.$$

Using relays, one can make the blocks u_{kl} controllable. Then the numbers a_{kl} will depend on the input signals σ.

From the pole 1'(2') the pulse of alternating current is again transmitted to the pole 1(2) after a delay $d_1(d_2)$. It is necessary to ensure synchronization of the pulse.

The state of the network at the instant t is the complex vector (i_1, i_2). Its transformation occurs via a complex matrix $\| a_{kl}(\sigma) \|$ (k, l = 1, 2). Thus, the network operates as a general linear automaton having the dimensionality 2 over the field K of complex numbers. A vector of state may serve as the network output (or it is necessary to use additional transforming blocks). Of course, the implementation of this network encounters certain engineering difficulties – it is necessary to ensure the accuracy of the amplitude and phase modulation. However, with a certain degree of approximation this network can be implemented even if it has a larger number of poles.

---

† Here analogies with quantum mechanics and the physics of elementary particles suggest themselves. E. Moore [18] indicated a certain analogy between experiments on automata and on particles in quantum mechanics.

On the other hand, the state of any physical system can be described by a ψ-function (i.e., a vector of Hilbert space).

Transformations of ψ-functions due to the effect of an external influence are usually linear. Thus, quantum-mechanical systems are countable-dimensional general linear automata (see [8, 7]).

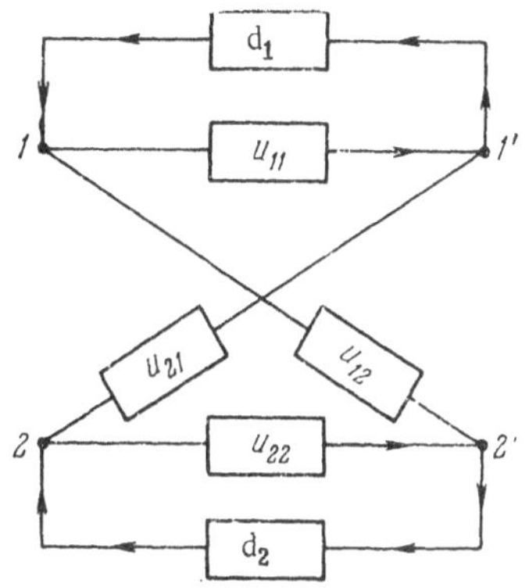

Fig. 1

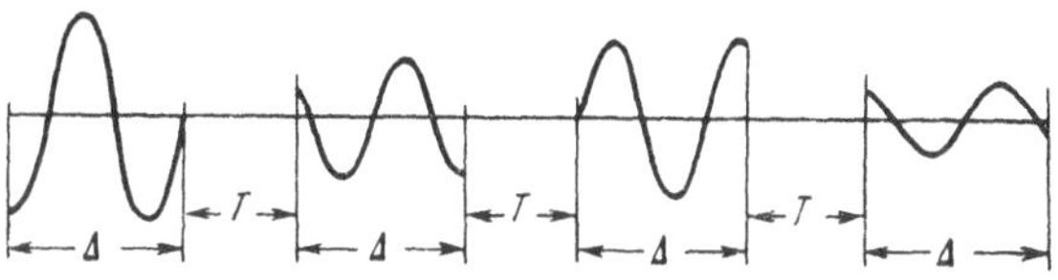

Fig. 2

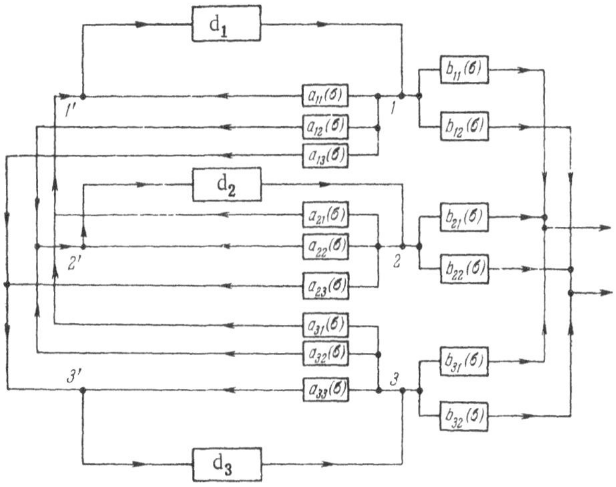

Fig. 3

It is not difficult to construct the analogous network for a general linear automaton having an arbitrary finite dimensionality over the field K.

The idea of using phase modulation for the transmission and manipulation of information in digital machines was stated by J. Neumann.

6. From the engineering standpoint, networks for any finite-dimensional general linear automaton over a finite field P may be fully implemented.

As an example, let us consider a general linear automaton of dimensionality 3 having an output of dimensionality 2.

For a general linear automaton of dimensionality 3 the network is shown in Fig. 3. The state of the general linear automaton $\mathfrak{A}$ consists of the vectors (v_1, v_2, v_3) (signals traveling along conductors may take a number of values corresponding to the number of elements in the field P). The elements a_{ij} (i, j = 1, 2, 3) carry out multiplication of the incoming signal by a_{ij} (for $a_{ij} = 0$ this element and the line passing through it may be dropped, while for $a_{ij} = 1$ only the element can be dropped; therefore, in fields having the characteristic 2, networks of autonomous general linear automata have an especially simple appearance). The elements a_{ij} may be made controllable (i.e., dependent on external inputs).

For fields having the characteristic 2 and k binary inputs $x_1, x_2, \ldots, x_k$, where the input signal $\sigma = (x_1, x_2, \ldots, x_k)$, the function $a_{ik}(\sigma) = a_{ik}(x_1, x_2, \ldots, x_k)$ will, in general, be an arbitrary logic-algebra function of $x_1, x_2, \ldots, x_k$, while for any finite field P of m elements the function $a_{ij}(\sigma)$ will be an m-valued logic function. In order to obtain the output vectors of the network, we require the additional blocks $b_{ik}(\sigma)$.

In this lies the difference between a general linear automaton and linear switching networks which use only $\oplus$ and delay elements in the binary case, and multiplying devices a_{ik} in the multivalued case; however, in this case the blocks a_{ik} do not depend on the input.

Finite-dimensioned general linear automata having a finite input alphabet over a finite field are finite automata. The advantage of representing them in the form of general linear automata lies in the fact that if one is able to code the states of a finite automaton (and there may be very many of them) by means of vectors over the field P (their dimensionality is of the order of $\log_m n$, where m is the number of elements of the field P, and n is the number of states of the finite automaton) and these vectors can be transformed linearly for transitions and outputs, it follows that in the network implementation it is possible to achieve a large economy of elements in comparison with the general case of a finite automaton.

Let us estimate the number of elements $\vee$, &, $\neg$ in the implementation of a general linear automaton over the field GF(2). Only the dependences of a_{ij} on the input symbols σ, coded by the collection $x_1, x_2, \ldots, x_k$ of elements from P, will be nonlinear (in the binary case $a_{ij}(x_1, x_2, \ldots, x_k)$ is an arbitrary logic-algebra function). However, for a small number of inputs and a large number of states the implementation of $a_{ij}(x_1, x_2, \ldots, x_k)$ requires comparatively few elements, while the remaining part of the network is implemented simply. At the same time for arbitrary coding of states in m-valued logic the number of elements will be considerably higher. A generalized linear automaton over the field P having the characteristic 2, which has the dimensionality n, $\tilde{k}$ binary inputs, $x_1, x_2, \ldots, x_{\tilde{k}}$ and m binary outputs requires n^2 blocks $a_{ij}(x_1, \ldots, x_{\tilde{k}})$ and nm blocks $b_{ij}(x_1, \ldots, x_k)$, each of which is implemented with a complexity not exceeding $\frac{(1+\varepsilon)2^{\tilde{k}}}{\tilde{k}}$ in the classical basis $\vee$, & and $\neg$; n + m adders modulo 2, each with n inputs, and n delay elements. Each adder is implemented with complexity Cn, where C is a constant. Thus, we find that it is sufficient to have

$$n(n+m)\left[(1+\varepsilon)\frac{2^{\tilde{k}}}{\tilde{k}}+C\right] \qquad (\varepsilon \to 0 \text{ for } k \to \infty)$$

elements $\vee$, & and $\neg$ and n delay elements [5]. However, when an arbitrary finite automaton having 2^n states, $\tilde{k}$ inputs, and m outputs is implemented, it is necessary to use at least

$$(1-\varepsilon)\left[m\,\frac{2^n \cdot 2m}{n+\tilde{k}}+n\,\frac{(2^{\tilde{k}}-1)\,2n}{\tilde{k}+n}\right]$$

elements $\wedge$, &, $\neg$ and n delay elements [5, Ch. 8], where $\varepsilon \to 0$ for $n + k \to \infty$. Moreover, it is considerably easier to analyze† n-dimensional general linear automata and arbitrary finite automata having 2^n states.

§ 3. Certain Possible Generalizations

1. Nondeterministic automata, which were considered by Rabin and Scott [21], are of interest. However, they do not fit directly into the scheme of a general linear automaton. Nevertheless, if instead of linear operations on vectors, which use the operations of the fields

†That is, to clarify the character of the mappings which they perform.

+ and ×, we consider the operations $\dot{+}$ and $\dot{\times}$ such that the operation $\dot{+}$ is commutative and associative, while $\dot{\times}$ is associative, and distributiveness holds:

$$(a \dot{+} b) \dot{\times} c = (a \dot{\times} c) \dot{+} (b \dot{\times} c) \tag{4}$$

and

$$c \dot{\times} (a \dot{+} b) = (c \dot{\times} a) \dot{+} (c \dot{\times} b), \tag{5}$$

then it follows that analogs of general linear automata may be considered over such an algebra; the states $\vec{v}$ and outputs $\vec{z}$ of these automata are declared to be vectors having coordinates that are elements of this algebra; the transition and output functions are determined by the matrices $\mathbf{A}(\sigma)$ and $\mathbf{B}(\sigma)$, where σ is the input symbol, in the conventional manner:

$$\vec{v}(t+1) = \vec{v}(t) \dot{\times} A(\sigma), \quad \vec{z}(t+1) = \vec{v} \dot{\times} B(\sigma),$$

while the multiplication of vectors by matrices and the multiplication of matrices in such an algebra is defined conventionally and has the properties of associativeness and distributiveness:

$$A \dot{\times} B \dot{\times} C = A \dot{\times} (B \dot{\times} C), \quad (A \dot{+} B) \dot{\times} C = (A \dot{\times} C) \dot{+} (B \dot{\times} C),$$
$$C \dot{\times} (A \dot{+} B) = (C \dot{\times} A) \dot{+} (C \dot{\times} B)$$

(see [1]).

Let us consider a Boolean algebra with $\vee$ instead of $\dot{+}$, and & instead of $\dot{\times}$, for which all of the required properties are fulfilled. Assume $\mathfrak{A}(S, F, \Sigma, U, \varphi, \psi)$ is a nondeterministic automaton [4, 21].

In the determination process the subsets $S = \{s_1, s_2, \ldots, s_n\}$ are taken as its state. Each subset $R \subseteq S$ is juxtaposed with a Boolean vector $\vec{r} = (r_1, r_2, \ldots, r_n)$, $r_i = 1$ being valid for $s_i \in R$ and $r_i = 0$ being valid for $s_i \bar{\in} R$. We define the matrix of transitions as a Boolean matrix $\mathbf{A}(\sigma)$, having placed $a_{ij}(\sigma) = 1$ when and only when transition to the state s_j is possible from the state s_i due to the action of σ. The output vector $\vec{b} = (b_1, b_2, \ldots, b_n)$, where $b_i = 1$ for $s_i \in F$ and $b_i = 0$ for $s_i \bar{\in} F$, while F is the set of isolated states; $\vec{b}$ is independent of σ (a Moore automaton). The output is binary (i.e., the input word is tagged or not), since we are speaking of the representability of events:

$$\vec{r}(t+1) = \vec{r}(t)\, A(\sigma), \tag{6}$$

$$y(t) = \vec{r}(t)\, \vec{b}, \tag{6'}$$
$$y(t) \ \text{output} \ (=0,1).$$

If the initial state was s_i, then the word $x = \sigma_1\sigma_2 \ldots \sigma_k$ will be tagged by the automaton $\mathfrak{A}$, if and only if

$$\vec{r}_F^{(i)} A(\sigma_1)\, A(\sigma_2) \ldots A(\sigma_n)\, \vec{b} = 1,$$

where $\vec{r}^{(i)}$ is a Boolean vector in which a single "one" appears in the i-th position.

An automaton whose operation can be described by Eqs. (6) and (6') (i.e., a Boolean automaton) adequately reflects the operation of the nondeterministic automaton $\mathfrak{A}$. Therefore, nondeterministic automata shall be called Boolean. Here, however, an arbitrary choice of the "basis" in the set of state vectors $\{\vec{r}\}$ is impossible, and this greatly complicates the finding of the structure of the Boolean automaton from an experiment on it. Therefore, gen-

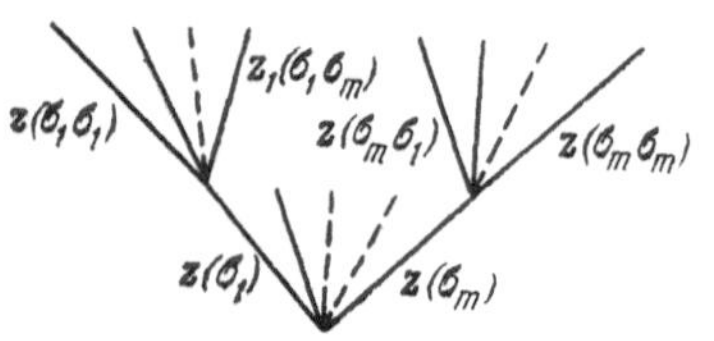

Fig. 4

eral linear automata are still the best generalizations of finite automata from the point of view of experiments.

2. Semilinear automata $\mathfrak{A}(M, \Sigma, Z, \varphi, R_i)$ are another generalization of general linear automata; these automata are distinguished by the fact that the output vector

$$\vec{z}(t) = (z_1(t), z_2(t), \ldots, z_m(t))$$

is defined by the polynomials $\{R_i\}$ ($i = 1, 2, \ldots, m$)

$$z_i(t) = R_{i,\sigma}[s_i(t), s_2(t), \ldots, s_n(t)],$$

where $R_{i,\sigma}$ is a polynomial of degree $g(i, \sigma)$ having coefficients that depend on i and the input σ. The transformations of the transitions

$$\vec{s}(t+1) = \varphi[\vec{s}(t), \sigma]$$

are linear as previously.

Here one may apply the linearization method [3] by taking vectors having coordinates corresponding to monomials of degree not exceeding $\max\limits_{i,\sigma} g(i, \sigma) = h$ from $s_1, s_2, \ldots, s_n$ (i.e., with coordinates of the form

$$q_{h_1 h_2 \ldots h_n} = s_1^{h_1} s_2^{h_2} \ldots s_n^{h_n},$$

where the number of each coordinate is determined by the collection $h_1, h_2, \ldots, h_n$, $h_1 + h_2 + \ldots + h_n \leq h$) instead of the vectors $(s_1, s_2, \ldots, s_n)$.

Since the vectors $\vec{s} = (s_1, s_2, \ldots, s_n)$ can be transformed linearly, it follows that the vectors $\vec{q}\,(q_{0,0}, \ldots, 0, \ldots, q_{h_1}, \ldots, h_n, \ldots)$ can similarly be transformed linearly, and the output $\vec{z}(t)$ depends linearly on $\vec{q}(t)$.

Automata for which the transformations φ_σ are any affine transformations of the space L(M):

$$\vec{s}(t+1) = \vec{s}(t) A(\sigma) + \vec{r}_1(\sigma),$$
$$\vec{z}(t+1) = \vec{s}(t) B(\sigma) + \vec{r}_2(\sigma)$$

also reduce to the case of a generalized linear automaton (it is sufficient to add one coordinate equal to 1 to the vector $\vec{s}_1$, and one row equal to $\vec{r}_1(\sigma)$ to the bottom of the matrix $\mathbf{A}(\sigma)$, and a column containing all 0 elements except the last element which is equal to 1 to the right of the matrix obtained, while to the matrix $\mathbf{B}(\sigma)$ it is required just to add the row $\vec{r}_2(\sigma)$ on the bottom.

§ 4. Experiments with General Linear Automata

The notion of experiments with automata was introduced by E. Moore [18] and was studied in papers by B. A. Trakhtenbrot, S. Ginsburg, T. Hibbard, A. A. Karatsuba, V. A. Dushskii, Yu. M. Borodyanskii, A. A. Muchnik, et al. [5, 17, 2, 9].

1. Since general linear automata are automata, the notions of equivalence and distinguishability of states, the notions of prime-uniform, branched, and multiple experiments [18], as well as the notion of distinguishability of automata, are automatically extended to them.

In the book [5] infinite trees (deterministic operators) are used to describe automata having the input alphabet Σ and the output alphabet U. Each string of inputs $\sigma(1), \sigma(2), \ldots, \sigma(t), \ldots$ corresponds to one and only one string of edges of the tree D with its beginning in the root of the tree. Each edge of the tree is assigned a corresponding output symbol $\vec{z}$. Assume that the input symbols will be vectors $\vec{z}$ over the field P. The linear combination $\alpha D_1 + \beta D_2$ of trees D_1 and D_2 $(\alpha, \beta \in P)$ is called the tree D whose edges are assigned the linear combinations $\alpha\vec{z}_1 + \beta\vec{z}_2$, where $\vec{z}_1$ and $\vec{z}_2$ are the vectors assigned to the corresponding edges D_1 and D_2. Thus, the set of all trees over the input alphabet Σ which have the output $\vec{z}$ will be a linear space over P. We shall call the dimensionality of the minimal hyperplane in this space, which contains all branches of the tree D, the *dimensionality of the tree D*. *The dimensionality of a certain set of trees* $\{D_i\}$ shall be defined as the dimensionality of the minimal hyperplane in the space over all trees which contains all branches of the trees from $\{D_i\}$.

It is obvious that each state of the general linear automaton $\mathfrak{A}$ $(M, \Sigma, Z, \varphi, \psi)$ corresponds to a certain tree D over Σ having outputs from Z (and the initial general linear automaton $\mathfrak{A}$ corresponds to one tree $D_{\mathfrak{A}}$). The dimensionality of the set of trees corresponding to the state of the general linear automaton does not exceed (as we shall see below) the dimensionality of the general linear automaton $\mathfrak{A}$. The converse is also true. Each tree D having the dimensionality n defines a certain initial general linear automaton having the dimensionality n. Its initial state corresponds to the root of D. Let us choose the basis in the set of branches of the tree D. Then each branch of D can be expressed as a linear combination of basis branches, and thereby the transformations of the transition and output will be defined. In exactly the same way the set of trees $\{D_i\}$ having the dimensionality n defines a general linear automaton having the dimensionality n.

Two states of the general linear automaton $\mathfrak{A}$ are called *equivalent* if identical trees correspond to them. Two general linear automata $\mathfrak{A}_1$ and $\mathfrak{A}_2$ are called equivalent if for each state of one general linear automaton one can find its equivalent state in the other, and vice versa. For a transition from the set of trees corresponding to $\mathfrak{A}$ to a general linear automaton, we may obtain a general linear automaton $\mathfrak{A}_1$ having a dimensionality smaller than the dimensionality of $\mathfrak{A}$ as a consequence of the splicing of equivalent states. Such a transition to a general linear automaton without equivalent states is called the operation of the *reduction* of the general linear automaton $\mathfrak{A}$. The reduced form of the complete general linear automaton $\mathfrak{A}$ is unique with accuracy up to isomorphism of the linear spaces L(M) which preserves the operators φ_σ and ψ_σ. However, the matrix form of a reduced general linear automaton $\mathfrak{A}$ always depends essentially on the choice of the basis L(M). Besides, the matrices $\mathbf{A}(\sigma)$ of the reduced general linear automaton go over into matrices of the form $\mathbf{C}^{-1}\mathbf{A}(\sigma)\mathbf{C}$ when the basis changes, where $\mathbf{C}$ is a certain nonsingular matrix that depends on the new basis and is independent of σ, while the matrices $\mathbf{B}(\sigma)$ go over into $\mathbf{C}^{-1}\mathbf{B}(\sigma)\mathbf{C}$, respectively.

One may consider the subtrees $D^{(1)}, D^{(2)}, \ldots$ of the tree D having the height h = 1, 2, ... (i.e., with h stories of edges).

The next theorem is an analogy of the Moore theorem on the length of a prime experiment† which distinguishes the state of the automaton.

Theorem 1. *For a general linear automaton $\mathfrak{A}$ of dimensionality n the states $\vec{v}_1$ and $\vec{v}_2$ are equivalent if and only if identical trees $D_1^{(n)}$ and $D_2^{(n)}$ of height n correspond to them (i.e., they are indistinguishable by any prime experiment of length n).*

†The corresponding theorem for probabilistic automata was proved independently by Carlyle [13, 14] and by the author (the results have not been published previously).

Proof. The necessity of this condition is obvious; let us prove its sufficiency. Let us consider the set M of states of the general linear automaton $\mathfrak{A}$. Let us assume that one can find two states $\vec{v}_1$ and $\vec{v}_2$ from M such that the trees of height n for them coincide, while in the (n + 1)-st story different outputs will be assigned to two identical edges. Let us prove that in this case the dimensionality Γ_M of the minimal hyperplane containing M will be no less than n + 1 [i.e., $r(\mathfrak{A}) \geqq n+1$]. For this purpose let us consider the hyperplane $\tilde{\Gamma}_M$ parallel to Γ_M passing through the origin (the point O; if $\vec{O} \in \Gamma_M$, then $\tilde{\Gamma}_M = \Gamma_M$).

$\tilde{\Gamma}_M$ is the minimal hyperplane which contains all differences of the vectors from M.

Let us consider the difference $\vec{v}_1 - \vec{v}_2$. It corresponds to the tree $D_1 - D_2$ for which the outputs are zeros on the first and second stories, while on the (n + 1)-st story there is at least one nonzero output $\vec{z}(n+1) = \vec{z}_1(n+1) - \vec{z}_2(n+1)$. Assume $\sigma_{i_1}, \sigma_{i_2}, \ldots, \sigma_{i_n}$ is a certain string of inputs that leads to a nonzero edge of the (n + 1)-st story of the tree $D_1 - D_2$.

Let us consider all branches (subtrees) of the tree $D_1 - D_2$

$$E_1 = D_1 - D_2, \quad E_2, \ldots, E_n, E_{n+1}$$

having roots at the vertices of this path. The k-th branch corresponds to the vertex $x = \sigma_{i_1}\sigma_{i_2}\ldots\sigma_{i_{k-1}}$ $(k = 1, 2, \ldots, n+1)$ for k = 1, $x = \Lambda$. Zeros appear in the k-th branch on the n − k + 1 lowest stories, while in the (n − k + 2)-nd story one can find a nonzero output. Therefore, all branches $E_1, E_2, \ldots, E_{n+1}$ are linearly independent; i.e., one can find n + 1 linearly independent vectors in the hyperplane $\tilde{\Gamma}_M \ni \vec{0}$. Therefore the dimensionality of $\tilde{\Gamma}_M$ is not less than n + 1, and consequently, $r(\mathfrak{A}) \geqslant n+1$, which contradicts the assumption of the theorem. Thus, if two trees D_1 and D_2 coincide at n stories, then they also coincide at the (n + 1)-st story, etc. The theorem has been proved.

Corollary 1 (E. Moore). For a finite automaton having n pairwise-distinguishable states any two states are distinguishable by a prime experiment of length n − 1.

Corollary 2 (E. Carlyle). For a probabilistic automaton with n states the necessary and sufficient condition for equivalence of two states (or the distributions of the initial states) is their indistinguishability by any experiment of length n − 1.

Corollary 3. If in an n-dimensional general linear automaton $\mathfrak{A}$ of the type I a tree of height n having a root in a certain state q is a linear combination of trees having a height n with roots in the states $q_1, \ldots, q_k$, then all of the infinite tree D_q is the same kind of linear combination of the trees $D_{q_1}, \ldots, D_{q_k}$.

For a general linear automaton of the type II, Corollary 3 is true if one considers linear combinations with the sum of the coefficients of 1, which we shall hereafter call normalized linear combinations.

The proof can be reduced to Theorem 1 if the general linear automaton $\mathfrak{A}$ is expanded to a complete general linear automaton $\bar{\mathfrak{A}}$, if $\mathfrak{A}$ is a general linear automaton of the I type, or to a weakly complete general linear automaton if $\mathfrak{A}$ is of the II type.

Then all linear combinations $\Sigma \alpha_i q_i$ for a general linear automaton $\mathfrak{A}$ of the type I and all linear combinations $\Sigma \alpha_i q_i$ having the condition $\Sigma \alpha_i = 1$ for a general linear automaton $\mathfrak{A}$ of the type II will be states of $\bar{\mathfrak{A}}$, which has a dimensionality equal to m.

Let us give another proof of Theorem 1, which has a clear geometric meaning and is closer to the ideas of E. Moore [18]. Instead of the general linear automaton $\mathfrak{A}$ we consider the expansion $\bar{\mathfrak{A}}$: a supplement if $\Gamma_M = L(M)$, or a weak supplement if $\Gamma_M \subset L(M)$; then we prove the theorem for $\bar{\mathfrak{A}}$, and consequently for $\mathfrak{A}$ also. Obviously, the dimensionality of $\bar{\mathfrak{A}}$ is equal to the dimensionality of $\mathfrak{A}$.

Assume that the automaton $\mathfrak{A}$ is initially in the state $\vec{s}$. Since we know nothing about the initial state, we may choose $\vec{s}$ to be any point from Γ_M. Each word transfers the state $\vec{s}$ to the state $\varphi_x(\vec{s})$. If the symbol σ is then applied to the input, the vector $\psi_\sigma[\varphi_x(\vec{s})]$ will appear at the output of the general linear automaton.

The value $\vec{u}$ of this vector $\psi_\sigma[\varphi_x(\vec{s})]$ will become known to us in the experiment $x\sigma$. Therefore, we obtain the linear equation

$$\psi_\sigma[\varphi_x(\vec{s})] = \vec{u}, \tag{7}$$

which isolates in $\Gamma(M)$ the hyperplane $\Gamma_{x\sigma}$ (which may coincide with $\Gamma(M)$) containing the state $\vec{s}$. If during the experiments (prime or multiple) the outputs of the general linear automaton $\mathfrak{A}$, corresponding to several words $x_1\sigma_1, x_2\sigma_2, \ldots, x_k\sigma_k$, becomes known to us, then this yields a system T of linear equations

$$\psi_{\sigma_i}[\varphi_{x_i}(\vec{s})] = \vec{u}_i \qquad (i = 1, 2, \ldots, k). \tag{8}$$

Such a system T likewise defines a certain hyperplane $\Gamma_T \subseteq \Gamma_M$. If the system T_2 derives from the system T_1, then $\Gamma_{T_1} \subseteq \Gamma_{T_2}$, i.e., a longer experiment defines a smaller hyperplane or the same one. For various values in the outputs $\vec{u}_i$ various hyperplanes parallel to each other will be obtained. Thus, just the inputs of the experiment define the direction of the hyperplanes $\Gamma(M)$ for us, while the experiment partitions $\Gamma(M)$ into hyperplanes parallel to Γ_T. (The difference between a uniform experiment which does not use information on the output of the automaton and is connected with a consideration of the partitions of the set of states, and a branched experiment in which the value of the output immediately isolates the subset of states to which the consideration is restricted during the subsequent course of the experiment resides precisely in this.)

Assume now that the general linear automaton has the dimensionality n, i.e., $r[\Gamma(M)] = n$. Let us assume further that $\Gamma(M) = \Gamma$ for the sake of brevity. The multiple experiment E_1 of length 1 isolates the hyperplane $\Gamma_1 \subseteq \Gamma$. The multiple experiment $E_k, E_{k-1}, \ldots, E_1$ (the length E_i is equal to i) respectively isolates the hyperplanes $\Gamma_k \subseteq \Gamma_{k-1} \subseteq \ldots \subseteq \Gamma_1 (\subseteq \Gamma)$.

Assume that this chain rigorously decreases to $k-1$, while $\Gamma_k = \Gamma_{k-1}$ (i.e., the experiment of length $k+1$ does not add new information on $\vec{s}$ in comparison with E_k). Let us prove that then $\Gamma_k = \Gamma_{k+1} = \Gamma_{k+2} = \ldots$; i.e., states which are not distinguishable by the multiple experiment E_k are not distinguishable at all. Let us take any input symbol σ and consider the hyperplane $\Gamma_k(\sigma)$ into which Γ_k makes the transition due to the action of φ_σ. $\Gamma_{k-1}(\sigma) \parallel \Gamma_{k-1}$, since if $\Gamma_{k-1}(\sigma) \nparallel \Gamma_{k-1}$ were to hold then the multiple experiment of length $k-1$ would partition $\Gamma_{k-1}(\sigma)$ into hyperplanes parallel to Γ_{k-1}. But then Γ_{k-1} – the prototype of $\Gamma_{k-1}(\sigma)$ for the mapping φ_σ – would be partitioned by an experiment of length k, but $\Gamma_k = \Gamma_{k-1}$; i.e., Γ_{k-1} is not partitioned. Since $\Gamma_k = \Gamma_{k-1}$, it follows that $\Gamma_k(\sigma) = \Gamma_{k-1}(\sigma)$ and $\Gamma_k(\sigma) \parallel \Gamma_{k-1}$. Therefore, $\Gamma_k(\sigma)$ is not partitioned by a multiple experiment of length k, and therefore Γ_k is not partitioned by an experiment of length $k+1$; i.e., $\Gamma_{k+1} = \Gamma_k$, etc., $\Gamma_k = \Gamma_{k+1} = \Gamma_{k+2} = \ldots$. The states that are not distinguishable by a multiple experiment of length k turn out to be indistinguishable altogether. The entire set $\Gamma(M)$ can be partitioned in this manner into sets of equivalent (indistinguishable) states that form hyperplanes obtained by parallel shifts of Γ_{k-1} (i.e., into contiguous classes Γ_M in Γ_{k-1}; these hyperplanes may also be points).

Now it remains for us to note that

$$n = r(\Gamma) > r(\Gamma_1) > r(\Gamma_2) > \ldots > r(\Gamma_{k-1}) = r(\Gamma_k) \geqslant 0.$$

From this it follows that $k \leq n$, i.e., states that are not distinguishable by experiments of length n are equivalent to r, etc.

From the proof given one may extract the following corollaries.

Colloary 4. If the distinguishable states $\vec{v}_1$ and $\vec{v}_2$ of the general linear automaton $\mathfrak{A}$ belong to the hyperplane Γ having the dimensionality n, which due to the action of each input σ goes over into the hyperplane $\Gamma(\sigma) \| \Gamma$, then these states are distinguishable by a prime experiment of length n, regardless of the dimensionality of the general linear automaton $\mathfrak{A}$.

Corollary 5. The state s of the general linear automaton $\mathfrak{A}$ is equivalent to the state q of the general linear automaton $\mathfrak{b}$ if and only if these states are indistinguishable by any experiment of length $r(\mathfrak{A})+r(\mathfrak{L})$.

For a proof it is sufficient to consider the direct sum $\mathfrak{A} \oplus \mathfrak{B}$ of the general linear automata $\mathfrak{A}$ and $\mathfrak{L}$. The dimensionality of $r(\mathfrak{A} \oplus \mathfrak{L})$ is equal to $r(\mathfrak{A})+r(\mathfrak{L})$. The states s and q are states of $\mathfrak{A} \oplus \mathfrak{B}$.

2. If states from Γ_{k-1} are identified in the latter proof of this theorem, then we obtain the reduced form of the general linear automaton $\mathfrak{A}$; i.e., we obtain a general linear automaton $\mathfrak{A}$ that is equivalent to the given one and has states which are all distinguishable from one another. Under these conditions the dimensionality of $\mathfrak{A}_1$ is equal to $n - r(\Gamma_{k-1})$ and to the dimensionality of the set of trees $\{D_s\}$ of the general linear automaton $\mathfrak{A}$. The set of states of $\mathfrak{A}_1$ will be the factor-hyperplane Γ_M/Γ_{k-1}.

The reduced form of $\mathfrak{A}_1$ is unique, since the partition of Γ_M into sets of equivalent states of the hyperplane Γ_{k-1} is unique. However, its matrix expression depends on the choice of the basis in L(M) and Z. Another expression for the reduced form of the automaton in the form of a graph of transitions-outputs is also possible, which is equivalent to the matrix expression. Assume that a basis in the space of trees D_s of the general linear automaton $\mathfrak{A}$ has been chosen – $D_1, D_2, \ldots, D_n$. Let us depict the trees $\{D_i\}$ by circle-vertices of a directed graph. n arrows issue from each vertex. These arrows connect it with the remaining vertices of the graph.

The arrow connecting the vertex D_i with D_j is assigned the expression

$$\pi_{ij} = \sum_{k=1}^{n} \sigma_k a_{ij}(\sigma_k),$$

whose meaning resides in the fact that the state s_i is caused by the action of the input σ_k to go over into the state $a_{ij}(\sigma_k) \cdot s_j$, where $a_{ij}(\sigma_k)$ is an element of the field P (Fig. 5). Each (i-th) circle-vertex of the graph is assigned an expression

$$\sum_{k=1,\, h=1}^{n,\, m} b_{ih}(\sigma_k)\vec{z}_h,$$

whose meaning resides in the fact that the general linear automaton, having received the signal σ_k at its input in the state D_i, produces the vector signal

$$\sum_h b_{ih}(\sigma_k)\vec{z}_h \quad (h = 1, 2, \ldots, p)$$

at its output (where p is the dimensionality and Z is the space of outputs), it being assumed that the basis $\vec{z}_1, \vec{z}_2, \ldots, \vec{z}_p$ has been chosen in Z. The matrices $\|\pi_{ij}\|$ and $\|b_{ih}(\sigma_k)\|$ are the matrices of transitions and outputs of $\mathfrak{A}$ (with allowance for all of the inputs σ).

3. However, let us return to experiments with general linear automata. An experiment is called uniform if it does not depend on the initial state (i.e., the next input signal is chosen independently of the output of the preceding portion of the experiment.

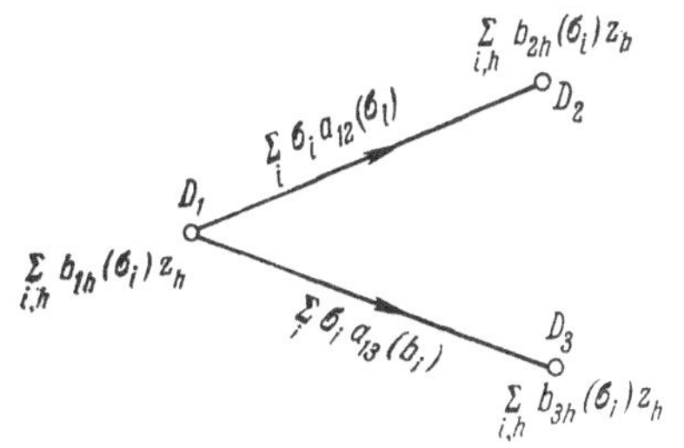

Fig. 5

Theorem 2. For any reduced general linear automaton $\mathfrak{A}$ having the dimensionality n there exists a uniform experiment E of length no greater than $n(n+1)/2$, which allows the state of $\mathfrak{A}$ at the end of the experiment to be established.

Proof. Since the general linear automaton $\mathfrak{A}$ is in reduced form and has the dimensionality n, it follows that the chain of hyperplanes (see the preceding theorem) $\Gamma_M \supset \Gamma_1 \supset \Gamma_2 \supset \ldots \supset \Gamma_k$ is not broken off before Γ_k, while Γ_k is a point (the initial state). Each hyperplane Γ_i is isolated from the results of a multiple experiment of length i. If multiple experiments are carried out in another state, the output will be different, and we shall obtain a different chain:

$$\Gamma_M \supset \Gamma_1' \supset \Gamma_2' \supset \ldots \supset \Gamma_k'.$$

Here Γ_k' consists solely of the new initial state of the experiments. Each hyperplane Γ_{i+1}' is obtained by a parallel shift of the corresponding hyperplane Γ_i', the inequality $r(\Gamma_{i+1}') < r(\Gamma_i')$ being valid, and the dimensionality $r(\Gamma_i) \leq n - i$, while $r(\Gamma_k) = 0$.

Thus, the multiple experiment E_1 of length 1 partitions $\Gamma = \Gamma_M$ into hyperplanes $\{\Gamma_1^*\}$, which are shifts of Γ_1; the multiple experiment of length 2 partitions each hyperplane Γ_1^* and the entire Γ_M into hyperplanes Γ_2^* which are shifts of Γ_2, and each hyperplane Γ_2^* and the entire hyperplane Γ_M is partitioned by a multiple experiment of length 3 into hyperplanes $\{\Gamma_3^*\}$ each of which is a parallel shift of Γ_3, etc.

From this it follows that there exists a simple uniform experiment of length 1 (i.e., an input symbol), $\mathscr{E}_1 = \sigma$, which partitions Γ into a set of parallel hyperplanes (shifts) $\{B_1^*\}$ that due to the action of $\mathscr{E}_1$ go over into the parallel hyperplanes $\{B_1^*(\mathscr{E}_1)\}$, while Γ_M goes over into $\Gamma(\mathscr{E}_1)$. The dimensionality of each $B_1^*(\mathscr{E}_1)$ is rigorously less than $r(\Gamma_M) = n$, i.e., it does not exceed $n-1$. Further, the multiple experiment E_2 of length 2 partitions Γ_M into parallel hyperplanes having a dimensionality not exceeding $n-2$. Therefore, if $B_1^*(\mathscr{E}_1)$ has the dimensionality $n-1$, the multiple experiment E_2 partitions $B_1^*(\mathscr{E}_1)$ into hyperplanes of dimensionality not exceeding $n-2$, and therefore one can also find a simple experiment $\mathscr{E}_2$ of length less than or equal to 2 that partitions each hyperplane $B_1^*(\mathscr{E}_1)$ and $\Gamma(\mathscr{E}_1)$ into parallel hyperplanes $B_2^*(\mathscr{E}_1)$ having a dimensionality not exceeding $n-2$. In this case, when the dimensionality of $B_1^*(\mathscr{E}_1)$ does not exceed $n-2$ one may assume that the input word $\mathscr{E}_2$ is empty, while $B_2^*(\mathscr{E}_1) = B_1^*(\mathscr{E}_1)$. Due to the action of $\mathscr{E}_2$ the set of hyperplanes $\{B_2^*(\mathscr{E}_1)\}$ goes over into the set of parallel hyperplanes $\{B_2^*(\mathscr{E}_1\mathscr{E}_2)\}$, while $\Gamma(\mathscr{E}_1)$ goes over into a hyperplane $\Gamma(\mathscr{E}_1\mathscr{E}_2)$ having a dimensionality less than or equal to $n-2$. If the dimensionality of $B_2^*(\mathscr{E}_1\mathscr{E}_2)$ is equal to $n-2$, then a certain simple experiment $\mathscr{E}_3$ of length no greater than 3 partitions each hyperplane $B_2^*(\mathscr{E}_1\mathscr{E}_2)$ into hyperplanes $\{B_3^*(\mathscr{E}_1\mathscr{E}_2)\}$ having a dimensionality not exceeding $n-3$, which due to the action of $\mathscr{E}_3$ go over into $\{B_3^*(\mathscr{E}_1\mathscr{E}_2\mathscr{E}_3)\}$, etc.

The sequence

$$\{B_1^*(\mathscr{E}_1)\},\ \{B_2^*(\mathscr{E}_1\mathscr{E}_2)\},\ \ldots,\ \{B_i^*(\mathscr{E}_1\mathscr{E}_2\ldots\mathscr{E}_i)\} \tag{9}$$

must be broken off no later than at the n-th place, since the dimensionality of the hyerplanes B_i^* decreases rigorously with increasing i, while the dimensionality of B_1^* is no greater than $n-1$. Assume that the last system of hyperplanes in the sequence (9) will be $\{B_k^*(\mathscr{E}_1\mathscr{E}_2\ldots\mathscr{E}_k)\}$. Each of the $B_k^*(\mathscr{E}_1\mathscr{E}_2\ldots\mathscr{E}_k)$ is a point, and their union forms $\Gamma(\mathscr{E}_1\mathscr{E}_2\ldots\mathscr{E}_k)$, i.e., it forms a hyperplane into which the hyperplane Γ_M goes over due to the action of the experiment

$\mathscr{E}_1\mathscr{E}_2 \ldots \mathscr{E}_k$. Thus, knowing the outcome of the experiment $\mathscr{E} = \mathscr{E}_1\mathscr{E}_2 \ldots \mathscr{E}_k$, we define the point-state into which the general linear automaton $\mathfrak{A}$ falls due to the action of the experiment $\mathscr{E}$. The length of $\mathscr{E}$ is estimated thus:

$$l(\mathscr{E}) \leqslant \sum_{i=1}^{k} l(\mathscr{E}_i) \leqslant \sum_{i=1}^{k} i = \frac{k(k+1)}{2} \leqslant \frac{n(n+1)}{2},$$

i.e., $l(\mathscr{E}) \leq n(n+1)/2$, which is what it was required to prove.

This theorem generalizes the theorems of T. Hibbard [17] and Karatsuba for finite automata, and its proof is even simplified so that it takes on a clear geometric meaning.

4. The general linear automaton $\mathfrak{A}(M, \Sigma, Z, \varphi, \psi)$ is called strongly connected if the dimensionality of the set of branches of the tree D_s is the same for all $s \in M$ and is equal to the dimensionality of $\mathfrak{A}$ itself.†

At first glance it seems that this definition derives directly from the definition of strong connectedness of automata according to Moore [18], extended to infinite automata. However, this is not so. Strong connectedness of the general linear automaton $\mathfrak{A}(V, M, \Sigma, Z, \varphi, \psi)$ differs from strong connectedness of $\mathfrak{A}$ treated as an infinite automaton. Let us present two simple examples:

1. Assume V is the real plane Oxy, while the set of states M is the ensemble of points of the straight line y = 1. $\Sigma = \{\sigma\}$; Z is the set of real numbers; the transformation φ_σ is the shift of the straight line y = 1: $x \to x + 1$ by the matrix $\begin{pmatrix} 1 & 1 \\ 1 & 0 \end{pmatrix}$, while the functional ψ: $\psi(x, y) = x$ and $\mathfrak{A}$ is a Moore general linear automaton.

Each state s = (x, 1) is transferred by the action of the word $u = \sigma^k$ into the state q = (x + k, 1), which differs from (x, 1), since $\psi(s) = x$, while $\psi(q) = x + k$. However, no word u_1 transfers q back to s. On the other hand, for each state s = (x, 1) the dimensionality of the set of branches of the tree $D_s = D_x$ (chains here) $\{x + k, x + k + 1, \ldots\}$ is unity, since the set of points $\{x + k, 1\}$ lies on the straight line y = 1,

$$D_{x+i} = D_x + i(D_{x+1} - D_x) = iD_{x+1} - (i-1)D_x,$$

while D_{x+1} and D_x are evidently linearly independent for all s. Therefore, $\mathfrak{A}$ is a strongly connected general linear automaton.

2. Another example of a general linear automaton is $\mathfrak{A}(V, M, \Sigma, Z, \varphi, \psi)$. V is again the real plane Oxy. M is the circle O(1) having its center at the origin O and a radius 1; $\Sigma = \{\sigma\}$.

The linear operator φ_σ performs rotation of the plane Oxy about O through the angle πa, where a is an irrational number $\psi(x, y) = x$. It is not difficult to see that the set of states which are accessible from any state s = (x, y) has the dimensionality 2, whereas from each state s – the points on O(1) – only those points q on O(1) are accessible which are such that the arc (s, q) is equal to $\pi r a - 2k\pi$, where r is a natural number and k is an integer.

† Here it may not be required that a path exist from each state s_i to any other state s_j, since in the overwhelming "majority of cases" (with the exception of a set of measure zero near P = D, K) the transition matrices $\mathbf{A}(\sigma)$ are such that each state s goes over into a linear combination with nonzero coefficients of other states. True, one could apply this conventional definition if generalized inputs were considered which are linear combinations of the input words (and not the input symbols) of the automaton. This latter definition is equivalent to the definition of strong connectedness of general linear automata that we adopted.

From strong connectedness of $\mathfrak{A}$ as an automaton there derives its strong connectedness as a general linear automaton.

Let us now consider the properties of strongly connected general linear automata.

Assume $\mathfrak{A}$ and $\mathfrak{B}$ are general linear automata having a general input alphabet and the output space Z. The general linear automaton $\mathfrak{A}$ is indistinguishable by the experiment $\mathscr{E}$ from the general linear automaton $\mathfrak{B}$, if for any state s of the general linear automaton $\mathfrak{A}$ there exists a state q of the automaton $\mathfrak{B}$, such that the results of the experiment $\mathscr{E}$ with the automaton $\mathfrak{A}$ in the initial state s coincides with the result of the experiment $\mathscr{E}$ with the automaton $\mathfrak{B}$ in the initial state q. In other words, the indistinguishability of $\mathfrak{A}$ from $\mathfrak{B}$ means that the set of outputs $\mathfrak{M}_{\mathscr{E}}(\mathfrak{A})$ of the experiment $\mathscr{E}$, conducted from various initial states $\mathfrak{A}$, is contained in the analogous set of outputs $\mathfrak{M}_{\mathscr{E}}(\mathfrak{B})$ of the automaton $\mathfrak{B}$.

The general linear automaton $\mathfrak{A}$ is indistinguishable from the general linear automaton $\mathfrak{B}$ if $\mathfrak{A}$ is indistinguishable from $\mathfrak{A}$ by any prime experiment.

The state s of the general linear automaton $\mathfrak{A}$ is equivalent to the state q of the general linear automaton $\mathfrak{B}$ if s and q are indistinguishable by any experiment. Let us recall that the general linear automaton $\mathfrak{A}$ is equivalent to the general linear automaton $\mathfrak{B}$, if for any state $\mathfrak{A}$ there exists an equivalent state of the general linear automaton $\mathfrak{B}$, and vice versa.

Theorem 3. *If a strongly connected general linear automaton $\mathfrak{A}$ is indistinguishable from the complete general linear automaton $\mathfrak{B} = (M, \Sigma, Z, \varphi, \psi)$, then for each state s of the automaton $\mathfrak{A}$ one can find an equivalent state q of the automaton $\mathfrak{B}$.*

Proof. For each state s_0 of the general linear automaton $\mathfrak{A}$ and a prime experiment (input word) $\mathscr{E}$ let us consider the set $M(\mathfrak{A}, \mathfrak{B}, s_0, \mathscr{E}) = \{q\}$ (more briefly, $M(\mathscr{E})$) of states of the general linear automaton $\mathfrak{B}$, which are such that the experiment begun in any state $M(\mathscr{E})$ has an outcome coinciding with the outcome of the experiment $\mathscr{E}$ on the automaton $\mathfrak{A}$ for the initial state s_0. It is not difficult to see that the set $M(\mathscr{E})$ is a hyperplane in the space $V_{\mathfrak{B}}$. For continuation of the experiment $\mathscr{E}$ the dimensionality of $M(\mathscr{E})$ may only decrease. Since $V_{\mathfrak{B}}$ has a finite dimensionality, there exists an experiment $\mathscr{E}^*$, such that for continuation of $\mathscr{E}^*$ the dimensionality of $M(\mathscr{E}^*)$ no longer decreases. The state $s_0\mathscr{E}^*$, into which the automaton $\mathfrak{A}$ goes over from s_0 at the end of the experiment $\mathscr{E}^*$, is equivalent to each state $M(\mathscr{E}^*)$. Actually, if there were to exist a certain experiment $\mathscr{E}_1$, distinguishing $s_0\mathscr{E}^*$ from $M(\mathscr{E}^*)$, then the set $M(\mathscr{E}^*\mathscr{E}_1)$ could be rigorously contained in $M(\mathscr{E}^*)$ and the dimensionality of $M(\mathscr{E}^*\mathscr{E}_1)$ would be lower than the dimensionality of $M(\mathscr{E}^*)$. Let us take a certain state $q_0 \in M(\mathscr{E}^*)$. Now for any state s of the general linear automaton $\mathfrak{A}$ one can indicate an equivalent state q of the general linear automaton $\mathfrak{B}$. For this purpose let us consider the set of states $\mathfrak{A}$ accessible from $s_0\mathscr{E}^*$. Its dimensionality r is equal to the dimensionality of $\mathfrak{A}$. Therefore, any state s of the automaton $\mathfrak{A}$ is a linear combination $r(\mathfrak{A})$ of states accessible from $s_0\mathscr{E}^*$, i.e., $s = \sum_{i=1}^{r(\mathfrak{A})} a_i s_i$, where $s_i = s_0\mathscr{E}^*\mathscr{E}_i \quad (i = 1, 2, \ldots, r(\mathfrak{A}))$.

The state

$$q = \sum_{i=1}^{r(\mathfrak{A})} a_i q_i$$

of the general linear automaton $\mathfrak{B}$ where $q_i = q_0\mathscr{E}^*\mathscr{E}_i \quad (i = 1, \ldots, r(\mathfrak{A}))$, exists by virtue of the completeness of $\mathfrak{B}$ and is equivalent to s, since $s_i = s_0\mathscr{E}^*\mathscr{E}_i$ is equivalent to $q_i = q_0\mathscr{E}^*\mathscr{E}_i$.

For the case in which the set of states of $\mathfrak{A}$ belongs to a certain hyperplane Γ that does not contain O (the origin of V) – i.e., when any state s of the automaton $\mathfrak{A}$ is a normalized linear combination $\sum a_i s_i$ ($\sum a_i = 1$) of the basis states s_i – it is sufficient to require weak completeness of the general linear automaton $\mathfrak{B}$ in the condition of Theorem 3.

Problem 1. The author does not know the degree to which one may be released from the requirement of completeness (or weak completeness) in the condition of Theorem 3.

If one considers simply infinite automata $\mathfrak{A}$ and $\mathfrak{B}$, where the automaton $\mathfrak{A}$ is strongly connected (in the general sense), then indistinguishability of $\mathfrak{A}$ from $\mathfrak{B}$ is insufficient for the validity of Theorem 3 of E. Moore [18], whose analog for a general linear automaton is the theorem that has been proved.†

Corollary 1. If two general linear automata $\mathfrak{A}$ and $\mathfrak{B}$ are strongly connected and $\mathfrak{A}$ is indistinguishable from $\mathfrak{B}$, then they are equivalent.

Corollary 2. If the general linear automata $\mathfrak{A}$ and $\mathfrak{B}$ are strongly connected and the state q of the automaton $\mathfrak{A}$ is equivalent to the state s of the automaton $\mathfrak{A}$, then the general linear automaton $\mathfrak{A}$ is equivalent to $\mathfrak{B}(\mathfrak{A} \sim \mathfrak{L})$.

For each general linear automaton $\mathfrak{A}$ there exists a unique general linear automaton $\mathfrak{A}_1$, which is equivalent to it and has pairwise-distinguishable states; this automaton is called the reduced form of the automaton $\mathfrak{A}$.

5. Let us now state the problem of experiments which allow the structure of a general linear automaton $\mathfrak{A}$ of the type I having the dimensionality n to be established. It is natural to restrict the analysis to the case of strongly connected general linear automata, since in other general linear automata there are states about which one may not obtain any information during an experiment (in finite automata and Markovian chains these states are called nonrecurrent). The class of all such general linear automata of the type I in reduced form over the field P having the input alphabet Σ and the output space Z is denoted by $T_{P,\Sigma,n,z}$.

Theorem 4. *Assume $\mathfrak{A}$ is a strongly connected general linear automaton of the type I over the field P having a dimensionality not exceeding n. A multiple experiment of length 2n begun at any state s of the general linear automaton $\mathfrak{A}$, allows the structure of $\mathfrak{A}$, to be established; i.e., it allows the transition and output operators of $\mathfrak{A}$ to be found.*‡

Proof. Let us find the basis of the general linear automaton $\mathfrak{A}$. Let us lexicographically number the vertices of the tree D_s of the automata $\mathfrak{A}$ which has the height 2n with its root in s.

Let us assume that $q_1 = s$. Let us prove by induction with respect to h that on the first h stories of the tree D_s for $h \le n_1$, where n_1 is the dimensionality of the general linear automaton $\mathfrak{A}$, one may choose no less than h linearly independent states. For h = 1 this is obvious. Assume this is true for a tree $D_s(h)$ having a height $h < n_1$. Let us assume that in the tree $D_s(h + 1)$ one cannot find $(h + 1)$ linearly independent states (i.e., the number of linearly independent states of $D_s(h + 1)$ is equal to h, and they are all situated in $D_s(h)$. Then each state

† The corresponding example was constructed in the paper by Ch. Facey [10]. The author introduced the notion of strong indistinguishability of automata. Ch. Facey proved the analog of the Moore theorem for strongly indistinguishable strongly connected infinite automata.

‡ It is natural that in general no experiment allows determination of the set M of states of the general linear automaton $\mathfrak{A}$; therefore the supplements of general linear automata are actually determined in Theorem 4 and the succeeding theorems in the experiments.

of the (h + 2)-nd story will be a linear combination of states from the first h + 1 stories, and therefore from just the h first stories (considering the linear dependence of the states of the (h + 1)-st story on the states of the first h stories), etc. Thus, we find that in D_s there are only h (and not n_1) linearly independent states, which is untrue. Therefore, in D_s(h + 1) there are at least h + 1 linearly independent states.

From what has been proved it follows that all n_1 linearly independent states of $\mathfrak{A}$ may be chosen to be on the first n_1 stories of D_s. Let us number all vertices of the tree D_s in the sequence of increasing stories, and lexicographically on one story. Let us choose the first n_1 linearly independent states in D_s to be on the first n_1 stories of D_s. Let us denote them by $q_1 = s_1, q_2, \ldots, q_{n_1}$. Any state q in the n_1 + 1 first stories is a linear combination $q_1, q_2, \ldots, q_{n_1}$. But for $q_1, q_2, \ldots, q_{n_1}$ and q we know the branches $D_{q_1}, D_{q_2}, \ldots, D_{q_n}$ and D_q of the tree D_s having a height n, since $n_1 < n$, while the length of a multiple experiment is equal to 2n. Recalling Theorem 1, we note that the expression for q may be found as a linear combination $\{q_i\}$ if D_{q_i} and D_q are known. In finding the expression for all states of the first n_1 + 1 stories in terms of $q_1, q_2, \ldots, q_{n_1}$, we thereby "close" the tree D_s in the graph of transitions-outputs and define the operators φ and ψ in the basis $q_1, q_2, \ldots, q_n$, which is what was required to be proved.

If we were to begin a multiple experiment in another state s* of the general linear automaton $\mathfrak{A}$, then, in general, we would obtain another basis $q_1^*, q_2^*, \ldots, q_{n_1}^*$ having the same dimensionality in accordance with the definition of a strongly connected general linear automaton, and the expression for φ and ψ in this new basis. Thus, we have defined the structure of a general linear automaton $\mathfrak{A}$ with an accuracy of up to the choice of the basis. Finding expressions for the state q of a general linear automaton $\mathfrak{A}$ as linear combinations of basis states is essentially the reduction of the general linear automaton $\mathfrak{A}$. Thus, if we were given a nonreduced general linear automaton $\mathfrak{A}$ having a dimensionality not exceeding n in Theorem 4, a multiple experiment of length 2n on it would allow us to find its reduced form $\mathfrak{A}_1$.

6. Let us now state the problem of constructing a simple experiment allowing the determination of the structure of any general linear automaton $\mathfrak{A}$ from $T_{P,\Sigma,n,Z}$. Let us note that the analogous problem for finite automata (the class $R_{n,m,p}$) is solved by E. Moore [18] with sorting of all automata from $R_{n,m,p}$. But in $T_{P,\Sigma,n,Z}$ with an infinite field P there are infinitely many elements. Therefore, the Moore method is not tenable here.

Hereafter the field P is fixed, and in place of $T_{P,\Sigma,n,Z}$ we simply write $T_{\Sigma,n,Z}$ or even T_n.

Theorem 5. *There exists a prime branching experiment E which allows the structure and terminal state of any general linear automaton $\mathfrak{A} \in T_n$ to be determined.*

Proof. Let us determine the structure of the general linear automaton $\overline{\mathfrak{A}}$ – the supplement of $\mathfrak{A}$.

Let us consider the experiment U_n, whose input consists of all strings of input letters $\sigma_1, \sigma_2, \ldots, \sigma_m$ of length n(n + 1)/2 taken in some sequence. The length U_n does not exceed $m^{\frac{n(n+1)}{2}} \cdot \frac{n(n+1)}{2}$ (it can be proved that the length U may be chosen equal to $m^{\frac{n(n+1)}{2}} + \frac{n(n+1)}{2} - 1$ [9]).

Such an input word U_n for each general linear automaton $\mathfrak{A} \in T_n$ allows its terminal state to be determined according to Theorem 2.

We take $F_1 = U_n$ to be the first place of the input of experiment E. Let us renumber all prime experiments of length 2n: $\alpha_1, \alpha_2, \ldots, \alpha_{m}^{2n}$. Let us carry out the experiment $E_1 = \alpha_1$ after

U_n. Assume that the experiment-input $F_1E_1F_2E_2 \ldots F_kE_k$ has been constructed, where $E_1 = \alpha_{i_1}$, $E_2 = \alpha_{i_2}$, $E_3 = \alpha_{i_3}, \ldots, E_k = \alpha_{i_k}$, while $F_i = U_n$ $(i = 1, 2, \ldots, k)$.

Let us place $F_{k+1} = U_n$. If the outcome of the experiment F_{k+1} is linearly independent of the outcome $F_1, F_2, \ldots, F_k$, then we take F_{k+1} to be α_1.

In the converse case we consider all different groups of experiments $(F_{j_1}, F_{j_2}, \ldots, F_{j_h})$ having linearly independent outcomes, on which the outcome F_{k+1} depends linearly. (Note that since the dimensionality of $\mathfrak{A}$ is equal to n, it follows that $h \leq n$, since the initial states of $F_{j_1}, F_{j_2}, \ldots, F_{j_h}$ must be linearly independent.) Let us place E_{k+1} equal to the first α_i whose outcome from the initial state (which is a final state for F_{k+1}) may not be determined from the previous part of the experiment (i.e., it may not be determined as a linear combination of outcomes for the words $E_{j_1} = \alpha_i, E_{j_2} = \alpha_i, \ldots, E_{j_h} = \alpha_i$ for a certain group $(F_{j_1}, F_{j_2}, \ldots, F_{j_h})$. In other words, let us assume that E_{k+1} is equal to a α_i such that additional information is obtained. The process is broken off when an $E_g = \alpha_i$ is determined such that for the outcome of the experiment $F_g = U_n$ the outcomes of all of the experiments α_i begun in the terminal state s_g of the experiment F_g may be determined from the experiment $E = F_1E_1F_2E_2 \ldots F_gE_g$.

In order to conclude the proof it remains to note that the outcomes of all experiments α_i having the length 2n are determined in the state s_g, and the structure of the general linear automaton $\mathfrak{A}$ may be determined in accordance with Theorem 4.

Let us note that each α_i is encountered in E no more than n times, and therefore it is necessary to take $g = n \cdot m^n$. Since $l(F_i) = \frac{n(n+1)}{2} m^{\frac{n(n+1)}{2}}$, while $l(E_i) = 2n$, it follows that $l(E) = g\left(\frac{n(n+1)}{2} m^{\frac{n(n+1)}{2}} + 2n\right)$. Hereafter we shall be able to construct a substantially shorter experiment $\mathfrak{E}$, which recognizes the structure of a general linear automaton $\mathfrak{A}$ for certain additional information on $\mathfrak{A}$.

For the time being we note that the uniform experiment U_{2n} allows any automaton $\mathfrak{A} \in T_n$ to be distinguished from any other $\mathfrak{B} \in T_n$. In fact, let us consider the direct sum $\mathfrak{A} \oplus \mathfrak{B}$, which is a general linear automaton of reduced form having the dimensionality 2n. The experiment U_{2n}, which is begun in a certain state $\mathfrak{A} \oplus \mathfrak{B}$, allows the terminal state $\mathfrak{A} \oplus \mathfrak{B}$ to be determined (and therefore the initial state also) along with that automaton from $\mathfrak{A}$ and $\mathfrak{B}$, to which it belongs (i.e., it allows $\mathfrak{A}$ to be distinguished from $\mathfrak{B}$). However, from this reasoning it does not follow that the experiment U_{2n} yields the algorithm for determining the structure $\mathfrak{A} \in T_n$, since here it is necessary to compare $\mathfrak{A}$ with an infinite set of general linear automata from T_n.

Problem 2. Does such an algorithm exist?

Problem 3. It is required to construct the shortest possible (uniform or nonuniform) experiment U_n, which allows the terminal state of any general linear automaton $\mathfrak{A} \in T_n$ to be determined.†

From E. Moore's example of a finite automaton having n states (the "secret lock" [18]), which evidently is a general linear automaton from T_{n-1}, there derives the lower bound for the length U_{n-1} equal to M^{n-1} (i.e., $l(U_n) \geq m^n$).

For finite automata from $R_{n,m,p}$ we were able to obtain an upper bound of order $n^2 \ln(ne) m^n$ for E_n [9]. However, our method of constructing U_n for $R_{n,m,p}$ rests on the sorting

† Beginning here, U_n denotes an arbitrary experiment with such a property, while U_n^{min} denotes the shortest experiment among U_n.

of automata from $R_{n,m,p}$, while the length U_n depends on the power of $R_{n,m,p}$; therefore, for the case of a general linear automaton with, in general, an infinite $T_{P,n,\Sigma,Z}$, this method is unsuitable.

An experiment U_n for $R_{n,m,p}$, which is longer than the experiment in [18] but does not have sorting of all automata from $R_{n,m,p}$, was constructed in the paper by Yu. M. Borodyanskii [2].

Let us now state the problem of constructing a simple uniform (nonbranching) experiment which allows the structure of any $\mathfrak{A} \in T_n$ to be determined.

Theorem 6. *There exists a uniform experiment F which allows the structure of any general linear automaton $\mathfrak{A} \in T_n$ to be determined.*

Proof. Assume x is an arbitrary word in the alphabet Σ, $x^k = \underbrace{x \cdot x \cdot \ldots \cdot x}_{k \text{ times}}$. Then, knowing the output corresponding to the word x^{2n} for the initial state s_0 of the n-dimensional general linear automaton $\mathfrak{B}$, we may determine the output of any word x^k for any k for the same initial state. Let us place $s_k = s_0 x^k$ (k = 1, 2, ...). Let us denote the outputs corresponding to the initial state s and the input string xxx ..., by y(s). Let us take the first state s_t in the string $\{s_k\}$ (k = 0, 1, 2, ...), which is such that $y(s_k)$ depends linearly on the preceding $y(s_i)$. It is obvious that $t \leq n$, since the maximum number of linearly independent states in $\{s_k\}$ does not exceed n. Two outputs y(s) and y(g) coincide if they coincide on the input word x^n. This is proved in the same way as Theorem 1 (for autonomous automata having an input consisting of the word x).

Assume $y_n(s)$ is the initial piece of y(s), which corresponds to the input word x^n. Let us represent $y_n(s_t)$ as the linear combination $\{y_n(s_i)\}$ (i = 1, 2, ..., t − 1). $y(s_t)$ is expressed by the same linear combination $\{y(s_i)\}$. Since we know the outputs $y_{2n}(s_n)$ $(1 \leq i < t)$, one can determine the output $y_{2n}(s_t)$; then all outputs $y_{3n}(s_i)$ $(i < t)$ become known, and the output $y_{3n}(s_t)$ can be determined, etc.

Let us place

$$F = (\alpha_1(\alpha_2(\alpha_3 \ldots (\alpha_{u-1}\alpha_u^{2n})^{2n} \ldots)^{2n})^{2n})^{2n},$$

where $\{\alpha_i\}$ (i = 1, 2, ..., $u = m^{2n}$) is the set of all words of length 2n in the alphabet Σ. Assume s is the terminal state of the experiment F. Finite pieces of F will be words of the form

$$\alpha_u^{2n},\ (\alpha_{u-1}\alpha_u^{2n})^{2n},\ (\alpha_{u-2}(\alpha_{u-1}\alpha_u^{2n})^{2n})^{2n},\ \ldots,\ (\alpha_1(\alpha_2 \ldots (\alpha_{u-1}\alpha_u^{2n})^{2n})^{2n})^{2n}.$$

Therefore, we may determine the outcomes of the experiments

$$\alpha_u,\ \alpha_{u-1}(\alpha_u)^{2n},\ \alpha_{u-2}(\alpha_{u-1},\ \alpha_u^{2n})^{2n},\ \ldots\ \alpha_1(\alpha_2 \ldots (\alpha_{u-1}\alpha_u^{2n})^{2n} \ldots)^{2n}.$$

from the state s taken as the original state. But thereby all prime experiments α_i of length 2n will be defined in s (i.e., the general linear automaton $\mathfrak{A}$ will be defined).

The length of the experiment F is fantastically great, but on the other hand the algorithm for determining the structure of $\mathfrak{A}$ from F, given in the notation presented above with parentheses, is fairly simple.

Theorem 7 (see below) allows a considerably shorter experiment than F to be constructed, but the algorithm reconstructing the structure of the general linear automaton from such an experiment turns out to be considerably more complex. Here we come up against a frequently encountered phenomenon: a reduction of the length of the calculation accompanies the complication of its program (see the collection of translations: "Problems of Mathematical Logic," Moscow, Mir (1970), Introduction).

Problem 4. It is required to construct the shortest possible uniform experiment which allows the structure of any general linear automaton $\mathfrak{A} \in T_n$ to be determined.

7. Usually, before an experiment is conducted with an automaton $\mathfrak{A}$, we have certain information available on it. Otherwise, as E. Moore [18] showed, it is impossible to determine either the subsequent behavior of the automaton or its structure. This information may be incorporated in the upper bound of the number of states of a finite automaton or the dimensionality of a general linear automaton, in the indication of strong connectedness of an automaton, the type of general linear automaton, the length of an experiment that establishes the final state of a general linear automaton, or directly in the indication in this experiment of the complete set Π of words $x_1, x_2, \ldots, x_t$ (i.e., a set such that the states $\vec{v}x_1, \vec{v}x_2, \ldots, \vec{v}x_t$ form a complete system of vector-states in $\mathfrak{A}$ for any state $\vec{v}$).

We shall consider a Moore general linear automaton for convenience.

The set of words $Y = \{y_1, \ldots, y_r\}$ is called the reference set for the general linear automaton $\mathfrak{A}$ if distinguishability (indistinguishability) of any two states of $\mathfrak{A}$ – $\vec{v}_1$ and $\vec{v}$ – may be established by a multiple experiment having the inputs $y_1, \ldots, y_r$.

Remark. It is obvious that for a complete general linear automaton the following statement holds: if the vector $\vec{u}$ whose coordinates u_i are the outputs of the states $\vec{v}y_i$ $(i = 1, \ldots, r)$ [i.e., $u_i = \vec{v}A(y_i)\vec{b}$] is a linear combination of vectors $\vec{u}^{(1)}, \ldots, \vec{u}^{(k)}$ corresponding to the states $\vec{v}^{(1)}, \ldots, \vec{v}^{(k)}$, then the vector $\vec{v}$ is the same linear combination of vectors $\vec{v}^{(1)}, \ldots, \vec{v}^{(k)}$. For a weakly complete general linear automaton the analogous statement is valid for normalized linear combinations.

The reference set Y is called the minimal reference set if none of the words $y_1, \ldots, y_r$ may be replaced by its own final piece in such a way that the resulting set of words remains a reference set.

Knowing the complete set Π and the reference set Y of a reduced strongly connected general linear automaton $\mathfrak{A}$, we may construct a uniform (i.e., independent of the initial state) multiple experiment which allows the structure of the general linear automaton $\mathfrak{A}$ to be established. Let us call the sum of the lengths of the words that determine the multiple experiment $\mathcal{E}$, its resultant length, and let us denote it by $L(\mathcal{E})$, while the number of words of the experiment shall be called its multiplicity $K(\mathcal{E})$. Let us use L(A) to denote the sum of the lengths of the set $A = \{a_1, \ldots, a_q\}$, while μA denotes the power of A.

Lemma 1. Assume $\mathfrak{A}$ is a complete connected general linear automaton $\mathfrak{A}$ in a reduced form having the input alphabet $\Sigma = \{\sigma_1, \ldots, \sigma_m\}$, $\Pi = \{x_1, \ldots, x_t\}$ is a complete set, and $Y = \{y_1, \ldots, y_r\}$ is the reference set of the words of $\mathfrak{A}$. Then there exists a uniform multiple experiment having the resultant length

$$L_{\Sigma, Y, \Pi} = (m+1)(tL(Y) + rL(\Pi)) + mrt$$

and the multiplicity $K_{\Sigma, Y, \Pi} = (m+1)\mu Y \cdot \mu\Pi$, which establishes the structure of $\mathfrak{A}$.

Proof. Let us consider a multiple experiment having the set of input words

$$\{x_i,\ y_j;\ x_i\sigma_k y_j\} \qquad (i=1, \ldots, t;\ j=1, \ldots, r;\ k=1, \ldots, m).$$

Assume $\vec{v}_0$ is a certain state of $\mathfrak{A}$. The states $\vec{v}_0x_1, \vec{v}_0x_2, \ldots, \vec{v}_0x_t$ form a complete system of vectors in the space of states of $\mathfrak{A}$. Knowing the outputs of the states $\vec{v}_0x_iy_j$ for all i, j, we may, according to the remark preceding the lemma, isolate the complete linearly independent subsystem of the system of vectors $\{\vec{v}_0x_i\}$ (i.e., the basis of the general linear automaton), while from Π we may isolate the basis set of words B. Knowing the output of the states $\vec{v}_0x_i\sigma_k y_j$ for all j, we may determine how the vector-state $\vec{v}_0x_i\sigma_k$ is expressed linearly in terms of the basis vector $\vec{v}_0x_i$, $x_i \in \Pi$, and thus determine which linear transformation carries out the entry of σ_k in the basis $\{\vec{v}_0x_i\}$, $x_i \in B$. Having performed this operation for all $\sigma_k \in \Sigma$, we determine the matrices of the transitions and outputs in the basis $\{\vec{v}_0x_i\}$ (i.e., the structure of the general linear automaton $\mathfrak{A}$).

The sum of the lengths of all words x_iy_j is equal to $tL(Y) + rL(\Pi)$, while the sum of the lengths of all words $x_i\sigma_k y_j$ is equal to

$$mtL(Y) + mrL(\Pi) + mrt.$$

Thus, the resultant length of a multiple experiment would be

$$(m+1)[tL(Y) + rL(\Pi)] + mrt,$$

while

$$K_{\Sigma, Y, \Pi} = (m+1)rt.$$

The lemma has been proved.

If we were to desire likewise to determine the initial state $\vec{v}_0$, then it would be required to perform experiments on the inputs y_j ($j = 1, \ldots, r$) having a total length $L(Y)$ in the state $\vec{v}_0$.

Let us note that for any general linear automaton $\mathfrak{A} \in T_n$ the complete and reference set of words is the set of all words having a length no longer than $n - 1$. Therefore the resultant length of a uniform multiple experiment for a general linear automaton of class T_n is $(2n-1)m^{2n-1}$; i.e., it is a multiple experiment of length 2n (if, according to the tradition going back to E. Moore, the length of an experiment is assumed to be the length of the chain of states corresponding to the input word, which is one greater than the input word). Since, however, one may select the reference and complete sets of words for a general linear automaton $\mathfrak{A} \in T_n$ having a considerably shorter resultant length (as we shall see further on, the first has a resultant length $< n^2/2$, while the second has a resultant length $< n^3/2$ – the number $t \le n^2$, $r \le n$), it follows that the estimates given by the lemma for this case will be:

$$L_{\Sigma, Y, \Pi} < (m+1)n^4 + mn^3 < (m+1)(n^4 + n^3),$$

while

$$K_{\Sigma, Y, \Pi} \leqslant (m+1)n^3.$$

As far as the lower bound for a universal uniform multiple experiment in the class T_n is concerned, constructions analogous to "the secret lock" may be used to prove that it coincides with the upper bound; i.e., the bound $(2n-1)m^{2n-1}$ for the class T_n cannot be improved.

Let us consider the case in which we know only the reference set of words $Y = \{y_1, \ldots, y_r\}$ for a complete general linear automaton $\mathfrak{A} \in T_n$, and this time let us construct a branching multiple experiment which determines the structure of $\mathfrak{A}$.

Lemma 2. There exists a branching† multiple experiment which allows determination of the structure of the complete general linear automaton $\mathfrak{A} \in T_{n,\Sigma}$ having the reference set $Y = \{y_1, \ldots, y_r\}$ of resultant length no greater than

$$mn\left(L(Y)+t+\frac{n-1}{2}\right)+L(Y)$$

and a power $K_{\Sigma, Y} \leqslant (m+1)nr$.

Proof. Initially let us carry out all prime experiments with inputs $y_1, \ldots, y_r$ and inputs $\sigma_k y_j$ ($k = 1, \ldots, m$; $j = 1, \ldots, r$). Let us assign the state $\vec{v}_0$ to the basis and use B_1 to denote the set $\{\vec{v}_0\}$. Among the states $\vec{v}_0\sigma_k$ ($k = 1, \ldots, m$) one can find at least one that is linearly independent of $\vec{v}_0$. Let us choose from among the states $\vec{v}_0, \vec{v}_0\sigma_1, \ldots, \vec{v}_0\sigma_m$ the maximal linearly independent system containing $\vec{v}_0$. This may be done, since we know the outputs of the states $\vec{v}_0\sigma_k y_j$ by virtue of the remark preceding Theorem 5 (on p. 199). Let us denote these vectors by $\vec{v}_1, \vec{v}_2, \ldots, \vec{v}_{i_1}$, their corresponding inputs by $\sigma_{k_1}, \sigma_{k_2}, \ldots, \sigma_{k_{i_1}}$, and the sets $\{\vec{v}_0, \ldots, \vec{v}_{i_1}\}$ by B_2. For each of these vector-states from $(B_2 \setminus B_1)$ we carry out all experiments with inputs $\sigma_k y_j$ and find from among

$$\vec{v}_0, \vec{v}_1, \ldots, \vec{v}_{i_1}, \vec{v}_i\sigma_k \qquad (i=0, 1, \ldots, i_1;\ k=1, \ldots, m)$$

the maximal linear independent system containing $\vec{v}_0, \vec{v}, \ldots, \vec{v}_{i_1}$. Among the vectors $\vec{v}_i\sigma_k$ one can find at least one vector that is linearly independent of the vectors $\vec{v}_0, \vec{v}_1, \ldots, \vec{v}_{i_1}$, provided only that these latter vectors do not form the basis for $\mathfrak{A}$ (i.e., if their number is less than n). Let us denote the vectors added to B_2 by $\vec{v}_{i_1+1}, \ldots, \vec{v}_{i_2}$, while the system of vectors obtained is denoted by B_3, etc., until at a certain step we obtain

$$B = B_q = \{\vec{v}_0, \vec{v}_1, \ldots, \vec{v}_{i_1}, \ldots, \vec{v}_{i_2}, \ldots, \vec{v}_{i_{q-1}} = \vec{v}_{n-1}\}.$$

For states from $(B_q \setminus B_{q-1})$ let us perform all experiments $\sigma_k y_j$. Thus, for each basis state $\vec{v}_i$ we will know how the vectors $\vec{v}_i\sigma_k$ are expressed in terms of $\vec{v}_i$ (i.e., we will define the transformations φ_{σ_k} of the vectors $\vec{v}_i$). The outputs in the states $\vec{v}_i$ are determinate, and therefore the structure of the initial state of $\mathfrak{A}$ can be determined.

Let us estimate the resultant length of the experiment. For each basis state we perform all experiments with inputs $\sigma_k y_j$, while for $\vec{v}_0$ we likewise perform all experiments y_j. The sum of the lengths of the corresponding words is

$$(mn+1)L(Y)+mnr.$$

Here we see that the number of all words in our experiment is equal to

$$mnr+r=(mn+1)r.$$

Moreover, the resultant length of the experiment includes the sum of the lengths of the words required to attain the state $\vec{v}_i$, multiplied by mr, since each word is encountered with mul-

†A branching multiple experiment, unlike a uniform multiple experiment, is constructed so that the next input word of this experiment is chosen as a function of the output of a general linear automaton in words defined earlier, while the input words are defined and applied to the input one after the other in one and the same initial state.

tiplicity mr. Since the length of a word leading from $\vec{v}_0$ to $\vec{v}_i \in (B_l \setminus B_{l-1})$ is equal to $l-1$, this sum is equal to

$$mr \sum_{l=1}^{q-1} l\mu(B_{l+1} \setminus B_l),$$

where μC denotes the power of the set C. Since all sets $(B_{l+1} \setminus B_l)$ and B_1 are not empty, it follows that $q \le n$, and the expression $\sum_{l=1}^{q-1} l\mu(B_{l+1} \setminus B_l)$ attains a maximum for $q = n$ when $\mu(B_{l+1} \setminus B_l) = 1$. This maximum is equal to

$$\sum_{l=1}^{n-1} l = \frac{n(n-1)}{2}.$$

Thus, we find that the resultant length of the experiment does not exceed

$$(mn+1)L(Y) + mnr + m\frac{n(n-1)}{2}r = mn\left(L(Y) + r + \frac{n-1}{2}\right) + L(Y),$$

which is what was required to be proved.

It is not difficult to see that the multiple experiment constructed in Lemma 2 is branching. For the chosen initial state $\vec{v}_0$ we constructed a certain set of words $\tilde{X}$ leading from $\vec{v}_0$ to the states $\vec{v}_0, \ldots, \vec{v}_i, \ldots, \vec{v}_{n-1}$ forming the basis in $\mathfrak{A}$. A knowledge of the number n is not required to construct the experiment here either, since the experiment stops as soon as it is discovered that all vectors $\vec{v}_i\sigma_k$ for all $\vec{v}_i$ from a certain set B_q can be expressed linearly in terms of $\vec{v}_i \in B_q$. However, the set of words $\tilde{X}$ may not be complete, since the system of vectors $\{\vec{v}x\}_{x\in\tilde{X}}$, being the basis for $\vec{v} = \vec{v}_0$, may not be such for another initial state $\vec{v} = \vec{v}_0'$. For example, assume that the general linear automaton $\mathfrak{A}$ having the input alphabet $\Sigma = \{0, 1\}$ is stipulated by the transition matrices

$$a_{ij}(0) = \begin{cases} 1, & \text{if } j \equiv i+1 \pmod{n}, \\ 0 & \text{otherwise} \end{cases}$$

(cyclic permutation),

$$a_{ij}(1) = \begin{cases} 1 & \text{for } i = j = 1, \\ 0 & \text{otherwise.} \end{cases}$$

The output vector is defined thus: $\vec{b} = (1, 0, \ldots, 0)$, i.e., in the state $\vec{v}_1$ the output is equal to 1, while in $\vec{v}_i$ $(i > 1)$ the output is 0. Obviously, all states of $\mathfrak{A}$ are distinguishable. Then the set of words $\tilde{X} = \{\Lambda, 0, 00, \ldots, 0^{n-1}\}$ for each basis state $\vec{v}_i$ will generate a basis $\{\vec{v}_j\}$, while it leaves the vector-state of the form $\vec{v} = (a, a, \ldots, a)$ as is (i.e., the set $\tilde{X}$ is complete, so to speak, r e l a t i v e t o any $\vec{v}_i$, but it is not complete for $\mathfrak{A}$). It can be proved that for completeness of the set $\tilde{X}$ of the input words of a complete general linear automaton $\mathfrak{A}$ in reduced form it is sufficient if the set of transition operators for the words φ_x is complete in the linear space generated by the transition operators φ_u for all words $u \in F(\Sigma)$. Therefore, one can always choose a complete set of words containing no more than n^2 words for a complete general linear automaton $\mathfrak{A}$ having the dimensionality n, since the dimensionality of the set of all n^2 matrices is $n \times n$. In the example given above, the dimensionality of the set of matrices $\| a_{ij}(u) \|$ for all words $u \in \{0, 1\}^*$ is equal to n^2.

D e f i n i t i o n. Let us call the set of words X *perfectly complete for the general linear automaton* $\mathfrak{A}$, if the set of all transition operators φ for $x \in X$ is complete in the linear state generated by the operators φ for all words $\in F(\Sigma)$. The perfectly complete set of words for a given general linear automaton $\mathfrak{A}$ can be found in the same way that we found the basis set of vectors in the proof of Lemma 2 (i.e., we being with Λ, then sort σ_k, which are all words of length 1, choosing from among them a maximal set such that the corresponding operators φ_Λ, φ_{σ_k} are linearly independent, etc.).

It can be shown that a perfectly complete set is likewise a reference set. The following problem develops.

Problem 5. Do n-dimensional complete general linear automata $\mathfrak{A}$ exist for which any complete set of words is perfectly complete? For which fields P may this hold?

Note that if the set of words X is such that for any word $x_1 \in X$ one may indicate a state $\vec{v}$ of the general linear automaton $\mathfrak{A}$ such that $\vec{v}x_1$ cannot be expressed linearly in terms of the remaining vectors $\vec{v}x$, where $x \in X$, then the set of operators $\varphi_x (x \in X)$ is linearly independent.

Finally, we take account of the fact that each of the sets of words: the complete set relative to the given initial state $\vec{v}$, the complete set for the general linear automaton $\mathfrak{A}$, and the perfectly complete set for the general linear automaton $\mathfrak{A}$ may be chosen so that

1) The minimality processes will be fulfilled; i.e., from the set one may not discard a single word in such a way that the set remains complete relative to $\vec{v}$, complete, and perfectly complete, respectively. We shall call such sets the basis set relative to $\vec{v}$, the basis set for the general linear automaton $\mathfrak{A}$, and the perfectly basis set, respectively.

2) Along with each word, the set X contains all of its initial pieces.

3) (The completeness criterion.) If a certain set of words R contains Λ and has the property that for each word $\tilde{x} \in R$ all states $\vec{v}\tilde{x}\sigma_k$ $(k = 1, \ldots, m)$ can be expressed linearly in terms of $\vec{v}\tilde{x}$ $(\tilde{x} \in R)$ for the state $\vec{v}$ of the general linear automaton $\mathfrak{A}$, it follows that the set R is complete relative to $\vec{v}$.

If the set R has the indicated property relative to each state $\vec{v}$, then the set is obviously complete.

4) (The criterion of perfect completeness.) If the set of words R contains Λ and for any word $\tilde{x} \in R$ the operators φ_{σ_k} or $\varphi_{\sigma_k \tilde{x}}$ $(k = 1, \ldots, m)$ of the transitions can be expressed linearly in terms of the operators φ_x $(x \in R)$, then R is a perfectly complete set.

The following method of constructing the complete set of words Π for a general linear automaton $\mathfrak{A}$ may be based on these criteria.

All words in $F(\Sigma)$ are ordered lexicographically.

Assume $x_1 = \Lambda$. Assume $x_1, \ldots, x_s$ are defined. x_{s+1} is equal to the first word, which is such that for a certain state $\vec{v}_i'$ the state $\vec{v}x_{s+1}$ cannot be expressed linearly in terms of the state $\vec{v}x_i$, $i = 1, \ldots, s$.

This process will be broken off, since the number of words x cannot be greater than n^2 because the system of operators $\{\varphi_{x_i}\}$ is obviously linearly independent. When the process breaks off, the set Π will have been constructed. The properties 1), 2), and 3) are fulfilled for it, and therefore it will be complete. Since for each word x and each state $\vec{v}$ the state $\vec{v}x$ can be expressed linearly in terms of the state $\vec{v}x_i$, where x_i lexicographically precede x (or coincide with x) while the set of all words of length not exceeding $n - 1$ form the complete set, it follows that the length of each word from Π does not exceed $n - 1$. Thus, in Π there are no more than n^2 words of length not exceeding $n-1$. Therefore, by virtue of the property 2)

$$L(\Pi) \leqslant n\frac{n(n-1)}{2} < \frac{n^3}{2}.$$

Problem 6. May a perfectly complete set R of words for $\mathfrak{A}$ be constructed in this way? It is required to give the upper bound of L(R).

Let us now consider the case in which we know the complete set $\Pi = \{x_1, \ldots, x_t\}$ for the general linear automaton $\mathfrak{A} \in T_n$. The reference set $y = \{y_1, \ldots, y_r\}$ is called the minimal reference set if none of the words $y_1, \ldots, y_r$ can be replaced by its own final piece in such a way that the resulting set of words remains a reference set. Here it is necessary for us to solve the dual problem to the one that we solved in the proof of Lemma 2, namely: to construct the minimal reference set of words.

Lemma 3. Assume that the complete general linear automaton $\mathfrak{A} \in T_{n,\Sigma}$ and $\Pi = \{x_1, \ldots, x_t\}$ is the complete set of words for $\mathfrak{A}$. Then there exists a branching multiple experiment that determines the structure of $\mathfrak{A}$ and has a resultant length no greater than

$$L_{\Sigma,\Pi} \leqslant m(n-1)\left(L(\Pi) + \frac{nt}{2}\right)$$

and a multiplicity $K_{\Sigma,\Pi} \leqslant (m+1)nt$.

Let us denote the initial-state vector by $\vec{v}_0$, while $\vec{v}_0 x_i$ is denoted by $\vec{v}_i$ ($i = 1, \ldots, n$).

We shall construct the set of reference words Y by steps: we carry out all experiments x_i ($i = 1, \ldots, t$). We place $Y_1 = \{\Lambda\}$; Λ is an empty word; $S_1 = \{\vec{u}\}$, where the vector $\vec{u} = (u_1, \ldots, u_t)$; $u_i = \psi(\vec{v}_i)$ is the output of the states $\vec{v}$, and ψ is the output operator of the general linear automaton $\mathfrak{A}$ (a Moore automaton!).

From among the vectors $\vec{u}(\sigma_1), \ldots, \vec{u}(\sigma_m)$, where $\vec{u}(\sigma_k) = (u_1(\sigma_k), \ldots, u_t(\sigma_k))$, while $u_i(\sigma_k) = \psi(\vec{v}_i\sigma_k)$ ($i = 1, \ldots, t$; $k = 1, \ldots, m$) we choose the maximal set R_1 of vectors which form (along with S_1) the system $R_1 = \{\vec{u}, \vec{u}(y_1), \ldots, \vec{u}(y_{j_1})\}$, where y_l denotes the corresponding letters σ_k.

Let us place $Y_2 = Y_1 \cup \{y_l\}$ ($l = 1, \ldots, j_1$), while $S_2 = S_1 \cup R_1$.

Further we consider the vectors $\vec{u}(\sigma_k y_l)$ ($l = 1, \ldots, j_1$; $k = 1, \ldots, m$), where $u_i(\sigma_k y_l) = \psi(\vec{v}_i\sigma_k y_l)$, and choose from among them the maximal set R_2 of vectors $\vec{u}(\sigma_k y_l)$, such that the union $S_3 = R_2 \cup S_2$ is a linearly independent system. The corresponding words $\{\sigma_k y_l\}$ are denoted by $y_{j_1+1}, \ldots, y_{j_2}$:

$$Y_3 = Y_2 \cup \{y_l\} \qquad (l = j_1 + 1, \ldots, j_2).$$

Then we consider the words $\sigma_k y_l$ ($j_1 < l \leqslant j_2$; $k = 1, \ldots, m$) and the vectors $\vec{u}(\sigma_k y_l)$, etc., until in a certain set S_q there turn out to be n linearly independent vectors. It can be proved that Y_i ($i = 1, \ldots, q$) is supplemented on each step, whence it follows that $q \leq n$. The corresponding set of words Y_q will be a minimal reference set. The point is that from the construction of Y_q it follows that for each $y_i \in Y_q$ ($i = 1, \ldots, j_q$) and each $\sigma_k \in \Sigma$ the vector $\vec{u}(\sigma_k y_i)$ can be expressed linearly in terms of the vectors $\vec{u}(y_l)$ ($l = 1, \ldots, j_q$), whence it follows that in each state $\vec{v}$ of the general linear automaton $\mathfrak{A}$, we may define the output on the word $\sigma_k y_i$ in this state knowing the outputs on all $y_l \in Y_q$. Further we define it the same way on all words of the form $\sigma_{k_1}\sigma_{k_2}y_i$, etc. In this way the outputs in the state $\vec{v}$ on all words $y \in \Sigma^*$ are defined.

In view of the arbitrariness of the choice of the state $\vec{v}$, it follows from this that Y_q is a reference set. The minimality of Y_q derives from the fact that for each $y_i \in Y_q$ $(i = 1, \ldots, j_q)$ the output in the state $\vec{v}y_i$ for an arbitrary state $\vec{v}$ may not be defined as a linear combination (independent of $\vec{v}$) of outputs of a general linear automaton in states $\vec{v}y_l$ $(l \neq i, y_l \in Y_q)$, while the set $Y_q \setminus \{y_i\}$ contains all intrinsic final pieces of the word y_i. Thus, y_i may not be replaced by any of its intrinsic final pieces. Thus, in the process of constructing Y_q all experiments $x_i y_l$, $x_i \sigma_k y_l$ $(i = 1, \ldots, t;\ l = 1, \ldots, j_q;\ k = 1, \ldots, m)$ are carried out; i.e., the structure of a general linear automaton $\mathfrak{A}$ is determined according to Lemma 1.

The reader will establish without particular difficulty the fact that the resultant length of the constructed experiment does not exceed

$$m(n-1)\left(L(\Pi) + \frac{tn}{2}\right),$$

while its multiplicity is

$$K_{\Sigma,\Pi} \leqslant (m+1)\,tn.$$

Now we shall formulate and give the outlines of the proofs of three theorems on prime experiments which allow the structure of a general linear automaton $\mathfrak{A} \in T_n$ to be established, provided that the input word E allowing the determination of the concluding state of the general linear automaton $\mathfrak{A}$ is known, along with either the complete or reference set of words $\mathfrak{A}$, or both. Let us begin with the latter case.

Theorem 7. If the complete general linear automaton $\mathfrak{A} \in T_n$, then, knowing E, the complete set of words Π, and the reference set Y, one may construct a prime branching experiment $\mathscr{E}$, that determines the structure of $\mathfrak{A}$ and has a length no greater than $n(K_{\Sigma,Y,\Pi}\, l(E) + L_{\Sigma,Y,\Pi})$, where $L_{\Sigma,Y,\Pi}$ and $K_{\Sigma,Y,\Pi}$ are the resultant length and power of the multiple experiment $\mathcal{Э}$, constructed in Lemma 1 for recognition of the structure of a general linear automaton, while l (E) is the length of the word E.

It is necessary to repeat the proof of Theorem 5 and Lemma 1 in their general features:

$$\mathscr{E} = EF_1EF_2EF_3E \ldots EF_q,$$

where $q \leq n(m+1)rt$. F_1 is the first word of the multiple experiment $\mathcal{Э}$. The next application of a word E allows determination of whether or not the terminal state of the experiment EF_1E coincides with the terminal state of E; if it coincides, then we assume that F_2 is equal to the second word of $\mathcal{Э}$. Otherwise, F_2 is equal to the first word of $\mathcal{Э}$, etc.

Remark 1. The construction of the experiment $\mathscr{E}$ does not require knowledge of the number n (the dimensionality of $\mathfrak{A}$) but is required for the upper bound in which n is included linearly.

Remark 2. As we shall see in the optimal selection of E, Y, Π,

$$l(E) < \frac{n^2}{2},\quad L(Y) < \frac{n^2}{2},\quad L(\Pi) < \frac{n^3}{2},\quad t \leqslant n^2,\quad r = n,$$

$$K_{\Sigma,Y,\Pi} \leqslant (m+1)\,n^3,\qquad L_{\Sigma,Y,\Pi} < (m+1)(n^4+n^3)$$

(see Lemma 1). whence

$$l(\mathscr{E}) < n\left[(m+1)\,n^3\frac{n^2}{2} + (m+1)(n^4+n^3)\right] < \frac{(m+1)(n^6+2n^5+2n^4)}{2}.$$

Theorem 8. If a complete $\mathfrak{A} \in T_n$, $Y = \{y_1, \ldots, y_r\}$ is the reference set for $\mathfrak{A}$, while E is a uniform terminal experiment† for $\mathfrak{A}$, then there exists a prime branching experiment $\mathscr{E}$, which establishes the structure of $\mathfrak{A}$, and has a length no greater than $n^3(m + 1)(l(E) + L(Y) + n) + n^3r$.

Proof. We shall construct the experiment $\mathscr{E}$ again according to steps:

$$\mathscr{E}_0 = \Lambda, \quad \mathscr{E}_1 = Ew_1,$$

where $w_1 = y_1$, $x_1 = \Lambda$, $\Pi_1 = \{x_1\}$, $W_1 = \{\Lambda\}$. Then assume the experiment $\mathscr{E}_p = Ew_1Ew_2 \ldots Ew_p$ has been constructed and the sets $\Pi_p = \{x_1, \ldots, x_s\}$ and $W_1, \ldots, W_p$ have been defined.

Assume $\vec{v}_0$ is the state of the general linear automaton $\mathfrak{A}$ at the beginning of the experiment $\mathscr{E}$. Here $\vec{u}_{p+1}$ denotes the state into which the general linear automaton $\mathfrak{A}$ has arrived at the end of the experiment $\mathscr{E}_pE$. From the output of the general linear automaton on the last piece E of the experiment E one can determine via which linear combinations of vectors $\vec{u}_1, \ldots, \vec{u}_p$ the vector $\vec{u}_{p+1}$ is represented (and whether it is represented at all). Let us define the set of words $\overline{W}_{p+1}$: $x \in \overline{W}_{p+1}$ if and only if there exists a linear combination $\sum_{i=1}^{I} \alpha_i \vec{u}_{l_i} = \vec{u}_{p+1}$, where $1 \leq l_i \leq p$, such that $x \in \bigcap_{i=1}^{I} W_{l_i}$. Let us consider the set of words x_iy_j and $x_i\sigma_k y_j$ ($i = 1, \ldots, s$; $j = 1, \ldots, r$), which is ordered as follows:

$$(x_{i_1}, y_{j_1}) \prec (x_{i_2}, y_{j_2}) \quad \text{for} \quad i_1 < i_2,$$

while if $i_1 = i_2$, then

$$x_iy_j \prec x_i\sigma_k y_j \quad \text{for} \quad j_1 < j_2 \quad \text{for all} \quad i, j, k;$$
$$x_i\sigma_{k_1}y_j \prec x_i\sigma_{k_2}y_j \quad \text{for} \quad k_1 < k_2;$$
$$x_{i_1}\sigma_k y_{j_1} \prec x_{i_2}y_{j_2}, \quad \text{if} \quad (x_{i_1}, y_{j_1}) \prec (x_{i_2}, y_{j_2}),$$

whence it follows from transitiveness that

$$x_{i_1}\sigma_{k_1}y_{j_1} \prec x_{i_2}\sigma_{k_2}y_{j_2}, \quad \text{if} \quad x_{i_1}y_{j_1} \prec x_{i_2}y_{j_2}.$$

If one can find just one input word x_iy_j or $x_i\sigma_k y_j$ which does not belong to $\overline{W}_{p+1}$, then we assume that w_{p+1} is equal to the first of such words in the sense of the order which has been introduced. Let us place

$$W_{p+1} = \overline{W}_{p+1} \cup \{w_{p+1}\}, \quad \mathscr{E}_{p+1} = \mathscr{E}_p w_{p+1}.$$

It is not difficult to establish the fact that W_{p+1} consists of all words for which the outputs in the state $\vec{u}_{p+1}$ have been determined from the experiment $\mathscr{E}_pEW_{p+1}$.

If after this the outputs on all input words x_iy_j and $x_i\sigma_k y_j$ have been defined in the state $\vec{u}_{p+1}$ (i.e., they belong to W_{p+1}), then we consider the vectors

$$\vec{u}_{p+1}x_i \text{ and } \vec{u}_{p+1}x_i\sigma_k \quad (i = 1, \ldots, s;\ k = 1, \ldots, m).$$

If the vectors $\vec{u}_{p+1}x_i\sigma_k$ can be expressed linearly in terms of the vectors $\vec{u}_px_i$ (and this can be found out from the outputs of the state $\vec{u}_{p+1}$ on the words $x_i\sigma_k y_j$ and x_iy_j, since $Y = \{y_1, \ldots, y_r\}$ is a reference set), while Π_p contains Λ and all the initial pieces of its own words, then from the criterion of completeness relative to a state the set Π_p is complete relative to $\vec{u}_p$.

†A terminal uniform experiment E is understood to be an input word which defines a terminal state.

From Π_p one may isolate the basis set of words Б relative to $\vec{u}_p$, while in the basis $\{\vec{u}_p x\}$ $(x \in Б)$ one can define the matrices of transitions $\mathbf{A}(\sigma_k)$ and outputs in the states

$$\{\vec{u}_p x\} \qquad (x \in Б,\ k = 1, \ldots, m),$$

i.e., one can determine the structure of the general linear automaton $\mathfrak{A}$. Then $\mathscr{E} = \mathscr{E}_p E w_{p+1}$.

<u>Remark.</u> If for $\vec{u}_{p+1}$ the experiment $\mathscr{E}_p E$ has been used to define the output on the words $x_i y_j$, $x_i \sigma_k y_j$ $(i = 1, \ldots, r;\ k = 1, \ldots, m)$ for those i such that all vectors $\vec{u}_{p+1} x_i \sigma_k$ can be expressed linearly in terms of the vectors $\vec{u}_p x_i$, then one can again find the basis among the vectors $\vec{u}_p x_i$ and go on to find the transition matrices $\mathbf{A}(\sigma_k)$ and outputs in this basis. In this case $\mathscr{E} = \mathscr{E}_p E$. One may avoid the sorting of all possible subsets Π_p with the object of clarifying their completeness relative to $\vec{u}_{p+1}$ by requiring that for $\vec{u}_{p+1}$ the output must be defined on all words $x_i y_j$ and $x_i \sigma_k y_j$, where $x_i \in \Pi_p$.

However, if a certain vector $\vec{u}_{p+1} x_l \sigma_k$ cannot be expressed linearly in terms of $\vec{u}_{p+1} x_i$ $(i = 1, \ldots, s)$, then assume x_{i_1} is equal to the least of such l, while k_1 is the least of such k for a given i_1, and place

$$x_{s+1} = x_{i_1} \sigma_{k_1}, \qquad \Pi_{p+1} = \{x_1, \ldots, x_{s+1}\},$$

$$w_{p+1} = x_{s+1} \sigma_1 y_1, \qquad W_{p+1} = \overline{W}_{p+1} \cup \{w_{p+1}\},$$

while $\mathscr{E}_{p+1} = w_{p+1} E$, etc.

Let us prove that this process will break off.

Actually, the set Π_p may not contain more than n^2 words: from the construction it follows that each word $x_i \in \Pi_p$ for $i > 1$ is such that for a certain p_i

$$\Pi_{p_i} \setminus \Pi_{p_i - 1} = \{x_i\} \text{ and } x_i = x_{i_1} \sigma_{k_1}$$

for certain $i_1 < i$ and k_1, and $\vec{u}_p x_i$ cannot be expressed linearly in terms of $\vec{u}_p x_l$ $(l < 1)$; from this it follows that the transition operator φ_{x_i} cannot be expressed linearly in terms of the operators φ_{x_l} $(l < i)$, and therefore the system of operators φ_{x_l} is linearly independent for $x_l \in \Pi_p$, while the dimensionality of the system of operators φ_x, $x \in \Sigma^*$, of an n-dimensional general linear automaton does not exceed n^2. Thus, the set Π_p is stabilized. $\Pi = \lim_{p \to \infty} \Pi_p$.

Assume $\Pi = \{x_1, \ldots, x_{\bar{s}}\}$. For such a word $x_i y_j$ and every word $x_i \sigma_k y_j$ $(i = 1, \ldots, \bar{s};$ $j = 1, \ldots, r,\ k = 1, \ldots, m)$ it is necessary to perform no more than n experiments $E x_i y_j$ or $E x_i \sigma_k y_j$, respectively. The total number of words $x_i y_j$ is $\bar{s} \cdot r \le n^2 r$, while the total number of words $x_i \sigma_k y_j$ is $\bar{s} \cdot m \cdot r \le n^2 m r$. Therefore, the length of the experiment $\mathscr{E}$ does not exceed

$$n\,[(n^2 r + n^2 m r)\, l(E) + (n^2 + n^2 m)\, L(Y) + L(\Pi)(m+1)\, r + n^2 r],$$

where $L(\Pi)$ is the sum of the lengths of the words $x \in \Pi$, while $n^2 r$ is the upper bound for the number of entries of σ_k in all words $x_i \sigma_k y_j$.

It remains for us to estimate $L(\Pi)$.

From the method of constructing the set Π it follows that along with each word $x \in \Pi$ all initial pieces of x are entered in Π.

If the word x_i is defined on the step p, then $x_i = x_{i_1} \sigma_k$, where $l < i_1$, and for the states $\vec{u}_p$ the outputs on all words of the form $x y_j$ are known from the preceding experiment, where $j = 1, \ldots, r$, and the length of x is less than the length of x_i. Otherwise, we would have taken the largest initial piece of x, which is the word $x_l \in \Pi$, to be such that $l < 1$ (in view of $\Lambda \in \Pi$ one can find such a piece). From the construction of the experiment $\mathscr{E}$ it follows that the outputs of all words $x_l \sigma_k y_j$ are defined for the state $\vec{u}_p$, and all vectors $\vec{u}_p x_l \sigma_k$ can be expressed linearly in terms of $\vec{u}_p x_1, \vec{u}_p x_2, \ldots, \vec{u}_p x_l$.

Otherwise, on the step p there would be added to Π_{p-1} one of the words $x_h \sigma_k$, where $h \leq l$, $1 \leq k \leq m$, which is shorter than x_i. But then the length of x_i may not be greater than $n-1$, since all words x of length $\leq n-1$ form a complete system for any state $\vec{u}$ of the general linear Moore automaton $\mathfrak{A}$, while if $x_i \geq n$, then all words x_f, $f < i$, would form a complete system for $\vec{u}_p$ which would contradict the condition: $\vec{u}_p x_i$ cannot be expressed linearly in terms of the vectors $\vec{u}_p x_f$. Thus, the length of each word $x_i \in \Pi$ is no greater than $n-1$, while their number is $\leq n^2$. From this $L(\Pi) < n^3$.† Thus, we obtain

$$l(\mathscr{E}) < n^3 (m+1) [l(E) + L(Y) + n] + n^3 r.$$

For the optimal choice of E and Y we obtain

$$l(\mathscr{E}) < (m+1) n^3 \left(\frac{n^2}{2} + \frac{n^2}{2} + n\right) + n^4 = (m+1)(n^5 + n^4) + n^4.$$

A paradoxical result is obtained: the bound $l(\mathscr{E})$ for given optimal E and Y turns out to be less than the bound $l(\mathscr{E})$ in Theorem 7 for optimal E, Y, and Π. However, this can be explained by the fact that during the process of constructing $\mathscr{E}$ in Theorem 8 we actually do not construct the entire complete (for the general linear automaton $\mathfrak{A}$) step Π, while in return the construction algorithm becomes more complicated.

Theorem 9. Assume that for a complete general linear automaton $\mathfrak{A} \in T_n$ we know E and $\Pi = \{x_1, \ldots, x_t\}$. Then a prime branching experiment $\mathscr{E}$ exists having the length

$$l(\mathscr{E}) < (m+1) n^2 \left[t l(E) + l(\Pi) + t \cdot \frac{n+1}{2} \right],$$

which allows the structure of the general linear automaton $\mathfrak{A}$ to be defined.

Proof. Let us place $F_1 = x_1$, $y_1 = \Lambda$, $Y_1 = \{y_1\}$, $\mathscr{E}_1 = EF_1$. Assume that the experiment $\mathscr{E}_p = EF_1EF_2 \ldots EF_p$ has been constructed (i.e., $F_1, F_2, \ldots, F_p$ have been defined), and assume $Y_p = \{y_1, \ldots, y_q\}$.

Assume $\mathfrak{A}$ has made the transition from the initial state $\vec{v}_0$ under the action of $\mathscr{E}_p E$ to the state $\vec{u}_p$ for which from a previous experiment the output has been defined on a certain set of words (perhaps empty) of the form $x_i y_j$ and $x_i \sigma_k y_j$ ($i = 1, \ldots, t$; $j = 1, \ldots, l$; $k = 1, \ldots, m$).

We shall order all words of this form as in the proof of Theorem 8.

For each word y and state $\vec{v}$ we use $\vec{w}(\vec{v}, y)$ to denote the vector having the coordinates $\psi(\vec{v}x_i y)$ ($i = 1, \ldots, t$). If in the state $\vec{u}_p$ the outputs are defined on all words $x_i y_j$, $x_i \sigma_k y_j$, then we see whether the vectors $\vec{w}(\vec{u}_p, \sigma_k y_j)$ for $k = 1, \ldots, m$; $j = 1, \ldots, q$ are expressed linearly in terms of $\vec{w}(\vec{u}_p, y_j)$ ($j = 1, \ldots, q$).

a) If this is so, then we assume $Y = Y_q$, and considering Y to be a reference set we define the matrix of transitions $\mathbf{A}(\sigma_k)$ in a basis which is first isolated from $\{\vec{u}_p x_i\}$ by means of Y (as in the proof of Theorem 8).

† By finer reasoning one may obtain the upper bound for $L(\Pi)$ equal to $n^3/2$.

b) If this is not so, then we take the least j_1 such that for a certain k the vector $\vec{w}(\vec{u}_p, \sigma_k y_{j_1})$ cannot be expressed linearly in terms of $\{\vec{w}(\vec{u}_p, y_j)\}$ (j = 1, ..., t). Assume k_1 is the least of such k for j_1. We add the word $\sigma_{k_1} y_{j_1}$ to Y_p, having placed

$$Y_{q+1} = \sigma_{k_1} y_{j_1} \text{ and } Y_{p+1} = \{y_1, \ldots, y_{q-1}\}.$$

In this case we place $F_{p+1} = x_1 \sigma_1 y_{q+1}$.

c) However, if in the state $\vec{u}_p$ the outputs were not defined from the experiment $\mathcal{E}_p E$ on all words $x_i y_j$ and $x_i \sigma_k y_j$ $(x_i \in \Pi, \sigma_k \in \Sigma, y_j \in Y_p)$, then we take the first of such words in our ordering and place F_{p+1} equal to this word, while $Y_{p+1} = Y_p$. If after this we turn out to be in situation a), then the transition and output matrices of the general linear automaton $\mathfrak{A}$ are defined.

Let us prove that the process breaks off. We use Y to denote the set of all y_j defined by our procedure. From the construction it follows that:

1) for any state $\vec{u}$ the vectors $\vec{w}(\vec{u}, y_j)$ (j = 1, 2, ...) are linearly independent;

2) along with each word y_j, the sets Y_p, containing the word y_j, also contain all of its "tails" (i.e., its final pieces). The i-th coordinate of the vector $\vec{w}(\vec{u}, y_j)$ is equal to

$$\psi(\vec{u} x_i y_j) = c_{ij}.$$

Since the rank of the system of vectors $\{\vec{u} x_i\}$ (i = 1, ..., t) is equal to n, it follows that the rank of the matrix $\| c_{ij} \|$ is likewise equal to n, and since all of its columns are the vectors $\vec{w}(\vec{u}, y_j)$, it follows that their number is equal to n. Thus, in Y there are no more than n words. Since each of the words $x_i y$, $x_i \sigma_k y$, where $x_i \in \Pi$, $\sigma_k \in \Sigma$, $y \in Y$, may be taken as F_p no more than n times, it follows that on some step the outputs on all words

$$x_i y, \ x_i \sigma_k y \ (x_i \in \Pi, \ \sigma_k \in \Sigma, \ y \in Y)$$

can be defined in the state $\vec{u}_p$ from the experiment $\mathcal{E}_p E$ or $\mathcal{E}_p E F_{p+1}$, while since the set Y may not be expanded further, the procedure breaks off and the structure $\mathfrak{A}$ is defined.

Computation shows that

$$l(\mathcal{E}) \leqslant (m+1) t \cdot \mu Y \cdot nl(E) + (m+1) \cdot \mu Y \cdot nl(\Pi) + (m+1) tnl(Y) + mtn \cdot \mu Y,$$

where $\mu Y = n$, $l(Y) \leq [n(n-1)]/2$, whence

$$l(\mathcal{E}) < (m+1) n^2 \left[tl(E) + l(\Pi) + t \cdot \frac{n+1}{2} \right],$$

which is what we were required to prove.

If E and Π are chosen optimally, i.e.,

$$l(E) < \frac{n^2}{2}, \quad l(\Pi) < \frac{n^3}{2}, \quad t \leqslant n^2,$$

then

$$l(\mathcal{E}) < (m+1) n^2 \left(\frac{n^4}{2} + n^3 + \frac{n^2}{2} \right) < (m+1) n^6 \ (n \geqslant 3).$$

Note that Lemmas 1-3 and Theorems 7-9 are valid for a general linear automaton $\mathfrak{A} \in T'_n$ belonging to the class of reduced strongly connected n-dimensional general linear automata of the II type if instead of completeness of the general linear automaton $\mathfrak{A}$ we require weak completeness of $\mathfrak{A}$, while linear dependence of the vectors is replaced by normalized linear dependence in the proofs.

One may also reject the requirement of completeness of a general linear automaton (or weak completeness), while in Lemmas 1-2 and Theorems 7-8 one may stipulate the set $\overline{Y}$, which is the reference set for the supplement (weak supplement) of the general linear automaton $\mathfrak{A} \in T_n$ (T'_n) in place of Y.

Assume now that we know only the experiment E which allows the terminal state of the investigated general linear automaton $\mathfrak{A} \in T_n$ to be established. The class of such general linear automata shall be denoted by $T_n(E)$. Then, taking account of the fact that the set of all words having a length no greater than n is complete in any n-dimensional general linear automaton, we obtain the following corollary.

Corollary. For any general linear automaton $\mathfrak{A} \in T_n(E)$ there exists a prime branching experiment F which establishes the structure of $\mathfrak{A}$, its length being

$$l(F) \leqslant (m+1)\, n^2\, (tl(E) + 2nt),$$

where $t = m^n$. (In order to define the output on all basis words it is sufficient for us to define it on all words of length n.)

8. Let us also note a number of cases with a priori information on the investigated automaton. Let us consider the case of a general linear automaton $\mathfrak{A}$ in reduced form having nondegenerate operators φ_{σ_i}, i.e., operators which yield a one-to-one transformation of the hyperfine Γ_M.

In the corresponding finite automata a permutation of states occurs due to the action of the input levels; therefore, S. Ginsburg called such automata permutation automata [16]. The class of such general linear automata shall be denoted by S_n. The intersection of the classes T_n and S_n shall be denoted by S'_n. Modifying the Yu. M. Borodyanskii method [3] (our proof is based on geometric notions), we prove the following theorem.

Theorem 10. The length of the shortest universal uniform experiment U(n) which establishes the terminal state of any general linear automaton $\mathfrak{A} \in S'_n$ does not exceed $(2^{n+1} - 2)(m+1)^n$. The length of the shortest uniform experiment which allows any general linear automaton $\mathfrak{A} \in S'_n$ to be distinguished from any other general linear automaton $\mathfrak{B} \in S'_n$, does not exceed $(2^{n+1} - 2)(m+1)^{2n}$.

Proof. Let us give the proof for a Mealy general linear automaton.

The essence of the Yu. M. Borodyanskii method consists in the abridged notation of certain words in the input alphabet Σ in the form of word codes in the expanded alphabet $\Sigma \cup \{\omega\}$.

Each code word of the form $\delta_1\omega\delta_2\omega\delta_3\omega \ldots \delta_k\omega\delta_{k+1}$, where $\delta_1, \delta_2, \ldots, \delta_{k+1}$ are words in the alphabet Σ, is decoded from the word $((\delta_1^2\delta_2)^2 \ldots \delta_k)^2\, \delta_{k+1}$ in the alphabet Σ (i.e., the symbol ω implies that it is necessary to repeat the preceding part of the word).

Assume that in the beginning the general linear automaton $\mathfrak{A}(\Gamma, \Sigma, Z, \varphi, \psi)$ may be in any state – any point of the hyperplane Γ. Then there exists an input σ_{i_1} such that the outcome of the experiment σ_{i_1} partitions Γ into parallel hyperplanes $\{\Gamma_1\}$, $\Gamma_1 = \Gamma(\sigma_{i_1})$, $r(\Gamma_1) < n$. Due to the action of the input σ_{i_1} these hyperplanes go over into the parallel hyperplanes

$\{\Gamma_1(\sigma_{i_1})\}$, where $r[\Gamma_1] = r[\Gamma_1(\sigma_{i_1})]$ in view of the nondegeneracy of the operators φ_σ, $\sigma \in \Sigma$. Let us assume that $E_1 = \sigma_{i_1}$, and the code $G_1 = \sigma_{i_1}$. Assume that we have constructed the experiment E_k with the code G_k, which partitions Γ into parallel hyperplanes $\{\Gamma(E_k)\}$, which the action of E_k causes to go over into the hyperplanes $\{\Gamma_k(E_k)\}$; under these conditions

$$r[\Gamma(E_k)] = r[\Gamma_k(E_k)] \leqslant n-k.$$

Let us consider two cases.

a) $\Gamma_k(E_k) \nparallel \Gamma(E_k)$. In this case the repeated application of E_k partitions the hyperplanes $\Gamma_k(E_k)$ into the hyperplanes $\Gamma_{k+1}(E_k)$, since all of Γ is partitioned into $\{\Gamma(E_k)\}$.

Under these conditions $\Gamma_{k+1}(E_k)$ is caused by the action of E_k to go over into $\Gamma_{k+1}(E_kE_k) = \Gamma_{k+1}(E_{k+1})$. Let us assume $E_{k+1} = E_kE_k$, $r[\Gamma_{k+1}(E_{k+1})] = r[\Gamma_{k+1}(E_k)] \leqslant n-k-1$, while we take the code G_{k+1} to be $G_k\omega$.

b) $\Gamma_k(E_k) \parallel \Gamma(E_k)$. In view of the coincidence of dimensionality, each hyperplane $\Gamma_k(E_k)$ coincides with a certain hyperplane $\Gamma(E_k)$.

Then 1) either $\Gamma_k(E_k)$ is a point and the experiment $U(n) = E_k$, since as a result of E_k the terminal state of $\Gamma_k(E_k)$ is defined as one of its points; or, 2) there exists a certain input $\sigma_{i'}$ that partitions $\Gamma_k(E_k)$ into $\Gamma_{k+1}(E_k)$ which due to the action of $\sigma_{i'}$ go over into $\Gamma_{k+1}(E_k\sigma_{i'})$. In this case we assume $E_{k+1} = E_k\sigma_i$, and the code $G_{k+1} = G_k\sigma_{i'}$. Here $r[\Gamma_{k+1}(E_k\sigma_{i'})] = r[\Gamma_{k+1}(E_k)] < r[\Gamma_k(E_k)] \leqslant n-k$, i.e., $r[\Gamma_{k+1}(E_{k+1})] \leqslant n-k-1$; 3) or for coincidence of the outputs of all points $\Gamma_k(E_k)$ there must exist on each input σ_i an input σ_{i*}, which causes $\Gamma_k(E_k)$ go over into the hyperplane $\Gamma_k(E_k\sigma_{i*}) \nparallel \Gamma_k(E_k)$. Otherwise, the output would be identical on each input word x for all points $\Gamma_k(E_k)$, and since $\mathfrak{A}$ is a reduced form it follows that $\Gamma_k(E_k)$ would be a point (but we have already considered this case).

Let us assume $E_{k+1} = E_k\sigma_{i*}E_k\sigma_{i*}$, while $G_{k+1} = G_k\sigma_{i*}\omega$. E_k partitions the hyperplane $\Gamma_k(E_k\sigma_{i*})$ into $\Gamma_{k+1}(E_k\sigma_{i*})$, since $\Gamma_k(E_k\sigma_{i*}) \nparallel \Gamma_k(E_k)$, and therefore $\Gamma_k(E_k\sigma_{i*}) \nparallel \Gamma(E_k)$; after all, $\Gamma_k(E_k)$ is a parallel shift of $\Gamma(E_k)$. Therefore, the hyperplane $\{\Gamma_{k+1}(E_k\sigma_{i*})\}$ and the hyperplane $\{\Gamma_{k+1}(E_k\sigma_{i*}E_k\sigma_{i*}) \underset{Df}{=} \Gamma_{k+1}(E_{k+1})\}$ have a dimensionality less than $r[\Gamma_k(E_k)]$ (i.e., not exceeding $n-k-1$). Thus, for a certain $k \leq n$ it turns out that $\Gamma_k(E_k)$ is a point, i.e., the terminal state of the experiment E_k can be defined. Therefore, one of the codes $x_1, x_2, \ldots, x_n$, where $x_j = \omega$, σ_{i_j} or $\sigma_{i_j}\omega$ $(j = 1, 2, \ldots, n)$, turns out to be the code of the experiment $E(x_1x_2 \ldots x_n)$, that determines the terminal state of $\mathfrak{A} \in T_n$. Carrying out all the experiments $E(x_1x_2 \ldots x_n)$ successively in any order for various allowed sets of values $x_1, x_2, \ldots, x_n$, we obtain the required experiment $U(n)$ which is universal for T_n.

Let us estimate the length of $U(n)$. We note only that in order to obtain the collection of experiments $\{E(x_1x_2 \ldots x_n)\}$ such that for each $\mathfrak{A} \in T_n$ a certain experiment determines the terminal state, it is sufficient to assign each x_j just two values ω and $\sigma_{i_j}\omega$, since the value σ_{i_j} is "absorbed" by $\sigma_{i_j}\omega$.

The total number of allowed sets $x_1, x_2, \ldots, x_n$ will thus be $(m+1)^n$.

In decoding the collection $x_1, x_2, \ldots, x_n$ the maximal length of $E(x_1x_2 \ldots x_n)$ will occur for $x_j = \sigma_{i_j}\omega$ $(j = 1, 2, \ldots, n)$. Under these conditions the length l_n of the experiment

$$E(x_1x_2 \ldots x_n) = (((\sigma_{i_1}^2\sigma_{i_2})^2\sigma_{i_3})^2 \ldots \sigma_{i_n})^2$$

is equal to $2l_{n-1} + 2$, $l_1 = 2$. From this we have

$$l_n = 2^n + 2^{n-1} + \ldots + 2^1 = 2^{n+1} - 2.$$

Therefore the length of U(n) does not exceed $(2^{n+1} - 2)(m + 1)^n$. The first statement of Theorem 10 has been proved.

Let us now consider the general linear automaton $\mathfrak{A} \in S'_n = T_n \cap S_n$, i.e., a strongly connected general linear automaton $\mathfrak{A}$ in reduced form with nondegenerate transition operators φ_σ. For any other general linear automaton $\mathfrak{B} \in S'_n$ let us consider the direct sum $\mathfrak{A} \oplus \mathfrak{B}$, which will be a general linear automaton from the class S_{2n} (only strong connectedness is lost). The experiment U(2n), which is universal for S_{2n}, allows the final state $\mathfrak{A} \oplus \mathfrak{B}$, to be established and thereby allows $\mathfrak{A}$ to be distinguished from $\mathfrak{B}$ (and $\mathfrak{B}$ from $\mathfrak{A}$). The length of U(2n) does not exceed

$$(2^{2n+1} - 2)(m + 1)^{2n}.$$

The theorem has been proved.

The basic idea running through the proofs of Theorems 1, 2, and 10 of this paper resides in the fact that the hyperplane Γ', which is parallel to the hyperplane Γ $(r(\Gamma) > r(\Gamma'))$, where Γ is determined by the output corresponding to the inputs σ_i, may be converted into the hyperplane $\Gamma'(x) \nparallel \Gamma$ by the word x of length $\leq r(\Gamma) - r(\Gamma')$.

In the case of fields of real numbers D and complex numbers K one may consider the measure in the space

$$\varphi_\Sigma \times \psi_\Sigma = \{\varphi_{\sigma_1} \times \varphi_{\sigma_2} \times \ldots \times \varphi_{\sigma_m} \times \psi_{\sigma_1} \times \psi_{\sigma_2} \times \ldots \times \psi_{\sigma_m}\},$$

which is the direct product of the spaces of the transition operators φ_{σ_i} (i = 1, 2, ...,m) and the output operators ψ_{σ_i} (i = 1, 2, ..., m). Each element $\varphi_\Sigma \times \psi_\Sigma$ for fixed V, Γ, Σ, and Z defines a general linear automaton $\mathfrak{A}(V, \Gamma, \Sigma, Z)$, while each element $\Gamma \times \varphi_\Sigma \times \psi_\Sigma$ defines an initial general linear automaton $\mathfrak{A}$. For "almost all" (in the sense of this natural measure) initial general linear automata $\mathfrak{A}$, coincidence of the trees D_{q1} and D_{q2} of the automata in any n edges is sufficient to ensure equivalence of any two states q_1 and q_2. Moreover, for "almost all" general linear automata $\mathfrak{A}$ any n vertices of the tree D_q will be linearly independent states. This derives from the fact that an algebraic surface defined by an equation in any n-dimensional space has measure zero in this space, while measure zero is invariant relative to the choice of the basis in the space.

From the indicated properties of "almost all" general linear automata over the fields D and K it likewise follows that the structure of "almost any" general linear automaton is determined by a multiple experiment (tree) having a length of the order of $\log_m n$, and any word of length n may serve as the input of a uniform experiment which determines the terminal state of a general linear automaton $\mathfrak{A}$. Moreover, any word of length n may form a basis (if one considers pieces of this word as basis words) in almost all general linear automata. In order to distinguish "almost any" general linear automaton $\mathfrak{A} \in T_n$ from "almost all" general linear automata $\mathfrak{B} \in T_n$ it is sufficient to carry out a uniform experiment with any input word having the length 2n. A prime branching experiment of length mn^2 which is constructed according to the method given in Theorem 7 allows simple (without sorting of general linear automata from T_n) establishment of the structure of "almost any" general linear automaton $\mathfrak{A}$.

These statements are also valid for finite probabilistic automata.

Here it is of interest to note that analogous properties hold for finite deterministic automata if the term "almost all" is understood in the sense of the tendency towards 1 of the fraction of (n, m, p)-automata having the given property out of all (n, m, p)-automata for $n \to \infty$ [6].

It would be of interest to consider the case of general linear automata over a field of rational numbers, having chosen a certain basis and considering the set $\mathfrak{M}_{n,k}$ of n-dimensional general linear automata with transition and output matrices whose elements are irreducible fractions p/q, where $|p| + |q| < k$.

Problem 7. How does the "majority" of general linear automata from $\mathfrak{M}_{n,k}$ behave in experiments? It is required to estimate the length of the experiments.

Problem 8. It is required to consider the case of the general linear automaton $\mathfrak{A}(V, M, \Sigma, Z, \varphi, \psi)$ over finite fields $P = GF(p^k)$. How does the length of the experiments depend on the dimensionality n, the numbers p and k, and the power M for "the majority" of automata $\mathfrak{A}$ over P?

I express thanks to the editor of the paper Yu. Ya. Breitbart for a number of comments.

Literature Cited

1. K. Berge, Theory of Graphs and Its Applications [Russian translation], IL, Moscow (1962), Ch. 14, pp. 150-152.
2. Yu. M. Borodyanskii, "Experiments with finite Moore automata," Kibernetika, No. 6, pp. 18-27, Kiev (1965).
3. A. Hill, Introduction to the Theory of Finite Automata [Russian translation], Mir, Moscow (1966).
4. V. M. Glushkov, Synthesis of Digital Automata, Moscow, Fizmatgiz (1962).
5. N. E. Kobrinskii and B. A. Trakhtenbrot, Introduction to the Theory of Finite Automata [in Russian], Fizmatgiz, Moscow (1961), Chap. II, V, and VII.
6. A. D. Korshunov, "On the degree of distinguishability of automata," in: Discrete Analysis, No. 10, Nauka, Novosibirsk (1967), pp. 39-60.
7. M. S. Lifshits, "On linear physical systems connected with the external world by communication channels," Izvestiya Akad. Nauk SSSR, 27:993-1030 (1963).
8. M. S. Lifshits, "Open systems and linear automata," Izvestiya, Akad. Nauk SSSR, 27:1215-1228 (1963).
9. A. A. Muchnik, "Length of an experiment for determining the structure of a finite strongly connected automaton," in: Systems Theory Research, Vol. 20, Consultants Bureau, New York (1971), p. 136.
10. Ch. Faisi, On the Distinguishability of Infinite Automata (see this volume, pp. 219-222).
11. M. L. Tsetlin, "On nonprimitive networks," in: Problemy Kibernetiki, Vol. 11, Fizmatgiz, Moscow (1958), p. 31.
12. G. Bacon, "Minimal-state stochastic finite-state systems," Trans. IEEE, CT-11:307-308 (1964).
13. J. W. Carlyle, "Reduced forms for stochastic sequential machines," J. Math. Annal. Appl., 7:167-175 (1963).
14. J. W. Carlyle, "On the external probability structure of finite-state channels," Inf. Contr., 7:385-397 (1964).
15. S. Even, Comments on the Minimization of Stochastic Machines, Res. Rep., SRRC-RR-64-50 (1964).
16. S. Ginsburg, "On the length of the smallest uniform experiment," ACM Journal, 5(3):266-280 (1968).
17. T. H. Hibbard, "Least upper bounds on minimal terminal-state experiments," ACM Journal, 8(4):601-612 (1961).
18. E. F. Moore, Gedanken-Experiments on Sequential Machines. Automata Studies, Princeton University Press (1956), pp. 129-153.
19. A. Paz, "Some aspects of probabilistic automata," Inf. Contr., 9(1):26-60 (1966).
20. M. O. Rabin, "Probabilistic automata," Inf. Contr., Vol. 6, No. 3 (1963).
21. M. O. Rabin and D. Scott, "Finite automata and their decision problems," IBM Res. Dev., 3(2):114-125 (1959).

22. G. N. Raney, Sequential Functions, ACM Journ., 5(2):177-180 (1958).
23. M. P. Schutzenberger, "On the definition of a family of automata," Inf. Contr., 4:245-270 (1961).
24. M. P. Schutzenberger, "Finite counting automata," Inf. Contr., 5:91-107 (1962).

DISTINGUISHABILITY OF INFINITE AUTOMATA†

Ch. Faisi

Moscow

1. Let $\mathfrak{A}=\{Q, X, Y, \varphi_{\mathfrak{A}}, \psi_{\mathfrak{A}}\}$ be a strongly connected synchronous automaton (i.e., the length of an output word is equal to the length of an input word), where Q is the set of states of the automaton, X the input alphabet, Y the output alphabet, $\varphi_{\mathfrak{A}}$ the transition function, and $\psi_{\mathfrak{A}}$ the output function. The automaton $\mathfrak{A}$ is either finite or infinite. Hence for any automaton $\mathfrak{B}=\{P, X, Y, \varphi_{\mathfrak{B}}, \psi_{\mathfrak{B}}\}$, whose input and output alphabets coincide with the corresponding alphabets of $\mathfrak{A}$, we have:

a) Either for any state q belonging to Q there exists an equivalent state p belonging to P,

$$\forall q \in Q \exists p \in P \forall a \mho X: \ \psi_{\mathfrak{A}}(q, a)=\psi_{\mathfrak{B}}(p, a);$$

($a \mho X$ signifies that a is a word of finite length in the alphabet X),

b) or no state q belonging to Q has an equivalent state in P:

$$\forall q \in Q \forall p \in P \exists a \mho X: \ \psi_{\mathfrak{A}}(q, a) \neq \psi_{\mathfrak{B}}(p, a).$$

P r o o f. Suppose that for a $q_0 \in Q$ there exists an equivalent state $p_0 \in P$:

$$\forall a \mho X: \ \psi_{\mathfrak{A}}(q_0, a)=\psi_{\mathfrak{B}}(p_0, a).$$

$\mathfrak{A}$ is strongly connected; hence for any $q_j \in Q$ there exists a word $b_j \mho X$ that carries $\mathfrak{A}$ from q_0 into q_j, and p_0 is equivalent to q_0; hence

$$\forall a: \ \psi_{\mathfrak{A}}(q_0, b_j a)=\psi_{\mathfrak{B}}(p_0, b_j a).$$

But

$$\psi_{\mathfrak{A}}(q_0, b_j a)=\psi_{\mathfrak{A}}(q_0, b_j)\psi_{\mathfrak{A}}(\varphi_{\mathfrak{A}}(q_0, b_j), a)=\psi_{\mathfrak{A}}(q_0, b_j)\psi_{\mathfrak{A}}(q_j, a),$$
$$\psi_{\mathfrak{B}}(p_0, b_j a)=\psi_{\mathfrak{B}}(p_0, b_j)\psi_{\mathfrak{B}}(\varphi_{\mathfrak{B}}(p_0, b_j), a)=\psi_{\mathfrak{B}}(p_0, b_j)\psi_{\mathfrak{B}}(p_{i(j)}, a),$$

where $p_{i(j)}=\varphi_{\mathfrak{B}}(p_0, b_j)$ is a state of $\mathfrak{B}$.

By our condition we have $\psi_{\mathfrak{A}}(q_0, b_j)=\psi_{\mathfrak{B}}(p_0, b_j)$; hence

$$\forall a \mho X: \ \psi_{\mathfrak{A}}(q_j, a)=\psi_{\mathfrak{B}}(p_{i(j)}, a)$$

the state $p_{i(j)}$ is equivalent to the state q_j.

† Original article submitted May 23, 1968.

2. Let $\mathfrak{A}$ and $\mathfrak{B}$ be infinite synchronous strongly connected automata, and $\mathfrak{A}$ cannot be distinguished from $\mathfrak{B}$ by any finite experiment, i.e.,

$$\forall q \in Q \forall a \mho X \exists p \in P: \; \psi_{\mathfrak{A}}(q, a) = \psi_{\mathfrak{B}}(p, a).$$

In this case it can happen that no $q \in Q$ has an equivalent $p \in P$.

Example. $X = Y = \{0, 1\}$, $Q = \{q_i, i = 0, \pm 1, \pm 2, \ldots\}$, $P = \{p_j, j = 0, \pm 1, \pm 2, \ldots\}$,

$$\varphi_{\mathfrak{A}}(q_i, 0) = q_{i-1}, \quad \varphi_{\mathfrak{B}}(p_j, 0) = p_{j-1},$$

$$\varphi_{\mathfrak{A}}(q_i, 1) = q_{i+1}, \quad \varphi_{\mathfrak{B}}(p_j, 1) = p_{j+1},$$

$$\psi_{\mathfrak{A}}(q_i, 0) = \psi_{\mathfrak{A}}(q_i, 1) = \begin{cases} 1 & \text{for } i = 0, \\ 0 & \text{for } i \neq 0, \end{cases}$$

$$\psi_{\mathfrak{B}}(p_j, 0) = \psi_{\mathfrak{B}}(p_j, 1) = \begin{cases} 1 & \text{for } j = \pm \frac{n(n-1)}{2}, \; n = 1, 2, \\ 0 & \text{for other j.} \end{cases}$$

The transition diagrams of these automata (the outputs for a given state are given in the circles) are:

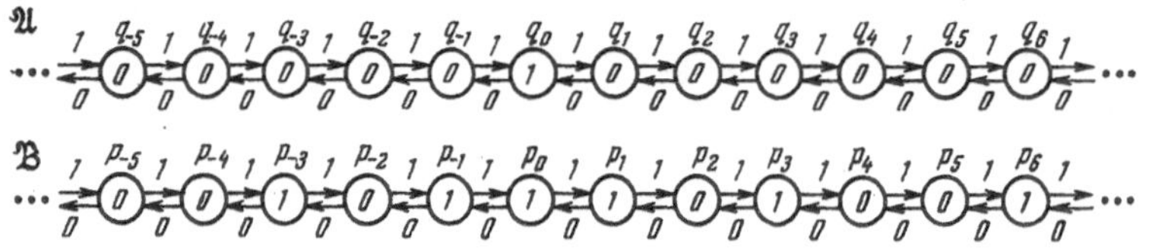

$\mathfrak{A}$ cannot be distinguished from $\mathfrak{B}$ by any finite experiment; suppose we are given an initial state q_i and an experiment a. In the operation of the automaton $\mathfrak{A}$ over this word there participates a finite number of states belonging to Q, more precisely, a "section" of states $[q_r, q_s]$, $r < s$, since $\mathfrak{A}$ can go over from one state to a neighboring (according to itş number) state. Suppose that these are the states from q_{i-m} to q_{i+n} $(n, m \geq 0)$.

a) $i - m > 0$ or $i + n < 0$. Then all these states yield zero. The transition function of $\mathfrak{B}$ is the same as the transition function of $\mathfrak{A}$; hence if we take p_j as the initial state, the operation of $\mathfrak{B}$ over a will involve the "second" from p_{j-m} to p_{j+n}. Let us take

$$j = \frac{(m+n+1)(m+n+2)}{2} + m + 1.$$

Then the states in operation will be those with numbers from

$$\frac{(m+n+1)(m+n+2)}{2} + m + 1 - m > \frac{(m+n+1)(m+n+2)}{2}$$

to

$$\frac{(m+n+1)(m+n+2)}{2} + m + 1 + n < \frac{(m+1+n)(m+n+2)}{2} + m + 2 + n = \frac{(m+n+2)(m+n+3)}{2},$$

i.e., these states likewise yield zeros alone.

b) $i - m \leq 0 \leq i + n$, i.e., $-n \leq i \leq m$. As the initial state let us take $p_{\frac{(n+m+1)(n+m+2)}{2} + i}$.

The states in operation will be those with numbers from

$$\frac{(m+n+1)(m+n+2)}{2} + i - m \geqslant \frac{(m+n+1)(m+n+2)}{2} - (n+m) >$$

$$> \frac{(m+n+1)(m+n+2)}{2} - (m+n+1) = \frac{(m+n+1)(m+n)}{2}$$

to

$$\frac{(m+n+1)(m+n+2)}{2}+i-n\leqslant\frac{(m+n+1)(m+n+2)}{2}+(m+n)<$$
$$<\frac{(m+n+1)(m+n+2)}{2}+(m+n+2)=\frac{(m+n+2)(m+n+3)}{2}.$$

Among these states, only $p_{\frac{(m+n+1)(m+n+2)}{2}}$ yields unity, whereas all the other states yield zero. Therefore the result will likewise coincide with the output of $\mathfrak{A}$. But, for example, for $q_0 \in Q$ there does not exist an equivalent state belonging to P: $\psi_{\mathfrak{A}}(q_0, 1^{\delta}) = 10^{\delta-1}$, whereas $\psi_{\mathfrak{B}}(p_j, 1^{\delta})$ terminates with 1, where for j ≠ 0, 1, 2, 3 we have $\delta = \frac{|j|(|j|-1)}{2} - j + 1$, since

$$\varphi_{\mathfrak{B}}(p_j, 1^{\delta-1}) = p_{\frac{|j|(|j|-1)}{2}};\ \psi_{\mathfrak{B}}(p_{\frac{|j|(|j|-1)}{2}}, 1) = 1,$$

and for j = 0, 1, 2, 3 we have

$$\delta = 7 - j:\ \varphi_{\mathfrak{B}}(p_j, 1^{\delta-1}) = \varphi_{\mathfrak{B}}(p_j, 1^{G-j}) = p_6,\ \psi_{\mathfrak{B}}(p_6, 1) = 1.$$

3. Theorem. Let $\mathfrak{A}$ be a strongly connected infinite synchronous automaton that cannot be distinguished by any infinite experiment (infinite sequence of input letters) from a synchronous automaton $\mathfrak{B}$ with a countable number of states (with the same input and output alphabets). Then for any state of $\mathfrak{A}$ there exists an equivalent state of $\mathfrak{B}$

$$\forall q \in Q \exists p \in P \forall a \mho X:\ \psi_{\mathfrak{A}}(q, a) = \psi_{\mathfrak{B}}(p, a).$$

Proof. Suppose this is not the case. But then (according to Sec. 1)

$$\forall q \in Q \forall p \in P \exists a \mho X:\ \psi_{\mathfrak{A}}(q, a) \neq \psi_{\mathfrak{B}}(p, a).$$

Let us number the states of $\mathfrak{A}$ and $\mathfrak{B}$: $P = \{p_1, p_2, p_3, \ldots\}$, $Q = \{q_1, q_2, q_3, \ldots\}$. Then we construct an infinite experiment x such that

$$\forall j:\ \psi_{\mathfrak{A}}(q_1, x) \neq \psi_{\mathfrak{B}}(p_j, x).$$

By $a_{i,j}$ we shall denote a word in X such that $\psi_{\mathfrak{A}}(q_i, a_{i,j}) \neq \psi_{\mathfrak{B}}(p_j, x)$ and write $x_1 = a_{1,1}$. By our condition we have $\psi_{\mathfrak{A}}(q_1, a_{1,1}) \neq \psi_{\mathfrak{B}}(p_1, a_{1,1})$.

If $\psi_{\mathfrak{A}}(q_1, a_{1,1}) \neq \psi_{\mathfrak{B}}(p_2, a_{1,1})$, then $x_2 = x_1$; if $\psi_{\mathfrak{A}}(q_1, a_{1,1}) = \psi_{\mathfrak{B}}(p_2, a_{1,1})$, we shall denote $\varphi_{\mathfrak{A}}(q_1, a_{1,1}) = q_{i_2}$, $\varphi_{\mathfrak{B}}(p_2, a_{1,1}) = p_{j_2}$, and $x_2 = x_1 a_{i_2, j_2}$.

Suppose that x_n has been constructed. If $\psi_{\mathfrak{A}}(q_1, x_n) \neq \psi_{\mathfrak{B}}(p_{n+1}, x_n)$, then $x_{n+1} = x_n$; but if $\psi_{\mathfrak{A}}(q_1, x_n) = \psi_{\mathfrak{B}}(p_{n+1}, x_n)$, we shall denote $\varphi_{\mathfrak{A}}(q_1, x_n) = q_{i_{n+1}}$, $\varphi_{\mathfrak{B}}(p_{n+1}, x_n) = p_{j_{n+1}}$ and write $x_{n+1} = x_n a_{i_{n+1}, j_{n+1}}$.

Let us show that $\psi_{\mathfrak{A}}(q_1, x_n) \neq \psi_{\mathfrak{B}}(p_k, x_n)$ for any k = 1, 2, ..., n. For k = 1 this is true.

Let k > 1. If $\psi_{\mathfrak{A}}(q_1, x_{k-1}) \neq \psi_{\mathfrak{B}}(p_k, x_{k-1})$, then $\psi_{\mathfrak{A}}(q_1, x_n) \neq \psi_{\mathfrak{B}}(p_k, x_n)$ (since x_n is a continuation of x_{k-1}, also $\psi_{\mathfrak{B}}(p_k, x_n)$ will be a continuation of $\psi_{\mathfrak{B}}(p_k, x_{k-1})$, and $\psi_{\mathfrak{A}}(q_1, x_n)$ will be a continuation of $\psi_{\mathfrak{A}}(q_1, x_{k-1})$; if their initial parts do not coincide, they will differ from one another).

If $\psi_{\mathfrak{A}}(q_1, x_{k+1}) = \psi_{\mathfrak{B}}(p_k, x_{k-1})$, then

$$\psi_{\mathfrak{A}}(q_1, x_n) = \psi_{\mathfrak{A}}(q_1, x_{k-1})\, \psi_{\mathfrak{A}}(q_{i_k}, a_{i_k, j_k}) \ldots,$$
$$\psi_{\mathfrak{B}}(p_k, x_n) = \psi_{\mathfrak{B}}(p_k, x_{k-1})\, \psi_{\mathfrak{B}}(p_{j_k}, a_{i_k, j_k}) \ldots$$

But $\psi_{\mathfrak{A}}(q_{i_k}, a_{i_k, j_k}) \neq \psi_{\mathfrak{B}}(p_{j_k}, a_{i_k, j_k})$.

Let us write $x = \lim_{n\to\infty} x_n$. Hence

$$\forall j:\ \psi_{\mathfrak{A}}(q_1, x) \neq \psi_{\mathfrak{B}}(p_j, x),$$

i.e., $\mathfrak{A}$ can be distinguished from $\mathfrak{B}$ by an experiment x. We have obtained a contradiction; this proves our assertion.

4. In the proof we used the fact that the states of $\mathfrak{B}$ can be numbered, i.e., that it has a countable number of states. If $\mathfrak{B}$ is also a strongly connected automaton, this condition will be satisfied.

It is not known whether there exists for any state of $\mathfrak{A}$ an equivalent state of $\mathfrak{B}$, if $\mathfrak{A}$ cannot be distinguished by any finite experiment from $\mathfrak{B}$, and $\mathfrak{B}$ cannot be distinguished by any finite experiment from $\mathfrak{A}$.

If the automata are not synchrönous, the assertion of Theorem 3 is not necessarily true. From the nondistinguishability by any infinite experiment does not yet follow the nondistinguishability by finite experiments.

Example. $\mathfrak{C} = \{Q, X, Y, \varphi_{\mathfrak{C}}, \psi_{\mathfrak{C}}\}$, $\mathfrak{D} = \{P, X, Y, \varphi_{\mathfrak{D}}, \psi_{\mathfrak{D}}\}$, $Q = \{q_i, i = 0, \pm 1, \pm 2, \ldots\}$, $P = \{p_j, j = 0, \pm 1, \pm 2, \ldots\}$, $X = \{0, 1\}$, $Y = \{1\}$. $\mathfrak{C}$ and $\mathfrak{D}$ have the same transition as $\mathfrak{A}$ and $\mathfrak{B}$ (this follows from Sec. 2)

$$\psi_{\mathfrak{C}}(q_i, 0) = \psi_{\mathfrak{C}}(q_i, 1) = \begin{cases} \Lambda & \text{for } i = \pm p, \\ 1 & \text{otherwise;} \end{cases}$$

$$\psi_{\mathfrak{D}}(p_j, 0) = \psi_{\mathfrak{D}}(p_j, 1) = \begin{cases} \Lambda & \text{for } i = \pm p,\ p > 2, \\ 1 & \text{for } i = \pm 2k,\ k = 0, 1, 2, \ldots, \\ 11 & \text{otherwise,} \end{cases}$$

where p is a prime, and Λ is the empty word. Any state of $\mathfrak{C}$ is distinguishable from any state of $\mathfrak{D}$, and any two states of $\mathfrak{C}$ (or $\mathfrak{D}$) are distinguishable from one another, but they yield the same output (consisting of infinitely many "ones") for any initial states and infinite sequences.

Literature Cited

1. A. A. Muchnik, "General linear automata," this volume, pp. 179-217.

PROGRAMMING

ON ALGORITHM SCHEMATA WHICH ARE DEFINED ON SITUATIONS †

R. I. Podlovchenko

Erevan

The notion of a memory consisting of individual cells and the notion of a memory state consisting of states of cells play a fundamental role in programming theory. At the same time, in certain branches of this theory, information on memory structure turns out to be redundant, and the memory state is perceived as a unified element. In this case memory may not be introduced at all, and the set of memory states may be considered as a set of abstract "situations."

The notion of a situation, which forms the basis for all the constructions in the present paper, is introduced as an indeterminate object and may certainly have interpretations differing from the notion of a memory state; however, all of the constructions are carried out in such a way that if a situation is understood to be a memory state, then we have both a ready device consisting of basic programming notions and a number of results which may be applied in the programming field.

Mappings of two types are constructed over a set of situations: actions and predicates on situations. If a situation is interpreted as a memory state, then an action will constitute an operator over a memory, and a predicate on situations will constitute a predicate over a memory.

The notion of a schema advanced in the paper is a generalization of the notion of a graph-scheme algorithm [2]. The construction of a schema is raised on a finite graph which may have several outputs for one input and is called a net; the vertices of the net are equipped with transformation functions (in this case a vertex is juxtaposed with an action) or recognition functions (and then the vertex is juxtaposed with a predicate on situations); both the links between the vertices of the nets and the objects juxtaposed with the vertices may be changed along with a change of situation; in particular, one and the same vertex of a schema (it is called a *vertex of the mixed type*) may be a transformer in certain situations and a recognizer in others.

In the majority of schemata defined on situations, an algorithm for executing the schema is introduced – an algorithm I which is applicable to any schema $\mathcal{E}$ in any situation ξ. The process of applying the algorithm I to the schema $\mathcal{E}$ in the situation ξ is called *the procedure for executing the schema $\mathcal{E}$ in the situation ξ* and can be reduced

†Original article submitted July 23, 1968.

to the parallel construction of two sequences: a sequence of situations and a sequence of vertices of the schema $\mathcal{S}$, which are tagged in these situations. The process of executing the schema $\mathcal{S}$ in the situation ξ may be infinite, finite without result, and finite with result; in the latter case the situation ξ is assigned to the definition domain of the schema $\mathcal{S}$.

The algorithmic properties of a schema are reflected by various of the characteristics. Among them the so-called value of the schema on the situation ξ occupies a special place; this characteristic carries information on the "history" of the execution of the schema which has been begun in the situation ξ, and it is used in introducing equivalence relations between schemata. For equivalence of two schemata it is required that their definition domains coincide and, in addition that the values of these schemata on each situation from their definition domain (weak equivalence) or on each possible situation (strong equivalence) coincide.

Among schemata we distinguish the following:

stationary in φ – the function juxtaposed with the vertex of the schema does not depend on the situation considered, and this applies to all vertices;

weakly nonstationary in φ – for nonstationarity the schema at the same time does not have vertices of the mixed type;

strongly nonstationary in φ – the schema does contain vertices of the mixed type;

stationary in φ – the arcs connecting the vertices of the schema remain unchanged (along with their markers) in all situations.

Stationary in φ and ψ simultaneously schemata are classified according to the structure of the net on which the schema is based; thus there appear:

connected schemata – the net of the schema is connected;

directionally connected schemata – the net of the schema is such that a path connecting the input with one of its outputs passes through every vertex of the schema;

linear schemata – the net of the schema has one output and is such that it allows the construction of an elementary path passing through all vertices of the schema.

The classification of schemata given in this paper yields a series of classes which are imbedded in one another. Theorems 1-7 establish sufficient attributes for a transition from a schema of one class to an equivalent schema of another (imbedded) class.

Examples illustrating the basic definitions and equivalent transformations of schemata are given in a separate section at the end of the paper. This emphasizes the particular character of the geometric interpretation of the structure of a schema and preserves the integrity of the exposition of the material of the paper.

§ 1. Auxiliary Notions

1.1. Let us give a series of definitions applying to the theory of groups (see [1]).

Assume $\mathfrak{A} = \{\alpha\}$ is an arbitrary set.

We shall say that the graph

$$\Gamma = \{\Pi(\alpha),\ \alpha \in \mathrm{a}\}$$

has been stipulated if a) the nonempty set $\mathrm{a} \subseteq \mathfrak{A}$ is given;

b) each element of $\alpha \in \mathrm{a}$ is juxtaposed with its set $\Pi(\alpha) \subseteq \mathrm{a}$ (in particular, $\Pi(\alpha)$ may be empty).

The elements of the set a are called vertices of the graph Γ.

1.2. An ordered pair (α, α'), where $\alpha \in \mathrm{a}$, $\alpha' \in \Pi(\alpha)$ shall be called an arc of the graph; concerning the arc (α, α') we say that it issues from the vertex α and enters the vertex α'; α and α' are called the beginning and end of the arc (α, α').

A path in the graph Γ is called a sequence of its arcs $\varkappa_1, \varkappa_2, \ldots, \varkappa_k$, such that the end of each preceding arc coincides with the beginning of the next one. If $\varkappa_i = (\alpha_i, \alpha_{i+1})$, $i = 1, 2, \ldots, k$, then the path, by definition, passes successively through the vertices

$$\alpha_1, \alpha_2, \ldots, \alpha_{k+1},$$

where α_1 is the beginning and α_{k+1} is the end of the path. A path is elementary if no vertex in it is encountered twice.

A contour is called a finite path in which the beginning and end coincide.

1.3. The edge of a graph Γ is called a set consisting of two of its vertices α_1 and α_2 (the so-called boundary vertices of the edge) such that either $\alpha_1 \in \Pi(\alpha_2)$ or $\alpha_2 \in \Pi(\alpha_1)$. A chain is a sequence of edges $\Delta_1, \Delta_2, \ldots, \Delta_k$, in which one of the boundary vertices of each edge Δ_i is also a boundary vertex for Δ_{i-1}, while the other is a boundary vertex for Δ_{i+1}.

A cycle is called a finite chain beginning and ending at one and the same vertex of the graph.

1.4. The graph

$$\Gamma = \{\Pi(\alpha), \alpha \in \mathrm{a}\}$$

is called a net and shall be written in the form

$$S = [\alpha_0, \mathrm{a}_w, \{\Pi(\alpha), \alpha \in \mathrm{a}\}],$$

if the following are true in it:

1) the vertex $\alpha_0 \in \mathrm{a}$, called the input of the net (it is such that exactly one arc of the graph issues from it and not a single arc of the graph enters it), is isolated.
2) The nonempty set $\mathrm{a}_w \subseteq \mathrm{a}$ is isolated whose elements are called outputs of the net and have the following property: not a single arc of the graph issues from them;
3) at least one arc of the graph issues from each graph vertex which does not coincide with α_0 and does not belong to the set a_w (such a vertex is called internal).

1.5. A graph is called connected if any two of its vertices may be connected with a chain.

A net is called directionally connected if, regardless of its vertex, there exists a path beginning at the input of the net which ends at one of its outputs and passes through this vertex.

A directionally connected net in which there are no samples is called a tree.

A net with one input is called linear if there exists an elementary path beginning at the input of the net and ending at its output, while passing through all vertices of the net.

§ 2. Basic Notions

2.1. Let us consider an arbitrary set $\Xi = \{\xi\}$, whose elements are called situations.

2.2. An action is called a partial mapping of the set Ξ in itself.

An action with an empty definition domain is called empty and is denoted by Λ.

Assume $\mathfrak{D} = \{A\}$ is the set of all possible actions.

2.3. Let us introduce the set $U = \{u\}$.

A predicate on situations is called a function which is defined on a certain subset of the set Ξ and maps this subset in the set U.

Assume $\mathfrak{R} = \{p\}$ is the set of all possible predicates on situations.

2.4. We shall call a schema a complex consisting of three mappings:

1. The mapping $\sigma(\xi)$ which juxtaposes each situation $\xi \in \Xi$ with its net; all juxtaposed nets, by stipulation, have the following in common: the set of vertices $\mathfrak{a}$, the inputs α_0, and the set of outputs $\mathfrak{a}_w$, and they differ from one another, perhaps, in the sets $\Pi(\alpha)$, $\alpha \in \mathfrak{a}$.

The vertices of the nets which are different from its inputs and outputs will be called internal; the net juxtaposed with the situation ξ shall be written in the form

$$[\alpha_0, \mathfrak{a}_w, \{\Pi(\alpha, \xi), \alpha \in \mathfrak{a}\}],$$

where $\Pi(\alpha, \xi)$ is the set of vertices juxtaposed with the vertex α in the situation ξ.

2. The mapping $\varphi(\alpha, \xi)$ which juxtaposes either an action A or a predicate p on situations with each pair (α, ξ), where α is an internal vertex of the schema, while ξ is an arbitrary situation.

3. The mapping $\psi(\alpha, \xi)$ which is defined for all vertices of the schema with the exception of its output and juxtaposes the pair (α, ξ) with an arbitrary mapping of the set U in the set $\Pi(\alpha, \xi)$, if $\varphi(\alpha, \xi)$ is a predicate, and the mapping of the set U in some particular element of the set $\Pi(\alpha, \xi)$, if $\varphi(\alpha, \xi)$ is an action.

If the vertex $\alpha' \in \Pi(\alpha, \xi)$ is the image of the element $u \in U$ in the mapping

$$\psi(\alpha, \xi): U \to \Pi(\alpha, \xi),$$

then the arc (α, α') is called a tagged element u.

For the schema we use the notation

$$\mathcal{Z} \equiv [\alpha, \mathfrak{a}_w, \{\Pi(\alpha, \xi), \varphi(\alpha, \xi), \psi(\alpha, \xi), \alpha \in \mathfrak{a}, \xi \in \Xi\}].$$

The vertex at which the arc issuing in the situation ξ from the vertex α_0 arrives shall be called the first arc and denoted by $\alpha_1(\xi)$.

2.5. The procedure of executing the schema $\mathcal{Z}$ in a stipulated situation ξ shall be understood to mean the procedure of joint construction of two sequences:

a) the sequence of situations

$$\xi_1, \xi_2, \ldots, \xi_l, \ldots, \tag{1}$$

b) the sequence of vertices of the schema $\mathcal{Z}$, which are tagged in the situation (1)

$$\alpha_1, \alpha_2, \ldots, \alpha_l, \ldots. \tag{2}$$

Let us define this procedure by induction.

The first step. Let us assume that $\xi_1 = \xi$, and let us tag the vertex $\alpha_1(\xi_1)$.

Assume that l steps ($l \geq 1$) have been taken, ξ_l is the situation constructed on the l-th step, and α_l is the vertex tagged on the l-th step (and, therefore, in the situation ξ_l).

On the l + 1-st step let us do the following:

I. Let us consider the vertex α_l.

If α_l is the output of the schema $\partial (\alpha_l \in \mathfrak{a}_w)$, then the process of executing the schema ∂ in the situation ξ is assumed to be completed with a result on the l-th step, and the situation ξ_l is considered to be terminal.

If the vertex α_l differs from the outputs of the schema $\partial (\alpha_l \bar{\in} \mathfrak{a}_w)$, then we go over to point II.

II. Let us consider the function φ at the vertex α_l and in the situation ξ_l; its value is either a certain action A or a certain predicate p on situations.

1. Assume $\varphi(\alpha_l, \xi_l) = A$ and the vertex $\alpha^* \in \Pi(\alpha_l, \xi_l)$ is the image of the set U in the mapping $\psi(\alpha_l, \xi_l)$; let us distinguish between the following cases:

a) the action A is applicable to the situation ξ_l; then we assume that

$$\xi_{l+1} = A(\xi_l), \qquad \alpha_{l+1} = \alpha^*,$$

after which the execution of the l + 1-st step is assumed to have been completed, and we go over to point I to execute the next step;

b) the action A is inapplicable to the situation ξ_l; in this case the process of executing the schema ∂ in the situation ξ is assumed to have been completed without result on the l-th step.

2. Assume $\varphi(\alpha_l, \xi_l) = p$; let us consider the following cases:

a) the predicate p is defined in the situation ξ_l, $p(\xi_l) = u^*$, and the mapping $\psi(\alpha_l, \xi_l)$ juxtaposes the element u* with the vertex $\alpha^* \in \Pi(\alpha_l, \xi_l)$; then we assume that

$$\xi_{l+1} = \xi_l, \qquad \alpha_{l+1} = \alpha^*,$$

consider the execution of the l + 1-st step to have been completed, and go over to point I to execute the next step;

b) the predicate p is not defined in the situation ξ_l; then the process of executing the schema ∂ and the situation ξ is assumed to have been completed without result on the l-th step.

The process of executing the schema ∂ in the situation ξ, as is evident from its description, may be either finite or infinite. If the process has been completed with a result and consequently leads to a terminal situation ξ', then the schema ∂ is assumed to be applicable to the original situation ξ, while the situation ξ is denoted by $\partial(\xi)$.

2.6. Assume Ξ_∂ is a set consisting of all such situations to which the schema ∂ is applicable.

Let us introduce the following characteristics of the schema.

The schema ∂ is juxtaposed with the action A_∂, which has the following properties:

1) the definition domain of the action A_∂ is the set Ξ_∂;

2) regardless of the situation $\xi \in \Xi_\partial$,

$$A_\partial(\xi) = \partial(\xi);$$

we shall speak of A_∂ as of an action realized by this schema ∂.

Assume $a_w = \{\alpha^{(1)}, \alpha^{(2)}, \ldots, \alpha^{(k)}\}$, $k \geq 1$. We shall say that the schema ∂ realizes the partitioning of the set Ξ_∂ into k disjoint subsets $\Xi_{\alpha^{(i)}}$, i = 1, 2, ..., k, which are defined as follows: the situation $\xi \in \Xi_\partial$ is assigned to the set $\Xi_{\alpha^{(i)}}$, if the procedure of executing the schema ∂ in the situation ξ is completed by tagging the output $\alpha^{(i)}$.

Assume the schema ∂ is executed in the situation ξ, and this leads to the construction of the sequences (1) and (2).

Let us consider the sequence

$$\bar{L}(\partial, \xi) = \varphi(\alpha_1, \xi_1),\ \varphi(\alpha_2, \xi_2),\ \ldots,\ \varphi(\alpha_k, \xi_k),\ \ldots;$$

its elements are actions or predicates which are juxtaposed with the tagged vertices of the schema ∂, the juxtaposition being in those situations in which these vertices were tagged during the execution of the schema ∂. From this it follows that: if the procedure of executing the schema ∂ in the situation ξ is finite with a result, then since the function φ is not defined at a single vertex of the set a_w (outputs of the schema), the sequence $\bar{L}(\partial, \xi)$ has a length one less than the sequence (1).

The value of the schema ∂ in the situation ξ shall be called a sequence

$$L(\partial, \xi) = \varphi(\alpha_{i_1}, \xi_{i_1}),\ \varphi(\alpha_{i_2}, \xi_{i_2}), \ldots, \varphi(\alpha_{i_t}, \xi_{i_t}), \ldots$$

of the sequence $\bar{L}(\partial, \xi)$, which is such that it contains all actions entered in $\bar{L}(\partial, \xi)$ and only those.

§ 3. Relations between Schemata. Classification of Schemata

3.1. Let us introduce a series of relations between schemata.

The schema ∂_1 is called the expansion of the schema ∂_2 if

$$\Xi_{\partial_1} \supseteq \Xi_{\partial_2},$$

and for any situation $\xi \in \Xi_{\partial_2}$

$$L(\partial_1, \xi) = L(\partial_2, \xi).$$

The schema ∂_1 and ∂_2 are called weakly equivalent if

$$\Xi_{\partial_1} = \Xi_{\partial_2},$$

and for any situation $\xi \in \Xi_{\partial_1}$

$$L(\partial_1, \xi) = L(\partial_2, \xi).$$

The schemata ∂_1 and ∂_2, are, by definition, strongly equivalent if, regardless of the situation $\xi \in \Xi$,

$$L(\partial_1, \xi) = L(\partial_2, \xi)$$

and, moreover,

$$\Xi_{\mathscr{E}_1} = \Xi_{\mathscr{E}_2}.$$

Weak equivalence of schemata derives from their strong equivalence; if each of the schemata $\mathscr{E}_1$ and $\mathscr{E}_2$ is an expansion of the other, then the schemata $\mathscr{E}_1$ and $\mathscr{E}_2$ are weakly equivalent.

The equivalence relation (both weak and strong) is reflective, symmetric, and transitive. The expansion relation is reflective and transitive.

3.2. The schemata

$$\mathscr{E}_1 = [\alpha_0', \mathfrak{a}_w', \{\Pi'(\alpha, \xi), \varphi'(\alpha, \xi), \psi'(\alpha, \xi), \alpha \in \mathfrak{a}', \xi \in \Xi\}]$$

and

$$\mathscr{E}_2 = [\alpha_0'', \mathfrak{a}_w'', \{\Pi''(\alpha, \xi), \varphi''(\alpha, \xi), \psi''(\alpha, \xi), \alpha \in \mathfrak{a}'', \xi \in \Xi\}]$$

are *isomorphic* if between the vertices of the sets $\mathfrak{a}'$ and $\mathfrak{a}''$ one may establish a one-to-one relationship σ, such that:

1) α_0' and α_0'' are corresponding vertices;

2) the sets $\mathfrak{a}_w'$ and $\mathfrak{a}_w''$ go over into each other;

3) regardless of $\xi \in \Xi$ and $\alpha \in \mathfrak{a}'$, the following statements are valid:

a) the sets $\Pi(\alpha, \xi)$ and $\Pi''(\sigma\alpha, \xi)$ consists of vertices that are pairwise corresponding to each other;

b) $\varphi'(\alpha, \xi) = \varphi''(\sigma\alpha, \xi)$;

c) the mappings

$$\psi'(\alpha, \xi) : U \to \Pi'(\alpha, \xi)$$

and

$$\psi''(\sigma\alpha, \xi) : U \to \Pi''(\sigma\alpha, \xi)$$

juxtapose each element $u \in U$ with corresponding (to each other) vertices of the sets $\Pi'(\alpha, \xi)$ and $\Pi''(\sigma\alpha, \xi)$.

The isomorphism relation of schemata is reflective, symmetric, transitive, and involves the relation of strong equivalence of schemata. It is obvious that isomorphic schemata realize one and the same operator and one and the same partitioning of the definition domain of the schemata.

3.3. Assume

$$\mathscr{E} = [\alpha_0, \mathfrak{a}_w, \{\Pi(\alpha, \xi), \varphi(\alpha, \xi), \psi(\alpha, \xi), \alpha \in \mathfrak{a}, \xi \in \Xi\}] -$$

is an arbitrary schema.

We use $\Pi^*(\alpha, \xi)$ to denote the image of the set U in the mapping

$$\psi(\alpha, \xi) : U \to \Pi(\alpha, \xi).$$

The schema $\mathcal{Z}$ is called **reduced in** ψ if for all α and ξ

$$\Pi^*(\alpha, \xi) = \Pi(\alpha, \xi).$$

If the schema $\mathcal{Z}$ is not reduced in ψ, then it may be juxtaposed with the schema

$$\mathcal{Z}^* = [\alpha_0, \mathfrak{a}_w, \{\Pi^*(\alpha, \xi), \varphi(\alpha, \xi), \psi(\alpha, \xi), \alpha \in \mathfrak{a}, \xi \in \Xi\}],$$

which a) will be reduced in ψ, and b) is strongly equivalent to the schema $\mathcal{Z}$.

Hereafter we shall consider only schemata which are reduced in ψ without making special mention of this each time.

3.4. We prefer classification of the schemata proper to classification of the vertices of a schema.

An internal vertex α of the schema $\mathcal{Z}$ is:

stationary in φ if the value of the function $\varphi(\alpha, \xi)$ is independent of the situation ξ, and nonstationary in φ otherwise;

stationary in ψ if the mappings $\psi(\alpha, \xi)$ do not change with a change in ξ, and nonstationary in ψ otherwise.

An interval vertex α is, by definition, **finite-valued** if the function $\varphi(\alpha, \xi)$ takes a finite number of values on the set Ξ. An internal vertex α of the schema $\mathcal{Z}$ is:

an A-vertex if the values of $\varphi(\alpha, \xi)$ in all situations $\xi \in \Xi$ are only actions;

a p-vertex if the values of $\varphi(\alpha, \xi)$ in all situations $\xi \in \Xi$ are only predicates;

a vertex of the mixed type if the values of $\varphi(\alpha, \xi)$ in certain situations are actions and in other situations are predicates.

3.5. Let us distinguish among the following classes of schemata.

The schema $\mathcal{Z}$ is called:

stationary in φ **(in** ψ**)** if all internal vertices of $\mathcal{Z}$ are stationary in φ (correspondingly in ψ), and nonstationary in φ (in ψ) otherwise;

stationary if $\mathcal{Z}$ is stationary in both φ and ψ.

Among schemata which are not stationary in φ we distinguish among:

weakly nonstationary schemata which do not contain the vertices of the mixed type;

strongly nonstationary schemata which contain such vertices.

Assume

$$\mathcal{Z} = [\alpha_0, \mathfrak{a}_w, \{\Pi(\alpha, \xi), \varphi(\alpha, \xi), \psi(\alpha, \xi), \alpha \in \mathfrak{a}, \xi \in \Xi\}]$$

is a schema which is stationary in ψ; consequently, the mapping $\Pi(\alpha, \xi)$, regardless of the vertex α, is independent of ξ if all situations $\xi \in \Xi$ are juxtaposed with one and the same net

$$S_{\mathcal{Z}} = [\alpha_0, \mathfrak{a}_w, \{\Pi_\xi^\prime(\alpha), \alpha \in \mathfrak{a}\}].$$

In §1, we define such properties of a net as connectedness, directional connectedness, linearity, and the property of the net making the net a tree.

If the net S_{∂} has one of the enumerated properties, then we shall agree to confer this property on the schema ∂ itself. Thus, in a set of schemata which are stationary in ψ there appear classes of connected, directionally connected, linear schemata, and tree-schemata.

3.6. A schema ∂ which is stationary in ψ (constructed on the net S_{∂}) shall be stipulated by agreement using the accepted method of depicting a net. For such a stipulation:

1) The vertices of the net are depicted by points of a plane; the input is marked in standard fashion with the index 0, and the index w is used to denote the outputs of the net.

2) The arcs of the net are depicted by directional pieces of curves which connect the vertices of the net.

3) The arcs issuing from the A-vertices do not carry tags.

4) The arcs issuing from the p-vertices are tagged by elements of the set U.

5) If ∂ is stationary in φ, then the vertices of the net are supplied with the designations of those actions and predicates which are juxtaposed with these vertices; for nonstationarity of ∂ in φ the mapping $\varphi(\alpha, \xi)$ must be stipulated separately.

Let us present examples of schemata of the simplest type which are stationary in ψ.

1.

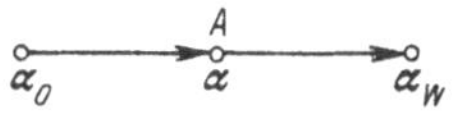

Here A is a certain action.

2.

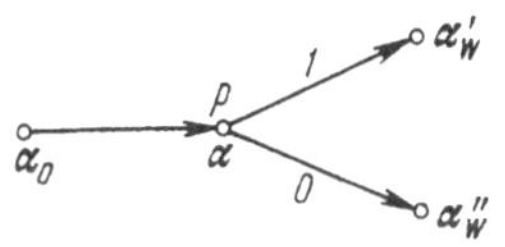

Here U = $\{0, 1\}$, and p is a certain predicate on situations.

§ 4. Operations on Schemata

4.1. In all the subsequent considerations:

1) all schemata are assumed to be reduced in ψ;

2) if a new schema is synthesized from several schemata, then the original schemta are assumed to have no common vertices.

In order for the first proposition to be valid, it is sufficient to go over from the original schema of transformations considered in subsection 3.3 to an equivalent schema; the transition from a schema to a schema isomorphic to it (and therefore equivalent to it) makes the second proposition valid also.

Besides these propositions, we assume that the following rule is operative: for transformations and synthesis of schemata, the vertices of the schemata are always taken to be inseparable from the functions juxtaposed with them (the arcs are inseparable from their tags); if the added vertex is an A-vertex, then nothing is said of the tags of the arc issuing from it.

4.2. The schema

$$\partial = [\alpha_0, \mathfrak{a}_w, \{\Pi(\alpha, \xi), \varphi(\alpha, \xi), \psi(\alpha, \xi), \alpha \in \mathfrak{a}, \xi \in \Xi\}]$$

is given whose outputs are assumed to be ordered $\mathfrak{a}_w = \{\alpha^{(1)}, \alpha^{(2)}, \ldots, \alpha^{(k)}\}$, and the schemata

$$\mathcal{Z}_i = [\alpha_0^{(i)}, \mathfrak{a}_w^{(i)}, \{\Pi^{(i)}(\alpha, \xi), \varphi^{(i)}(\alpha, \xi), \psi^{(i)}(\alpha, \xi), \alpha \in \mathfrak{a}^{(i)}, \xi \in \Xi\}]$$

in an amount equal to the number of outputs of the schema $\mathcal{Z}$ (i = 1, 2, ..., k).

Let us construct the schema $\mathcal{Z}^*$, which we shall call the $\mathcal{Z}$-composition of the schemata $\mathcal{Z}_1, \mathcal{Z}_2, \ldots, \mathcal{Z}_k$. Let us begin by giving a substantive description of the procedure for constructing it.

1. We combine all of the vertices of the schemata $\mathcal{Z}, \mathcal{Z}_1, \mathcal{Z}_2, \ldots, \mathcal{Z}_k$ into the set.

2. The input of the future schema is designated to be the vertex α_0, and the outputs are designated to be all outputs of the schemata $\mathcal{Z}_1, \mathcal{Z}_2, \ldots, \mathcal{Z}_k$.

3. In each situation $\xi \in \Xi$ we execute the following:

a) we consider the set containing all arcs of the schemata $\mathcal{Z}, \mathcal{Z}_1, \ldots, \mathcal{Z}_k$;

b) each arc arriving at the output $\alpha^{(i)}$, i = 1, 2, ..., k, of the schema $\mathcal{Z}$, is directed to the first vertex $\alpha_1^{(i)}(\xi)$ of the schema $\mathcal{Z}_i$, with no alteration of the tag of the arc itself;

c) we discard the outputs $\alpha^{(i)}$ of the schema $\mathcal{Z}$ and the inputs of all schema $\mathcal{Z}_i$, i = 1, 2, ..., k.

Now the $\mathcal{Z}$-composition of the schema $\mathcal{Z}_1, \mathcal{Z}_2, \ldots, \mathcal{Z}_k$ will be defined formally; it will be the schema

$$\mathcal{Z}^* \equiv (\mathcal{Z}, \mathcal{Z}_1, \mathcal{Z}_2, \ldots, \mathcal{Z}_k) \equiv [\alpha_0^*, \mathfrak{a}_w^*, \{\Pi^*(\alpha, \xi), \varphi^*(\alpha, \xi), \psi^*(\alpha, \xi), \alpha \in \mathfrak{a}^*, \xi \in \Xi\}],$$

in which:

1) the input α_0^* is the input α_0 of the schema $\mathcal{Z}$;

2) the outputs constitute the set

$$\mathfrak{a}_w^* = \bigcup_{i=1, 2, \ldots, k} \mathfrak{a}_w^{(i)};$$

3) the set of internal vertices is obtained by a union of the sets of the internal vertices of the schemata $\mathcal{Z}, \mathcal{Z}_1, \mathcal{Z}_2, \ldots, \mathcal{Z}_k$;

4) for all internal vertices α of the schema $\mathcal{Z}^*$ and all situations ξ we have

$$\varphi^*(\alpha, \xi) = \begin{cases} \varphi(\alpha, \xi), & \text{if} \quad \alpha \in \mathfrak{a}, \\ \varphi^{(i)}(\alpha, \xi), & \text{if} \quad \alpha \in \mathfrak{a}^{(i)}, \quad i = 1, 2, \ldots, k; \end{cases}$$

5) regardless of the vertex α of the schema $\mathcal{Z}^*$ and the situation $\xi \in \Xi$,

a) if $\alpha \in \mathfrak{a}^{(i)}$, then $\Pi^*(\alpha, \xi) = \Pi^{(i)}(\alpha, \xi)$, $\psi^*(\alpha, \xi) = \psi^{(i)}(\alpha, \xi)$;

b) however, if $\alpha \in \mathfrak{a}$, then $\Pi^*(\alpha, \xi)$ is obtained from $\Pi(\alpha, \xi)$ by replacement of each vertex $\alpha^{(i)}$ in the latter by the vertex $\alpha_1^{(i)}(\xi)$, $i = 1, 2, \ldots, k$ (here $\alpha_1^{(i)}(\xi)$ is the first vertex of the schema $\mathcal{Z}_i$ and the situation ξ); analogously, the mapping $\psi^*(\alpha, \xi)$ is obtained from the mapping

$$\psi(\alpha, \xi): U \to \Pi(\alpha, \xi),$$

provided only that in the set $\Pi(\alpha, \xi)$ each vertex $\alpha^{(i)}$ is replaced by $\alpha_1^{(i)}(\xi)$, $i = 1, 2, \ldots, k$.

4.3. The vertex α of the schema ∂ is called the p r e o u t p u t of the schema ∂ in the situation ξ if the set $\Pi(\alpha, \xi)$ contains just one output of the schema ∂.

We shall say that

∂ i s a s c h e m a o f t h e t y p e A, if:

1) all of its preoutputs are A-vertices which are stationary in ψ;

2) not one output has two preoutputs (i.e., for any preoutputs α' and α'' the inequality $\alpha' \neq \alpha''$ entails the equation $\Pi(\alpha') \cap \Pi(\alpha'') = \phi$);

∂ i s a s c h e m a o f t h e t y p e p, if:

1) all of its preoutputs are p-vertices which are stationary in ψ;

2) all arcs issuing from each preoutput lead only to outputs of the schema;

3) not one output of the schema has two preoutputs.

Let us introduce the following operations:

1) s u b s t i t u t i o n o f t h e s c h e m a ∂_1 o f t h e t y p e A i n t o t h e s c h e m a ∂_2 i n p l a c e o f i t s A - v e r t e x α^*;

2) s u b s t i t u t i o n o f t h e s c h e m a ∂_1 o f t h e t y p e p i n t o t h e s c h e m a ∂_2 i n p l a c e o f i t s p - v e r t e x α^*.

The result of executing each of these operations is a new schema which is denoted by ∂^* in both cases.

Let us describe the substance of the procedure of synthesizing a schema ∂^*.

1. Let us combine all vertices of the schemata ∂_1 and ∂_2 into one set.

2. The input of the featured schema is assumed to be the input α_0'' of the schema ∂_2, while its outputs are assumed to be all outputs of the schema ∂_2.

3. In each situation $\xi \in \Xi$ we do the following:

a) we combine all arcs of the schemata ∂_1 and ∂_2 into one set;

b) each arc arriving at the vertex α^* is directed to the first vertex $\alpha_1'(\xi)$ of the schema ∂_1, without changing the tags of this arc;

c) we eliminate the input of the schema ∂_1 and all of the outputs together with the arcs arriving at them;

d) if from the vertex α^* there issues an arc (α^*, α), then from each preoutput of the schema ∂_1 we bring out an arc to the vertex α and tag it with the same elements of the set U which were used to tag the arc (α^*, α); after this the arc (α^*, α) is discarded; the described procedure is carried out until the set of arcs issuing from α^* becomes empty;

e) we discard the vertex α^*.

The resultative schema ∂^* may also be stipulated formally. If

$$\partial_1 = [\alpha_0', \mathrm{a}_w, \{\Pi'(\alpha, \xi), \varphi'(\alpha, \xi), \psi'(\alpha, \xi), \alpha \in \mathrm{a}', \xi \in \Xi\}],$$

$$\mathcal{Z}_2 = [\alpha_0'', \mathbf{a}_w'', \{\Pi''(\alpha, \xi), \varphi''(\alpha, \xi), \psi''(\alpha, \xi), \alpha \in \mathbf{a}'', \xi \in \Xi\}],$$
$$\mathcal{Z}^* = [\alpha_0^*, \mathbf{a}_w^*, \{\Pi^*(\alpha, \xi), \varphi^*(\alpha, \xi), \psi^*(\alpha, \xi), \alpha \in \mathbf{a}^*, \xi = \Xi\}],$$

then

1) $\alpha_0^* = \alpha_0''$;

2) $\mathbf{a}_w^* = \mathbf{a}_w''$;

3) the set of internal vertices of the schema $\mathcal{Z}^*$ is obtained by combining the sets of internal vertices of the schemata $\mathcal{Z}_1$ and $\mathcal{Z}_2$ and eliminating the vertex α^*;

4) for all internal vertices α of the schema $\mathcal{Z}^*$ and all situations ξ we have

$$\varphi^*(\alpha, \xi) = \begin{cases} \varphi'(\alpha, \xi), & \text{if} \quad \alpha \in \mathbf{a}', \\ \varphi''(\alpha, \xi), & \text{if} \quad \alpha \in \mathbf{a}''; \end{cases}$$

5) regardless of the vertex α of the schema $\mathcal{Z}^*$ and the situation $\xi \in \Xi$:

a) if α is not a preoutput of the schema $\mathcal{Z}_1$, then

$$\Pi^*(\alpha, \xi) = \begin{cases} \Pi'(\alpha, \xi), & \text{if} \quad \alpha \in \mathbf{a}', \\ \Pi''(\alpha, \xi), & \text{if} \quad \alpha \in \mathbf{a}'' \text{and } \alpha^* \overline{\in} \Pi''(\alpha, \xi), \\ \{\alpha_1'(\xi)\} \cup \Pi''(\alpha, \xi)\{\alpha^*\}, & \text{if} \quad \alpha \in \mathbf{a}'', \quad \alpha^* \in \Pi''(\alpha, \xi); \end{cases}$$

$$\psi^*(\alpha, \xi) = \begin{cases} 1)\ \psi'(\alpha, \xi), & \text{if} \quad \alpha \in \mathbf{a}'; \\ 2)\ \psi''(\alpha, \xi), & \text{if} \quad \alpha \in \mathbf{a}'' \text{and} \alpha^* \overline{\in} \Pi''(\alpha, \xi), \\ \text{3) however, if } \alpha \in \mathbf{a}'' \text{and } \alpha^* \in \Pi''(\alpha, \xi)\text{, then the} \\ \quad \text{mapping } \psi^*(\alpha, \xi) \text{ is obtained from the} \\ \quad \text{mapping } \psi''(\alpha, \xi)\text{: } U \to \Pi''(\alpha, \xi) \text{ by replacement of the} \\ \quad \text{vertex } \alpha^* \text{ by the vertex } \alpha_1'(\xi) \text{ in } \Pi''(\alpha, \xi); \end{cases}$$

b) if α is a preoutput of the schema $\mathcal{Z}_1$, then

$$\Pi^*(\alpha, \xi) = \Pi''(\alpha, \xi), \qquad \psi^*(\alpha, \xi) = \psi''(\alpha, \xi).$$

4.4. The operation of splicing the outputs α' and α'' of the schema $\mathcal{Z}$, which leads to the schema $\mathcal{Z}^*$, is defined only according to intent.

In constructing the schema $\mathcal{Z}^*$ we do the following:

1. We take the set of vertices of the schema $\mathcal{Z}$.

2. The input of the schema $\mathcal{Z}^*$ is assumed to be the input of the schema $\mathcal{Z}$, while the outputs of the schema $\mathcal{Z}^*$ are assumed to be all outputs of the schema $\mathcal{Z}$, with the exception of α''.

3. In each situation $\xi \in \Xi$:

a) we scan the arcs of the schema $\mathcal{Z}$;

b) each arc arriving at the output α'' is directed to the output α';

c) we discard the vertex α''.

It can easily be verified that the resultative schema $\mathcal{Z}^*$ will be strongly equivalent to the schema $\mathcal{Z}$, although it will also realize a new partitioning of the set $\Xi_{\mathcal{Z}}$ into subsets.

§ 5. Nonstationary Schemata and Their Transformations

5.1. Let us specify the sets $D \subseteq \mathfrak{D}$ and $R \subseteq \mathfrak{R}$ of actions and predicates on situations.

The schema $\mathcal{C} = [\alpha_0, \mathfrak{a}_\omega, \{\Pi(\alpha, \xi), \varphi(\alpha, \xi), \psi(\alpha, \xi), \alpha \in \mathfrak{a}, \xi \in \Xi\}]$ is called a schema over D and R if

$$\varphi(\alpha, \xi) \in D \cup R,$$

regardless of the situation ξ from the internal vertex α of the schema $\mathcal{C}$.

$\mathcal{C}$ will be called a schema over R if, regardless of the situation ξ and the interval vertex α of the schema $\mathcal{C}$,

$$\varphi(\alpha, \xi) \in R.$$

It can easily be seen that the operator realized by a schema over R is unitary.

5.2. Assuming D and R to be nonempty sets, let us consider the class of all possible schemata over D and R; we establish a series of results.

Theorem 1. *A strongly nonstationary schema $\mathcal{C}$ may be transformed into a weakly nonstationary schema which is equivalent*† *to it if regardless of the partitioning of Ξ into two subsets one can find a schema over R which realizes this partitioning.*

Proof. Assume $\mathcal{C}$ is a strongly nonstationary schema; this means that among its internal states one can find at least one vertex of a mixed type. Obviously, for proof of the theorem it is sufficient to advance a method of transforming the schema $\mathcal{C}$ into a schema $\mathcal{C}_1$, equivalent to it in which the number of vertices of a mixed type is one less than it is in the schema $\mathcal{C}$. Let us present such a method.

Assume α is a vertex of the mixed type in the schema $\mathcal{C}$, and Ξ'_α is the set of all such situations (and only those) in which actions are juxtaposed with the vertex α; therefore, the set

$$\Xi''_\alpha = \Xi / \Xi'_\alpha$$

combines in itself all such situations (and only them) in which predicates are juxtaposed with the vertex α.

In accordance with the proposition of the theorem one can find a schema $\bar{\mathcal{C}}$ over R which realizes the partitioning of Ξ into Ξ'_α and Ξ''_α. The schema $\bar{\mathcal{C}}$, just as any schema over R, will be weakly nonstationary.

Without loss of generality of the considerations, let us assume that $\bar{\mathcal{C}}$ has a total of two outputs α' and α'', and

$$\Xi_{\alpha'} = \Xi'_\alpha, \quad \Xi_{\alpha''} = \Xi''_\alpha.$$

(If this is not so and other outputs are present in the schema $\bar{\mathcal{C}}$, then they are juxtaposed with empty sets of situations; in this case it is sufficient to carry out the operations of placing these outputs with, for example, the output α' in the final number, and we arrive at a schema equivalent to the original one and having the property postulated above.) Let us stipulate the following schemata which are stationary in ψ:

† Here and below equivalence shall be understood to mean strong equivalence.

$$\varphi'(\gamma', \xi) = \begin{cases} \varphi(\alpha, \xi), & \text{if } \xi \in \Xi'_\alpha, \\ A & \text{otherwise;} \end{cases}$$

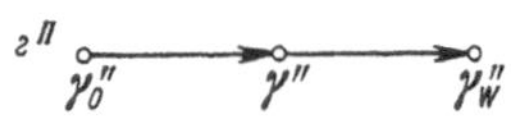

$$\varphi''(\gamma'', \xi) = \begin{cases} \varphi(\alpha, \xi), & \text{if } \xi \in \Xi''_\alpha, \\ p & \text{otherwise;} \end{cases}$$

here A and p represent an action and predicate which are arbitrary but specified. Both schemata, as is easily seen, are weakly nonstationary.

Let us compose the $\bar{\mathcal{G}}$ -composition of the schemata $\mathcal{G}'$ and $\mathcal{G}''$:

$$\mathcal{G}''' = (\bar{\mathcal{G}}, \mathcal{G}', \mathcal{G}''),$$

while assuming that in the schema $\bar{\mathcal{G}}$ the first output is the output α'. The schema $\mathcal{G}'''$ will be weakly nonstationary.

Finally, let us synthesize the schema $\mathcal{G}_1$, while performing the following operations.

1. We combine all vertices of the schemata $\mathcal{G}$ and $\mathcal{G}'''$ into one set.

2. The input of the schema $\mathcal{G}_1$ is designated to be the input of the schema $\mathcal{G}$, while the outputs are designated to be all outputs of the schema $\mathcal{G}$.

3. In each situation ξ we do the following:

a) we combine all arcs of the schemata $\mathcal{G}$ and $\mathcal{G}'''$ into one set;

b) each arc arriving at the vertex α of the schema $\mathcal{G}$, is directed to the first vertex of the schema $\mathcal{G}'''$, without altering the tags of this arc;

c) assume that in the situation ξ the vertex α is juxtaposed with an operator; therefore, a single arc (α, α') issues from it; then from the preoutput γ' of the schema $\mathcal{G}'''$ we bring out the arc to the vertex α', while from the preoutput γ'' we bring out an arc to the same vertex while tagging the latter arc with all elements of the set U†; after this the arc (α, α') is discarded;

d) assume that the vertex α is juxtaposed with a predicate; then from the preoutput γ' we bring out the arc to the same vertex; then we proceed as follows: if the arc (α, α') issues from the vertex α, then from the preoutput γ'' we bring out an arc to the vertex α' and label it with the same elements of the set U that were used to label the arc (α, α'); after this the arc (α, α') is discarded; the described procedure is repeated until the set of arcs issuing from α becomes empty;

e) we discard the vertex α, and the input and outputs of the schema $\mathcal{G}'''$.

The results of our operation will be a schema in which the number of the vertices of the mixed type is actually one less than it is in the schema $\mathcal{G}$. The theorem has been proved.

5.3. Theorem 2. A schema $\mathcal{G}$ which is nonlinear in φ may be transformed into a schema which is stationary in φ and is equivalent to it if:

a) all internal vertices of the schema $\mathcal{G}$ are finite-valued;

b) regardless of the finite partitioning of the set Ξ, one can find a schema over R which is stationary in φ and realizes this partitioning.

† Since the schema $\mathcal{G}_1$ is reduced in ψ, it follows that it is superfluous to tag the single arc issuing from the A-vertex.

Proof. Since condition b) makes the statement of Theorem 1 valid, it is sufficient to consider the case in which ∂ is a weakly nonstationary schema. Therefore, in the schema ∂ each internal vertex is either an A-vertex or a p-vertex, and at least one of them is nonstationary in φ.

The theorem will be proved if a method is given for constructing a schema equivalent to the schema ∂ and containing vertices which are nonstationary in φ and are one less in number than those contained in the schema ∂. Let us propose such a method.

Assume α is a vertex which is nonstationary in φ and belongs to the schema ∂. For α let us construct the schema ∂', having the following properties:

1) ∂' is stationary in φ;

2) ∂' represents a schema of the type A if α is an A-vertex, and a schema of the type p if α is a p-vertex.

By virtue of the second property the schema ∂' may be substituted into the schema ∂ instead of the vertex α; assume that the result of the schema is ∂''. The schema ∂' can be constructed so that it is calculated to make ∂'' equivalent to the schema ∂. The fact that in the schema ∂'' the number of vertices which are nonstationary in φ will be one less than in the schema ∂ derives from stationarity in φ of the schema ∂'.

As has already been noted, in designing the schema ∂', it is necessary to distinguish whether or not the vertex α is an A-vertex or a p-vertex. We shall carry out all the necessary constructions only in the first case on the basis of the fact that the second case is completely analogous to the first.

Thus, α is an A-vertex. Let us use the proposition concerning the finite-valuedness of α, and let us write out all values of the function $\varphi(\alpha, \xi)$, $\xi \in \Xi$,

$$A_1, A_2, \ldots, A_m.$$

Let us use Ξ_i to denote the set of all those situations (and only those) for which

$$\varphi(\alpha, \xi) = A_i, \quad i = 1, 2, \ldots, m.$$

It can easily be seen that

$$\Xi_i \cap \Xi_j = \phi, \quad \text{if} \quad i \neq j, \qquad \bigcup_{i=1, 2, \ldots, m} \Xi_i = \Xi.$$

Therefore, according to proposition b) one can find a scheme $\bar{\partial}$ over R which is stationary in φ and realizes the partitioning of the set Ξ into Ξ_i, i = 1, 2, ..., m. Without loss of generality, we shall assume (see subsection 5.2) that the schema $\bar{\partial}$ has m outputs $\alpha^{(1)}, \alpha^{(2)}, \ldots, \alpha^{(m)}$, and that $\Xi_{\alpha^{(i)}} = \Xi_i$, i = 1, 2, ..., m.

For each $i \in \{1, 2, \ldots, m\}$ we construct the stationary (in φ and in ψ) schema ∂_i:

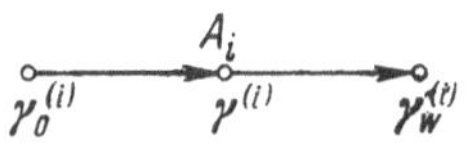

(i.e., for all $\xi \in \Xi$ $\varphi(\gamma^{(i)}, \xi) = A_i$).

Making use of the fact that we have already ordered the outputs of the schema $\bar{\mathcal{Z}}$, we construct the $\bar{\mathcal{Z}}$ -composition of the schemata $\mathcal{Z}_1, \mathcal{Z}_2, \ldots, \mathcal{Z}_m$.   And this will be the schema $\mathcal{Z}'$.

The theorem has been proved.

5.4. T h e o r e m 3.   A s c h e m a $\mathcal{Z}$ w h i c h i s s t a t i o n a r y i n φ m a y b e t r a n s f o r m e d i n t o a s t a t i o n a r y s c h e m a e q u i v a l e n t t o i t i f

a) e a c h p r e d i c a t e j u x t a p o s e d w i t h a p-v e r t e x o f t h e s c h e m a i s f i n i t e-d i m e n s i o n e d;

b) r e g a r d l e s s o f the f i n i t e p a r t i t i o n i n g o f t h e s e t Ξ, o n e c a n f i n d a s t a t i o n a r y s c h e m a o v e r R w h i c h r e a l i z e s t h i s p a r t i t i o n i n g.

P r o o f.   In just the same way as in the proof of Theorems 1 and 2, let us indicate the method of transforming the schema $\mathcal{Z}$ into a schema $\mathcal{Z}'$, which is equivalent to it and has a smaller number than $\mathcal{Z}$, of vertices which are nonstationary in ψ. This will prove the theorem.

Let us consider the vertex α of the schema $\mathcal{Z}$; which is nonstationary in ψ; this vertex will be an A-vertex or a p-vertex, and it will be stationary in φ for both cases.

Assume α is an A-vertex; then in each situation ξ exactly one arc issues from it. Let us use $\Xi(\alpha, \alpha')$, $\alpha' \in a$, to denote the set of all situations (and only them) such that $\Pi(\alpha, \xi) = \alpha'$.

From among the sets $\Xi(\alpha, \alpha')$, $\alpha' \in a$, let us isolate all nonempty words:

$$\Xi(\alpha, \alpha_1), \quad \Xi(\alpha, \alpha_2), \ldots, \Xi(\alpha, \alpha_m).$$

It is obvious that their number m exceeds 1, and in aggregate they have the properties

$$\Xi(\alpha, \alpha_i) \cap \Xi(\alpha, \alpha_j) = \phi \quad \text{for} \quad \alpha_i \neq \alpha_j,$$
$$\bigcup_{i=1, 2, \ldots, m} \Xi(\alpha, \alpha_i) = \Xi.$$

Consequently, by assumption b) one can find a stationary schema over R (we shall denote it by $\mathcal{Z}$), which realizes the partitioning of the set Ξ into $\Xi(\alpha, \alpha_i)$, $i = 1, 2, \ldots, m$. Without loss of generality, it may be assumed that $\bar{\mathcal{Z}}$ has the vertices $\alpha^{(1)}, \alpha^{(2)}, \ldots, \alpha^{(m)}$ as its outputs and

$$\Xi_{\alpha^{(i)}} = \Xi(\alpha, \alpha_i), \quad i = 1, 2, \ldots, m.$$

Having the schemata $\mathcal{Z}$ and $\bar{\mathcal{Z}}$, available, we synthesize the schema $\mathcal{Z}'$. For this purpose:

1. We combine α vertices of the schemata $\mathcal{Z}$ and $\bar{\mathcal{Z}}$ into one set.

2. The input and outputs of the schema $\mathcal{Z}'$ are designated as the input and output of the schema $\mathcal{Z}$.

3. In each situation ξ we execute the following operations:

a) we combine the arcs of the schemata $\mathcal{Z}$ and $\bar{\mathcal{Z}}$ into one set;

b) we discard the arc issuing from the vertex α;

c) we bring out an arc from α to the first vertex of the schema $\bar{\mathcal{Z}}$;

d) for each i = 1, 2, ..., m all arcs arriving at the output $\alpha^{(i)}$ of the schema $\bar{\mathcal{Z}}$, are directed to the vertex α_i of the schema $\mathcal{Z}$;

e) we eliminate the input and outputs of the schema $\bar{\mathcal{Z}}$.

The resultative schema $\mathcal{Z}'$ is our desired schema for the case in which α is an A-vertex.

Assume α is a p-vertex, and that the predicate p juxtaposed with α takes the values

$$u_1, u_2, \ldots, u_t$$

[we use proposition (a)].

Let us specify a certain u_i $(1 \le i \le t)$; we use $\Xi_i(\alpha, \alpha')$ to designate the set of all those situations ξ (and only them) for which the following statement is valid: in the situation ξ the arc (α, α') carrying the tag u_i issues from the vertex α. Among the sets $\Xi_i(\alpha, \alpha')$, $\alpha' \in \mathbf{a}$, let us enumerate all nonempty ones:

$$\Xi_i(\alpha, \alpha_{i1}), \quad \Xi_i(\alpha, \alpha_{i2}), \ldots, \Xi_i(\alpha, \alpha_{im_i}).$$

It is clear that

$$\Xi_i(\alpha, \alpha_{ij}) \cap \Xi_i(\alpha, \alpha_{ik}) = \phi, \quad \text{if} \quad j \neq k,$$
$$\bigcup_{j=1, 2, \ldots, m_i} \Xi_i(\alpha, \alpha_{ij}) = \Xi.$$

Consequently, according to proposition b) one can find a stationary schema $\bar{\mathcal{Z}}_i$ over R which realizes the partitioning of Ξ into $\Xi_i(\alpha, \alpha_{ij})$, j = 1, 2, ..., m. It can be assumed that $\gamma_{i1}, \gamma_{i2}, \ldots, \gamma_{im_i}$ are all outputs of the schema $\bar{\mathcal{Z}}_i$, and

$$\Xi_{\gamma_{ij}} = \Xi_i(\alpha, \alpha_{ij}), \quad j = 1, 2, \ldots, m_i.$$

Assume that for each u_i we have found the schema $\bar{\mathcal{Z}}_i$ over R having the property described above. Let us synthesize from the schemata $\mathcal{Z}$ and $\bar{\mathcal{Z}}_i$, i = 1, 2, ..., t, the schema $\mathcal{Z}'$, which gives the proof of the theorem. For this purpose:

1) we combine all vertices of the schemata $\mathcal{Z}$ and $\bar{\mathcal{Z}}_1, \bar{\mathcal{Z}}_2, \ldots, \bar{\mathcal{Z}}_t$ into one set;

2) the input and outputs of the schema $\mathcal{Z}'$ are called the input and outputs of the schema $\mathcal{Z}$;

3) in each situation ξ we do the following:

a) we combine the arcs of the schemata $\mathcal{Z}$ and $\bar{\mathcal{Z}}_1, \bar{\mathcal{Z}}_2, \ldots, \bar{\mathcal{Z}}_t$ into one set;

b) we discard all arcs issuing from the vertex α;

c) for each i = 1, 2, ..., t we bring out an arc tagged by the element u_i from the vertex α and direct it to the first vertex of the schema $\bar{\mathcal{Z}}_i$;

d) for each i = 1, 2, ..., t and each j = 1, 2, ..., m we execute the following operation: all arcs arriving at the output γ_{ij} of the schema $\bar{\mathcal{Z}}_i$ are directed to the vertex α_{ij};

e) we discard all inputs and outputs of the schemata $\bar{\mathcal{Z}}_i$, i = 1, 2, ..., t.

The theorem has been proved.

§ 6. Stationary Schemata

6.1. Let us show that for a nonempty set R one can construct a stationary schema over R whose definition domain is empty.

Actually, let us consider a schema $\mathcal{Q}_\Lambda$

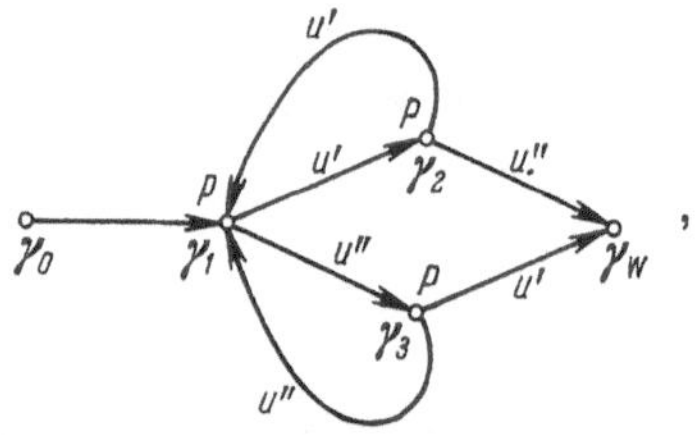

in which all internal vertices (γ_1, γ_2, γ_3) are juxtaposed with one and the same predicate p, while the sets $U' \subseteq U$ and $U'' \subseteq U$ (each of them combines those elements of the set U which tag one and the same arc of the schema) satisfy the conditions

$$U' \cup U'' = U, \quad U' \cap U'' = \phi.$$

It can easily be checked that when this schema is executed in any situation ξ the sequence of vertices we have mentioned has one of the two possible forms:

1) $\gamma_1, \gamma_2, \gamma_1, \gamma_2, \ldots, \gamma_1, \gamma_2, \ldots$,

2) $\gamma_1, \gamma_3, \gamma_1, \gamma_3, \ldots, \gamma_1, \gamma_3, \ldots$;

the validity of our statement derives from this.

Note that the schema $\mathcal{Q}_\Lambda$ is directionally connected.

6.2. For stationary schemata over nonempty sets E and R the following theorems are valid.

Theorem 4. *An arbitrary stationary schema $\mathcal{Q}$ may be transformed into its equivalent connected schema.*

Proof. Assume

$$\mathcal{Q} = [\alpha_0, \mathfrak{a}_w, \{\Pi(\alpha), \varphi(\alpha), \psi(\alpha), \alpha \in \mathfrak{a}\}]$$

is a stationary schema. Let us consider the set $\mathfrak{a}^* \subseteq \mathfrak{a}$, which consists of all such vertices α of the schema $\mathcal{Q}$ (and only them) for which one can construct a chain beginning at the vertex α_0 and ending at α. Let us distinguish between the cases:

1) $\mathfrak{a}^*$ contains just one output of the schema $\mathcal{Q}$;

2) $\mathfrak{a}^*$ does not contain a single output of the schema $\mathcal{Q}$.

In the first case one may choose the schema

$$\mathcal{Q}^* = [\alpha_0, \mathfrak{a}_w^*, \{\Pi(\alpha), \varphi(\alpha), \psi(\alpha), \alpha \in \mathfrak{a}^*\}],$$

as the desired connected schema, where $\mathfrak{a}_w^* = \mathfrak{a}^* \cap \mathfrak{a}_w$; $\mathcal{Q}^*$ is obviously equivalent to the schema $\mathcal{Q}$.

In the second case the definition domain of the schema $\mathcal{Q}$ will be empty, and therefore the connected schema $\mathcal{Q}_\Lambda$ will be weakly equivalent to the schema $\mathcal{Q}$.

6.3. Theorem 5. *For each connected schema $\mathcal{Q}$ one may construct a directionally connected schema which is an expansion of the schema $\mathcal{Q}$.*

Proof. Let us construct the set $a^* \subseteq a$, while including in it all such vertices $\alpha \in a$, through which at least one path passes which connects the input of the schema $\mathfrak{A}$ with one of its outputs.

If the set a^* is empty,† then the schema $\mathfrak{A}$ realizes an empty action, and therefore the schema $\mathfrak{A}_\Lambda$ which is weakly equivalent to it may be considered as an expansion of the schema $\mathfrak{A}$.

Assume a^* is not empty; let us construct the required schema $\mathfrak{A}^*$;

1) the set a^* is considered to be the set of vertices of $\mathfrak{A}^*$;

2) the input of the schema $\mathfrak{A}^*$ is called the input of the schema $\mathfrak{A}$; the outputs of $\mathfrak{A}^*$ are called those outputs of the schema $\mathfrak{A}$, which are contained in a^*, and in addition the vertex α_w^*, where $\alpha_w^* \bar{\in} a$;

3) we preserve all of the arcs of the schema $\mathfrak{A}$ (with their tags) which issue from the vertices of the set a^* and arrive at the vertices of this same set;

4) each arc issuing from the vertex $\alpha \in a^*$ and arriving at a vertex which is not contained in a^*, shall be directed to a complementary output α_w^* of the schema $\mathfrak{A}^*$, without changing the tag of this arc.

The theorem has been proved.

6.4. Theorem 6. Each connected schema $\mathfrak{A}$ may be transformed into a directionally connected schema $\mathfrak{A}^*$ which is weakly equivalent to it.

For the proof it is sufficient to carry out the following constructions:

1) combine the set of vertices a^* with the vertices of the schema $\mathfrak{A}_\Lambda$;

2) designate the input of the schema $\mathfrak{A}^*$ as the input of the schema $\mathfrak{A}$, the outputs of the schema $\mathfrak{A}^*$ as all those outputs of the schema $\mathfrak{A}$, which are contained in the set a^*, and the output γ_w of the schema $\mathfrak{A}_\Lambda$;

3) combine all arcs of $\mathfrak{A}$, which issue from the vertices a^* and arrive at the vertices of the same set and the arcs of the schema $\mathfrak{A}_\Lambda$ into one set;

4) direct each arc issuing from a vertex $\alpha \in a^*$ and arriving at a vertex which is not contained in a^*, to the first vertex of the schema $\mathfrak{A}_\Lambda$;

5) discard the input γ_0 of the schema $\mathfrak{A}_\Lambda$ [along with the arc (γ_0, γ_1)].

6.5. Theorem 7. Each directionally connected schema $\mathfrak{A}$ having one output may be transformed into a linear schema which is equivalent to it, provided only that the set R contains an everywhere-definite predicate which does not take even one value $u_0 \in U$.

† This case will occur if $\mathfrak{A}$ is a schema

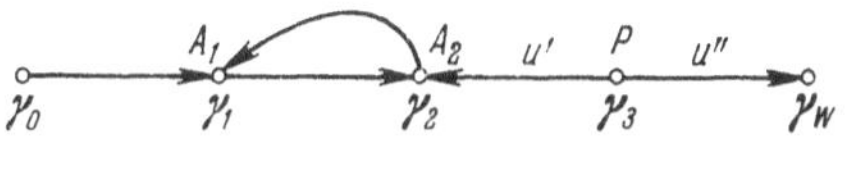

;

here A_1 and A_2 are actions, and p is a predicate on situations; $U' \subseteq U$ and $U'' \subseteq U$ satisfy the conditions

$$U' \cup U'' = U, \quad U' \cap U'' = \phi.$$

Proof. Assume

$$\mathscr{P} = [\alpha_0, \alpha_w, \{\Pi(\alpha), \varphi(\alpha), \psi(\alpha), \alpha \in \mathbf{a}\}]$$

is a directionally connected schema with one input, and

$$S = [\alpha_0, \alpha_w, \{\Pi(\alpha), \alpha \in \mathbf{a}\}]$$

is the net of this schema.

If ρ is a certain elementary path in the net S, then each vertex belonging to the path ρ and differing from its beginning and end is called internal.

The number of arcs forming the path ρ is called its length.

Two paths of the set S, by definition, do not cross if not a single internal vertex of one path is an internal vertex of the other path.

Two paths having identical beginnings and identical ends are called contiguous.

Assume K is the set of elementary paths of the net S which are disjoint. The elementary path ρ which does not cross any other path of the set K and is such that the beginning and ends of ρ belong to certain paths from K is called based on K.

Let us show that for the net S one can construct a sequence of elementary paths

$$K = \rho_0, \rho_1, \rho_2, \ldots, \rho_m$$

which is such that

a) no paths ρ_i and ρ_j, $i \neq j$, cross;

b) for any $i = 1, 2, \ldots, m$, the path ρ_i is based on the set

$$\{\rho_0, \rho_1, \ldots, \rho_{i-1}\};$$

c) each vertex of the net S belongs to at least one path of the set K.

Let us begin the construction of the sequence K with the selection of the path ρ_0.

As ρ_0 we shall take any elementary path on the net S, which connects the input α_0 with the output α_w.

Let us use $\mathbf{a}_0$ to denote the set of vertices belonging to the path ρ_0.

If $\mathbf{a}_0 = \mathbf{a}$, then the path ρ_0 exhausts the desired sequence K.

Let us assume that $\mathbf{a}_0 \neq \mathbf{a}$, and let us note that the vertices of the set $\mathbf{a}_0$ are naturally ordered in the path ρ_0.

Let us use K_1^* to denote the set of all elementary paths of length greater than unity which are based on the path ρ_0. It is evident that K_1^* is not empty if $\mathbf{a}_0 \neq \mathbf{a}$.

Let us introduce partial order in the set K_1^*. We shall say that $\rho' \in K_1^*$ precedes $\rho'' \in K_1^*$ $(\rho' < \rho'')$, if one of the following two conditions holds:

1) the beginning of the path ρ' has a lower number than the beginning of the path ρ'';

2) the numbers of the beginnings of the paths ρ' and ρ'' coincide, but the number of the end of the path ρ' is lower than the number of the end of the path ρ''.

Let us consider the maximum sequence of paths from K_1^*

$$\rho_1, \rho_2, \ldots, \rho_l, \tag{3}$$

which satisfies the conditions:

1) $\rho_1 \prec \rho_2 \prec \ldots \prec \rho_l$;

2) ρ_1 is a minimal element of the partially ordered set K_1^* ;

3) for all $i = 1, 2, \ldots, l$ the path ρ_i does not cross any of the paths $\rho_1, \rho_2, \ldots, \rho_{i-1}$.

It is evident that in the general case the sequence (3) may be constructed ambiguously, but its length l remains unchanged.

Let us attach the following in order at the right to the sequence (3):

all paths contiguous with ρ_1;

all paths contiguous with ρ_2;

. . . .

all paths contiguous with ρ_l

(the order established among the paths which are contiguous with a certain ρ_i, $1 \le i \le l$ may be arbitrary); the sequence of paths obtained shall be denoted by $K_1^!$.

Finally, to the sequence K_0 consisting of the path ρ_0 we attach the sequence $K_1^!$ at the right, and use K_1 to denote the resultative sequence. It is obvious that it satisfies the two requirements imposed on the sequence K.

Let us use $\mathfrak{a}_1$ to denote the set obtained by the union of all vertices through which paths from K_1 pass.

If $\mathfrak{a}_1 = \mathfrak{a}$, then the sequence K_1 is the one desired.

If $\mathfrak{a}_1 \neq \mathfrak{a}$, then we shall construct a sequence of paths K_2 in a manner completely analogous to the manner in which the sequence K_1 was constructed. For this purpose it is first necessary to order the vertices of the set $\mathfrak{a}_1$. Note that $\mathfrak{a}_1$ is derived from $\mathfrak{a}_0$ by adding internal vertices of all paths of the sequence $K_1^!$. The vertices of the set $\mathfrak{a}_0$ have already been ordered; let us assume that they are precursors of all the remaining vertices of the set $\mathfrak{a}_1$ we also order these remaining ones.

If the vertices belong to the same path, then we preserve the order of their sequence in the path.

Of two vertices belonging to different paths we call the precursor that one which is contained in the path having the lower number.

Thus, the set $\mathfrak{a}_1$ is ordered, and one may go over to constructing the set K_2^* and ordering its elements.

The procedure described terminates as soon as for a certain $i \ge 1$ the set of vertices $\mathfrak{a}_i$ coincides with $\mathfrak{a}$.

Having available the sequence K, let us carry out the following transformation of the schema $\mathcal{C}$.

Assume p is a predicate on situations having a definition domain Ξ, which does not take the value u_0.

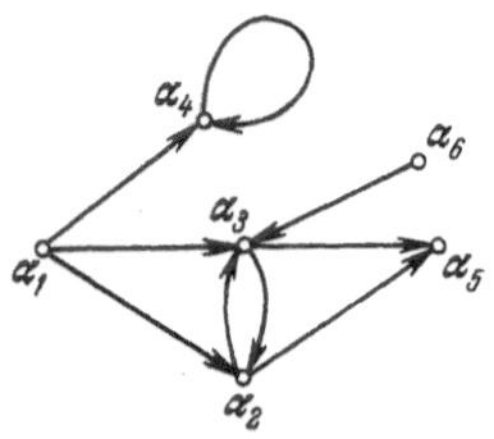

Fig. 1

1. To the set of vertices a we add the vertices

$$\beta_0, \beta_1, \ldots, \beta_m,$$

each of which is juxtaposed with a predicate p.

2. The input and output of the future schema remain α_0 and α_w.

3. For each i = 0, 1, ..., m we do the following:

a) all arcs arriving at the end of the path ρ_i are directed to the vertex β_i without any change of their tags;

b) from the vertex β_i we bring out two arcs: one, which is tagged by the element u_0, is directed to the first internal vertex of the path ρ_{i+1} (for i = m it is directed to the vertex α_w); the second, which is tagged by all of the remaining elements in the set U, is directed to the end of the path ρ_i. It is easy to verify the fact that the designed schema is linear. Actually, assume the path ρ_i passes through the vertices

$$\alpha_{i0}, \alpha_{i1}, \alpha_{i2}, \ldots, \alpha_{it_i}, \quad i=0, 1, 2, \ldots, m.$$

Then the sequence of vertices

$$\begin{aligned} \alpha_0=\alpha_{00}, \alpha_{01}, \alpha_{02}, \ldots, \alpha_{0,t_0-1}, \beta_0, \\ \alpha_{11}, \alpha_{12}, \ldots, \alpha_{1,t_1-1}, \beta_1, \\ \ldots\ldots\ldots\ldots \\ \alpha_{m1}, \alpha_{m2}, \ldots, \alpha_{m,t_m-1}, \beta_m, \alpha_w=\alpha_{0t_0}, \end{aligned}$$

which contains all the vertices of the schema, constitutes an elementary path. The equivalence of the original and resultative schema is obvious. The theorem has been proved.

§ 7. Examples

7.1. As has already been assumed in subsection 3.6, the vertices of a graph are depicted by points on a plane and the arcs are depicted by directional curves connecting the vertices of the graph.

Figure 1 depicts the graph $\Gamma = \{\Pi(\alpha)\ \alpha \in \mathsf{a}\}$, where

$$\begin{gathered} \mathsf{a}=\{\alpha_1, \alpha_2, \alpha_3, \alpha_4, \alpha_5, \alpha_6\}, \\ \Pi(\alpha_1)=\{\alpha_2, \alpha_3, \alpha_4\}, \ \Pi(\alpha_2)=\{\alpha_3, \alpha_5\}, \\ \Pi(\alpha_3)=\{\alpha_2, \alpha_5\}, \\ \Pi(\alpha_4)=\{\alpha_4\}, \quad \Pi(\alpha_5)=\phi, \quad \Pi(\alpha_6)=\{\alpha_3\}. \end{gathered}$$

The graph Γ has nine arcs; here the arcs (α_1, α_2), (α_1, α_3), and (α_1, α_4) issue from the vertex α_1, the arc (α_4, α_4) issues from the vertex α_4, and no arc issues from the vertex α_5; at least one arc of the graph enters all vertices except α_1 and α_6.

The sequence of arcs

$$(\alpha_6, \alpha_3), \quad (\alpha_3, \alpha_2), \quad (\alpha_2, \alpha_3), \quad (\alpha_3, \alpha_5);$$

may serve as an example of a path in the graph Γ; this path passes through the vertices α_6, α_3, α_2, α_3, α_5 and contains the contour (α_3, α_2), (α_2, α_3). Assume (α_1, α_2), (α_2, α_3), (α_3, α_5) is elementary.

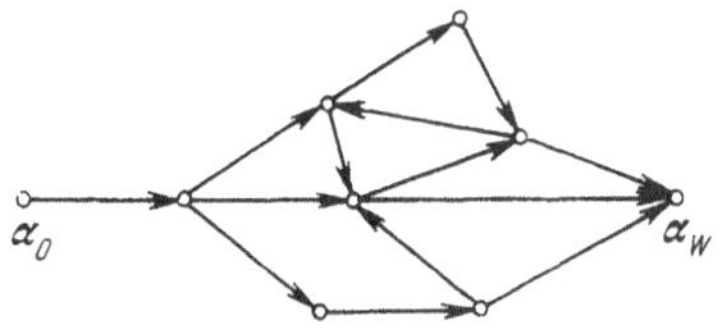

Fig. 2

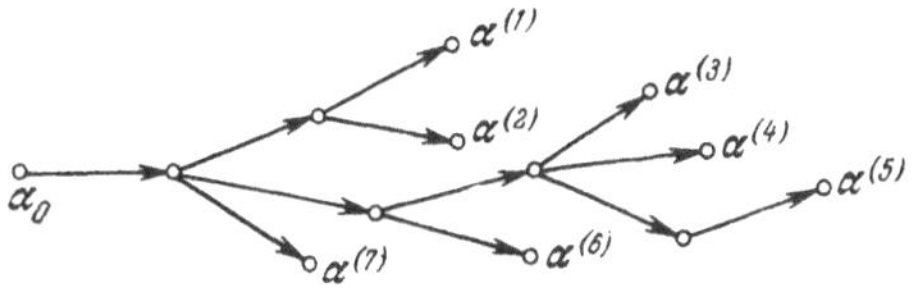

Fig. 3

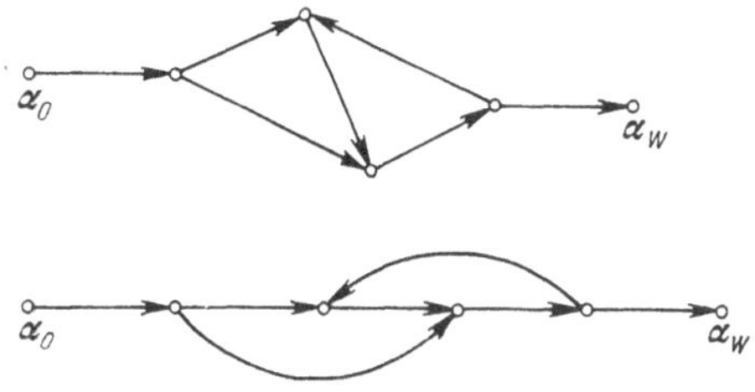

Fig. 4

There are eight edges in the graph Γ: the vertices α_2 and α_3 are connected by one edge $\Delta = \{\alpha_2, \alpha_3\}$.

The graph Γ is connected, since any two of its vertices may be connected by a chain. For example, the vertices α_6 and α_4 are connected by the chain

$$\{\alpha_6, \alpha_3\}, \quad \{\alpha_3, \alpha_1\}, \quad \{\alpha_1, \alpha_4\}, \quad \{\alpha_4, \alpha_4\},$$

while the vertices α_6 and α_6 are connected by the chain

$$\{\alpha_6, \alpha_3\}, \quad \{\alpha_3, \alpha_6\}.$$

If in the graph Γ the vertex α_6 (not a single arc of the graph enters it, and exactly one arc issues from it) is called the input, while the vertex α_5 (not a single arc of the graph issues from it) is called the output, then we obtain the net

$$S = [\alpha_6, \alpha_5, \{\Pi(\alpha), \alpha \in \mathfrak{a}\}];$$

the vertices α_1, α_2, α_3, and α_4 are internal.

The net S will not be directionally connected; actually no such path exists which would begin at the input, end at the output, and pass through the vertex α_1.

We present examples of a directionally connected net (Fig. 2), a net-tree (Fig. 3), and a linear net (Fig. 4); each of the nets is stipulated geometrically; only the input and output of the net are provided with designations. In Fig. 4, one and the same linear net is given in two depictions; the second of them clarifies the appearance of the term "linear."

7.2. Let us present an example of a schema on situations.

Assume the mapping $\sigma(\xi)$ is such that it partitions the set of all situations Ξ into four subsets

$$\Xi^{(1)}, \ \Xi^{(2)}, \ \Xi^{(3)}, \ \Xi^{(4)},$$

which are disjoint, and juxtaposes all situations of an individual set $\Xi^{(i)}$ with one and the same (but its own for each set) net S_i, i = 1, 2, 3, 4.

Let us recall that the set of vertices of a schema, and likewise its input and output, do not depend on the situation considered. Depicting each vertex of a schema by a point on a plane, we shall agree to preserve their mutual arrangement on the plane unchanged, regardless of the net associated with the schema which we might be describing. Then it is sufficient to give the designation of the vertices themselves on just one of the nets.

Assume that in our case

$$\mathfrak{a} = \{\alpha_0, \alpha_1, \alpha_2, \alpha_3, \alpha_w\},$$

where α_0 is the input, and α_w is the output of the schema, while Fig. 5 provides descriptions of all nets S_i, i = 1, 2, 3, 4.

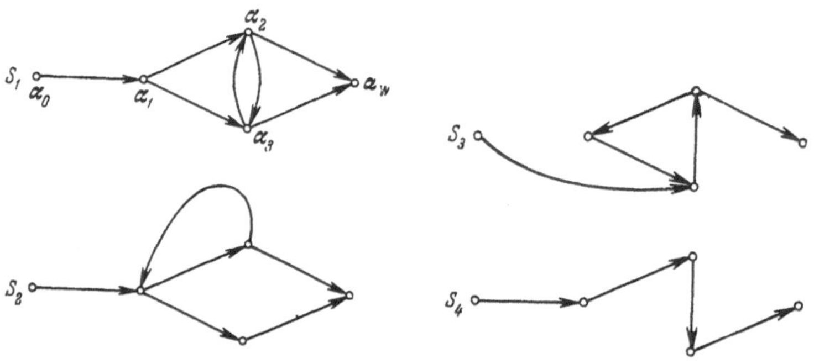

a b

Fig. 5

The mapping $\varphi(\alpha, \xi)$ is constructed as follows: each of the sets $\Xi^{(1)}$ and $\Xi^{(3)}$ is partitioned into two nonintersecting subsets

$$\Xi^{(1)} = \Xi_1 \cup \Xi_2, \quad \Xi^{(3)} = \Xi_4 \cup \Xi_5,$$

and then the entire set Ξ will constitute the sum of six disjoint subsets

$$\Xi_1, \ \Xi_2, \ \Xi_3 = \Xi^{(2)}, \ \Xi_4, \ \Xi_5, \ \Xi_6 = \Xi^{(4)}.$$

Let us assume that for each internal vertex α of the schema $\mathfrak{S}$ the following holds: the image of the pair (α, ξ) in the mapping $\varphi(\alpha, \xi)$ is determined solely by the choice of the vertex α and by which subset Ξ_i the situation ξ belongs to (i.e., the image of the pair (α, ξ) does not change while the situation ξ traverses the subset Ξ_i).

Consequently, the mapping $\varphi(\alpha, \xi)$ may be described thus: on each of the subsets Ξ_i, $1 \le i \le 6$, it is required to consider the net juxtaposed with the situations of this subset, and each internal vertex of the net is tagged with the action or predicate juxtaposed with it. Such a description of $\varphi(\alpha, \xi)$ is given in Fig. 6a; the symbols A_1, A_2, A_3 denote certain actions, while the symbols p, q, r denote certain predicates on situations.

Let us now postulate that the set U consists of two elements 0 and 1, and let us make use of the already available partitioning of Ξ into the subsets Ξ_i, i = 2, 2, ..., 6 for stipulating the mappings $\psi(\alpha, \xi)$.

Let us suppose that $\psi(\alpha, \xi)$ is determined solely by the choice of the vertex α and by those of the sets Ξ_i to which the considered situation ξ belongs. Then it is sufficient to use the elements 0 and 1 to tag the arcs of the nets juxtaposed with the set Ξ_i in order to describe the mappings $\psi(\alpha, \xi)$ acting under these conditions in accordance with the definition given in 2.4. Such tagging of the arcs is displayed in Fig. 6a.

Thus, Fig. 6a contains complete information on the schema $\mathfrak{S}$. Note that the first vertex of the schema $\mathfrak{S}$ is α_3 if the situation ξ belongs to the set $\Xi^{(3)}$ and α_1 in all remaining cases.

7.3. In executing the schema $\mathfrak{S}$ in a stipulated situation ξ it is necessary in the first place to establish which of the sets Ξ_i the situation ξ belongs to.

Assume, for example, $\xi \in \Xi_1$; then the following cases are possible:

1) the process of executing the schema $\mathfrak{S}$ is completed without result on the first step, since $\xi \notin \Xi_p$, where Ξ_p is the definition domain of the predicate p; the value of the schema $\mathfrak{S}$ in this case will be an empty sequence;

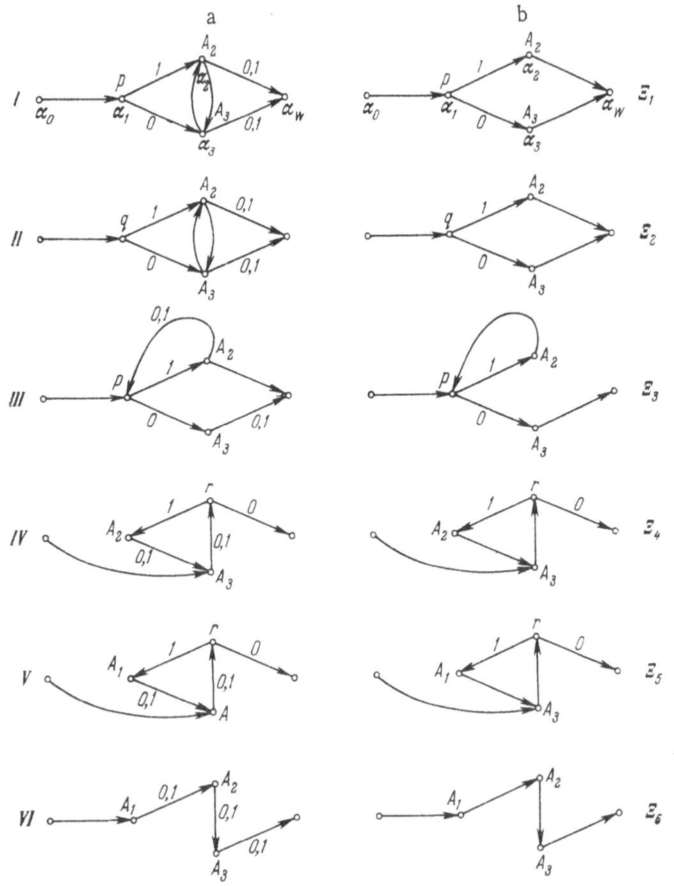

Fig. 6

2) the process of executing the schema $\mathfrak{A}$ is completed without result on the second step; this occurs when $\xi \in \Xi_p$ and one of the two following conditions holds:

a) $p(\xi) = 1$ and $\xi \bar{\in} \Xi_{A_2}$;

b) $p(\xi) = 0$ and $\xi \bar{\in} \Xi_{A_3}$;

here Ξ_{A_2} and Ξ_{A_3} denote the definition domains of the actions A_2 and A_3; the value of the schema $\mathfrak{A}$ in the case a) will be the action A_2, while in the case b) it will be the action A_3;

3) the process of executing the schema $\mathfrak{A}$ is terminated with a result on the third step, the sequences (1) and (2) having the following form:

$$\left.\begin{array}{ll} \text{a) } \xi,\ \xi,\ A_2(\xi); & (1) \\ \quad \alpha_1,\ \alpha_2,\ \alpha_w; & (2) \end{array}\right\} \text{ for } p(\xi) = 1;$$

$$\left.\begin{array}{ll} \text{b) } \xi,\ \xi,\ A_3(\xi); & (1) \\ \quad \alpha_1,\ \alpha_3,\ \alpha_w; & (2) \end{array}\right\} \text{ for } p(\xi) = 0;$$

in case a) the terminal situation is $A_2(\xi)$, and the value of the schema consists of the action A_2; in case b) the situation $A_3(\xi)$ is the terminal one, and the value of the schema consists of the action A_3.

However, if the original situation $\xi \in \Xi_6$ and the superposition

$$A_3(A_2(A_1(\xi)))$$

is meaningful, then the process of executing the schema $\mathcal{Q}$ in the situation ξ leads to the construction of the sequences

$$\xi, \quad A_1(\xi), \quad A_2(A_1(\xi)), \quad A_3(A_2(A_1(\xi))); \tag{1}$$

$$\alpha_1, \quad \alpha_2, \quad \alpha_3, \quad \alpha_w; \tag{2}$$

here the terminal situations will be $A_3(A_2(A_1(\xi)))$, while the value of the schema will be the sequence of actions

$$A_1, \; A_2, \; A_3.$$

The process of executing the schema $\mathcal{Q}$ may also be infinite. Actually, suppose

a) $\Xi_{p1} \cap \Xi_3 \neq \phi$, where Ξ_{p1} is the set of all those situations on which the predicate p takes the value 1;

b) the action A_2 maps the set $\Xi_{p1} \cap \Xi_3$ in itself. Then for execution of the schema $\mathcal{Q}$ in the situation $\xi \in \Xi_{p1} \cap \Xi_3$ we shall inspect the vertices α_1 and α_2 endlessly, going over from α_1 to α_2 and returning from α_2 to α_1. The value of the schema $\mathcal{Q}$ in this case will be the infinite sequence

$$A_2, \; A_2, \; \ldots, \; A_2, \; \ldots$$

In order to describe the definition domain $\Xi_{\mathcal{Q}}$ of the schema $\mathcal{Q}$ and the action $A_{\mathcal{Q}}$ realized by the schema, it is necessary to concretize the choice of the actions A_1, A_2, A_3 and the predicates p, q, r. One thing is obvious here: since the schema $\mathcal{Q}$ has one output α_w, the partitioning of $\Xi_{\mathcal{Q}}$ realized by it consists of just this set $\Xi_{\mathcal{Q}}$.

7.4. Let us use the classification of schema introduced in § 3, and let us define the type of schema $\mathcal{Q}$.

First of all let us note that the schema $\mathcal{Q}$ is not reduced in ψ. Actually, in situations $\xi \in \Xi^{(1)}$ the arcs (α_2, α_3) and (α_3, α_2), which are not tagged by the elements of the set U, issue from the vertices α_2 and α_3, while in situations $\xi \in \Xi^{(2)}$ the untagged arc (α_2, α_w) issues from the vertex α_2.

Figure 6b gives the schema $\mathcal{Q}_1$, reduced in ψ, which differs from the schema $\mathcal{Q}$ solely by the elimination of the arcs enumerated above. The schema $\mathcal{Q}_1$ is obviously equivalent (strongly) to the schema $\mathcal{Q}$.

Let us go over to the characteristic of the vertices of the schema $\mathcal{Q}_1$. Of the three internal vertices of the schema $\mathcal{Q}_1$ only the vertex α_3 is stationary in φ: in all situations $\xi \in \Xi$ it is juxtaposed with the action A_3 (i.e., α_3 is an A-vertex); the vertices α_1 and α_2 are of the mixed type. All three vertices α_1, α_2, α_3 are finite-valued and nonstationary in ψ.

Thus, the schema $\mathcal{Q}_1$ is nonstationary in ψ and strongly nonstationary in φ, which allows the transformations described in Theorems 1-3 to be demonstrated on it.

Note that the schema $\mathcal{Q}_1$ is stipuled by six schemata I-VI, each of which is reduced in ψ, stationary in φ and ψ, and desribes the schema $\mathcal{Q}_1$ on one of the sets Ξ_i, i = 1, 2, ..., 6.

In describing schemata which are reduced in ψ let us agree to leave arcs issuing from A-vertices and from the input without tags, as has already been done in Fig. 6b.

7.5. Postulating the schema $\mathfrak{d}_1$ to be a schema over the sets D and R, we shall to the degree necessary impose on R the requirements which ensure fulfillment of the presumptions of Theorems 1-3.

Let us first carry out the transition from the schema $\mathfrak{d}_1$, which is strongly nonstationary in φ, to its equivalent schema $\mathfrak{d}_2$, which is weakly nonstationary in φ. (It is obvious that in order to describe the latter we shall require no more than six schemata which are stationary in φ and ψ and are relegated to the sets Ξ_i, i = 1, 2, ..., 6.)

For this purpose let us consider the mapping $\varphi(\alpha, \xi)$ in vertices of the mixed type

$$\varphi(\alpha_1, \xi) = \begin{cases} p, & \xi \in \Xi_1 \cup \Xi_3, \\ q, & \xi \in \Xi_2, \\ A_2, & \xi \in \Xi_4, \\ A_1, & \xi \in \Xi_5 \cup \Xi_6; \end{cases}$$

$$\varphi(\alpha_2, \xi) = \begin{cases} r, & \xi \in \Xi_4 \cup \Xi_5, \\ A_2, & \xi \in \Xi_1 \cup \Xi_2 \cup \Xi_3 \cup \Xi_6. \end{cases}$$

In order for the vertices α_1 and α_2 to be replaced by vertices of the type A and p, it is necessary to realize partitioning of Ξ into sets $\Xi_1 \cup \Xi_2 \cup \Xi_3$ and $\Xi_4 \cup \Xi_5 \cup \Xi_6$ in one case, and partitioning of Ξ into $\Xi_4 \cup \Xi_5$ and $\Xi_1 \cup \Xi_2 \cup \Xi_3 \cup \Xi_6$ in the other case.

Let us postulate that the predicates

$$\mu_1(\xi) = \begin{cases} 0, & \xi \in \Xi_1 \cup \Xi_2 \cup \Xi_3, \\ 1, & \xi \in \Xi_4 \cup \Xi_5 \cup \Xi_6; \end{cases}$$

$$\mu_2(\xi) = \begin{cases} 0, & \xi \in \Xi_4 \cup \Xi_5, \\ 1, & \xi \in \Xi_1 \cup \Xi_2 \cup \Xi_3 \cup \Xi_6 \end{cases}$$

belong to the set R, and let us construct weakly stationary in φ (and stationary in ψ) schemata $\mathfrak{d}'$ and $\mathfrak{d}''$.

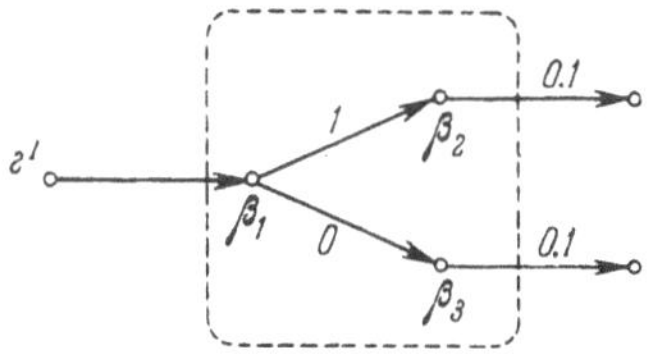

$$\varphi'(\beta_1, \xi) = \mu_1;$$

$$\varphi'(\beta_2, \xi) = \begin{cases} A_2, & \xi \in \Xi_4, \\ A_1, & \xi \in \Xi_5 \cup \Xi_6, \\ A, & \xi \in \Xi_1 \cup \Xi_2 \cup \Xi_3; \end{cases} \qquad \varphi'(\beta_3, \xi) = \begin{cases} p, & \xi \in \Xi_1 \cup \Xi_3, \\ q, & \xi \in \Xi_2, \\ \lambda, & \xi \in \Xi_4 \cup \Xi_5 \cup \Xi_6; \end{cases}$$

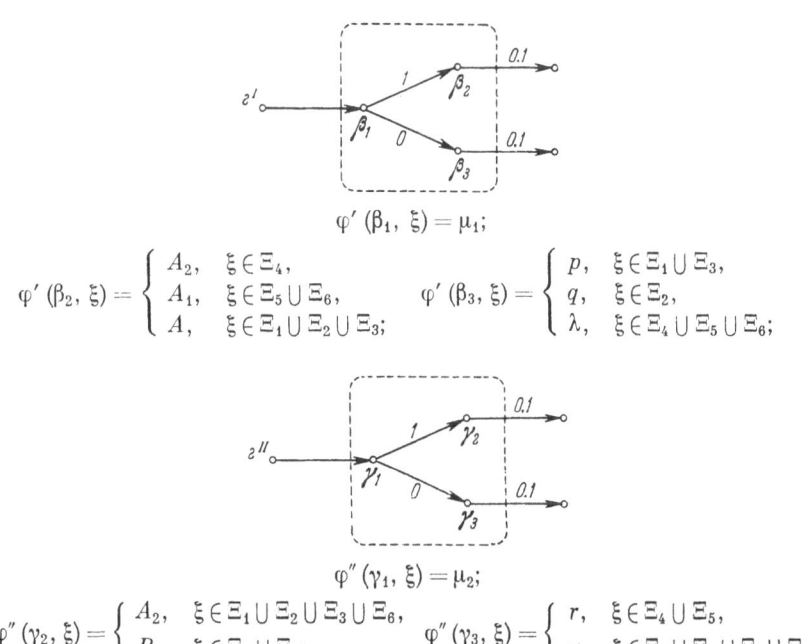

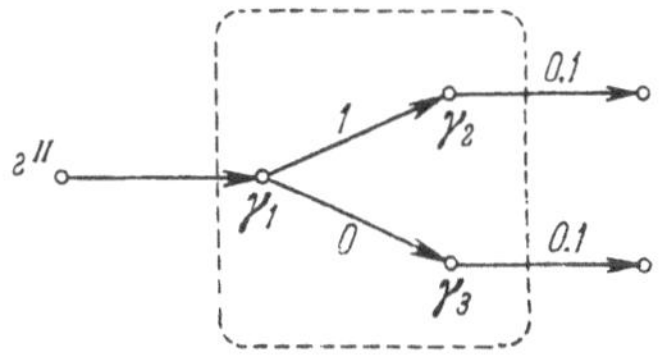

$$\varphi''(\gamma_1, \xi) = \mu_2;$$

$$\varphi''(\gamma_2, \xi) = \begin{cases} A_2, & \xi \in \Xi_1 \cup \Xi_2 \cup \Xi_3 \cup \Xi_6, \\ B, & \xi \in \Xi_4 \cup \Xi_5; \end{cases} \qquad \varphi''(\gamma_3, \xi) = \begin{cases} r, & \xi \in \Xi_4 \cup \Xi_5, \\ \tau, & \xi \in \Xi_1 \cup \Xi_2 \cup \Xi_3 \cup \Xi_6. \end{cases}$$

Here A and B are arbitrarily chosen actions, while λ and τ are arbitrarily chosen predicates.

Making use of an intentional language, it may be said that the schema $\mathcal{S}'$ executes the functions of the vertex α_1, while the schema $\mathcal{S}''$ executes the functions of the vertex α_2.

The desired schema $\mathcal{S}_2$ can be synthesized from the schemata $\mathcal{S}_1$, $\mathcal{S}'$ and $\mathcal{S}''$ according to the rules expounded in subsection 5.2. In order to trace more easily how this synthesis is achieved, rectangles are drawn around those vertices and their connecting arcs in the schemata $\mathcal{S}'$ and $\mathcal{S}''$ which must enter into the composition of the schema $\mathcal{S}_2$; these same vertices and arcs, which are already incorporated in the schema $\mathcal{S}_2$, are enclosed in rectangles there too (see Fig. 7). Considering the schemata I-VI describing $\mathcal{S}_2$, and comparing them with the schemata I-VI describing $\mathcal{S}_1$, it can easily be established that the rectangles take the place of the vertices α_1 and α_2.

7.6. The schema $\mathcal{S}_2$ is weakly stationary in φ. Let us transform it into the schema $\mathcal{S}_3$ which is equivalent to it and stationary in φ.

For this purpose we must consider all vertices of the schema $\mathcal{S}_2$ which are nonstationary in φ. In order to reduce their number we place

$$B = A_2, \ \tau = r;$$

then in the schema $\mathcal{S}_2$ only two vertices β_2 and β_3 will be nonstationary in φ. If we take

$$A = A_1, \ \lambda = p,$$

then this reduces the set of values taken by the function φ_2 (we denote the mapping of φ in the schema $\mathcal{S}_2$ this way) at the vertices β_2 and β_3. Thus,

$$\varphi_2(\beta_2, \xi) = \begin{cases} A_2, & \xi \in \Xi_4, \\ A_1, & \xi \bar{\in} \Xi_4; \end{cases}$$

$$\varphi_2(\beta_3, \xi) = \begin{cases} q, & \xi \in \Xi_2, \\ p, & \xi \bar{\in} \Xi_2. \end{cases}$$

We shall postulate that the set R contains the predicates μ_3 and μ_4 defined by the equations

$$\mu_3(\xi) = \begin{cases} 0, & \xi \in \Xi_4, \\ 1, & \xi \bar{\in} \Xi_4; \end{cases}$$

$$\mu_4(\xi) = \begin{cases} 0, & \xi \in \Xi_2, \\ 1, & \xi \bar{\in} \Xi_2. \end{cases}$$

Then the stationary (in φ and ψ) schemata $\mathcal{S}'''$ and $\mathcal{S}^{IV}$ will execute the same functions as do the vertices β_2 and β_3.

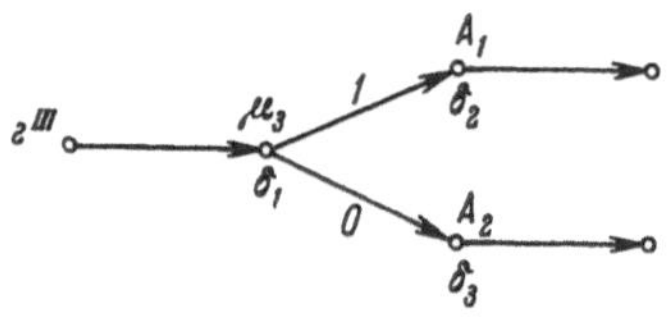

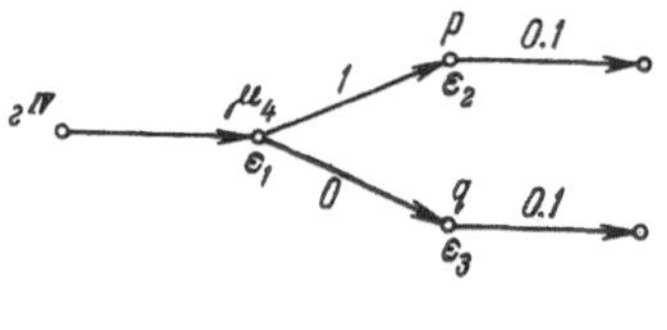

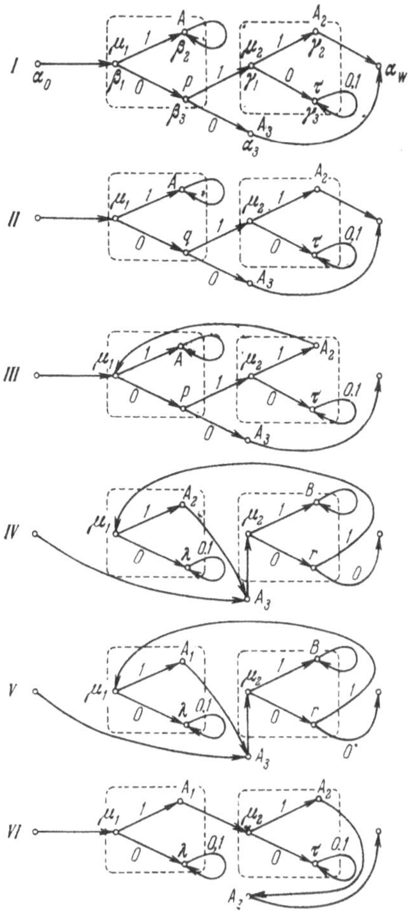

Fig. 7

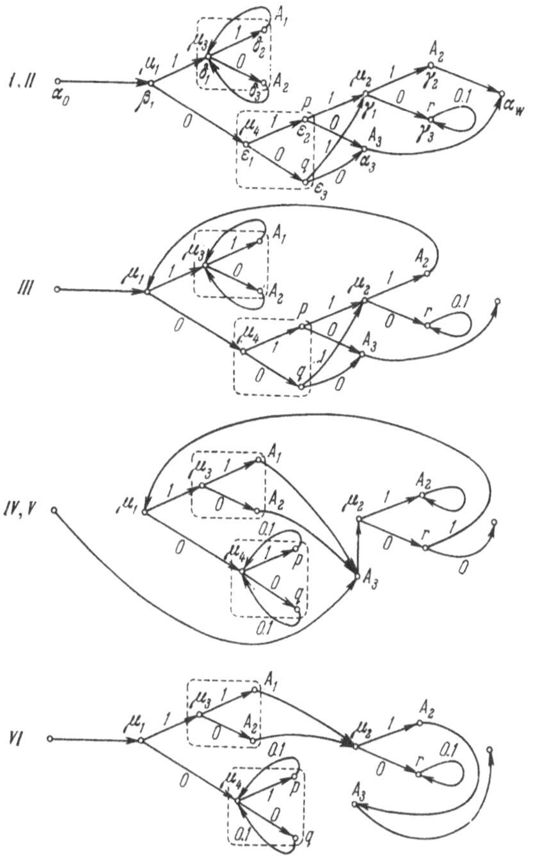

Fig. 8

Substituting the schema $\mathfrak{Z}'''$ (type A) for the A-vertex β_2, and the schema $\mathfrak{Z}^{\mathrm{IV}}$ (type p) for the p-vertex β_3 in the schema $\mathfrak{Z}_2$, we shall obtain the schema $\mathfrak{Z}_3$ which is stationary in φ (Fig. 8). $\mathfrak{Z}_3$ can be described by means of stationary (in φ and ψ) schemata of four types; the rectangles in them denote the places occupied in the schema $\mathfrak{Z}_2$ by the vertices β_2 and β_3.

7.7. In the schema $\mathfrak{Z}_3$ four vertices are stationary in ψ

$$\beta_1,\ \delta_1,\ \varepsilon_1,\ \gamma_1,$$

and the remaining ones are nonstationary in ψ.

From the A-vertices δ_2, δ_3, γ_2, and α_3 one arc issues in each situation ξ; this arc is directed as follows:

$$\text{from } \delta_2 \text{ and } \delta_3 \text{ to } \begin{cases} \delta_1 & \text{for} \quad \xi \in \Xi_1 \cup \Xi_2 \cup \Xi_3, \\ \alpha_3 & \ldots \quad \xi \in \Xi_4 \cup \Xi_5, \\ \gamma_1 & \ldots \quad \xi \in \Xi_6; \end{cases} \tag{4}$$

$$\text{from } \gamma_2 \text{ to } \begin{cases} \alpha_w & \text{for} \quad \xi \in \Xi_1 \cup \Xi_2, \\ \beta_1 & \ldots \quad \xi \in \Xi_3, \\ \gamma_2 & \ldots \quad \xi \in \Xi_4 \cup \Xi_5, \\ \alpha_3 & \ldots \quad \xi \in \Xi_6; \end{cases} \tag{5}$$

$$\text{from } \alpha_3 \text{ to } \begin{cases} \alpha_w & \text{for} \quad \xi \in \Xi_1 \cup \Xi_2 \cup \Xi_3 \cup \Xi_6, \\ \gamma_1 & \text{for} \quad \xi \in \Xi_4 \cup \Xi_5. \end{cases} \tag{6}$$

From the p-vertices ε_2, ε_3, and γ_3 arcs issue in each situation ξ which are tagged by the elements 0 and 1; an arc having the tag 0 travels

$$\text{from } \varepsilon_2 \text{ and } \varepsilon_3 \text{ to } \begin{cases} \alpha_3 & \text{for} \quad \xi \in \Xi_1 \cup \Xi_2 \cup \Xi_3, \\ \varepsilon_1 & \text{for} \quad \xi \in \Xi_4 \cup \Xi_5 \cup \Xi_6; \end{cases} \tag{7}$$

$$\text{from } \gamma_3 \text{ to } \begin{cases} \gamma_3 & \text{for} \quad \xi \in \Xi_1 \cup \Xi_2 \cup \Xi_3 \cup \Xi_6, \\ \alpha_w & \text{for} \quad \xi \in \Xi_4 \cup \Xi_5; \end{cases} \tag{8}$$

an arc with the tag 1 travels

$$\text{from } \varepsilon_1 \text{ and } \varepsilon_2 \text{ to } \begin{cases} \gamma_1 & \text{for} \quad \xi \in \Xi_1 \cup \Xi_2 \cup \Xi_3, \\ \varepsilon_1 & \text{for} \quad \xi \in \Xi_4 \cup \Xi_5 \cup \Xi_6; \end{cases} \tag{9}$$

$$\text{from } \gamma_3 \text{ to } \begin{cases} \gamma_3 & \text{for} \quad \xi \in \Xi_1 \cup \Xi_2 \cup \Xi_3 \cup \Xi_6, \\ \beta_1 & \text{for} \quad \xi \in \Xi_4 \cup \Xi_5. \end{cases} \tag{10}$$

An arc issues from the input α_0 to the vertex β_1,

$$\text{if } \xi \in \Xi_1 \cup \Xi_2 \cup \Xi_3 \cup \Xi_6,$$
$$\text{and to the vertex } \alpha_3, \quad \text{if } \xi \in \Xi_4 \cup \Xi_5. \tag{11}$$

In order to realize the partitionings which we require of the set Ξ into subsets (these partitionings are represented by the descriptions (4)-(11)), we postulate that the predicate

$$\mu_5(\xi) = \begin{cases} 0, & \xi \in \Xi_1 \cup \Xi_2, \\ 1, & \xi \in \Xi_3 \cup \Xi_4 \cup \Xi_5 \cup \Xi_6 \end{cases}$$

belongs to the set R.

Let us construct the following stationary schemata over R:

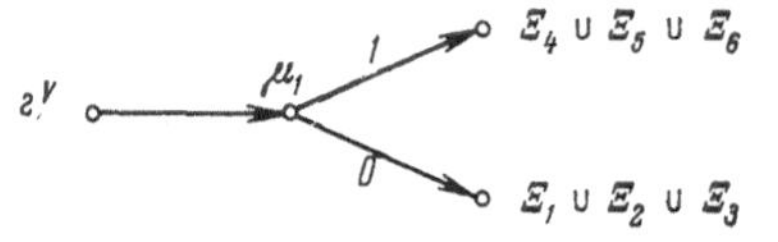

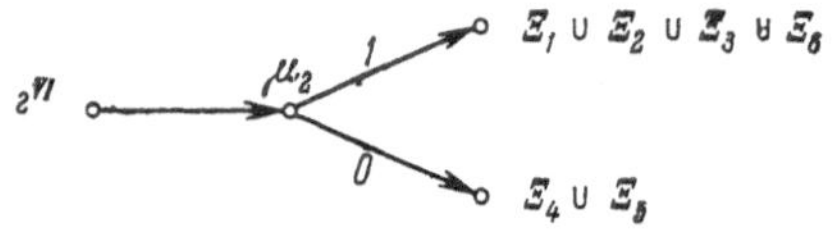

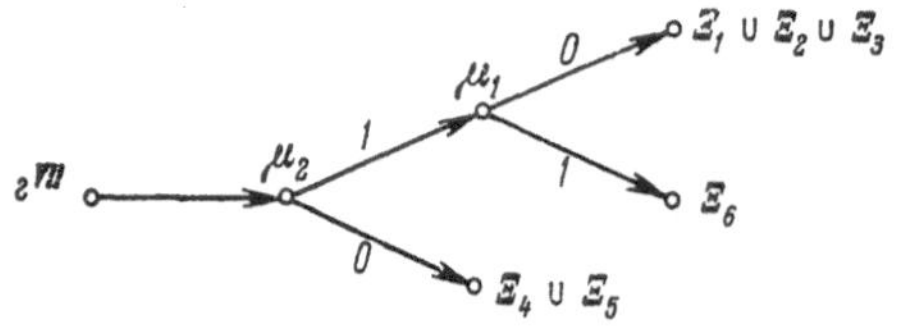

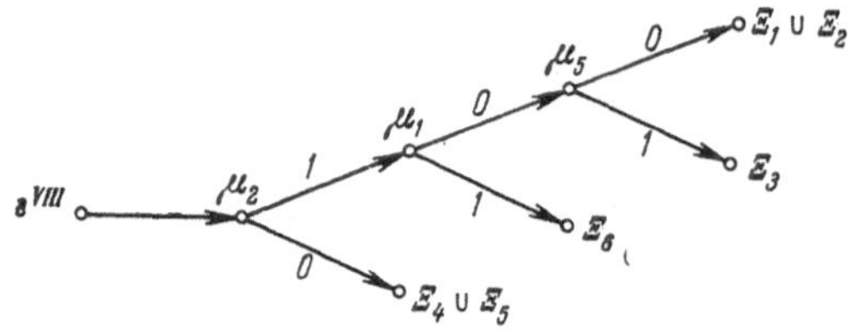

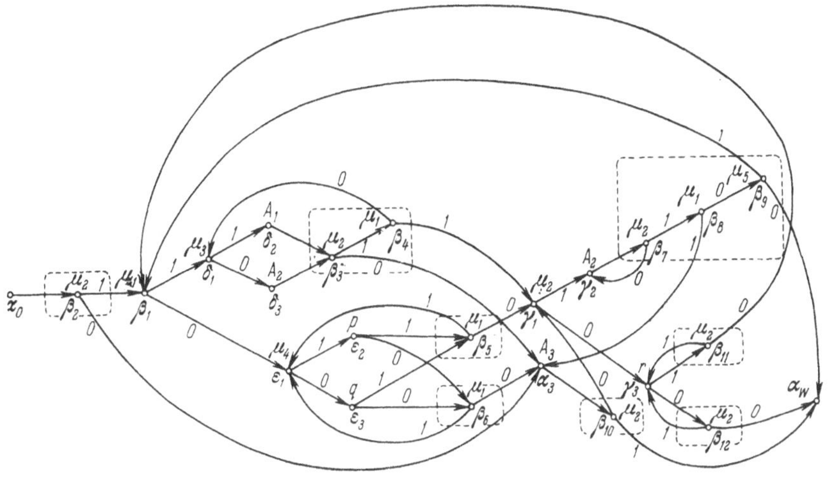

Fig. 9

Here each of the outputs of the schema is tagged by a subset consisting of all those situations in which we arrive at this output if we execute the schema.

It can easily be verified that the schema $\mathfrak{A}^{\mathrm{V}}$ realizes the parititioning specified in the description (7) and (9), the schema $\mathfrak{A}^{\mathrm{VI}}$ realizes the partitioning specified in (6), (8), (10), and (11), the schema $\mathfrak{A}^{\mathrm{VII}}$ realizes the partitioning specified in (4), and, finally, the schema $\mathfrak{A}^{\mathrm{VIII}}$ realizes the partitioning specified in the description (5).

The transformation of the schema $\mathfrak{A}_3$ into a schema $\mathfrak{A}_4$ which is equivalent to it and is stationary (in φ and ψ) consists of a series of transformations which eliminate nonstationarity in ψ of the individual vertices of the schema $\mathfrak{A}_3$.

Let us consider, for example, how nonstationarity of the vertex δ_2 is eliminated. The schema $\mathfrak{A}^{\mathrm{VII}}$ is taken which realizes the partitioning required in this case; the input of the schema $\mathfrak{A}^{\mathrm{VII}}$ is made to coincide with the actual vertex δ_2; that output of the schema $\mathfrak{A}^{\mathrm{VII}}$, which is tagged by the set $\Xi_1 \cup \Xi_2 \cup \Xi_3$, is made to coincide with the vertex δ_1 (in the schema $\mathfrak{A}_3$ the arc from δ_2 arrived at it in the situations $\xi \in \Xi_1 \cup \Xi_2 \cup \Xi_3$); the output of the schema $\mathfrak{A}^{\mathrm{VII}}$, tagged with the set $\Xi_4 \cup \Xi_5$, is made to coincide with the vertex α_3 (the arc from δ_2 in the situations $\xi \in \Xi_4 \cup \Xi_5$) arrived at it); finally, the remaining third output of the schema $\mathfrak{A}^{\mathrm{VII}}$, having the tag Ξ_6, is made to coincide with γ_1 (the arc from δ_2 in the situations $\xi \in \Xi_6$) arrived at this vertex in the schema $\mathfrak{A}_3$).

Figure 9 shows the resultative schema $\mathfrak{A}_4$. The vertices of the schemata $\mathfrak{A}^{\mathrm{V}} - \mathfrak{A}^{\mathrm{VIII}}$ which are included in it are enclosed in rectangles; it is easy to discern the fact that arcs (one or two; the latter holds in those cases in which the schema is used twice while being included in the composition of $\mathfrak{A}_4$ at only one spot) enter each such rectangle from vertices which were previously nonstationary in ψ.

7.8. The schema $\mathfrak{A}_4$ is a directionally connected schema having one output. We shall assume that the predicate

$$\mu_6(\xi) \equiv 0$$

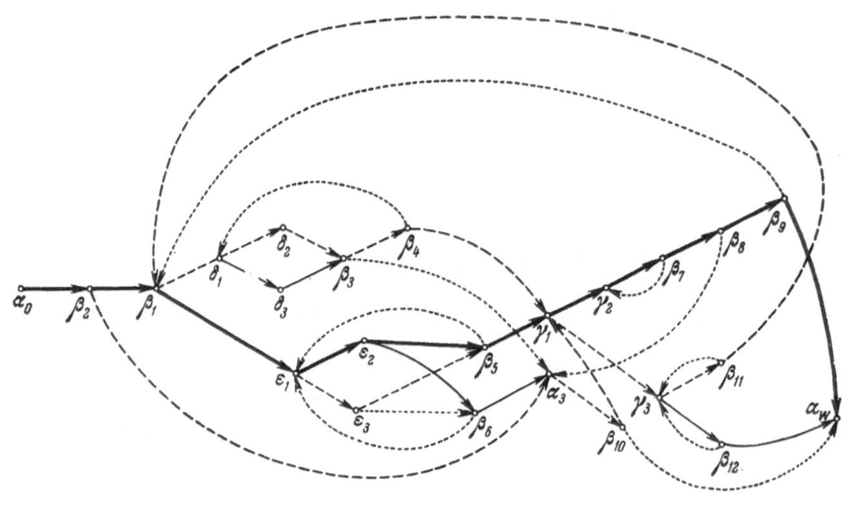

Fig. 10

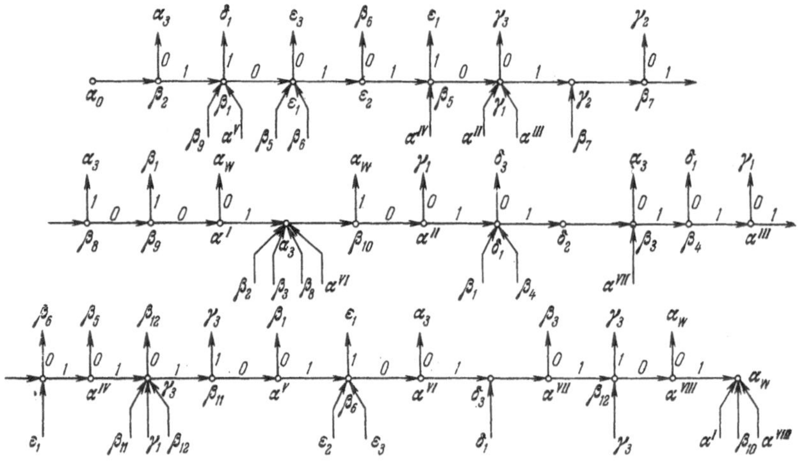

Fig. 11

is contained in the set R over which the schema $\mathcal{E}_4$ is considered (this predicate does not take the value 1). Then the premises of Theorem 7 will be fulfilled; i.e., the possibility develops of transforming the schema $\mathcal{E}_4$ into a linear schema equivalent to it.

Let us realize such a transformation.

Assume S_4 is a net on which the schema $\mathcal{E}_4$ has been constructed. For S_4 we must construct the sequence

$$K = \rho_0, \rho_1, \ldots, \rho_m$$

of elementary paths which are disjoint and are such that for each i ($1 \leq i \leq m$), the path ρ_i is based on the set of paths

$$\{\rho_0, \rho_1, \ldots, \rho_{i-1}\};$$

Each vertex of the net S_4 must belong to at least one of the paths of the sequence K.

The construction of the sequence K has been demonstrated in Fig. 10. As ρ_0 we have chosen the path which passes through the vertices

$$\alpha_0,\ \beta_2,\ \beta_1,\ \varepsilon_1,\ \varepsilon_2,\ \beta_5,\ \gamma_1,\ \gamma_2,\ \beta_7,\ \beta_8,\ \beta_9,\ \alpha_w;$$

this path is denoted in Fig. 10 by the heavy lines.

The elementary paths which are based on ρ_0 are marked on Fig. 10 by dashed lines; these are

$$\rho_1=\beta_2,\ \alpha_3,\ \beta_{10},\ \gamma_1,\quad \rho_2=\beta_1,\ \delta_1,\ \delta_2,\ \beta_3,\ \beta_4,\ \gamma_1,\ \rho_3=\varepsilon_1,\ \varepsilon_3,\ \beta_5,\ \rho_4=\gamma_1,\ \gamma_3,\ \beta_{11},\ \beta_1.$$

(Here and further on a path is stipulated by the sequence of vertices through which it passes.)

It is obvious that the paths ρ_1, ρ_2, ρ_3, and ρ_4 are disjoint and are connected by successor relations

$$\rho_1 \prec \rho_2 \prec \rho_3 \prec \rho_4.$$

We leave it to the reader to check the fact that the sequence ρ_1, ρ_2, ρ_3, and ρ_4 is the maximal one among sequences of elementary paths having the properties noted.

The paths

$$\rho_0,\ \rho_1,\ \rho_2,\ \rho_3,\ \rho_4$$

fail to contain only three vertices of the net S_4: δ_3, β_6, β_{12}. Through these vertices we draw the paths

$$\rho_5=\delta_1,\ \delta_3,\ \beta_3,\quad \rho_6=\varepsilon_3,\ \beta_6,\ \alpha_3,\quad \rho_7=\gamma_3,\ \beta_{12},\ \alpha_w.$$

The paths ρ_5, ρ_6, ρ_7 are disjoint, are based on the set of paths $\{\rho_0, \rho_1, \rho_2, \rho_3, \rho_4\}$, and are in the relation

$$\rho_5 \prec \rho_6 \prec \rho_7;$$

they are shown in Fig. 10 by the thin solid lines; all paths of length 1 are drawn in Fig. 10 using dotted lines.

The sequence K has been constructed and has the following form:

$$K=\rho_0,\ \rho_1,\ \rho_2,\ \rho_3,\ \rho_4,\ \rho_5,\ \rho_6,\ \rho_7.$$

Let us use $\mathcal{Z}_5$ to denote the linear schema which is equivalent to the schema $\mathcal{Z}_4$ and is obtained from $\mathcal{Z}_4$ by the transformations described in subsection 6.5; S_5 denotes the net on which the schema $\mathcal{Z}_5$ is constructed.

The net S_5 includes all vertices of the net S_4 and has vertices added in quantity equal to the number of elements in the sequence K. Each additional vertex is juxtaposed with a predicate μ_6; the vertices which go over into S_5 from the net S_4 are juxtaposed with the same actions and predicates as in the schema $\mathcal{Z}_4$. Therefore, if the net S_5 is constructed and the arcs issuing from its p-vertices are equipped with the tags 0 and 1, then we obtain complete information on the schema $\mathcal{Z}_5$.

The net S_5 with the tagged arcs is depicted in Fig. 11.

In describing the net S_5 we isolated that elementary path (we shall call it the principal one) which begins at the input α_0, ends at the output α_w, and passes successively through all vertices of the net. All branchings from the principal path are given by arcs whose beginning and end belong to the principal path. Each such arc in Fig. 11 is represented by two arrows:

1) one issues from the vertex which is the beginning of the arc considered, and it is tagged by the symbol of the end-vertex of the arc; 2) the other arrives at the vertex which is the end of the arc, and it is labeled by the symbol of the beginning vertex of the arc.

7.9. Let us sum up. In order to demonstrate transformations of schemata used in the proof of Theorems 1-3 and 7 we chose a schema $\mathcal{Z}$ with one output, which was strongly nonstationary in φ and nonstationary in ψ.

The fulfillment of a portion of the premises of the theorems mentioned was ensured by the choice of constructions of the schema $\mathcal{Z}$; in order to fulfill the remaining premises it was necessary to include the predicate $\mu_1, \mu_2, \ldots, \mu_6$ in the set R (over which the schema $\mathcal{Z}$ was considered).

The schema $\mathcal{Z}$ which was not reduced in ψ was first replaced by the schema $\mathcal{Z}_1$, which was equivalent to it and was reduced in ψ; strong nonstationarity in φ and nonstationarity in ψ carried over to $\mathcal{Z}_1$ from the schema $\mathcal{Z}$.

By equivalent transformations, the schema $\mathcal{Z}_1$ was converted into the schema $\mathcal{Z}_2$, which was weakly nonstationary in φ; the schema $\mathcal{Z}_2$ was converted into the schema $\mathcal{Z}_3$, which was stationary in φ, but nonstationary in ψ; the schema $\mathcal{Z}_3$ was converted into the schema $\mathcal{Z}_4$, which was stationary in φ and ψ; finally, the schema $\mathcal{Z}_4$ was converted into the linear schema $\mathcal{Z}_5$.

Literature Cited

1. C. Berge, Theory of Graphs and Its Application [Russian translation], IL, Moscow (1962).
2. L. A. Kaluzhnin, "On algorithmization of mathematical problems," in: Problemy Kibernetiki, Vol. 2, Fizmatgiz, Moscow (1959), pp. 51-68.
3. R. I. Podlovchenko, "On transformations of program schemata and their application in programming," in: Problemy Kibernetiki, Vol. 7 [in Russian], Fizmatgiz, Moscow (1962), pp. 161-188.
4. H. Thiele, Wissenschaftstheoretische Untersuchungen in Algorithmischen Sprachen, I, VEB Deutscher Verlag der Wissenschaften, Berlin (1966).

CONTROL PROCESSES IN LIVING ORGANISMS

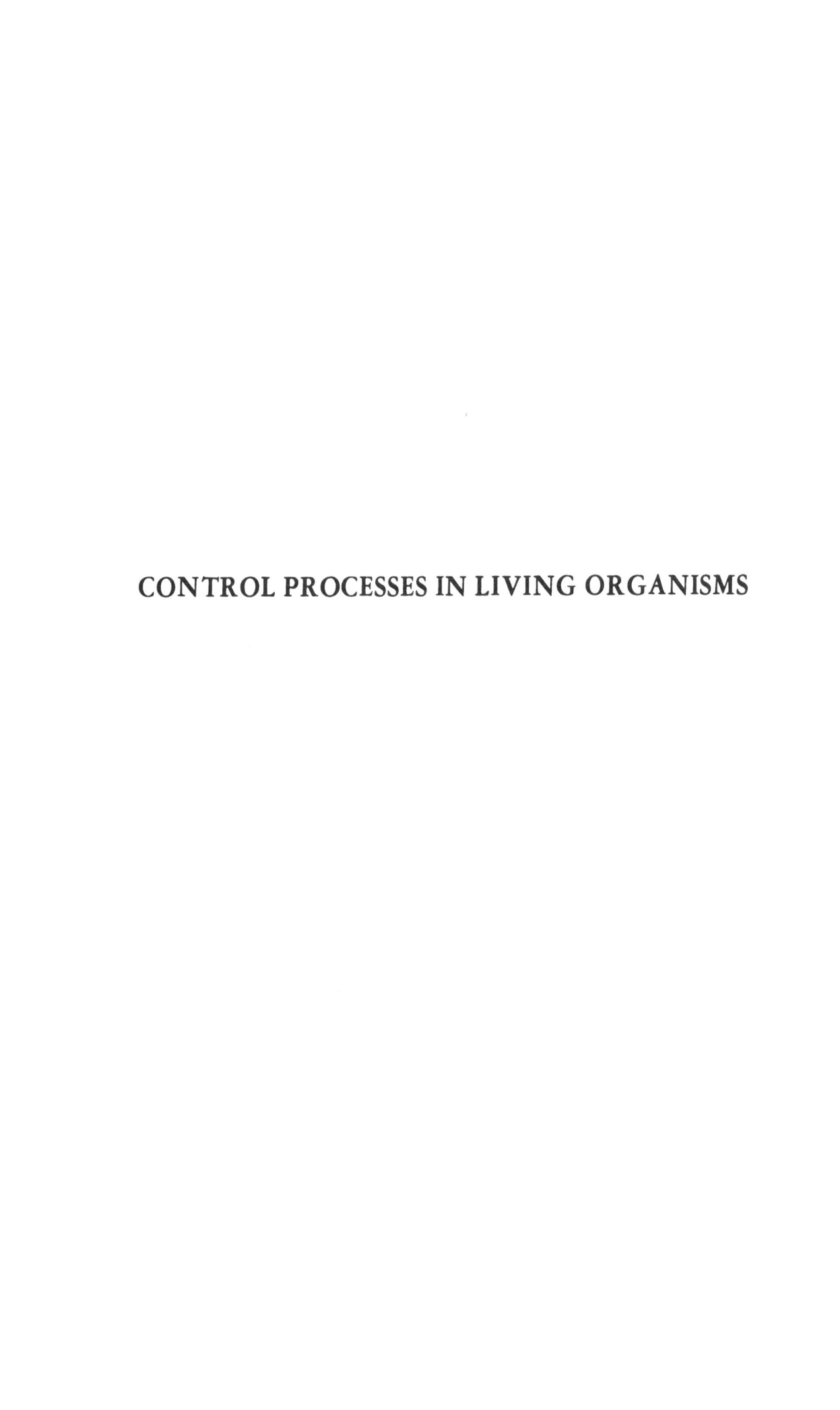

ON THE PROBLEM OF MODELING FOR AN EVOLUTIONARY PROCESS WITH REGARD TO METHODS OF SELECTION. II†

T. I. Bulgakova, O. S. Kulagina, and A. A. Lyapunov

Moscow and Novosibirsk

The present work is an immediate continuation of our previous paper [1]. The basic problem studied is that of determining a more detailed statistical mechanism of divergence form and in studying the formation of a genetically isolated "invariant" of a group in a limiting population under different mechanisms of choice affecting the population.

§ 1. Types of Choices

In the work [1] we described one kind of choice (a "one-sided" choice) and represented results of experiments for the population No. 6, belonging to zones of "stable instability" (see [2]).

In the present work we examine three types of choices, and respectively, three series of experiments. As initial populations we select the populations Nos. 6, 8, and 9 of [2]. Populations Nos. 6 and 8 refer to the zone of "stable instability" for a lack of choice, i.e., for which, under identical conditions in distinct realizations the changes occur in the same fashion: we are to give random numbers for the groups and the collections with the greatest or least isolation of one from the other (for characterizing such populations we will use the word "unstable"). Population No. 8 in the experiments without choice has a tendency to disintegrate into isolated groups (such populations will be called "disintegrating"). In addition to the populations described in [2] we describe a population No. 10, which in the experiment without choice is merged into one group (such a population is called "contracting"). The population No. 10 has the following characteristics: the number of elements in the initial population $N_0 = 150$; the number of genes in the genetic type, influencing the restrictions on the process, $n = 18$; the number of support trains $M_0 = 25$; for each support train for $q = 5$ genotypes, the distance from support to spread is not greater than $l = 4$. Definitions of all terms can be found in [2].

In the first series of experiments the choice of organisms was that described in [1]. Exactly as in [1] the advantage in survivial is given to the genotypes which are nearer to an "ideal" (an "ideal" genotype is supposed to consist of a unit). For the formation of a descendant the number α is calculated by the formula $\alpha = \frac{1}{2} - \beta_0 \frac{r_m - r}{r_m} + \gamma$, where r is the length

†Original article submitted August 22, 1968.

of the geonotype of this descendant with respect to the "ideal" genotype; r_m is the maximal possible distance from the "ideal" genotype (this "ideal" genotype consists only of the unit, up to $r_m = n$, where n is a number and $(r_m - r)$ is the number of units in it); β_0 is a positive constant for every experiment; γ is a variable, regulating the size of the population in such a way that the size is not overly dependent on its initial value. The correction γ is calculated in the following way. By definition the interval of time indicates the size of the population N and its deviation from the initial value $\Delta N = N - N_0$. If $\Delta N > \delta$, then $\gamma = +\gamma_0$ (γ_0 and δ are fixed positive numbers); if $(-\Delta N) > \delta$; then $\gamma = -\gamma_0$, if $|\Delta N| < \delta$, $\gamma = 0$.

The value α is equal to the random number ξ, chosen from the interval [0, 1] and subordinated to this interval by the rule of distribution with constant density. If $\xi > \alpha$, then the descendant survives and determines in this way its location in the population. If $\xi \leq \alpha$, then the descendant will not survive. A choice of this type is called a one-sided choice. We recall that in an experiment without choice for the determination of the survival of the descendant the random number ξ is equal to 1/2.

In the second series of experiments the choice was organized in the following way. We introduce two "ideal" genotypes, with one consisting only of zero, and the other, only of one. The random variable ξ is chosen, as in the first series of experiments, equal to α, while α is selected by the formula

$$\alpha = \begin{cases} \frac{1}{2} - \beta_0 \frac{r_0 - r}{r_0} + \gamma, & \text{if} \quad r \leqslant r_0; \\ \frac{1}{2} - \beta_1 \frac{r - r_0}{r_m - r_0} + \gamma, & \text{if} \quad r > r_0, \end{cases}$$

where r is the former distance of the genotype from the unit, $0 \leq r_0 \leq r_m$; $r_m = n$, β_0 and β_1 are fixed positive numbers for every experiment. Therefore this choice gives a preference for the survival of those genotypes which are close to one of the "ideal" genotypes, and the intermediate genotypes have a smaller probability of survivial. This choice will be called two-sided.

In the third series of experiments the preference is given to those genotypes having r_0 zeros and $(r_m - r_0)$ units and such that α is determined by the formula

$$\alpha = \begin{cases} \frac{1}{2} - \beta_0 \frac{r}{r_0} + \gamma, & \text{if} \quad r \leqslant r_0; \\ \frac{1}{2} - \beta_0 \frac{r_m - r}{r_m - r_0} + \gamma, & \text{if} \quad r > r_0. \end{cases}$$

Thus for r_0 close to $r_m/2$, the preferred genotypes have approximately equal numbers of zeros and units. This choice is called central.

The fourth series of experiments was carried out because in the basic paper [1] very limited experimental material is found, while because of this, most subsequent papers bear the character of preliminary information. At the same time, experiments on the components of the I-series of the present paper entirely confirm the results of [1]. The series of experiments II and III in some other arrangements also verify the existing phenomena described in [1].

§ 2. Description of the Experiments

In [1] and [2] the details were given of a model population and a model process of reproduction which we will not repeat. We recall only that one act of the process, corresponding to unit time, consists of selecting a pair of separate entities, capable of having descendants, and forming four offspring; after this we determine the survival of each of the offspring, as described in §1, and the survival of the offspring distributed in the population. As in [1] and

[2], we will determine the process of evolution for an initial population with a known structure. The structure is characterized by the collection of groups in the population, and by their size (i.e., the collection of elements found in every group), and the order of linkage of the groups in the family (by the number of isolated groups, the number of groups, found in every family, and the size of the families).

Since the choice is carried out in such a way that preference is given to those elements having certain defined numbers of units in the genotype, it is generally found that the evolution of the population bears the following characteristics.

1. The mean number of units in the genotype of the population $E_{av}=\frac{\sum_{m=1}^{N} e_m}{N}$, where e_m is the unit set in the genotype with number m, m = 1, 2, ..., N.

2. The mean number of units in the genotype of the i-th group, $E_i = \frac{\sum_{m_i}^{N_i} e_{mi}}{N_i}$, where N_i is the number of elements in the i-th group.

3. The mean number of units in the genotype of the maximal group, for which $N_{i_0} = \max_i N_i$, is denoted by E_{i_0}.

If the characteristics described at the points 1, 2, 3, are calculated by a machine, then the characteristics described at the points 4 and 5 are calculated automatically.

4. The mean number of units in the genotypes of the k-th group is E_k (in the present experiments k denotes the assumed values 0, 1, 2, 3).

If k = 1, then E_k is given by the formula $\frac{E_{av}N - \sum_{i_1}^{I_1} E_{i_1}N_{i_1}}{N_k}$, where $N_k = N - \sum_{i_1}^{I_1} N_{i_1}$, and i_1 is the number of isolated groups.

If k = 0, then the quantity E_k is not calculated; if k > 1, then E_k is calculated approximately by the mean of the number of units in the groups found in the k-th family.

5. The mean number of units in the genotypes of the maximal family for which $N_{k_0} = \max_k N_k$, is denoted by E_{k_0}.

Every characteristic E is calculated with an accuracy to the first decimal place.

The results of experiments are presented with the aid of tables and graphs. In every experiment we present the distance between elements of the group ρ = 5. In each of the experiments we enumerate the group, calculate E_{av}, E_i, and construct four graphs: the first time for the original population is given by means of all 960 units of time, which corresponds to a change of 13 generations. Thus, as in [2], the population is traced through a duration of 2880 units of time, i.e., for 40 generations. The computation of the correction γ for every experiment is carried out for every unit of time.

Each of the tables corresponds to the determinations of a series of experiments and characterizes the structure of the finite population, obtained from the initial population after 2880 time steps.

In Table 1 we present the results of experiments for the one-sided choice, in Table 2, for the two-sided, and in Table 3, for the central choice. For every type of choice we carry

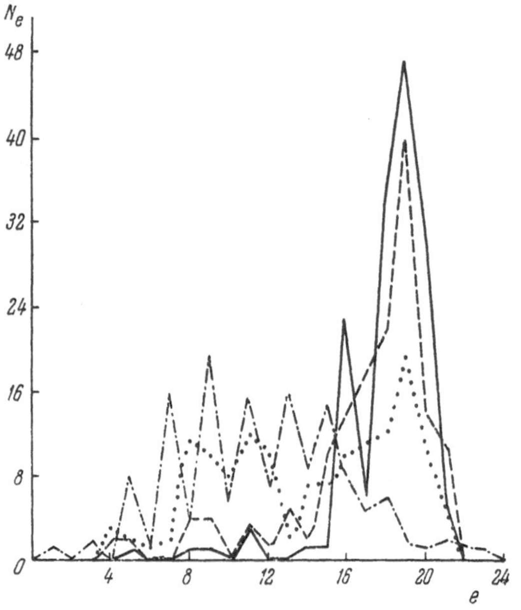

Fig. 1

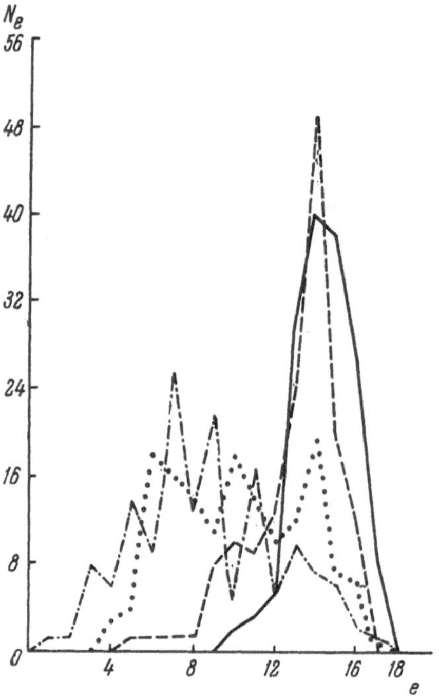

Fig. 2

out the experiments with four different populations. In each series of experiments with one and the same population differing pairwise by means of the order of the formation of the pair of genotypes, given descendants, and values of the parameters of the choice β, γ.

We now explain some of the graphs of the tables.† In graph 3 for every population we introduce the number n of genes in the genotypes of the population; in graph 5, the numbers of finite populations. In graph 9 we introduce the number of groups found in each of the families. For miscellaneous families the numbers are separated by the sign "/." In graph 10 we introduce the number of elements in the family and in the isolated groups, as well as the number of elements in the families contained in the brackets. In graph 14 we find the mean number of units E_k in the genotypes of the family (included in brackets) and in the isolated group. For this the order of the groups and families are the same as that of graph 10. In graphs 13, 15, 16, whenever we indicate the number of units, we introduce the mean number of units, calculated as on p. 264. For the remaining graphs of the tables an explanation is unnecessary.

Additional tables refer to the graphs of $N_e = f(e)$, where e is the number of units in the genotypes, which can be taken as a discrete value between 0 and n. N_e is the number of critical points in the population, whose genotypes contain e units. The graphs are found not for every experiment, but only for certain characteristic cases. The remaining cases are qualitatively similar to the previous ones. In each figure we present four graphs: the first for the initial time (denoted by lines and points), the second for the population obtained after 960 times units (the graph denoted by dots), the third, obtained after 1920 time units (denoted by points), the fourth for the final population (denoted by means of lines).

By means of the graphs and tables it is possible to observe that in the case of a "one-sided" choice (Figs. 1 and 2) the principle part of the population is displaced by the aspects of the "ideal" genotype. For this "compressed" population (No. 10) an "ideal" displacement dictates the choice of one family (see Fig. 2, experiments 18, 19). The "unstable" populations

† The enumeration of the experiments is that of [1].

TABLE 1

Experiment No.	Population No.	n	Parameters of choice β_0, γ	No. of elements N	No. of groups I	No. of isolated groups I_1	No. of families	Structure of families in groups	Structure of population	No. of elements in max. family Nk_0	No. of elements in max. group Ni_0	No. of units in population E_{av}	Structure of population in units	No. of units in max. family E_{k_0}	No. of units in max. group E_{i_0}
10	6	24	0.3 0.2	148	11	5	2	4/2	(130), (7) 3, 2, 4 1, 1	130	121	18.4	(19.1); (15.5) 9.0; 15.0; 13.3; 8.0; 12.0	19.1	19.1
11	6	24	0.3 0.1	151	10	4	1	6	(125), 2, 20, 1, 3	125	111	18.1	(18.9); 8.5; 16.0; 5.0; 11.0	18.9	18.9
12	9	30	0.2; 0.1	130	12	12	—	—	1, 105, 4, 7, 1, 1, 4, 2, 1, 1, 2, 1	—	105	25.1	16.0; 27.1; 9.0; 19.5; 21.0; 17.0; 21.0; 13.0; 18.0; 14.0; 19.0; 20.0	—	27.1
13	9	30	0.2; 0.1	135	8	6	1	2	(105), 226, 1, 2, 1, 3, 1	105	104	24.8	(25.9); 23.0; 19.0; 10.0; 16.0; 16.0; 18.9	25.9	25.9
14	9	30	0.2; 0.1	136	13	2	1	11	(130), 4, 2	130	87	24.7	(25.4); 9.0; 13.0	25.4	25.9
15	8	28	0.2; 0.1	145	25	5	1	20	(130), 8, 1, 2, 3, 1	130	43	22.3	(22.5); 8.5; 14.0; 19.0; 10.0; 16.0	22.5	21.9
16	8	28	0.2; 0.1	144	14	1	1	13	(139), 5	139	62	23.2	(23.4); 17.8	23.4	24.4
17	8	28	0.2; 0.1	144	17	3	1	14	(137), 4, 1, 2	137	62	22.9	(23.4); 13.0; 19.0; 13.0	23.4	24.0
18	10	18	0.2; 0.1	159	22	—	1	22	(159)	159	81	13.5	(13.5)	13.5	14.3
19	10	18	0.2; 0.1	153	14	—	1	14	(153)	153	103	14.4	(14.4)	14.4	14.8

(Nos. 6 and 8) arise in the following manner: we select a family which including the basic part of the population, and strongly displaced in quality, which in turn dictates the choice of this part of the population divided into isolated groups, which are generally not displaced (the "invariant" form) (see Fig. 1, experiments 10, 11, 15, 16, 17). It can be noted that the non-evolution of a group arises if its genotypes are found at a very small distance from each other and often are simply equal.

TABLE 2

Experiment No.	Population No.	n	Parameters of choice β_1, γ, r_0	No. of elements N	No. of groups I	No. of isolated groups I_1	No. of families	Structure of families in groups	Structure of population	No. of elements in max. family N_{k_0}	No. of elements in max. family N_{i_0}	No. of units in population E_{av}	Structure of population in units	No. of units in max. family E_{k_0}	No. of units in max. group E_{i_0}
20	6	24	0.2; 0.2; 0.1; 12	147	4	2	1	2	(111), 40, 6	111	110	8.4	(2.5); 22.0; 18.0	2.5	2.5
21	6	24	0.2; 0.2; 0.1; 12	161	11	4	1	7	(153), 2, 4, 1, 1	153	121	21.4	(22.5); 9.0; 4.0; 14.1; 10.0	22.5	22.1
22	6	24	0.2; 0.2; 0.1; 12	152	14	4	1	10	(134), 1, 8, 6, 3	134	120	4.3	(2.5); 12.0; 17.0; 17.0; 9.0	2.5	2.5
23	6	24	0.2; 0.2; 0.1; 12	151	7	—	2	5/2	(131); (20)	131	126	19.9	(21.1); (3.1)	21.1	21.1
24	6	24	0.2; 0.2; 0.1; 12	149	15	1	2	10/4	(98), (50), 1	98	71	9.5	(2.8); (21.3); 14.0	2.8	2.5
25	9	30	0.2; 0.2; 0.1; 12	130	7	7	—	—	25, 57, 29, 17, 1, 2, 3	—	57	5.9	7.0; 2.5; 7.0; 9.0; 27.0; 21.0; 7.0	—	2.5
26	9	30	0.2; 0.2; 0.1; 12	131	8	7	1	2	(78), 34, 3, 7, 7, 1, 1	78	50	11.3	(3.1); 26.0; 9.0; 25.0; 9.0; 8.0; 6.0	3.1	4.0
27	9	30	0.2; 0.2; 0.1; 12	131	7	7	—	—	89, 20, 7, 1, 1, 1, 12	—	89	21.5	28.4; 6.1; 9.8; 13.0; 22.0; 14.0; 4.0	—	28.4
28	9	30	0.2; 0.2; 0.1; 12	132	3	3	—	—	19, 112, 1	—	112	25.7	4.1; 29.4; 8.0	—	29.4
29	9	30	0.2; 0.2; 0.1; 12	139	4	4	—	—	66, 71, 1, 1	—	71	14.4	26.9; 3.0; 12.0; 9.0	—	3.0
30	8	28	0.2; 0.2; 0.1; 12	157	14	—	1	14	(157)	157	111	3.5	(3.5)	3.5	2.7
31	8	28	0.2; 0.2; 0.1; 12	143	12	5	1	7	(125), 12, 2, 1, 2, 1	125	107	22.8	(24.8); 6.8; 11.0; 7.0; 17.0; 13.0	24.8	24.8
32	8	28	0.2; 0.2; 0.1; 15	152	8	4	1	4	(138), 11, 1, 1, 1	138	124	6.3	(4.2); 26.0; 20.0; 10.0; 22.0	4.2	4.3
33	8	28	0.2; 0.2; 0.1; 12	150	16	4	2	2/10	(13), (122), 1, 11, 2, 1	122	80	8.6	(25.8); (5.9); 13.0; 12.0; 6.0; 7.0	5.9	6.0
34	8	28	0.2; 0.2; 0.1; 12	146	4	4	—	—	138, 4, 1, 13	—	138	2.8	2.0; 25.0; 24.0; 7.0	—	2.0
35	8	28	0.3; 0.2; 0.2; 15	166	12	4	1	8	(111), 50, 1, 3, 1	111	94	17.3	(25.4); 1.0; 7.9; 8.0; 7.0	25.4	25.4
36	8	28	0.3; 0.2; 0.2; 15	216	5	3	1	2	(189), 25, 1,1	189	188	23.1	(25.9); 2.0; 13.0; 18.0	25.9	26.0
37	8	28	0.2; 0.2; 0.1; 15	152	2	2	—	—	150, 2	—	150	3.0	2.7; 24.5	—	2.7
38	8	28	0.2; 0.2; 0.1; 15	145	4	4	—	—	130, 12, 1, 2	—	130	2.1	1.0; 10.0; 19.0; 22.0	—	1.0
39	10	18	0.2; 0.2; 0.1; 9	153	8	—	2	2/6	(90), (63)	90	89	10.8	(15.8); (2.9)	15.8	15.8

TABLE 3

Experiment No.	Population No.	n	Parameters of choice β_0, γ, r_0	No. of elements N	No. of groups I	No. of isolated groups I_1	No. of families	Structure of families in groups	Structure of population	No. of elements in max. family N_{k_0}	No. of elements in max. group N_{i_0}	No. of units in population E_{av}	Structure of population in units	No. of units in max. family E_{k_0}	No. of units in max. group E_{i_0}
40	6	24	0.2; 0.1; 12	168	15	6	3	2/4/3	(18), (65), (17), 39, 11, 10, 2, 4, 2	65	58	11.9	(10.9); (11.4); (11,1); 12.5; 15.0; 12.0; 16.0; 13.0; 7,0	11.4	11.4
41	6	24	0.2; 0.1; 12	149	13	11	1	2	(84), 12, 3, 7, 17, 19, 1, 1, 2, 1, 1, 1	84	81	12.0	(11.5); 10.0; 12.0; 11,7; 11.7; 12.9; 18,0; 10.0; 14.0; 17.0; 14.0; 8.0	11,4	11.5
42	6	24	0.2; 0.1; 12	158	9	5	1	4	(76), 3, 46, 29, 3, 1	76	46	11.9	(11.3); 16.0; 12.4; 11.1; 12.0; 11.0	11.3	12.4
43	6	24	0.2; 0.1; 12	151	27	5	1	22	(132), 11, 1, 2, 1, 4	132	38	11.7	(11.5); 14.0; 12.0; 14,0; 14.0; 11.0	11.5	12.1
44	6	24	0.2; 0.1; 12	158	8	6	1	2	(67), 53, 11, 7, 15, 4, 1	67	65	11.7	(12.0); 11.7; 12.0; 13.1; 10.8; 9.0; 13.0	12.0	11.9
45	9	30	0.2; 0.1; 15	153	10	8	1	2	(30), 28, 56, 4, 13, 2, 18, 1, 1	30	56	15.0	(14.7); 16.7; 15.0; 13.0; 16.0, 16.0; 13.1; 12.0; 7.0	14.7	15.0
46	9	30	0.2; 0.1; 15	140	13	6	1	7	(100), 20, 2, 2, 11, 2, 3	100	84	14.0	(14.1); 13.0; 15.0; 13,0; 15,0; 12.0; 16.0	14.1	14.1
47	9	30	0.2; 0.1; 15	130	15	15	—	—	25, 19, 39, 1, 2, 7, 1, 4, 23, 1, 1, 1, 4, 1, 1	—	39	15.3	16.0; 15.1; 16.0; 22.0; 11.0; 12.0; 9.0; 12.0; 15.0; 12.0; 19.0; 16.0; 17.0; 14.0; 13.0	—	16.0
48	9	30	0.2; 0.1; 15	137	10	10	—	—	27, 21, 38, 7, 35, 1, 2, 4, 1, 1	—	38	15.0	15.0; 14.0; 15.0; 16.0; 15.9; 9.0; 15.0; 15.0; 12.0; 11.0	—	15.0
49	9	30	0.2; 0.1; 15	130	10	8	1	2	(64), 31, 19, 9, 2, 1, 1, 1, 2	64	63	14.8	(13.8); 15.1; 15.0; 16.0; 16.0; 16.0; 17,0; 22,0; 17.5	13.8	13.9
50	8	28	0.2; 0.1; 12	153	22	5	1	17	(105), 38, 4, 4, 1, 1	105	38	18.3	(18.7); 18.5; 24.0; 11,3; 13.0; 15.0	18.7	18.5
51	8	28	0.2; 0.1; 15	161	13	6	1	7	(113), 37, 1, 1, 3, 5, 1	113	82	14.5	(14,1); 15,0; 17.0; 8.0; 21.0; 13.0; 10.0	14.1	14.1
52	8	28	0.2; 0.1; 15	147	14	10	1	4	(64), 33, 20, 9, 1, 3, 10, 2, 3, 1, 1	64	51	15.5	(16,3); 16,1; 13.5; 11.5; 25.0; 18.7; 16.0; 8.0; 14.0; 10.0; 16.0	16.3	16.3
53	8	28	0.2; 0.1; 15	170	10	5	2	3/2	(48), (15), 102, 1, 2, 1, 1	48	102	14.9	(14.5); (15.2); 15.0; 8.0; 18.0; 10.0; 25.0	14.5	15.0
54	8	28	0.2; 0.1; 15	152	9	2	3	3/3	(48), (93), 8, 2, 1	93	88	14.8	(14.4); (14.7); 18.0; 18.0; 11.0	14.7	14.7
55	10	18	0.2; 0.1; 9	163	26	—	1	26	(163)	163	67	9.0	(9.0)	9.0	8.5
56	10	18	0.2; 0.1; 9	155	33	—	1	33	(155)	155	35	8.7	(8.7)	8.7	9.0

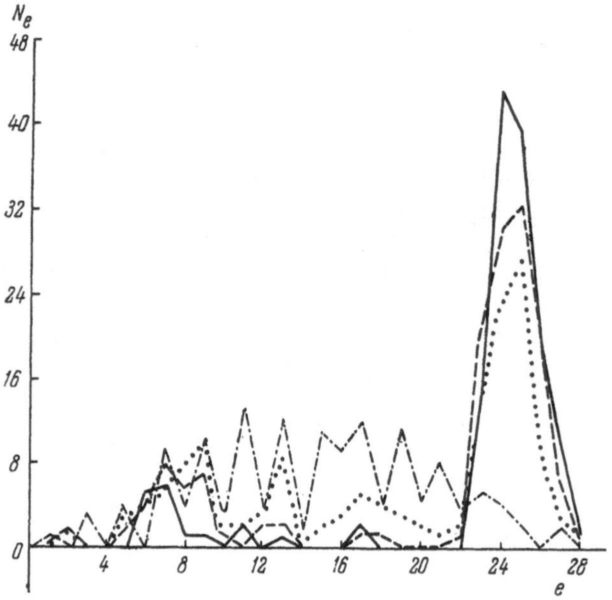

N_e
48
40
32
24
16
8
0
4
8
12
16
20
24
28
e

Fig. 3

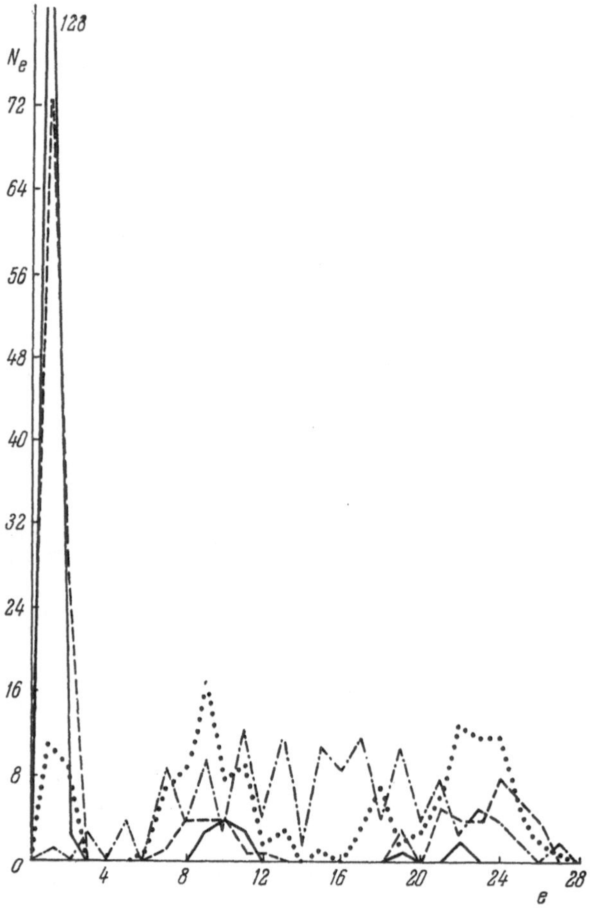

128
N_e
72
64
56
48
40
32
24
16
8
0
4
8
12
16
20
24
28
e

Fig. 4

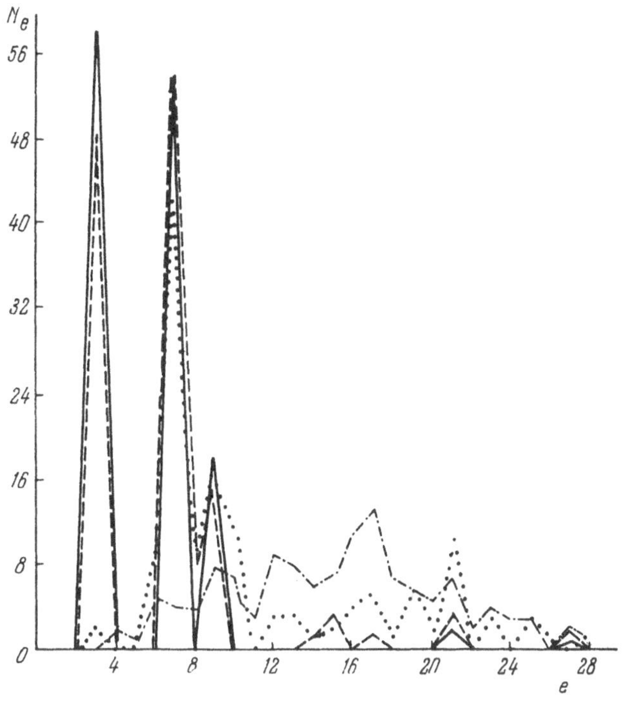

Fig. 5

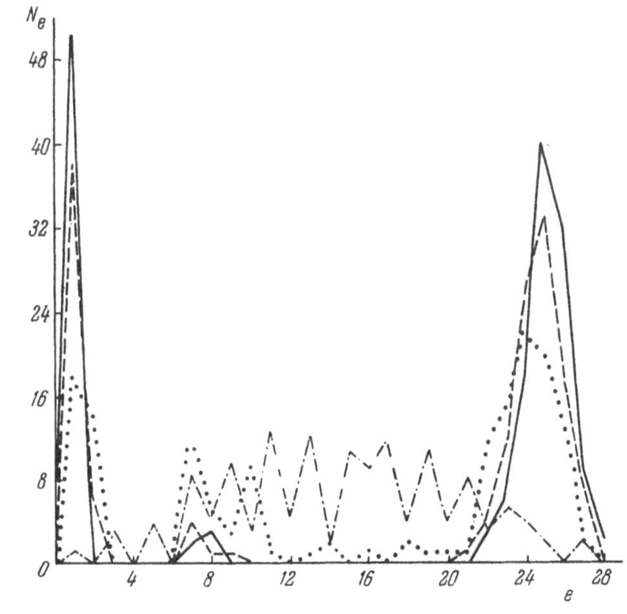

Fig. 6

In Fig. 1 the peak of the value e = 16 corresponds to the group of slow evolution, while the non-extreme peaks correspond to the values e = 11, 8-9, and 5 correspond to the non-evolutionary "invariant" forms.

Reducing the population (No. 9) for a one-sided choice into different cases without choice, we obtain a self-similar instability. In this way we distinguish a larger part (a group or a

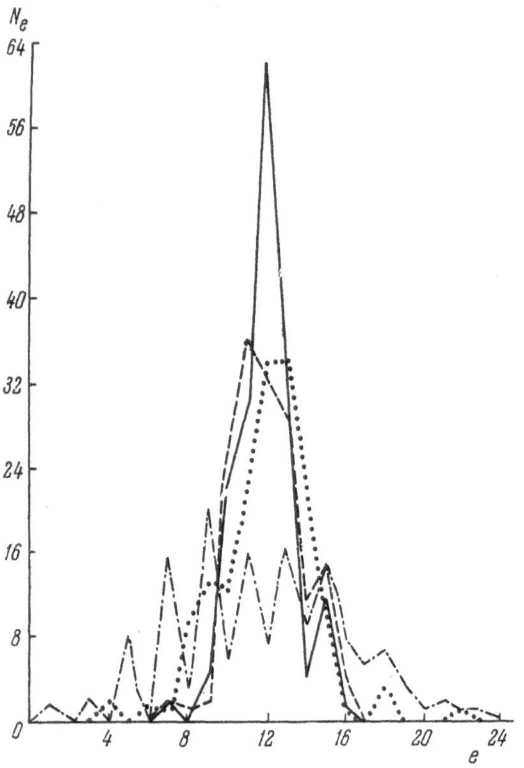

Fig. 7

family) which is strongly displaced towards the direction of the "ideal genotype," and moreover, which generates through a slow evolutionary process into a small non-evolutionary group (experiments 12, 13, 14).

For the two-sided choice in the various experiments, we obtain results of three types. In one case we obtain a situation reducing to the one-sided choice: the separating basic mass of the population is displaced to a genotype consisting only of units (Fig. 3, experiments 21, 23, 27, 28, 31, 36). In the other cases the basic mass of the population is displaced to a genotype consisting only of zeros (Fig. 4, experiments 22, 30, 32, 33, 34, 37, 38). In this case the basic part of the population is generally represented by one family, while sometimes (for the "reduction" of the population) it separates into roughly equal numbers of elements of isolated groups, with every displacement in one direction but with different orders (see Fig. 5, experiment 25).

Finally, we consider the case when the population divides into two approximately equal numbers of parts, which are displaced in different directions (see Fig. 6, experiments 20, 24, 26, 29, 35, 39).

We will generally speak of the greater part of the population. In the majority of cases we may speak of the basic isolated mass of the group, which is displaced into one or another small group or is not displaced at all (an "invariant" form).

It may be noted that a faster disintegration of the population into parts than noted, in which its parts undergo evolution, is dictated by the method choice.

For the central choice of the population we have a displacement dictated by the choice, i.e., E_{av} becomes approximately equal to $(n - r_0)$. From the graphs (Fig. 7) it is possible to note one peak, obtained at the point $(n - r_0)$. However, this does not mean that the population combines to form one group, since it may instead evolve into various different genotypes, consisting of $(n - r_0)$ units and r_0 zeros. In the general case the basic part of the population is divided into approximately equal numbers of isolated groups, divided into sufficiently remote pairs (experiments 40-54). Exceptions are the "contractible" populations, which in this case give one family (and do not give isolated groups) (experiments 55, 56).

In every figure we can see that the first and second graphs (i.e., the graphs of the initial population and the population after 960 units of time), as well as the second and third graphs, differ pairwise much more than the third and fourth graphs, although each graph is formed 960 unit time steps after the previous one. This means that the drift of the population slows down with increasing time. Clearly, the larger the variety in the genotypes, the faster will be the evolution of the population.

§ 3. General Conclusions

On the basis of the experiments performed, we have arrived at the following general conclusions.

1. In the problem on the divergence and evolution of biological organisms, an evolutionary process is obtained of such a character that some small genotype descendants will not be given, while large descendants have a behavior indicative of Markov processes, i.e., in essence the fluctuation of genotypes of the population. Thus it is not at all necessary to differentiate between special geographically or biologically isolated mechanisms in order to perform a natural taxonometric study of new organisms. It is highly probable that because of this the general rhythm of evolutionary processes is accelerated, and that it is necessary to consider the theme of this evolution.

2. Repeated experiments with the same initial population and probability characterization of the process of selection and with only different random numbers (chosen from one and the same distribution) leads to completely different realizations of the process of evolution up to that which in the case of the two-sided choice for one realization results in a drift of genotypes of the population to one side; for two realizations, into two sides; and for three, two genotypically isolated pairwise populations, from which one genotypically drifts into another, and pair into pair. This shows to what extent a random element may affect the qualitative characterization of the course of a random process. Hence, it is clear that it is inappropriate, in order to explain the qualitative and macroscopic details of the evolutionary process, to look for special determining properties of an essential nature. On the contrary, the representation of very important features of this process requires the knowledge of the character of the four kinetic evolutionary processes, which can be determined by means of random properties as well as the properties of this process.

3. We call attention to a common result of our experiments, in which for an unknown structure the choice of the character of the genotype dictates the evolution of the population with the passage of time.

Thus the population No. 8 (see Fig. 4) goes through a process of the following form. If we observe its changes after 960 time steps, then it seems that it divides into two parts, which will be evolving in different forms. But then the character of the drift is known, and the part of the population which is close to the genotype consisting of one unit is almost invariant (see also p. 270 on the retardation of the drift). This indicates the great role of the fluctuations in the progress of evolution of the process as well as the fact that we must deal with great care with the history of known genotypes of a population under known conditions of choice.

4. In the present experiments it was assumed that the establishment of the invariant forms results in an evolutionary impasse when the population is quickly evolving under known sufficiently abruptly changing conditions of choice. This collection of invariant forms turns out to be highly genetically homogeneous. The case of strong fluctuations and the evolution of invariant forms, if observed for various types of choices, is essentially seen to be of a peculiar character and is caused by only exceptionally posed experiments. Great interest has arisen in connection with the extent to which the appearance of spreading is natural, and the genotype structure of the resulting groups.

5. Our results have attracted the attention of investigators towards aspects of the important problems of the mathematical theory of evolution of populations in the following form. Thus special interest has arisen in the cybernetic zones in time, which are characterized in the following way. The fate of the genotype population in a small interval of time is entirely in

agreement with the description of elementary probability-combinatorical methods. Similarly, over long time intervals it is well described by asymptotic methods. However, there exists an extension of the time interval in which neither combinatorical nor asymptotic methods are applicable to the problem. In this case it is necessary to combine analytical methods with sufficiently accurate estimates for the experiment. This zone of time has been referred to as the cybernetic zone by Yu. I. Zhuravleb.

Literature Cited

1. T. I. Bulgakova, O. S. Kulagina, and A. A. Lyapunov, "The Question of Modeling the Evolutionary Process with Selection Taken into Account. I., Systems Theory Research, Vol. 20, Consultants Bureau, New York (1971), p. 22.
2. O. S. Kulagina and A. A. Lyapunov, "On modeling the evolutionary process," Problemy Kibernetiki, Vol. 16, Nauka, Moscow (1966).

ON THE DYNAMICS AND CONTROL OF THE AGE STRUCTURE OF A POPULATION†

L. R. Ginzburg

Leningrad

The problem of time dynamics of the size of natural populations occupies one of the most important places in biological literature [1-9, 11]. Many investigators isolate varied mechanisms for regulating the size of natural populations, which operate at both the inter-population and intrapopulation levels. For many reasons, some of which will become clear below, the study of the dynamics of the overall number of a population without allowance for its age structure is hardly satisfactory. Besides being of purely ecological interest, problems of the dynamics of the age composition of a population have practical significance in connection with a number of applied problems of controlling the dynamics of natural and artificial populations.

In the first section of the present paper the mathematical model of the dynamics of the age structure of a population in a stationary medium is considered, and one of the possible hypotheses explaining the fluctuations in population size with time is discussed. In the second section a generalization of this model for the case of nonstationary external conditions is considered. In the third section the constructed model is used as the basis for stating certain control problems.

Dynamics of the Population Size in a Stationary Medium

Various approaches exist which explain the periodic character of the variation of population size. The best known and most widely used mathematical concept is the Lotka–Volterra concept [10-13] whose meaning is that the fluctuation of population size in time is caused by predator–prey relations. The equations

$$\left.\begin{aligned} \frac{dN_1}{dt} &= \alpha_1 N_1 - \beta_1 N_1 N_2, \\ \frac{dN_2}{dt} &= -\alpha_2 N_2 + \beta_2 N_1 N_2 \end{aligned}\right\} \qquad (1)$$

serve as the foundation of the mathematical model of this theory; here N_1 is the size of the prey population; N_2 is the size of the predator population; α_1, α_2 are coefficients which characterize the intrinsic birth and death rates of both populations; β_1, β_2 are coefficients which describe the interpopulational interaction.

†Original article submitted June 4, 1968.

This model has undergone various kinds of criticism and refinement [14, 15, 16]. In particular, A. N. Kolmogorov noted the low accuracy of the system (1) in view of the fact that the derivatives dN_1/dt and dN_2/dt have meaning only for fairly long time intervals.

Moreover, the experimental material obtained up to the present by biologists allows the statement to be made that factors operating at an intrapopulational level have an important significance in the dynamics of population size [1-5, 8, 9].

We have made an attempt at constructing a mathematical model which describes regular fluctuations of population size on the basis of a nonuniform age distribution of the individuals in the population (this holds in the overwhelming majority of natural populations). This fact makes necessary a study of the dynamics of populations with allowance for the age composition. As will be shown below, consideration of the age composition of a population improves the accuracy of the model and reveals qualitatively new effects connected with an explanation of the fluctuating processes in the dynamics of population growth.

Let us go over to a construction of the mathematical model for the growth of a certain population with allowance for the age composition. Related problems were considered in [17, 18, 20, 21] in connection with the growth of human population and the age composition of the growing population.

In order to simplify the mathematical reasoning we shall consider the growth of a unisex population in a stationary medium. All of what follows will be generalized for the case of a bisexual population in a nonstationary medium. In certain places those mathematical proofs that are not essential to an understanding of the biological meaning of the constructions of the model will be omitted.

We shall call the function of two variables u(x, t) the a g e d e n s i t y (or simply the d e n s i t y) of the number of individuals of a given age at time t if for any ages a and b the number of individuals having an age from the interval $[a, b]$ is expressed as

$$\int_a^b u(x, t)\, dx.$$

Obviously, the function $\tilde{u}(x, t)$, which is defined by the equation

$$\tilde{u}(x, t) = \frac{u(x, t)}{\int_0^\infty u(x, t)\, dx},$$

may be treated as the density of the probability that an individual chosen at random has an age not exceeding x at time t.

Let us assume (again only for simplicity) that the population is totally isolated (i.e., there is no exchange of migrants with other populations of the same type).

The basic system of equations which is satisfied by the function u(x, t) has the form

$$\left.\begin{aligned} &\frac{\partial u}{\partial t} + \frac{\partial u}{\partial x} = -d(x)\, u, \\ &u(0, t) = \int_0^\infty b(x)\, u\, dx, \\ &u(x, 0) = g(x). \end{aligned}\right\} \tag{2}$$

Here d(x) is the mortality rate of individuals of age x; b(x) is the birth rate for parents of age x; g(x) is the initial numerical distribution by age.

The first equation of the system (2) describes the population loss at various ages as a result of natural mortality and as a result of natural aging of individuals with the passing of time. The second equation expresses the number of newborn at time t, summed over all ages of the parents. The third equation of the system (2) constitutes the initial condition. In deriving the equations the assumption of a linear dependence of birth and mortality on numerical size was used. This hypothesis corresponds to a situation in which there is no noticeable effect of limiting factors such as, for example, shortage of food and restricted living space. The fact that the coefficients b(x) and d(x) are independent of time is what designates the medium as stationary.

Let us now investigate the system of equations (2). The general solution of the first equation has the form

$$u(x, t) = \Omega(t-x)\, e^{-\int_0^x d(\xi)d\xi}, \tag{3}$$

where Ω is an arbitrary function.

Having substituted (3) into the second and third equations of the system (2), we obtain

$$\Omega(t) = \int_0^\infty b(x)\, e^{-\int_0^x d(\xi)d\xi}\, \Omega(t-x)\, dx, \tag{4}$$

$$\Omega(-x) = g(x)\, e^{\int_0^x d(\xi)d\xi}. \tag{5}$$

Let us introduce the notation

$$\left.\begin{aligned} K(x) &\equiv b(x)\, e^{-\int_0^x d(\xi)d\xi},\\ \varphi(t) &\equiv g(t)\, e^{\int_0^t d(\xi)d\xi}. \end{aligned}\right\} \tag{6}$$

Equations (4) and (5) take the following form for the conditions (6):

$$\left.\begin{aligned} \Omega(t) &= \int_0^\infty K(x)\,\Omega(t-x)\,dx,\\ \Omega(-t) &= \varphi(t). \end{aligned}\right\} \tag{7}$$

We shall seek the solution of Eq. (7) in the form $\Omega(t) = e^{zt}$; then we obtain the basic equation for z:

$$F(z) = \int_0^\infty K(x)\, e^{-zx}\, dx - 1 = 0. \tag{8}$$

It may be shown that the function F(z) is an integer function of the complex variable z having an order of growth which is higher than that of a power law. Consequently, it has an infinite countable set of zeros on the complex plane [19], and therefore one may attempt to find the solution of Eq. (7) in the form

$$\Omega(t)=\sum_{i=1}^{\infty} C_i e^{z_i t}, \tag{9}$$

where z_i are the roots of the function F, and C_i are constant numbers. Let us rewrite Eq. (7) in the form

$$\Omega(t)=\int_0^t K(x)\,\Omega(t-x)\,dx+\int_t^{\infty} K(x)\,\Omega(t-x)\,dx. \tag{10}$$

In the second integral the argument of the function Ω is nonpositive, and therefore from the second condition (7) we obtain

$$\int_t^{\infty} K(x)\,\Omega(t-x)\,dx=\int_t^{\infty} K(x)\,\varphi(x-t)\,dx\equiv f(t). \tag{+}$$

Now we apply the Laplace transform to Eq. (10) with allowance for (+); then,

$$\widetilde{\Omega}(z)=\frac{\widetilde{f}(z)}{1-\widetilde{K}(z)}.$$

Here Ω, $\widetilde{f}$, and $\widetilde{K}$ denote the Laplace transforms of the functions Ω, f, and K. According to the expansion theorem from operational calculus, the original of the function $\widetilde{\Omega}$ is determined according to the equation

$$\Omega(t)=\sum_{i=1}^{\infty} \operatorname*{res}_{z_i}\Omega(z)\,e^{z_i t},$$

where $\operatorname*{res}_{z_i}\Omega$ are the residues of the complex-variable function Ω of the corresponding poles, while z_i are the roots of the equation

$$\widetilde{K}(z)=1.$$

The latter, as can easily be seen, coincides with Eq. (8).

Thus, the coefficients C_i may be calculated explicitly from the stipulated functions φ(t) and K(x) as residues of the function of the complex variable (3) at the corresponding poles.

Let us note certain properties of the solution obtained. First of all let us note that among z_i there is one and only one real number. Actually, by virtue of the positiveness of the kernel K(x) the function F is monotonic on the real axis and takes all possible real values from −1 to ∞. Consequently, it has one and only one real root.

Moreover, it can easily be shown that for all the remaining complex roots of the function F(z) the condition

$$\operatorname{Re} z_i \leqslant \lambda \tag{11}$$

is fulfilled, where λ is the sole real root of Eq. (8). Actually, from (8) it follows that

$$1=\left|\int_0^\infty K(x)\,e^{-z_i x}\,dx\right|\leqslant\int_0^\infty K(x)\,e^{-(\mathrm{Re}\,z_i)x}\,dx. \tag{12}$$

By virtue of the monotonicity of F on the real axis, Re $z_i \leq \lambda$, i.e., the statement (11) has been proved.

Thus, in Eq. (9) one can isolate the principal term $e^{\lambda t}$, which has the principal effect on the dynamics of population size for larget t:

$$\Omega(t)=Ce^{\lambda t}+\sum_{i=2}^{\infty}C_i e^{z_i t}. \tag{13}$$

All terms of the sum in Eq. (13) describe the time fluctuation of population size. The frequencies of the fluctuations are the imaginary parts of the roots z_i of the function F(z). Thus, the solutions of the system of equations (2) may be oscillatory in character, which is particularly noticeable for small t. In the case $\lambda = 0$, which corresponds to equality of the average birth and mortality rates, undamped fluctuations of the population size u(x, t) with time may be observed.

Let us consider the problem of the frequencies of the size fluctuations of a natural population with time. It is clear that high frequencies are not realized in practice, and therefore it is of interest to attempt to estimate the lowest frequencies corresponding to the most noticeable fluctuations having the maximum period. Of course, for a stipulated function K(x) the frequencies may be calculated numerically with any degree of accuracy, but we shall attempt to estimate them, having made the simplest assumptions concerning the function K(x).

a) Assume that the reproductive age is concentrated in the interval [A, B], i.e.,

$$K(x)=0,\qquad \{x\colon x<A,\ x>B\}; \tag{14}$$

then Eq. (8) takes the form

$$\left.\begin{aligned}&\int_A^B K(x)\,e^{-\alpha x}\cos\omega x\,dx=1,\\&\int_A^B K(x)\,e^{-\alpha x}\sin\omega x\,dx=0,\end{aligned}\right\} \tag{15}$$

where

$$\alpha=\mathrm{Re}\,z,\qquad \omega=\mathrm{Im}\,z. \tag{16}$$

From the second equation of the system (15) it follows that $\sin\omega x$ changes sign on the interval [A, B]; therefore, for the maximum period T of the fluctuations one may obtain the estimate

$$T\leqslant 2B; \tag{17}$$

i.e., the maximum period of the fluctuations of population size is a quantity of the order of the lifetime of one generation.

b) Assume that the reproductive age is concentrated at one point M (i.e., the birth rate has a δ-shaped form). This case may provide a good description of the situation which holds for several species of fish for which total mortality of the parent individuals occurs immediately after spawning. Moreover, let us assume that $\lambda = 0$; therefore,

$$\int_0^\infty K_0\delta(x-M)\,dx = 1. \tag{18}$$

Equation (18) has the following form for $K(x) = K_0\delta(x-M)$:

$$e^{-zM} = 1. \tag{19}$$

Taking account of (18) and making use of the expansion of the difference of the exponent into an infinite product, we obtain

$$ze^{-\frac{1}{2}Mz}\prod_{k=1}^{\infty}\left[1+\frac{M^2z^2}{4k^2\pi^2}\right]=0. \tag{20}$$

In this case the entire spectrum of frequencies can be determined from the exact formula

$$\omega_k = \frac{2k\pi}{M} \qquad (k=1, 2, \ldots), \tag{21}$$

while the possible periods can be determined respectively from the equation

$$T_k = \frac{M}{k}. \tag{22}$$

Thus, in this case the most noticeable will be fluctuations having the period $T_1 = M$. In this particular case the final solution of the problem has an especially simple form:

$$u(x, t) = \left[\sum_{k=1}^{\infty} C_k \cos\frac{2\pi k}{M}(t-x)\right] e^{-\int_0^x d(\xi)d\xi}, \tag{23}$$

where C_k are the coefficients of the expansion of the function $\varphi(t)$ into a Fourier series in cosines.

Let us attempt to compare the fluctuation frequencies obtained approximately from the model constructed above with the actual fluctuation frequencies cited in the literature.

Let us consider, for example, the fluctuations of the population size of foxes in Canada on the Labrador Peninsula. According to the data given by Elton [6] the average period of population-size fluctuations has been approximately four years during the past 100 years. In order of magnitude this approximately corresponds to the period of fluctuations obtained theoretically [see (17)]. The fluctuation frequencies obtained in the preceding section likewise correspond to the data on the population fluctuation of field voles and lemmings [3-6]. For these species the fluctuation period is equal to three to four years. The population-size fluctuations of the salmon family with a period equal to the reproduction age (M) are well known. The accuracy of the observations is so low that there is no need to carry out exact calculations. The order of the compared quantities is evidence of the fact that the model given may be used for a rough prediction of the size of natural populations. Of course, the actual fluctuations are considerably more complex than those obtained from the model. This is natural, since the basic hypothesis on which the model is founded is stationarity of the medium, and this hypothesis is naturally violated under actual conditions.

In conclusion let us again compare the model advanced for the dynamics of population size in the present paper with the model developed by V. Volterra. However paradoxical this might be, it is nevertheless true that in certain cases the objections to a model of the predator-prey type lies in the fact that it is too general; i.e., the fluctuation frequencies which derive from it may be the most varied. Actually, however, an amazing similarity of the frequencies of the observed fluctuations [3-6] can be detected for one and the same species under different conditions. What has been said in no way denigrates the role of the interaction model of the predator–prey type in the explanation of many ecological facts. However, the thought comes to mind that practical oscillations of population sizes may in many cases be explained by simpler notions (for example, those which are presented in the present paper).

The Dynamics of Population Size under Nonstationary External Conditions

In order to introduce nonstationarity of the external conditions into the model considered, it is sufficient to suppose that the birth and death rates depend on time.

The system of equations describing the dynamics of the population size will be of the form

$$\frac{\partial u}{\partial t}+\frac{\partial u}{\partial x}=-d(x,t)\,u, \tag{24}$$

$$u(0,t)=\int_0^\infty b(x,t)\,u\,dx. \tag{25}$$

The general solution of Eq. (24) has the following form:

$$u(x,t)=\Omega(t-x)\,e^{-D(x,t)}, \tag{26}$$

where Ω is an arbitrary function, while

$$D(x,t)=\int_0^x d(\xi,t-x+\xi)\,d\xi. \tag{27}$$

Having substituted the general solution (26) into Eq. (25) along with the initial condition (2), we obtain

$$\Omega(t)=\int_0^\infty b(x,t)\,e^{-D(x,t)}\Omega(t-x)\,dx, \tag{28}$$

$$\Omega(-x)=g(x)\,e^{D(x,0)}. \tag{29}$$

Let us note that since the reproductive age is positive, the integration in Eq. (28) may be assumed to extend from a certain $A > 0$ to ∞, rather than from zero. Moreover, let us introduce the notation:

$$\left.\begin{aligned}K(x,t)&\equiv b(x,t)\,e^{-D(x,t)},\\ \tilde{\varphi}(t)&=g(t)\,e^{D(t,0)}.\end{aligned}\right\} \tag{30}$$

In the new notation the integral equation (28) and the initial condition (29) take the following form:

$$\Omega(t)=\int_0^\infty K(x,t)\,\Omega(t-x)\,dx, \tag{31}$$

$$\Omega(-t)=\tilde{\varphi}(t). \tag{32}$$

The function Ω is known to us for negative values of the argument. It is necessary to recover it from Eq. (31) for positive values of the argument. Since x varies from A to ∞ in the integration in (31), it follows that the argument of the function Ω varies within the limits

$$-\infty \leqslant t - x \leqslant t - A.$$

If one considers the functions Ω on the interval $0 \leq t \leq A$, then the argument $\Omega(t-x)$ is negative, and therefore the right side of Eq. (31) is known to us. Thus, for $0 \leq t \leq A$ we have

$$\Omega(t) = \int_A^\infty K(x, t)\, \tilde{\varphi}(x-t)\, dx.$$

Now the function Ω is known to us on the interval $-\infty \leq t \leq A$. Having repeated the same procedure for the interval $A \leq t \leq 2A$, we obtain the function Ω for $-\infty \leq t \leq 2A$. Continuing this process ad infinitum, we find the function Ω for all values of the argument.

Let us determine the sequence of functions $\omega_i(t)$ by means of the recurrence formula

$$\left.\begin{aligned} &\omega_{i+1}(t) = \int_A^\infty K(x, t)\, \omega_i(t-x)\, dx, \\ &\omega_0(t) = \tilde{\varphi}(t) \quad \text{for} \quad t \leqslant 0; \end{aligned}\right\}$$

then the solution of Eq. (31) for the initial condition (32) will have the following form:

$$\Omega(t) = \begin{cases} \omega_0(t) & \text{for} \quad -\infty \leqslant t \leqslant 0, \\ \omega_1(t) & \text{for} \quad 0 \leqslant t \leqslant A, \\ \dots\dots\dots\dots, \\ \omega_i(t) & \text{for} \quad (i-1)A \leqslant t \leqslant iA, \\ \dots\dots\dots\dots \end{cases} \tag{33}$$

The solution of Eq. (31) may be carried out by another method also. Let us partition the integration interval in Eq. (31) into two intervals from 0 to t and from t to ∞. Then by virtue of the initial condition (32) we obtain

$$\Omega(t) = \int_0^t K(x, t)\, \Omega(t-x)\, dx + \int_t^\infty K(x, t)\, \varphi(x-t)\, dx.$$

This equation may be solved by the method of successive approximations.

The following fact may likewise be established using the method of successive approximations.

Assume that the nonstationary birth and mortality rates for all t are confined between certain stationary values, i.e.,

$$\left.\begin{aligned} b_1(x) \leqslant b(x, t) \leqslant b_2(x), \\ d_1(x) \geqslant d(x, t) \geqslant d_2(x). \end{aligned}\right\} \tag{34}$$

Then the corresponding solutions of the three systems of equations having the coefficients $b_1(x)$, $d_1(x)$; $b(x, t)$, $d(x, t)$; $b_2(x)$, $d_2(x)$ satisfy the inequality

$$u_1(x, t) \leqslant u(x, t) \leqslant u_2(x, t). \tag{35}$$

Making use of the knowledge of the analytic solution of the stationary problem, one can indicate the boundaries within which the solution of the nonstationary problem lies. The asymptotic estimate

$$C_1 e^{\lambda_1 t} e^{-\lambda_1 x - \int_0^x d_1(\xi) d\xi} \leqslant u(x, t) \leqslant C_2 e^{\lambda_2 t} e^{-\lambda_2 x} e^{-\int_0^x d_2(\xi) d\xi}$$

has an especially simple form. This estimate is especially useful in view of the fact that in many cases the exact values of the coefficients are unknown. In this case the possibility presents itself of giving a bilateral estimate of the solution.

The final exact solution of the problem is given by Eq. (26), where it is determined by Eq. (33) or Eq. (34).

Let us now consider the case of a bisexual population with allowance for separate dynamics of the population sizes of the sexes. By analogy, the dynamics equations will have the following form:

$$\left.\begin{aligned} &\frac{\partial u}{\partial t} + \frac{\partial u}{\partial x} = -d_u(x, t) u, \\ &\frac{\partial v}{\partial t} + \frac{\partial v}{\partial x} = -d_v(x, t) v, \\ &u(0, t) = \int_0^\infty b_u(x, t) u \, dx, \\ &v(0, t) = \int_0^\infty b_v(x, t) u \, dx, \\ &u(x, 0) = g_u(x), \\ &v(x, 0) = g_v(x). \end{aligned}\right\} \tag{36}$$

Here u(x, t) is the population density of females; v(x, t) is the population density of males; d_u, d_v, b_u, b_v are respectively mortality and birth rates of females and males.

Let us introduce the notation:

$$\left.\begin{aligned} &D_u(x, t) \equiv \int_0^x d_u(\xi, t - x + \xi) \, d\xi, \\ &D_v(x, t) \equiv \int_0^x d_v(\xi, t - x + \xi) \, d\xi, \\ &K_u(x, t) \equiv b_u(x, t) e^{-D_u(x, t)}, \\ &K_v(x, t) \equiv b_v(x, t) e^{-D_u(x, t)}, \\ &\varphi_u(t) \equiv g_u(t) e^{D_u(t, 0)}, \\ &\varphi_v(t) \equiv g_v(t) e^{D_v(t, 0)}. \end{aligned}\right\} \tag{37}$$

In this notation we obtain

$$\left.\begin{aligned} u(x, t) = \Omega_u(t - x) e^{-D_u(x, t)}, \\ v(x, t) = \Omega_v(t - x) e^{-D_v(x, t)}, \end{aligned}\right\} \tag{38}$$

where Ω_u and Ω_v can be determined from the equations

$$\Omega_u(t)=\int_0^\infty K_u(x, t)\,\Omega(t-x)\,dx, \tag{39}$$

$$\Omega_v(t)=\int_0^\infty K_v(x, t)\,\Omega_u(t-x)\,dx, \tag{40}$$

$$\Omega_u(-t)=\varphi_u(t), \tag{41}$$

$$\Omega_v(-t)=\varphi_v(t). \tag{42}$$

It is evident that Eq. (39), which describes the dynamics of the female population size, can be isolated for the condition (41). This equation coincides completely with the corresponding equation for the unisex case. And this should be expected, since the dynamics of a bisexual population is determined in the final analysis by the dynamics of the female population size.

Equation (39) is solved for condition (41) by the method described above, while the function Ω_v is found simply by substituting Ω_u into Eq. (40).

Finally, the dynamics of the populations of the sexes in a bisexual population can be described by Eqs. (38), where the functions Ω_u and Ω_v are determined from the rule described above.

On Controlling the Dynamics of the Sizes of Natural and Artificial Populations

In the previous sections it was shown how, knowing the birth rate b(x, t), the mortality rate d(x, t), and the initial age distribution g(x) of the population size, we may predict the population size as a function of time with allowance for the age composition. It is of interest to state the problem of controlling population size for a certain purpose. In the case of combating the populations of various species of pets such a purpose may be to maintain the population size at a certain fixed level, while in the case of the exploitation of a school of fish (a natural population) or a herd of large horned cattle (an artifical population) the purpose of the control may be to optimize a certain economic criterion.

The determination of the required characteristics of the population, such as birth rates, mortality rates, etc. (we are speaking, of course, of natural populations), causes great difficulties in solving problems of this kind. The problem of the means of control available to us is likewise important. Thus, the age composition of caught fish is regulated by the size of the net meshes, but the relationship between age and fish size is, generally speaking, statistical. This requires consideration of additional information associated with the corresponding statistical characteristics.

Frequently (as, for example, in the case of exploitation of a herd of large horned cattle) we have the possibility of externally adding a certain quantity of individuals to the population. This provides additional possibilities for control. In this case the system of equations takes on a somewhat different form:

$$\frac{\partial u}{\partial t}+\frac{\partial u}{\partial x}=-d(x, t)\,u+w(x, t),$$

$$u(0, t)=\int_0^\infty b(x, t)\,u\,dx+w(0, t),$$

which, in general, does not increase the difficulty of solution. The function w describes the rate of artificial influx or removal of individuals having the age x from the population at time t.

However, we shall for the time being neglect this possibility and consider problems of controlling isolated populations [$w(x, t) = 0$]. We shall assume that control constitutes an additional mortality rate $\mu(x, t)$ (the catching of fish, the slaughtering of cattle) in such a way that the mortality equation with control takes the form

$$\frac{\partial u}{\partial t} + \frac{\partial u}{\partial x} = -[d(x, t) + \mu(x, t)]\,u.$$

First of all let us consider the problem of determining the necessary ecological characteristics (i.e., using control-theory terminology, let us consider the problem of observability and identification). For this purpose let us recall the integral equation for the function $\Omega(t)$ of the number of newborn at time t (for simplicity we shall deal with the stationary characteristics $b(x)$ and $d(x)$):

$$\Omega(t) = \int_0^t K(x)\,\Omega(t-x)\,dx + \int_t^\infty K(x)\,\varphi(x-t)\,dx.$$

If we can measure the number of newborn in the population over a generation (i.e., for a change in t from 0 to S), then we obtain the integral equation

$$\int_t^\infty K(x)\,\varphi(x-t)\,dx = l(t),$$

from which, knowing $K(x)$, one can determine the initial age composition $\varphi(x)$ of the population, or, knowing $\varphi(x)$, one can determine the function $K(x)$ (i.e., a certain relationship between the birth and mortality rates). Then, knowing the birth rate $b(x)$, which is usually known more accurately than the mortality rate, one may determine the mortality rate $d(x)$ from the formula

$$d(x) = \frac{d}{dx} \ln \frac{b(x)}{K(x)}.$$

Actually, the initial distribution $\varphi(x)$ may be obtained by observing not only the function $\Omega(t)$ but also the dynamics of the population size of any fixed age, and even the dynamics of the overall population size. The latter is especially interesting, since the overall population size $N(t) = \int_0^\infty u(x, t)\,dx$ is the parameter that probably is most accessible to observation.

The second problem, naturally, is the problem of the attainability of a certain age structure $G(x)$ by means of our control $\mu(x, t)$ (i.e., using control-theory terminology, the problem of controllability).

In many cases of practical interest we are actually concerned with the problem of the attainability of a certain optimal stationary age structure $G(x)$, but for stationarity of a certain distribution it is required that the condition

$$G(0) = \int_0^\infty b(x)\,G(x)\,dx$$

be fulfilled, or, recalling that in the stationary case

$$G(x) = G(0)\,e^{-\int_0^x [d(\xi)+\mu(\xi)]\,d\xi}, \tag{43}$$

we obtain the condition

$$\int_0^\infty K(x)\, e^{-\int_0^x \mu(\xi)\,d\xi}\, dx = 1 \tag{44}$$

for the control. This equation is a constraint on the control $\mu(x)$ in the case of a stationary policy. We may state that all distributions $G(x)$ representable in the form (43), where $\mu(x)$ satisfies the condition (44), are stationary and attainable.

With this we shall end our brief discussion of controllability, observability, and identification problems, which of course require special consideration in view of their importance to practical problems.

As an example let us consider the problem of optimizing the age structure of a herd of large horned cattle. The exploitation of such a herd is connected with obtaining two basic forms of production – milk and meat, whose intensified production produces an obvious contradiction between them. Therefore, the age structure of the herd should be optimized on the basis of a certain resultant criterion. Such a criterion may be, for example, the income obtained by the farm from the exploitation of the herd. Let us begin by dwelling on the problem of the optimal stationary structure of the herd. In this case the equations take the simple form:

$$\left.\begin{aligned} \frac{du}{dx} &= -[d_u(x) + \mu_u(x)]\, u, \\ \frac{dv}{dx} &= -[d_v(x) + \mu_v(x)], \\ u(0) &= \int_0^\infty b_u(x)\, u\, dx, \\ v(0) &= \int_0^\infty b_v(x)\, u\,(dx). \end{aligned}\right\} \tag{45}$$

The income from the exploitation of the herd is made up of the income obtained from the sale of milk and meat less the expenditures required to maintain a herd of a stipulated size.

We shall not enter into detail in the present paper (i.e., we shall not write out the dependences of the milk productivity of the cattle and of the slaughtering rate on age, etc.), but it is completely clear that the optimality criterion will be a linear functional of the distributions $u(x)$ and $v(x)$, which may be treated as the independent variables instead of $\mu_u(x)$ and $\mu_v(x)$. The constraints are stipulated by Eqs. (45) and the limitations on feed which are stipulated by inequalities of the form:

$$\int_0^\infty L_u(x)\, u(x)\, dx \leqslant L_{u0}; \quad \int_0^\infty L_v(x)\, v(x)\, dx \leqslant L_{v0}, \tag{46}$$

where $L(x)$ is the amount of feed required for one individual of age x, and L_0 is the overall amount of feed of the given kind.

Thus, after quantization the problem of optimizing the age structure of a stationary herd reduces to a conventional problem in linear programming. The initial experience in solving this kind of problem for one of the state farms of the Leningrad Region shows that the income may be increased by 5 to 7% compared with the actually existing income by realizing the optimal age structure.

The problem of the optimal process of transition from a stipulated nonoptimal age structure to the optimal stationary age structure is more complex. This problem can be solved numerically by one of the existing methods.

The problem of optimizing the age structure of a herd for planned growth of its size is of great practical significance. First of all, having calculated the growth index λ_0 for $\mu_u = 0$ as the root of the equation

$$\int_0^\infty K(x)\, e^{-\lambda x}\, dx = 1,$$

we may establish the upper bound of the population growth rate in the form

$$N(t) \leqslant N_0 e^{\lambda_0 t}.$$

For optimization of some economic criterion one may pose the additional condition

$$\lambda = \lambda^* \leqslant \lambda_0,$$

which will guarantee the required planned growth of the herd size.

It is not difficult to see that all of the problems indicated may be related with equal success to the dynamics of a school of fish or other natural populations.

As a second example of the application of the mathematical model considered, let us dwell on the problem of planning the intensity with which agricultural pests are combatted. In this case one can state the problem of maintaining the population size at a certain stationary level. In the general statement of the problem the population size N(t) can be expressed by the equation

$$N(t) = \int_0^\infty \Omega(t-x)\, e^{-D(x,\, t)} e^{-\int_0^x \mu(\xi,\, t-x+\xi)\, d\xi}\, dx.$$

Our problem is to choose the function $\mu(x, t)$ in such a way that $N(t) \equiv N_0$, the function $\Omega(t)$ likewise being a complex functional of $\mu(x, t)$ – the solution of the corresponding integral equation with a kernel that depends on $\mu(x, t)$. Under these conditions some economic criterion may be optimized. The basic difficulty in solving the problem stated lies in the absence or inaccuracy of information on the natural birth and mortality rates for different ages.

Thus, in the present paper we have considered the mathematical model of the dynamics of the age composition of unisex and bisexual populations in stationary and nonstationary media. Based on the model constructed, problems of the control of population-size dynamics by choosing the mortality rate as a function of age and time have been discussed.

Literature Cited

1. S. P. Naumov, "General regularities of governing the population size of a species and its dynamics," in: Investigation of the Causes and Regularities of the Dynamics of the Population Size of the White Rabbit in Yakutia, Izd. AN SSSR (1960).
2. L. Z. Kaidanov, "On the problem of the role of behavior as a factor in microevolution," in: Issledovaniya po Genetike, Vol. 3, Izd. LGU (1967).
3. T. V. Koshkina, "Population density and its significance in regulating the population size of the red field vole," Byull. MOIP, Otdel. Biol., Vol. 20, No. 1 (1965).

4. T. V. Koshkina, "On periodic variations of the population size of field voles," Byull. MOIP, Otdel. Biol., Vol. 21, No. 3 (1966).
5. T. V. Koshkina, "Population control of rodents," Byull. MOIP, Vol. 22, No. 6 (1967).
6. C. S. Elton, Voles, Mice and Lemmings, Clarendon Press, Oxford (1942).
7. C. S. Elton and M. Nicholson, "The ten-year cycle in numbers of lynx," J. Animal Ecol., Vol. 11 (1942).
8. V. C. Wynne-Edwards, Animal Dispersion in Relation to Cosial Behavior, London (1962).
9. I. I. Christian, Endocrine Adaptive Mechanisms and the Physiological Regulation of Population Growth, London (1963).
10. V. Volterra, Lecons sur la Théorie Mathématique de la Lutte Pour la Vie, Paris (1931).
11. R. N. Chapman, J. Animal Ecol., London (1931).
12. U. D'Ancona, The Struggle for Existence, Leiden (1954).
13. A. Y. Lotka, Essays on Growth and Form, Clarendon Press, Oxford (1945).
14. I. A. Poletaev, "On the mathematical models of elementary processes and biogeocenoses," in: Problemy Kibernetiki, Vol. 16, Nauka, Moscow (1966).
15. T. I. Éman, "On certain mathematical models of biogeocenoses," in: Problemy Kibernetiki, Vol. 16, Nauka, Moscow (1966).
16. W. R. Utz and P. E. Waltman, "Periodicity and boundedness of solutions of the generalized differential equation of growth," Bulletin of Mathematical Biophysics, Vol. 25 (1963).
17. R. A. Fisher, The Genetical Theory of Natural Selection, Clarendon Press, Oxford (1930).
18. P. A. P. Moran, The Statistical Processes of Evolutionary Theory, Clarendon Press, Oxford (1962).
19. B. Ya. Levin, The Distribution of the Roots of Integer Functions, Gostekhizdat (1956).
20. Boyarskii (ed.), A Demography Course, Moscow (1967).
21. T. Harris, Theory of Branching Random Processes, Mir (1966).

ON THE CONTROL OF CARDIAC RHYTHM†

Yu. A. Vlasov and A. T. Kolotov

Novosibirsk

The importance of controlled variation of cardiac rhythm need hardly be stressed. A successful solution of this problem would make it possible to control a number of pathological conditions such as auricular and ventricular flutter, extrasystole, and high- or low-frequency rhythms.

Unfortunately, in spite of the fact that investigations in this direction continue for a relatively long time, practical solutions have been obtained in only a few most simple cases such as, for example, increasing the heart rate by means of independent cardiac stimulation.

The present article is an attempt to analyze with the aid of a model the effectiveness of control intervention in reducing the rate of spontaneous heart contractions.

The problem of changing the cardiac rhythm can be approached from two fundamentally different directions. To the first belong methods of changing the spontaneous activity of the automatic cardiac nodes (by various pharmacological means or by acting directly on the nervous system). Subsequent evolution of the excitation process proceeds without further intervention. In contrast, the second approach presupposes active intervention into the excitation process. As tools of such intervention serve various electrical stimulation devices that are being intensively developed in recent times.

In our discussion we shall deal mainly with the second approach.

It is well known that any spontaneous or induced extra contraction of the heart (extrasystole) can be followed by a prolonged (compensating) pause as a result of the fact that the next pulse arriving from the rhythm source is blocked by refractive cells. The first attempt to use this mechanism for clinical reduction of the rate of heart contraction has been made as recently as in 1963-1964 [3]. A decisive role in this lag is played apparently by the fact that most physicians associate extrasystole with a pathological condition, and this hindered the attempts to use extrasystole for slowing down the rhythm. Unfortunately, the method of paired cardiac stimulation especially developed for this purpose, in which regular application of a pair of stimuli to the heart produces extrasystole, does not provide reliable reduction of the rate of heart contraction. This circumstance forced experimenters to resort to various modifications of this method (such as varying the number and shape of the applied pulses, changing the location of the stimulating electrodes, etc.). In the best case, the paired stimulation method ensures an approximately twofold reduction of the rhythm rate.

†Original article submitted July 23, 1968.

At present, there is no comprehensive theory concerning the interaction among cellular elements of the heart which could serve as a basis for developing reliable methods of controlling the rate of heart contractions. In such a situation it is quite important to analyze systematically all the factors relating to this problem.

I. Problem Formulation

We shall base our discussion on the model described in [1].

Consider a connected net of cells T which has two poles: A, the net input, and B, the net output.

If to the input A we apply pulses so that the time interval Θ between two conseutive pulses is longer than the refractivity period of the cells ($\Theta > n$), the output pulses at B will be of the same periodicity. (If desirable, the pole A can be assumed to be capable of periodical self-excitation.)

The discussed problem can now be formulated as follows: how can the rate of the output signals of the Net T be reduced without changing the rate of input signals? (It is assumed that the net structure is fixed and that only stimuli are allowed to act on the net.)

It should be noted that the pole B cannot be completely blocked, i.e., its excitation cannot be discontinued after a certain finite time interval.

In fact, otherwise the set of all cells of the net T would be divided into two disjoint subsets $\mathfrak{M}$ and $\mathfrak{N}$, where $\mathfrak{M}$ is the set of all blocked cells and $\mathfrak{N}$ is the set of all remaining cells. None of these sets is empty as the pole B belongs to the first set and the pole A to the second. In virtue of its connectivity the net T will necessarily contain a pair of adjoining cells a_i and a_j (directly connected one with the other) such that $a_i \in \mathfrak{N}$ and $a_j \in \mathfrak{M}$. But in such a case the cell a_i would be excited an infinite number of times. Consequently, after a certain finite time interval the cell a_i will force the cell a_j to fire as soon as the latter turns into a quiescent state, i.e., the cell a_j cannot be blocked, which contradicts the condition of its choice.

We have answered the problem stated under the following assumption: for any section of the net one can find a stimulus such that its region of application coincides with the given section; in other words, every cell of the net can serve as its input. Let the sequence of stimuli applied to the net T be called *the experiment* $\varepsilon(T)$ *on the net T*.

It is clear that an arbitrary experiment $\varepsilon(T)$ can change the sequence of output signals.

II. Solution

We will show that the pause between two consecutive stimulations of the pole B is limited to a certain value independent of the experiment.

Let T* be an arbitrary connected subnet of the net T such that both the poles A and B are contained in it. For any experiment $\varepsilon(T)$ on the net T in which the maximum pause at the output of this net (i.e., the maximum interval between two consecutive stimulations of the pole B) is equal to σ, we can indicate a certain experiment $\varepsilon^*(T^*)$ on the net T* that gives a maximum pause $\sigma^* \geq \sigma$. (It is assumed that the input and output of the net T* are the same poles A and B, and that A receives pulses with an initial periodicity Θ.)

In fact, such an experiment $\varepsilon^*(T^*)$ can be realized by retaining all stimuli that affect the subnet T* in the original experiment $\varepsilon(T)$ and replacing all stimuli from the direction $T \setminus T^*$ with T* equivalent stimuli. Hence, in particular, follows that any upper bound of attainable pauses at the output of the net T* is at the same time the upper bound of the attainable pauses in T.

Let $L = \{a_1, a_2, \ldots, a_l\}$, where $a_1 = A$ and $a_l = B$, be the shortest chain of cells connecting both poles, and let l be the length of this chain. Considering L as an autonomous net, we shall apply to A pulses with a periodicity Θ. Let the arbitrary instant t_0 of stimulation of the pole B be taken as the origin, and let $t_0 = 0$. Any cell a $(1 \leq i \leq l-1)$ being at the instant t in an excited state will force the cell a_{i+1} to fire at the next $(t+1)$-st instant provided the cell a_{i+1} has not been in a refractive phase at the instant t.

Thus if even a single cell a_i $(1 \leq i \leq l-1)$ is excited in the interval $[t_1, t_2]$, then at least one excitation of the cell a_{i+1} will take place in the interval $[t_1 - (n-1), t_2 + 1]$, where n is the period of refractivity and 1 is the magnitude of the latent period of reaction.

Reasoning as above we arrive at the following conclusion: if the pole A (cell a_1) is excited only once in the time interval $[t_1, t_2]$, then:

a_2 will be excited at least once in the interval $[t_1 - (n-1), t_2 + 1]$,

a_3 will be excited at least once in the interval $[t_1 - 2(n-1), t_2 + 2]$,

. . . .,

and, finally, the pole B (cell a_l) will be excited at least once in the time interval $[t_1 - (l-1)(n-1), t_2 + l - 1]$.

It now remains to choose the appropriate values of t_1 and t_2. Since the pole A receives external pulses with a period Θ, it must be excited at least once during any time interval of the form $[t, t + \Theta + n - 1]$. Thus, we take $t_2 = t_1 + \Theta + n - 1$. To find t_1, note that if the interval $[t_1 - (l-1)(n-1), t_2 + l - 1]$ does not contain the point $t_0 = 0$, then any excitation of the pole B within this interval will be distinct from the initial stimulation. We can thus assume $t_1 - (l-1)(n-1) = 1$.

Thus, $t_1 = (l-1)(n-1) + 1$ and $t_2 = t_1 + \Theta + n - 1$.

Hence, the pole B must be excited at least once within the time interval $[t_1 - (l-1)(n-1), t_2 + l - 1] = [1, (l-1)(n-1) + 1 + \Theta + n - 1 + l - 1] = [1, ln + \Theta]$, i.e., the pole B will be stimulated a second time not later than the instant $t = ln + \Theta$.

Now, regarding the chain L as $T^* \subset T$, we can make use of the previous remark. Thus, the quantity $ln + \Theta$ is the upper bound of the admissible pauses also for the original net T.

Thus, if a certain experiment $\varepsilon(T)$ on the net T reduces the rate of output signals, then the initial pause equal to Θ can be lengthened by not more than ln, where l is the length of the shortest chain of cells that joins both poles, and n is the refractive period of the cells.

The established upper bound is an attainable one, i.e., it is possible to devise an experiment $\varepsilon(T)$ in which the maximum pause at the output of the net is lengthened by exactly ln as compared with the initial pause Θ. However, if Θ, l, and n are arbitrary, such an experiment, generally speaking, cannot guarantee constant pauses between the output pulses since some pauses are liable to be lengthened at the expense of others. We shall thus consider the case in which the pole A is stimulated at a high rate (i.e., Θ is, roughly speaking, identical with n) and l is sufficiently long.

The experiment proposed below ensures significant reduction of the rate of output signal of the net T while at the same time keeping a constant interval between consecutive excitations of the pole B. (Sacrificing the constancy of the intervals, it is possible by means of a slight modification to obtain the maximum pause $ln + \Theta$ at the output.)

Let us associate the number 1 with the pole A. With each cell in the neighborhood of the pole A (i.e., with each cell directly connected with A) let us associate the number 2. Further, if the number i − 1 has been used already, we shall assign the number i to all those unnumbered cells that belong to the neighborhood of at least one cell with the number i − 1. Thus, the cell a has the number i assigned to it if there is a chain that joins this cell with the pole A and contains exactly i cells, but there is no connecting chain containing less cells.

The set of cells to which the number i is assigned is called the i-th layer and denoted by P_i. Clearly, $B \in P_l$.

The totality of cells belonging to all layers whose number is lower than k + 1 is called the domain R_k, i.e., $R_k = \bigcup_{i=1}^{k} P_i \ (k = 1, 2, \ldots)$.

We will successively narrow down the region of application of stimuli exciting at first the domain R_l (completely), and then R_{l-1}, R_{l-2}, and so on down to the domain R_1 whose only point is the pole A. Each time we shall delay as far as possible the application of the next stimulus, but under the condition, however, that at the instant of excitation of the stimulated domain R_k there is around it a barrier of refractivity formed by the layer P_{k+1} as a result of the preceding stimulus. (For this the time interval between two consecutive stimuli should be equal to n − 1.) As soon as the domain of application of stimuli contracts to a single point, we, after waiting for a time equal to the period of refractivity, once again cover the entire domain R_1. This procedure is then repeated.

It is evident that the proposed experiment guarantees at the output of net T a constant pause $\sigma = l(n-1)$, i.e., the pause is shorter than the maximum pause only by l, which is equal to the time of propagation of the excitation wave from the pole A to the pole B.

III. Interpretation

The concept of refraction used above is local in nature since it relates to an individual cell and not to the heart as a whole. This property makes it possible in principle to exploit the inhomogeneity of the different parts of the heart in respect to refractivity.

One of the foregoing remarks leads to the following conclusion: the atrioventricular A−B node cannot be blocked by any stimulations, i.e., it is impossible to secure complete rest of the A−B node with the periodic activity of the sinus node remaining unchanged. Thus, either we allow some pulse to pass from the sinus node to the A−B node, or pulses of spontaneous origin due to the blocking stimulation will break through to the A−B node in the course of interception of pulses from the sinus node.

It should be kept in mind that a spontaneous periodic excitation that suppresses the desired effect arises in the A−B node when the pause between two transmitted pulses exceeds some critical length. This can be compared with the shift of the rhythm carrier leading to the center of the second-order automatism that takes place in the case of a complete transverse heart block caused by anatomical interruption of the conducting tract. In fact, in the considered situation we also deal with a block but of an entirely functional character.

As an experimental test of the results obtained with the described model we can conceive the following experiment which can be realized in two variants: 1) on a complete working organ and 2) on an isolated muscular strip from the heart wall.

1. On the external surface of the heart auricles let us mark out a sufficiently large number of zones imbedded one into the other so that the maximal zone borders on the A−B node and the sinus node lies at the center of the minimal zone (the most inner one); this is in practice the region of the ostium of the superior vena cava. Each marked zone is fitted with separate electrodes arranged around its perimeter. The first stimulus is applied at once to all

electrodes. After a certain time interval, somewhat shorter than the refractivity period, the second stimulus is applied to all electrodes except those around the maximal zone, etc. After a number of steps, equal to the number of zones marked out, the stimulus is applied to the electrodes of the minimal zone containing the sinus node. The entire procedure is then repeated.

2. In the second version we isolate from the heart wall a muscular strip so that it contains no cells capable of spontaneous excitation (cells of the conductive system). On the strip we select two poles to one of which we apply stimulating pulses of a constant frequency whose passage is recorded at the other pole. From here on the experiment is conducted as in the preceding version.

In the first version of the experiment we should expect spontaneous excitation of the center of the second-order automatism. In the second case, we should obtain a maximum pause between two consecutive excitations of the output pole of the strip. In both versions we neglect the thickness of the myocardium of the auricles as well as the thickness of the myocardial strip and regard both of them as flat muscular layers.

By interrupting in the model experiment the sequence of stimuli at an appropriate instant we can obtain practically any retardation from the minimum possible to the maximum obtainable, i.e., practically any frequency both above and below the frequency of spontaneous excitation in the sinus node. Thus, we now deal with a case in which the heart rhythm can be varied at will in any desired direction. We wish once more to stress that the proposed concept of reducing the heart rhythm is based on multiplying the unit delay by a factor which is a multiple of the number of cells in the shortest chain between the sinus node and the atrioventricular node (i.e., using the local property of refractivity).

Literature Cited

1. A. T. Kolotov, An Automatic Model of the Heart, in: Systems Theory Research, Vol. 20, Consultants Bureau, New York (1971), p. 210.
2. P. F. Cranefield, "The force of contraction of extrasystoles and the potentiation of force of the postextrasystolic contraction: a hystorical review," Bull. N. Y. Acad. Med., 41(5):419 (1965).
3. J. E. Lopez, A. Edelist, and L. N. Katz, "Reducing heart rate of the dog by electrical stimulation," Circul. Research, 15:414 (1964).

BRIEF COMMUNICATIONS

A NOTE ON DETERMINISTIC LINEAR LANGUAGES†

A. Ya. Dikovskii

Novosibirsk

§ 1. Basic Concepts

Definition. A finite automaton with two tapes (in short, a 2-K-automaton) is specified by an ordered sextupole $S=\langle K, \Sigma \cup \{\#\}, q_1, q_0, \varepsilon, \delta\rangle$, where: 1, 2) K and Σ are finite sets (of states and input symbols), and $\# \bar{\in} \Sigma$ (a right boundary marker); 3) $q_1 \in K$ (an initial state); 4) $q_0 \in K$ (a terminal state); 5) $\varepsilon \bar{\in} \Sigma \cup \{\#\}$ (an auxiliary symbol); 6) δ is a mapping of the set $(\Sigma \cup \{\#\} \cup \{\varepsilon\} \times K \times (\Sigma \cup \{\#\} \cup \{\varepsilon\}) - \{\varepsilon\} \times K \times \{\varepsilon\}$ into the set of all subsets of K.‡

Definition. Any ordered triple α belonging to the set $\Omega = (\Sigma^* \{\nabla\} \Sigma^* \{\#\}) \times K \times (\Sigma^* \{\nabla\} \Sigma^* \{\#\})$ §), where $\nabla \bar{\in} \Sigma \cup \{\#\} \cup \{\varepsilon\} \cup K$ (the indicator of the location of the reading head) is called a configuration of the 2-K-automaton S.

The mapping δ induces the following "$\vdash$" relation on the set Ω:

$$(1)\ (x_1\nabla a x_2, q, y_1\nabla b y_2) \vdash (x_1 a \nabla x_2, q', y_1 b \nabla y_2), \quad \text{if} \quad q' \in \delta(a, q, b);$$
$$(2)\ (x_1\nabla a x_2, q, y_1\nabla y_2) \vdash (x_1 a \nabla x_2, q', y_1 \nabla y_2), \quad \text{if} \quad q' \in \delta(a, q, \varepsilon);$$
$$(3)\ (x_1\nabla x_2, q, y_1\nabla b y_2) \vdash (x_1 \nabla x_2, q', y_1 b \nabla y_2), \quad \text{if} \quad q' \in \delta(\varepsilon, q, b)$$

for any $q, q' \in K$; $x_1, y_1 \in \Sigma^*$; $x_2, y_2 \in \Sigma^*\{\#\}$; $a, b \in \Sigma \cup \{\#\}$.

Let $\alpha, \beta \in \Omega$. We shall write $\alpha \models \beta$, if there exist $\alpha_1, \alpha_2, \ldots, \alpha_k \in \Omega$ such that $\alpha_1 = \alpha$, $\alpha_k = \beta$, and $\alpha_i \vdash \alpha_{i+1} (1 \leqslant i < k)$.

Definition. A string pair $\langle x, y\rangle \in \Sigma^* \times \Sigma^*$ is allowed by a 2-K-automaton S if $(\nabla x\#, q_1, \nabla y\#) \models (x\#\nabla, q_0, y\#\nabla)$. The set $L(S) = \{x\hat{y} \mid \langle x, y\rangle$ is allowed by a 2-K-automaton S$\}$¶ is called a language allowed by a 2-K-automaton S. The set $L^2(S) = \{\langle x, y\rangle \mid \langle x, y\rangle$ is allowed by a 2-K-automaton S$\}$ is called an event representable in S.

Definition. A context-free grammar†† $\Gamma = \langle V, V_1, \Pi p, P\rangle$ is said to be linear (and the language generated by it is called a linear language) if the scheme P contains only rules of the form $A \to aBb$, $A \to aB$, $A \to Bb$, $A \to a$, where $A, B \in V_1$ and $a, b \in V$.

†Original article submitted June 27, 1968.

‡ Sometimes the inclusion $q' \in \delta(\xi, q, \eta)$ will be written in the form of an instruction $(\xi, q, \eta) \to q'$.

§ Let X and Y be sets of strings. $XY \overset{df}{=} \{xy \mid x \in X, y \in Y$, xy being a concatenation of x and y$\}$.

$X^0 \overset{df}{=} \{\Lambda\}$, Λ being the empty string; $X^{i+1} \overset{df}{=} X^i X$; $X^* \overset{df}{=} \bigcup_{i=0}^{\infty} X^i$.

¶ $\hat{\Lambda} \overset{df}{=} \Lambda$, $\widehat{xa} \overset{df}{=} a\hat{x}$, $\hat{L} \overset{df}{=} \{y \mid \hat{y} \in L\}$, i.e., $\hat{x}$ is an inversion of the word x.

†† The principal concepts of the theory of grammars are assumed known.

Theorem 1. A language L is linear if and only if there exists a 2-K-automaton S such that L = L(S) (see [3], Theorem 11).

Definition. A diagram of a 2-K-automaton $S=\langle K, \Sigma \cup \{\#\}, q_1, q_0, \varepsilon, \delta\rangle$ is defined as a directed graph D_S of the following form:

1) The vertices of D_S are states of K; 2) the vertex q is connected with the vertex q' by an

a) arc $q \xrightarrow[b]{a} q'$, if $q' \in \delta(a, q, b)$;

b) arc $q \xrightarrow[\Lambda]{a} q'$, if $q' \in \delta(a, q, \varepsilon)$;

c) arc $q \xrightarrow[b]{\Lambda} q'$, if $q' \in \delta(\varepsilon, q, b)$;

3) The vertices of D_S are connected by arcs only by virtue of a)-b).

In D_S let us consider a path from the vertex q to the vertex q'

$$\pi(q, x, y, q') = q \xrightarrow[b(1)]{a(1)} q(1) \xrightarrow[b(2)]{a(2)} q(2) \to \ldots \to q(n-1) \xrightarrow[b(n)]{a(n)} q' \dagger$$

(it is possible that some vertices q(i) and q(j) coincide with one another or with q or q', and that some a(i) and b(j) are equal to Λ) such that $a(1)a(2) \ldots a(n) = x$ and $b(1)b(2) \ldots b(n) = y$. By W(q, x, y, q') we shall denote the set of all paths π(q, x, y, q'). Moreover, let $I(D_S)$ be the set of all pairs of strings $\langle x, y\rangle$ such that $x, y \in \Sigma^*$ and suppose that in D_S there exists a path $\pi(q_1, x\#, y\#, q_0)$. It is evident that $L^2(S) = I(D_S)$.

Definition. A 2-K-automaton $S=\langle K, \Sigma \cup \{\#\}, q_1, q_0, \varepsilon, \delta\rangle$ is said to be deterministic if the following conditions hold:

1) If $\mathfrak{A} \to q$ and $\mathfrak{A} \to q'$ are instructions of S, then q = q';

2) if Q_1, Q_2, and Q_3 are sets of states encountered in the left-hand sides of instructions of S of the form $(a, q, b) \to q'$, $(a, q, \varepsilon) \to q'$, $(\varepsilon, q, b) \to q'$, then Q_1, Q_2, and Q_3 will be pairwise disjoint.

Definition. A language is said to be 2-K-deterministic if it is allowed by a deterministic 2-K-automaton.

We have the following evident

Lemma 1. Let S be a 2-K-automaton and D_S its diagram. Then:

1) If

$$\pi(q, x_1, y_1, q') = q \begin{smallmatrix} a(1) \\ b(1) \end{smallmatrix} q(1) \begin{smallmatrix} a(2) \\ b(2) \end{smallmatrix} q(2) \ldots q(n-1) \begin{smallmatrix} a(n) \\ b(n) \end{smallmatrix} q'$$

and

$$\pi(q', x_2, y_2, q'') = q' \begin{smallmatrix} a'(1) \\ b'(1) \end{smallmatrix} q'(1) \begin{smallmatrix} a'(2) \\ b'(2) \end{smallmatrix} q'(2) \ldots q'(m-1) \begin{smallmatrix} a'(m) \\ b'(m) \end{smallmatrix} q''$$

are paths in D_S, then

$$\pi(q, x_1x_2, y_1y_2, q'') = \pi(q, x_1, y_1, q')\,\pi(q', x_2, y_2, q'') =$$
$$= q \begin{smallmatrix} a(1) \\ b(1) \end{smallmatrix} q(1) \begin{smallmatrix} a(2) \\ b(2) \end{smallmatrix} q(2) \ldots q(n-1) \begin{smallmatrix} a(n) \\ b(n) \end{smallmatrix} q' \begin{smallmatrix} a'(1) \\ b'(1) \end{smallmatrix} q'(1) \begin{smallmatrix} a'(2) \\ b'(2) \end{smallmatrix} q'(2) \ldots q'(m-1) \begin{smallmatrix} a'(m) \\ b'(m) \end{smallmatrix} q''$$

will also be paths in D_S;

† In the following we shall drop the arrows.

2) if S is a deterministic 2-K-automaton, any nonempty set $W(q, x, y, q')$ will be a one-element set.

Corollary. Any 2-K-deterministic language is a well-defined language.†

Let $\mathscr{K}$ be the class of all CF-languages, $\mathscr{K}^D$ the class of all CF-languages allowed by deterministic automata with a storage memory, $\mathscr{L}^0$ the class of all well-defined linear CF-languages, $\mathscr{L}$ the class of all linear CF-languages, and $\mathscr{L}^D$ the class of all 2-K-deterministic languages.

§ 2. Relations between the Classes $\mathscr{L}$, $\mathscr{L}^D$, $\mathscr{L}^0$ and $\mathscr{K}^D$

Let us consider a linear language $L=\{xc\hat{x}y \mid x\in\Sigma^*,\ c\bar{\in}\Sigma,\ y\in(d\Sigma_1^*\cup\{\Lambda\}),\ \Sigma=\Sigma_1\cup\{d\},\ d\bar{\in}\Sigma_1,\ |\Sigma|\geqslant 3\}$ ‡

Lemma 2. Any 2-K-automaton S that allows a language L satisfies the condition $\exists k>0\forall x_1,\ x_2,\ x,\ y\,[x=x_1x_2\,\&\,((\langle x_1, \hat{y}xc\hat{x}_2\rangle\in L^2(S)\vee\langle xcx_2, yx_1\rangle\in L^2(S))\supset l(x_2)\leqslant k]$.

Proof. Suppose that there exists a 2-K-automaton S with a diagram D_S that allows a languate L such that one of the following conditions holds:

a) There exists a sequence of natural numbers $k_1 < k_2 < k_3 < \ldots < k_n < \ldots$, such that $\forall i>0\exists x_1,\ x_2,\ x,\ y\,[x=x_1x_2\,\&\,l(x_2)=k_i\,\&\,\langle x_1, \hat{y}xc\hat{x}_2\rangle\in L^2(S)]$;

b) there exists a sequence of natural numbers $l_1 < l_2 < \ldots < l_j < \ldots$, such that $\forall j>0\exists x_1,\ x_2,\ x,\ y\,[x=x_1x_2\,\&\,l(x_2)=l_j\,\&\,\langle xc\hat{x}_2, \hat{y}x_1\rangle\in L^2(S)]$. Let us assume that Condition (b) holds (the simpler case (a) can be analyzed in a similar way). We shall consider the set $F=\{\langle xc\hat{x}_2, \hat{y}x_1\rangle\}$ of all pairs of strings yielded by this condition. It is easy to see that for a sufficiently large j_0 there exists a pair $\langle x^0c\hat{x}_2^0, \hat{y}^0x_1^0\rangle\in F$ and a path $\pi^0(q_1, x^0c\hat{x}_0^2\#, \hat{y}^0x_1^0\#, q_0)\in W(q_1, x^0c\hat{x}_2^0\#, \hat{y}^0x_1^0\#, q_0)$ such that $\pi^0(q_1, x^0c\hat{x}_2^0\#, \hat{y}^0x_1^0\#, q_0)=q_{b(1)}^{a(1)}q(1)_{b(2)}^{a(2)}q(2)\ldots q(\gamma-1)_{b(\gamma)}^{a(\gamma)}q(\gamma)\ \ldots\ q(\varkappa-1)_{b(\varkappa)}^{a(\varkappa)}q(\varkappa)\ \ldots\ q(\lambda-1)\,{}_{b(\lambda)}^{a(\lambda)}q(\lambda)\ \ldots\ q(m-1)_{b(m)}^{a(m)}q_0$ and that $a(\gamma)=c,\ q(\varkappa-1)=q(\lambda),\ a(\varkappa)a(\varkappa+1)\ldots a(\lambda)\neq\Lambda$ and $l(x_2^0)=lj_0$ (otherwise the lengths of strings x_2 such that $\langle xcx_2, yx_1\rangle\in F$, are bounded by the number of states of the 2-K-automaton S). Let us consider the following paths and strings: $q_{1b(1)}^{a(1)}q(1)\ldots q(\varkappa-1)\overset{df}{=}\bar{\pi}^1$, $q(\varkappa-1)_{b(\varkappa)}^{a(\varkappa)}q(\varkappa)\ \ldots\ q(\lambda)\overset{df}{=}\bar{\pi}^2$, $q(\lambda)\ldots q(m-1)_{b(m)}^{a(m)}q_0\overset{df}{=}\bar{\pi}^3$, $a(\gamma+1)\ldots a(\varkappa-1)\overset{df}{=}\bar{u}_1$, $a(\varkappa)\ldots a(\lambda)\overset{df}{=}z$, $a(\lambda+1)\ldots a(\varphi-1)\,b(\psi-1)\ldots b(\lambda+1)\overset{df}{=}\bar{v}$ (where φ and ψ are specified by the relations $a(\varphi)=b(\psi)=\#$), $b(\lambda)\ldots b(\varkappa)\overset{df}{=}z'$ and $b(\varkappa-1)\ldots b(1)\overset{df}{=}\bar{u}_2$. It is evident that $\pi^0(q_1, x^0c\hat{x}_2^0\#, \hat{y}^0x_1^0\#, q_0)=\bar{\pi}^1\bar{\pi}^2\bar{\pi}^3$ and $x^0c\hat{x}^0y^0=x^0c\bar{u}_1z\bar{v}z'\bar{u}_2$. Furthermore, for any $t>0$ we have in D_S the path $\underbrace{\bar{\pi}^1\bar{\pi}^2\bar{\pi}^2\ldots\bar{\pi}^2}_{t\text{ times}}\bar{\pi}^3$ (Lemma 1 of Sec. 1), and hence for any $t>0$ we have $x^0c\bar{u}_1(z)^t\,\bar{v}\,(z')^t\,\bar{u}_2\in L$. Let us note that $\bar{v}z'\bar{u}_2=wy^0$ and $\bar{u}_1zw=\hat{x}^0$ for an appropriate string w. Since $l(z)>0$, there exists a t_0 such that $l[(z)^{t_0}]>l(x^0)$. Let $y^0\in d\Sigma_1^*$ (the case $y^0=\Lambda$ is trivial). For an appropriate string $\bar{w}$ we then have the relation $\bar{v}(z')^{t_0}\bar{u}_2=\bar{w}\bar{y}^0$, where $\bar{y}^0\in d\Sigma_1^*$. On the other hand, since $x^0c\bar{u}_1(z)^{t_0}\bar{v}(z')^{t_0}\bar{u}_2\in L$, it follows that $\bar{u}_1(z)^{t_0}=\hat{x}^0\bar{\bar{w}}$ for an

† A CF-grammar Γ is said to be well-defined if for any string belonging to L(Γ) there exists a unique left-side derivation (i.e., a derivation in which the rule is applied at each step to the leftmost auxiliary symbol). A CF-language is said to be well-defined if there exists a well-defined CF-grammar generating this language.

‡ $|\Sigma|$ is the number of elements of the set Σ.

appropriate string $\bar{\bar{w}}$. But in this case we have $\overline{\overline{ww}y^0} \in d\Sigma_1^*$, which cannot be the case, since $l(\overline{w\bar{\bar{w}}}) > 0$ and $\bar{y}^0 \in d\Sigma_1^*$. This completes the proof of the lemma.

Lemma 3. If there exists a deterministic 2-K-automaton S that allows a language L, it is possible to construct from it effectively another, equivalent, deterministic 2-K-automaton $\bar{S}$ that satisfies the additional condition $\forall_x^\infty \forall_y [xc\hat{x}y \in L \leftrightarrow \langle xc, \bar{y}x\rangle \in L^2(\bar{S})]$.†

Outline of Proof. By A we shall denote the set of paths $\pi(q_1, z_1\#, z_2\#, q_0)$ in the diagram D_S of the automaton S that satisfy the condition: $\exists x, y, w\,[xc\hat{x}y = z_1\hat{z}_2 \,\&\, w\hat{z}_2 = y \,\&\, l(w) > 0]$. By B we shall denote the set of other paths in D_S leading from q_1 to q_0. By reasoning in the same way as in Lemma 2, we can show that the set X_A is finite. On the basis of the diagram D_S it is easy to construct a diagram D_S' in which any c-A-path (i.e., part of an A-path of the form $q_{b(\alpha)}^{a(\alpha)}q' \ldots q_0$, where either $a(\alpha) = c$, or $b(\alpha) = c$) is incident to any c-B-path only at the vertex q_0. Indeed, D_S does not contain any fragment of the form

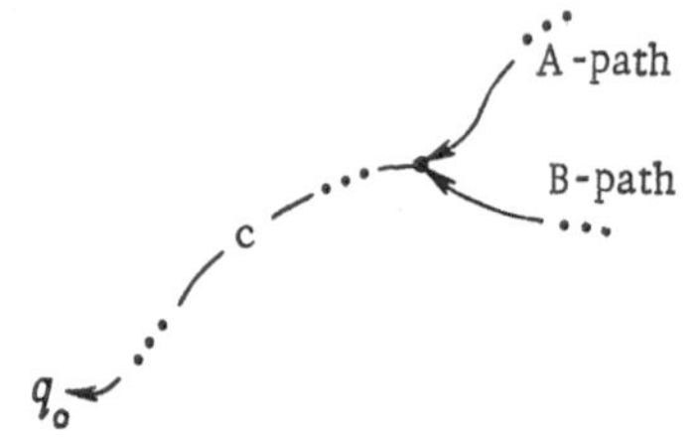

(by the definition of an A-path), and no fragment of the form

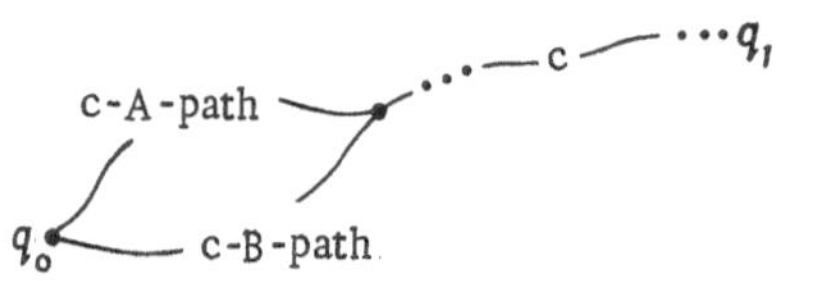

where the path from q_1 to c is traversed by a sufficiently long string x (since as a result of the deterministic property the paths from q to q_0 are distinct and the traversal of one of them does not violate the "mirror image" with respect to c). If the fragment D_S has the form

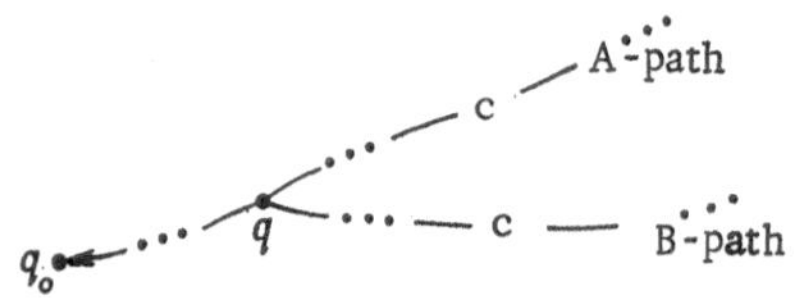

the common section from q to q_0 must be "split," by adding a limited number of new vertices. By reasoning in the same way, we can construct on the basis of the diagram D_S' a diagram D_S'' in which any two c-B-paths are incident only to the vertex q_0. By virtue of Lemma 2 the c-B-paths D" do not contain cycles. In this case the c-B-path $q(1)_{b(1)}^{a(1)} q(2)_{b(2)}^{a(2)} q(3) \ldots q(i)_{b(i)}^{a(i)} \ldots q(j)_{b(j)}^{a(j)} q_0$, where (for definiteness) $a(1) = c$, $a(i) = b(j) = \#$, can be replaced by a path $q(1)_{\bar{b}(1)}^{\bar{a}(1)} p(2)_{\bar{b}(2)}^{\bar{a}(2)} p(3) \ldots p(l)_{\bar{b}(l)}^{\bar{a}(l)} q_0$, where the vertices p(2), ..., p(l) are new vertices, $\bar{a}(l) = \bar{b}(l) = \#$, $\bar{a}(1) = c$, $\bar{a}(2) = \bar{a}(3) = \ldots = \bar{a}(l-1) = \Lambda$ and $\bar{b}(1)\bar{b}(2) \ldots \bar{b}(l-1) = b(1)b(2) \ldots b(j-1)\,a(i-1) \ldots a(3)\,a(2)$. The diagram $D_{\bar{S}}$ obtained by this transformation is the diagram of the sought-for 2-K-auto-

† $\forall_x^\infty P(x) \overset{df}{=} P(x)$ holds for all x, with the possible exception of finitely many.

maton $\bar{S}$ (since the condition $\forall y\,[xcxy \in L \leftrightarrow \langle xc, \hat{y}x\rangle \in L^2(\bar{S})]$ is not satisfied by strings belonging to a finite set).

<u>Theorem 2.</u> There does not exist a deterministic 2-K-automaton that allows a language L.

<u>Proof.</u> Suppose that there exists a deterministic 2-K-automaton S that allows a language L. On its basis let us construct a deterministic 2-K-automaton $\bar{S}$ as in Lemma 3. Let us consider a sufficiently long string $y \in \Sigma_1^* d$; $yc\hat{y}\hat{y} \in L$. Then $W(q_1, yc\#, yy\#, q_0) \neq \phi$. Next, $yc\hat{y} \in L$, hence $W(q_1, yc\#, y\#, q_0) \neq \phi$. Since in $D_{\bar{S}}$ any B-path terminates with an arc $q \overset{\#}{\underset{\#}{\to}} q_0$, this being the only arc originating at q, we have

$$\pi(q_1, yc\#, y\#, q_0) = \pi(q_1, yc, y, q)\, q^{\#}_{\#}\, q_0.$$

It follows from Lemma 1 of Sec. 2 that the path

$$\pi(q_1, yc\#, yy\#, q_0) = \pi(q_1, yc, y, q)\,\pi' q_0$$

is unique in the set $W(q_1, yc\#, yy\#, q_0)$. Yet at the vertex q there originates a unique arc $q \overset{\#}{\underset{\#}{\to}} q_0$ with $y \neq \Lambda$. Therefore, $yc\hat{y}\hat{y} \bar{\in} L$, which contradicts our assumption.

Let us go over to relations between the classes $\mathcal{L}$, $\mathcal{L}^D$, $\mathcal{L}^0$ and $\mathcal{K}^D$. We can present the following table (the symbols

$$\supset, \subset, \infty,$$

standing at the intersection of a row of class X and of a column of class Y signify

$$X \supsetneq Y, \quad X \subsetneq Y;$$
$$X \cap Y \neq \phi \;\&\; X - Y \neq \phi \neq Y - X$$

respectively):

	$\mathcal{L}$	$\mathcal{L}^D$	$\mathcal{L}^0$	$\mathcal{K}^D$
$\mathcal{L}$				
$\mathcal{L}^D$	$\subset$ [1]			
$\mathcal{L}^0$	$\subset$ [2]	$\supset$ [3]		
$\mathcal{K}^D$	∞ [4]	∞ [5]	∞ [6]	

Relation 2 follows from the fact that the linear language

$$L' = \{x \mid x = a^n b^k a^n b^l \quad \text{or} \quad x = a^k b^m a^l b^m;\ k, l, m, n \geqslant 1\}$$

is essentially undefined (i.e., there does not exist a well-defined CF-grammar generating it) [2]. The inclusions 1 and 3 follow from Theorem 2.† Relations 4, 5, and 6 follow from the

† The inclusion 1 follows also from the corollary of Lemma 1 and from the fact that the language L' is essentially undefined. Our attention was drawn to this fact and to relation 3 by A. V. Gladkii.

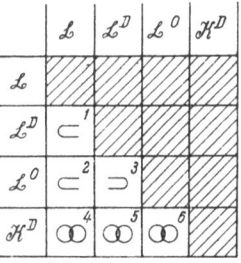

fact that the language $L'' = \{x\hat{x} \mid x \in \Sigma^*, \Sigma$ contains at least two symbols$\}$ belongs to class $\mathcal{L}^D$ and does not belong to $\mathcal{K}^D$ (the latter is proved in [1] and [4]).

The author expresses his deep gratitude to A. V. Gladkii for valuable remarks.

Literature Cited

1. S. Ginsburg and S. Greibach, "Deterministic context-free languages," Information and Control, 9:6 (1966).
2. R. Parikh, "Language-generating devices," RLE Quart. Progr. Rept., MIT, Cambridge, Mass, No. 60, pp. 199-212.
3. A. L. Rosenberg, "A machine realization of the linear context-free languages," Information and Control, 10(2):175-188 (1967).
4. A. Ya. Dikovskii, "Relations between the class of all context-free languages and the class of deterministic context-free languages," Algebra i Logika, 7(3):23-37 (1968).

NONRECURRENT CODES WITH MINIMAL DECODING COMPLEXITY †

A. A. Markov

Gorki

Let U be a finite system of distinct words in the alphabet A, $\mathfrak{A}$ a free semigroup over A, [U] a subsemigroup of $\mathfrak{A}$, generated by the set U, and λ the empty word. By $\|X\|$ we shall denote the number of elements of the set X, and by $|x|$ the length of the word x.

To the system of words U we shall assign a finite rooted digraph $\Gamma(U)$ with a set of vertices V formed by some suffixes of the words of U, the set of edges E, and a function φ with values ± 1 defined on E:

1°. $\lambda \in V$ and is a root of $\Gamma(U)$.

2°. If $u = u'a$ $(u, u' \in U, \ a \in \mathfrak{A}, \ a \neq \lambda)$, then $u' \in V, \ a \in V$ and $(\lambda, u'), (u', a) \in E$.

3°. If $a \in V$ and $aa' \in U (aa' \in \mathfrak{A}, a \neq \lambda, a' \neq \lambda)$, then $a' \in V$ and $(a, a') \in E$.

4°. If $a \in V$ and $a = ua' \ (u \in U, \ aa' \in \mathfrak{A}, \ a \neq \lambda, \ a' \neq \lambda)$, then $a' \in V$ and $(a, a') \in E$.

5°. The graph $\Gamma(U)$ contains only the vertices and edges that can be obtained by Rules 1°-4°. If the pair (a, a') can be obtained both by 3° and by 4°, the set E will contain the pair of edges (a, a'), and we shall distinguish between them.

6°. For $e \in E$ we shall write

$$\varphi(e) = \begin{cases} +1 \text{ if e has been obtained by } 2°\text{-}3°, \\ -1 \text{ if e has been obtained by } 4°. \end{cases}$$

The graph $\Gamma(U)$ is a slight modification of the construction proposed for the first time in [1] (see also [7]) for studying variable-length code systems that do not have the prefix property (so-called nonrecurrent codes). The purpose of the improved version proposed by us here is to enrich and sharpen the information content of the graph assigned to a system of words. In the same way as in [1], we can show that for $a \in \mathfrak{A} \setminus U \ (a \neq \lambda)$ we have

(1) $a \in V \leftrightarrow$ there exists a relation $u_1, \ldots, u_k a = u_1', \ldots, u_e'$, where $u_i, \ u_i' \in U, \ |a| < |u_e'|$.

In the following we shall consider only complete independent systems of words. Let us recall that a system of words U is said to be independent if any word in $\mathfrak{A}$ can be represented as a union of words in U in at most a unique manner, and it is said to be complete if for any $v \in \mathfrak{A} \setminus U$ the system $\{U \cup v\}$ is no longer independent [2]. The following property is characteristic for independent system of words (see [7], [1], and [5]).

†Original article submitted December 10, 1968.

(2) For any path $\lambda, v_1, v_2, \ldots, v_k, \ldots$ in the graph $\Gamma(U)$ we have $v_i \in U$ if and only if $i = 1$.
If an independent system U is complete, it will satisfy (as is shown in [2]) the conditions

(3) $(\exists u \in [U])\ (\forall a \in \mathfrak{A})\ (\exists a' \in \mathfrak{A})\ uaa' \in [U]$,

(4) $\sum_{u \in U} \|A\|^{-|u|} = 1$,

each of the conditions (3) and (4) implying not only that U is independent, but also that it is complete. If (3) holds for a given $u = u_0$, we shall refer to u_0-completeness of U.

Let $I^+(U)$ be the number of edges $e \in E$ with $\varphi(e) = 1$, and $I^-(U)$ the number of edges with $\varphi(e) = -1$.

As a quantitative measure of the information about the system U contained in the graph $\Gamma(U)$ we shall take $I(\Gamma \mid U) = I^+(U) + I^-(U)$, i.e., the number of edges of the graph $\Gamma(U)$. The quantity $I(\Gamma \mid U)$ represents a certain aspect of the complexity of decoding if the system is taken as a code. More precisely, it characterizes in general the diversity of problematic situations that can occur in sequential decoding of messages. In particular, in accordance with intuition we have $I(\Gamma \mid U) = 0$ if and only if $\Gamma(U) = \{\lambda\}$, i.e., in the case of recurrent (prefix) coding [3, 4, 6]. We shall henceforth assume that $I(\Gamma \mid U) \neq 0$, thus confining ourselves to nonrecurrent codes.

The redundancy of information in $\Gamma(U)$ will be defined as $I(\Gamma \mid U) - \|U\|$. This is due to the following reasons. Any edge $e \in E$ contains together with $\varphi(e)$ information about one (and only one) word in U which will be denoted by $\mu(e)$. More precisely,

$$\mu(e) = \begin{cases} aa', & \text{if } e = (a, a') \text{ and } \varphi(e) = +1, \\ a'', & \text{if } e = (a, a'),\ \varphi(e) = -1 \text{ and } a = a''a'. \end{cases} \tag{5}$$

Let us show that if the system U is complete, μ will map E onto the entire system U, and hence

$$I(\Gamma \mid U) - \|U\| \geqslant 0.$$

Moreover, in this case we have

(6) $U = \mu(E^+)$, where $E^+ = \{e \mid e \in E,\ \varphi(e) = +1\}$.

Indeed, we know [6] that if $\Gamma(U) \neq \{\lambda\}$, then

(7) $(\exists \alpha \in \mathfrak{A})(\forall \alpha' \in \mathfrak{A})\ \alpha\alpha' \notin [U]$ and the equation $\alpha = u\alpha''$ cannot hold for any $u \in U$. In the case of u_0-completeness of U for any $u \in U$ and any n of the form (3), there exists an $a \in \mathfrak{A}$, such that $u_0 u^n \alpha a \in [U]$, where α is selected on the basis of (7). In this case we shall have for some m and a' the relation $u_0 u^m a' = v_1 \ldots v_k = v \in [U]$, where $u = a'a''$, $|a'| < |v_k|$, $u_0 u^m \in [U]$. Now we have $a' \in V$ by virtue of (1), whereas 3° and 6° yield $e = (a', a'') \in E$ with $\varphi(e) = +1$. Hence, we conclude that $I(\Gamma \mid U) \geq I^+(U) \geq \|U\|$. But for constructing U on the basis of $\Gamma(U)$ we must know only $\|U\|$ edges, i.e., a number $I(\Gamma \mid U) - \|U\|$ of edges is redundant for this purpose. In particular, $I^-(U)$ negative edges can be always regarded as redundant.

Our objective is now to study a class of complete independent systems of words such that $I(\Gamma \mid U) - \|U\| = 0$.

If W is a system of words in the alphabet B and $\|W\| = \|A\|$, whereas τ is a one-to-one mapping of A onto W, we shall write for any words $a_1, a_2 \in \mathfrak{A}$ the relation $\tau(a_1 a_2) = \tau(a_1)\tau(a_2)$ and for the set of words U we shall write $\tau(U) = \{\tau(u) \mid u \in U\}$. In [5], the system $\tau(U)$ is called a composition of the systems U and W. By a^* we shall denote the inversion of the word a, i.e., if $a = a_1 \ldots a_k$ $(a_i \in A)$, then $a^* = a_k \ldots a_1$. Hence, $U^* = \{u^* \mid u \in U\}$. Let K_{ij} be a complete system of words in the alphabet $B_{ij} = \{1, 2, \ldots, i + j\}$ consisting of words 1, 2, ..., i of length 1 and of words (i + p)q of length 2 for all p = 1, 2, ..., j; q = 1, 2, ..., i + j. We shall show that if the system U is complete and $I(\Gamma \mid U) = \| U \|$, then the set $V \setminus \{\lambda\}$ will be a complete prefix system and for some i and j there exists a mapping τ of the alphabet B_{ij} onto V such that $U = \tau(K^*_{ij})$.

By virtue of (6) we have $\varphi(e) = +1$ for any $e \in E$, and the mapping μ in (5) will be a bijection. Any word $u \in U$ can be uniquely represented in the form $v_i v_j$, where $v_i, v_j \in V$, $(v_i, v_j) \in E$. Let us show that none of the words $\{V \setminus \lambda\}$ is the initial section of another word of this set. Let us assume the contrary: $v_i, v_j \in V$ and $v_i = v_j a$. In this case the vertex v_i in $\Gamma(U)$ must be a dead end; otherwise we would have for some $u \in U$ a relation $v_i a' = u$ of type 3° and $\mu(v_i, a') = \mu(v_j, aa')$, which is impossible. Let $(v_k, v_i) \in E$ (such an edge must exist in view of (1) if $v_i \neq \lambda$, with $v_k \neq \lambda$) and $v = \mu(v_k, v_i)$. Let us consider the system $U' = \{U \cup v\alpha\}$, where α has been selected for U on the basis of (7). Without loss of generality it can be assumed that $|\alpha| > |u|$ for any $u \in U$. It is evident that $\Gamma(U)$ is a subgroup of $\Gamma(U')$; hence if $\Gamma(U')$ contains any vertices or edges other than V and E, the former must appear according to 2°-4° for $u = v\alpha$ or $u' = v\alpha$. Let us consider these possibilities.

2°. $u = u'a$, $u' \neq v\alpha$, and hence $v\alpha = u'a$. This relation has a unique solution $u' = v$, $a = \alpha$, since otherwise we would have $\mu(\lambda, v) = \mu(v_k, v_i)$, or $u'a' = v$ and $\mu(u', a') = \mu(v_k, v_i)$. Thus according to 2° we must include in V' the vertices V and α, and in E' the edges (λ, v) and (v, α) and only them.

3°. $aa' = v\alpha$, $a, a' \neq \lambda$, $a \in V$. We have $aa'' = v$ (if $a = va''$, then $\varphi(aa'') = -1$ in $\Gamma(U)$, which is impossible). But $\mu(a, a'') = \mu(v_k, v_i)$, whence $a = v_k$. Hence according to 3° we must add $v_i\alpha \in V'$ and the edge $(v_k, v_i\alpha)$ in E'.

Rule 4° is evidently not applicable, and all the added vertices are dead-end vertices; therefore $\Gamma(U')$, as well as $\Gamma(U)$, have Property (1). But this is impossible, since U is complete by assumption. The obtained contradiction proves that the system $V \setminus \{\lambda\}$ is a prefix system.

Thus, $U = (U \cap V) \cup (u_{ij} = v_i v_j)$, $v_i, v_j \in V$. It follows directly from the completeness of U that V is a complete prefix system. Let $\sigma(v_i) = i$, $\sigma(V \setminus \{\lambda\}) = B$. Hence there exists a system W consisting of words of length not greater than 2 in the alphabet B and such that $U = \sigma^{-1}(W)$. It follows from the latter formula that W is independent and that if U is u_0-complete, then W will be $\sigma^{-1}(u_0)$-complete. For concluding the proof we must convince ourselves that the only complete systems W with $|w| \leqslant 2$ $(w \in W)$ and i words of length 1 are K_{ij} and K^*_{ij}. Indeed, let $W = \{1, 2, \ldots, i, 1A_1, 2A_2, \ldots, mA_m\}$, where $A_i \subset A$, $m = \| A \|$.

By virtue of (4) we have $\sum_i m^{-e_i} = \frac{i}{m} + \frac{\sum_j \| A_j \|}{m^2} = 1$, whence $\sum_{j=1}^{m} \| A_j \| = m(m-i)$. The condition of independence of W can be formulated in terms of a binary relation R on B: $jRk \leftrightarrow (jk) \in W$. W is independent if and only if $R^n \cap I \times I = \phi$, for any n; here $I = \{1, 2, \ldots, i\}$. Suppose that n_1 symbols of A are to the left with respect to I, whereas n_2 symbols are to the right. Since by virtue of the independence of W there are no cycles passing through I, we have $\sum_j \| A_j \| \leqslant (m-i)i + n_1^2 + (m-i-n_1)^2$. Hence we obtain $n_1 \geq m - i$. But $n_1 + n_2 + i = m$, $n_1, n_2 \geq 0$, and we have $n_1 = m - i$, $n_2 = 0$ (or conversely), and these cases correspond to $W = K_{ij}$, or $W = K^*_{ij}$. By taking $\sigma^{-1} = \tau$, we obtain $U = \tau(K^*_{ij})$, which completes the proof.

Literature Cited

1. A. A. Markov, "Alphabet coding," Dokl. Akad. Nauk SSSR, 132(3):521-523 (1960).

2. A. A. Markov, "Completeness condition for nonuniform codes," Problemy Kibernetiki, 9:327-331, Fizmatgiz, Moscow (1963).
3. A. A. Markov, "Nonrecurrent coding," Problemy Kibernetiki, 8:169-186, Fizmatgiz, Moscow (1962).
4. B. Mandelbrot, "On recurrent noise limiting coding," Symposium on Information Networks, Polytechnic Institute of Brooklyn (1955).
5. M. Nivat, "Elements de la theorie generale des codes," Automata Theory, Academic Press, New York-London (1966), pp. 278-294.
6. E. N. Gilbert and E. F. Moore, "Variable-length binary encodings," BSTJ, 38(4):933-967 (1959).
7. A. A. Sardinas and G. W. Patterson, "A necessary and sufficient condition for unique decomposition of coded messages," Conv. Rec., Trans. IRE, IT-8:104-108 (1953).

REALIZATION OF DISJUNCTIONS AND CONJUNCTIONS IN MONOTONIC BASES†

É. I. Nechiporuk

Leningrad

This note belongs to a series of papers devoted to finding "nonlinear" lower bounds for the complexity of Boolean functions [1-5].

In the note we show that in some monotonic bases it is possible to realize a disjunction and a conjunction of n arguments with a complexity of order n^C, where C is an arbitrarily large constant.

We shall consider the realization of functions

$$D_n = x_1 \vee x_2 \vee \ldots \vee x_n, \quad K_n = x_1 x_2 \ldots x_n$$

by superpositions in a basis consisting of one monotonic function

$$\varphi_{l,m} = \bigvee_{i=1}^{l} \mathop{\&}_{j=1}^{m} x_{i,j}, \quad \text{where } l \geqslant 2,\ m \geqslant 2.$$

Each of these bases generates all the monotonic functions apart from the constants 0 and 1. It is easy to see that in the basis $\{\varphi_{l,m}, 0, 1\}$ the functions D_n and K_n are realized with a complexity‡ of order n. Yet we have the following theorems:

<u>Theorem 1.</u> $L_{\varphi_{l,m}}(D_n) \asymp n^{\frac{\log m}{\log l}+1}$.

<u>Theorem 2.</u> $L_{\varphi_{l,m}}(K_n) \asymp n^{\frac{\log l}{\log m}+1}$.

1°. We shall represent a superposition by a tree such that: Each internal node is an occurrence of a base element $\varphi_{l,m}$; each terminal node is an occurrence of the argument; each edge will be numbered by a pair (i, j) indicating the input of a base element. The occurrences of a superposition will be split into tiers (see Fig. 1). The number of the tier of an occurrence is the number of edges connecting the occurrence with an output element. The number of the last tier (it contains only arguments) will be denoted by P and called the depth of a superposition. If for any p, $0 \le p \le P-2$, the p-th tier contains $(lm)^p$ base elements, the superposition is said to be complete.

† Original article submitted October 31, 1969.

‡ The complexity of a superposition is defined as the number of base elements in it. By $L_{\varphi_{l,m}}(f)$ we denote the minimum number of base elements $\varphi_{l,m}$ sufficient for realizing a Boolean function f.

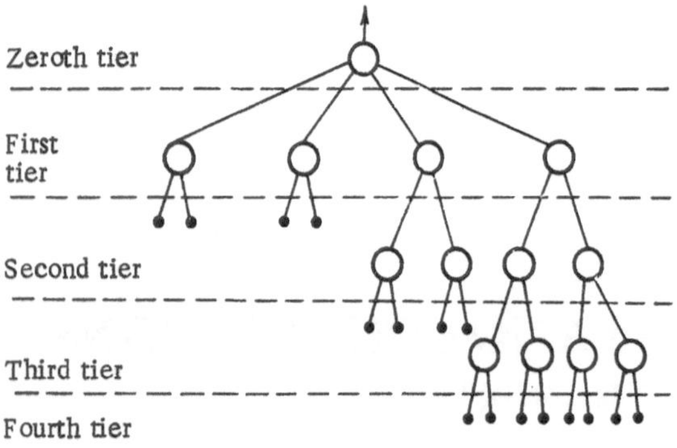

Fig. 1

2°. Proof of Theorem 1. The upper bound is reached on complete superpositions of depth $]\log_l n[$.

The lower bound. We shall transform the given superposition that realizes the disjunction D_n, without increasing its complexity, into an equivalent complete superposition.

1) If a given superposition has "superfluous" letters other than the letters $x_1, x_2, \ldots, x_n$, we shall replace each such letter by one of the letters $x_1, x_2, \ldots, x_n$.

2) Let us consider the subformulas

$$\bigvee_{i=1}^{l} \mathop{\&}_{j=1}^{m} A_{i,j} \tag{1}$$

in the order of decreasing tiers.† For each i we shall select from the subformulas $A_{i,1}$, $A_{i,2}, \ldots, A_{i,m}$ the most economical one and replace all the subformulas $A_{i,j}$ by this subformula. The function realized in this way can only increase; but it does not, since the "superfluous" letters have been eliminated.

After all these replacements have been effected, each subformula will realize a disjunction.

3) For each p, $1 \le p \le P$, we shall assign to a subformula‡ of the p-th tier an array of pairs

$$(i_1, j_1), \ (i_2, j_2), \ \ldots, \ (i_p, j_p),$$

corresponding to the edges connecting the given subformula with an output element. This array will be called the record of the subformula. Subformulas which have the same arrays $(i_1, i_2, \ldots, i_p)$ in the record, are said to be monoconjunctive. It is evident that monoconjunctive subformulas located in the p-th tier consists of m^p subformulas.

If a superposition has letters x_k $(1 \le k \le n)$ in the p-th tier and subformulas A which are not arguments in the (p + 1)-st tier, we shall replace in the p-th tier each of the m^p monoconjunctive letters x_k by the subformula A, and in the (p + 1)-st tier each of the m^{p+1} monoconjunctive subformulas A by the letter x_k.

After all these replacements, the superposition will be uniform with a depth not less than $]\log_l n[$. Hence,

$$L_{\varphi_{l,m}}(D_n) \geqslant (lm)^{]\log_l n[-2} \gtrsim n^{\frac{\log m}{\log l}+1}.$$

3°. Proof of Theorem 2. The upper bound is reached on complete superpositions of depth $]\log_m n[$.

† At first we shall consider all the subformulas of the (p − 1)-st tier, then all the subformulas of the (P − 2)-nd tier, etc., up to the zeroth tier.

‡ The subformula can be one of the letters $x_1, x_2, \ldots, x_n$.

The lower bound. We shall transform the given superposition that realizes the conjunction K_n, without increasing its complexity, into an equivalent complete superposition.

1) We eliminate the "superfluous" letters as in 2° above.

2) We consider the subformulas (1) in the order of decreasing tiers. In (1) we select an i such that the set of subformulas

$$A_{i,1}, A_{i,2}, \ldots, A_{i,m}$$

is the most economical. In (1) we then replace for each k, $k \neq i$, the subformula $A_{k,1}$ by $A_{i,1}$, $A_{k,2}$ by $A_{i,2}$, ..., $A_{k,m}$ by $A_{i,m}$. The function realized in this way can only decrease, but it does not, since the "superfluous" letters have been eliminated.

After all these replacements each subformula realizes a conjunction.

3) As in 2° we assign to each subformula its record. Subformulas which have the same arrays $(j_1, j_2, \ldots, j_p)$ in their record are said to be monodisjunctive. It is evident that monodisjunctive subformulas are identical, and that each group of monodisjunctive subformulas located in the p-th tier consists of l^p subformulas.

If a superposition has letters x_k $(1 \leq k \leq n)$ in the p-th tier and subformulas A which are not arguments in the (p + 1)-st tier, we shall replace in the p-th tier each of the l^p monodisjunctive letters x_k by a subformula A, and in the (p + 1)-st tier each of the l^{p+1} monodisjunctive subformulas A by the letter x_k.

After all these replacements, the superposition will be uniform with a depth not less than $]\log_m n[$. Hence,

$$L_{\varphi_{l,m}}(K_n) \geqslant (lm)^{]\log_m n[-2} \gtrsim n^{\frac{\log l}{\log m}+1}$$

Corollary. For $\frac{\log m}{\log l} \neq \frac{\log t}{\log s}$ the bases $\varphi_{l,m}$ and $\varphi_{s,t}$ are incommensurable (for the definition see [6]).

Literature Cited

1. B. A. Subbotovskaya, "Realization of linear functions by formulas in the bases ∨, &, –," Dokl. Akad. Nauk SSSR, 136(3):553-555 (1961).
2. A. A. Markov, "Minimal gate-contact networks for monotonic symmetrical functions," Problemy Kibernetiki, Vol. 8, 117-121, Fizmatgiz, Moscow (1962).
3. R. E. Krichevskii, "A minimal circuit with make contacts for a Boolean function of n arguments," Diskretnyi Analiz, No. 5, pp. 89-92, Novosibirsk (1965).
4. É. I. Nechiporuk, "A Boolean function," Dokl. Akad. Nauk SSSR, 169(4):765-766 (1966).
5. É. I. Nechiporuk, "On a Boolean matrix," Systems Theory Research, Vol. 21, Consultants Bureau, p. 236.
6. B. A. Subbotovskaya, "Comparison of bases in the realization of functions of the algebra of logic by formulas," Dokl. Akad. Nauk SSSR, 149(4):784-787 (1963).

CIRCUITS TO RAISE RELIABILITY †

M. M. Rokhlina

Moscow

Let us consider circuits of functional elements that in a certain sense can be said to "vote." Voting consists of the following: in a circle of radius k the set (0, 0, ..., 0) realized by the circuit of the function takes on the value 0, while in a circle of the same radius the set (1, 1, ..., 1) takes the value 1; in this case we say that the circuit "corrects k errors." This type of circuit must be used in the synthesis of self-correcting circuits made of functional elements [1]. The circuits considered here are built of $\mathscr{E}_\mu$, elements, realizing the function $h_\mu (x_1, \ldots, x_{\mu+1}) = \bigvee_{i=1}^{\mu+1} x_1x_2 \ldots x_{i-1}x_{i+1} \ldots x_{\mu+1}$. From the set of all circuits that correct k errors we separate the subset of circuits in which the output of each element, different from the output of the whole circuit, is connected to one input of a certain element.

<u>Note 1.</u> The circuits just defined above are isomorphous to the equations in a base consisting of one function $h_\mu (x_1, \ldots, x_{\mu+1})$. All subsequent conclusions concerning the complexity of such circuits can be formulated as conclusions concerning the complexity of the corresponding equations if the complexity of the equation is defined as the number of symbols of the base function entering into it.

We will show that the minimum complexity of circuits with tree-like shapes correcting k errors is no less than $k^{\log_2 (\mu+1)}$.

As an application of this result for any constant c, note the incomplete base B_c and the "effectively defined" sequence of functions $f_n^c (x_1, \ldots, x_n)$, expressed in base B_c such that the order of complexity of the equation when f_n^c is realized in this base is no less than n_c.

Let us introduce certain concepts and terminology.

A function is said to satisfy the condition $\langle A^\mu\rangle$, $\mu \geqslant 2$, if any μ of the sets on which the function returns to one have a common unit component [2]. Obviously:

a function satisfying condition $\langle A^\mu\rangle$, cannot take the value 1 on opposing sets. (1)

Clearly, function $h_\mu (x_1, \ldots, x_{\mu+1}) = \bigvee_{i=1}^{\mu+1} x_1x_2, \ldots x_{i-1}x_{i+1} \ldots x_{\mu+1}$ satisfies condition $\langle A^\mu\rangle$.

We shall designate by $l_0(\tilde{\alpha})$ the number of zeros in the set $\tilde{\alpha}$, and by $l_1(\tilde{\alpha})$ the number of ones in set $\tilde{\alpha}$. Obviously:

$$h_\mu(\tilde{\alpha}) = 0 \text{ if and only if } l_0(\tilde{\alpha}) \geqslant 2. \tag{2}$$

† Original article submitted April 15, 1968.

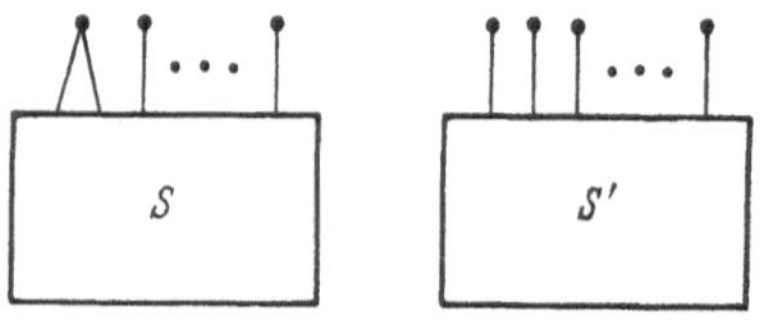

Fig. 1

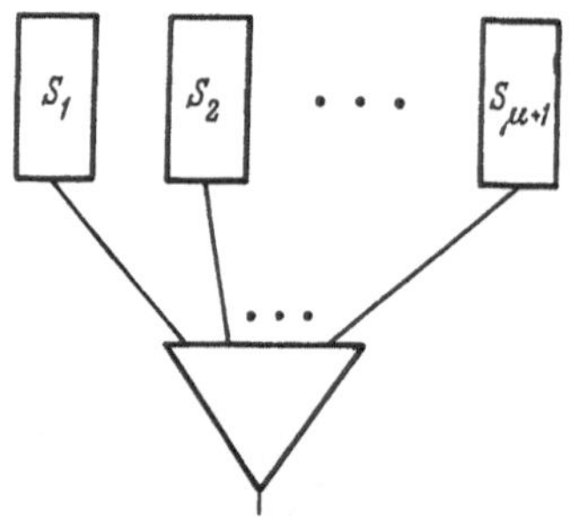

Fig. 2

We shall call the function of algebraic logic $g(x_1, \ldots, x_n)$ the k-function ($k = 1, 2, \ldots$), if it satisfies the following conditions:

$$\text{if } l_0(\tilde{\alpha}) \leqslant k, \text{ then } g(\tilde{\alpha}) = 1, \quad \text{if } l_1(\tilde{\alpha}) \leqslant k, \text{ then } g(\tilde{\alpha}) = 0$$

(the values of the function on the remaining sets are insignificant). Obviously, if $g(x_1, \ldots, x_n)$ is the k-function, then it is also an l-function for any l less than k. From this discussion and from [1] it follows that for each k there exists a k-function realized by a circuit of $\mathscr{E}_\mu$ elements.

The complexity of the circuit S we shall define by the number of elements $\mathscr{E}_\mu$ in it [symbol $L(S)$]. Let $L^\mu(k) = \min L(S)$, where the minimum is taken over all the circuits S realizing the k-function. From the definition of the k-function it follows that

$$\text{if } k' \leqslant k'', \text{ then } L^\mu(k') \leqslant L^\mu(k''). \tag{3}$$

Circuit S is called nonrepetitive if to each circuit input is connected no more than the input from one element while different variables are ascribed to all the inputs. We designate by $L_0^\mu(k)$ the minimum of the numbers $L(S)$ over all the nonrepetitive circuits S realizing the k-function.

It is true that

$$L^\mu(k) = L_0^\mu(k). \tag{4}$$

In fact, the inequality $L^\mu(k) \leq L_0^\mu(k)$ is obvious. On the other hand, let S be a circuit realizing the k-function and such that $L(S) = L^\mu(k)$. If S is a nonrepetitive circuit, then we separate the identified inputs of the elements forming the circuit inputs (Fig. 1). We obtain a nonrepetitive circuit S'. Clearly, $L(S) = L(S')$. And S' also realizes a certain k-function, so $L(S') \geq L_0^\mu(k)$ and $L^\mu(k) \geq L_0^\mu(k)$. Thus we have shown the validity of (4). From here it follows that the number of inputs for the minimum circuit realizing a certain k-function may not be fixed.

We shall designate the tree-type circuits connected to the inputs of the output elements of circuit S the *major subcircuits of circuit S* (Fig. 2).

Lemma 1. *Let S be a nonrepetitive circuit and $S_1, \ldots, S_{\mu+1}$ be its major subcircuits. The following two conditions are equivalent:*

$$K(S) \geqslant k; \tag{I}$$

for any i and j such that $i \neq j$ and $1 \leq i, j \leq \mu + 1$, it is true that

$$K(S_i) + K(S_j) \geqslant k - 1. \tag{II}$$

Proof. We designate by F (or F_i, respectively, $1 \leq i \leq \mu + 1$) the function realized by circuit S (S_i, respectively).

(I) → (II). Let $K(S) \geq k$. It is sufficient to prove one of the relationships of condition (II), such as $K(S_1) + K(S_2) \geq k - 1$. Assume that it is not true, i.e.,

$$K(S_1) + K(S_2) < k - 1. \tag{5}$$

From the definition of $K(S_i)$ it follows that for each of the subcircuits S_i ($i = 1, 2$) there exists one input set $\tilde{\alpha}_i$ satisfying one of the following conditions:

a) $l_0(\tilde{\alpha}_i) = K(S_i) + 1$ and $F_i(\tilde{\alpha}_i) = 0$,

b) $l_1(\tilde{\alpha}_i) = K(S_i) + 1$ and $F_i(\tilde{\alpha}_i) = 1$.

Further, if one set $\tilde{\alpha}_i$ satisfies condition b), then by virtue of (1), the set $\bar{\tilde{\alpha}}_i$ opposite to it satisfies condition a). Thus, there exist two sets, $\tilde{\alpha}_1'$ and $\tilde{\alpha}_2'$, satisfying condition a) for $i = 1, 2$.

Then $F(\tilde{\alpha}_1', \tilde{\alpha}_2', 1, \ldots, 1) = h_\mu(0, 0, 1, \ldots, 1) = 0$. But from a) and (5) it follows that $l_0(\tilde{\alpha}_1', \tilde{\alpha}_2', 1, \ldots, 1) \leqslant k$, and this contradicts the fact that F is a k-function.

(II) → (I). Let $\tilde{\alpha}$ be an arbitrary set and $\tilde{\alpha}_i$ its part applied to the inputs of the major subcircuit S_i ($1 \leq i \leq \mu + 1$). Let $l_0(\tilde{\alpha}) \leqslant k$. We will show that $F(\tilde{\alpha}) = 0$. Assume that $F(\tilde{\alpha}) \leqslant k$. Then from property (2) it follows that no less than two values of i fulfill $F_i(\tilde{\alpha}_i) = 0$. Let $F_1(\tilde{\alpha}_1) = 0$ and $F_2(\tilde{\alpha}_2) = 0$. Then from the definition of $K(S_1)$ and $K(S_2)$ it follows that $l_0(\tilde{\alpha}_1) \geqslant K(S_1) + 1$, and $l_0(\tilde{\alpha}_2) \geqslant K(S_2) + 1$. Since circuit S is nonrepetitive, from the latter two inequalities and by virtue of condition (II) it follows that $l_0(\tilde{\alpha}) \geqslant l_0(\tilde{\alpha}_1) + l_0(\tilde{\alpha}_2) \geqslant k + 1$, and this contradicts the fact that $l_0(\tilde{\alpha}) \leqslant k$.

Now let $l_1(\tilde{\alpha}) \leqslant k$. Let us show that $F(\tilde{\alpha}) = 0$. Assume that $F(\tilde{\alpha}) = 1$. From property (2) it follows that then no less than μ values of i satisfy $F_i(\tilde{\alpha}_i)$. Assume that these values of i are $1, 2, \ldots, \mu$. Considering only subcircuits S_1 and S_2, as above, we contradict the fact that $l_1(\tilde{\alpha}) \leqslant k$.

Lemma 1 is proven.

We introduce the notation:

$\mathfrak{A}_k$ is a set of sets of numbers $(k_1, \ldots, k_{\mu+1})$ satisfying condition (II) of Lemma 1.

$\mathfrak{B}_k$ is a set of sets of numbers $(k_1, \ldots, k_{\mu+1})$ satisfying the conditions:

$$k_1 + k_2 = k - 1, \qquad k_1 \leqslant k_2 = k_3 = \ldots = k_{\mu+1}.$$

<u>Lemma 2.</u> $L_0^\mu(k) = 1 + \min\limits_{(k_1, \ldots, k_{\mu+1}) \in \mathfrak{B}_k} \{\sum\limits_{i=1}^{\mu+1} L_0^\mu(k_i)\}$.

<u>Proof.</u> First we show that

$$L_0^\mu(k) = 1 + \min_{(k_1, \ldots, k_{\mu+1}) \in \mathfrak{A}_k} \{\sum_{i=1}^{\mu+1} L_0^\mu(k_i)\}. \tag{6}$$

Let S be a nonrepetitive circuit realizing a certain k-function and such that $L(S) = L_0^\mu(k)$. Further, let $S_1, \ldots, S_{\mu+1}$ be its major subcircuits and $k_i = K(S_i)$ ($1 \leq i \leq \mu + 1$). Then by virtue of lemma 1 $(k_1, \ldots, k_{\mu+1}) \in \mathfrak{A}_k$ and furthermore $L(S_i) \geq L_0^\mu(k)$. Therefore,

$$L_0^\mu(k) = L(S) = 1 + \sum_{i=1}^{\mu+1} L(S_i) \geqslant 1 + \sum_{i=1}^{\mu+1} L_0^\mu(k_i) \geqslant 1 + \min_{\mathfrak{A}_k} \sum_{i=1}^{\mu+1} L_0^\mu(k_i).$$

On the other hand, let $(k_1^0, \ldots, k^0_{\mu+1})$ be a set on which we reach $\min\limits_{\mathfrak{A}_k} \sum\limits_{i=1}^{\mu+1} L_0^\mu(k_i)$. Let S be a nonrepretitive circuit whose major subcircuits $S_1, \ldots, S_{\mu+1}$ (they are nonrepetitive) have the properties:

S_i realizes a certain k_i^0-function and $L(S_i) = L_0^\mu(k_i^0)$.

Then by virtue of Lemma 1, circuit S realizes a certain k-function and

$$1+\min_{\mathfrak{A}_k} \sum_{i=1}^{\mu+1} L_0^\mu(k_i) = 1 + \sum_{i=1}^{\mu+1} L_0^\mu(k_i^0) = 1 + \sum_{i=1}^{\mu+1} L(S_i) = L(S) \geqslant L_0^\mu(k).$$

Thus, (6) is proven. Further, it is obvious that $\mathfrak{A}_k \supseteq \mathfrak{B}_k$. From this it follows that

$$\min_{\mathfrak{A}_k} \sum_{i=1}^{\mu+1} L_0^\mu(k_i) \leqslant \min_{\mathfrak{B}_k} \sum_{i=1}^{\mu+1} L_0^\mu(k_i).$$

The opposite inequality is easily derived from (3), and Lemma 2 is proven.

We define an auxiliary function $f(k)$ as follows:

$$\left.\begin{aligned} &f(0) = 0, \\ &f(2s+1) = (\mu+1) f(s) + 1, \qquad s \geqslant 0, \\ &f(2s+2) = \mu f(s+1) + f(s) + 1, \quad s \geqslant 0. \end{aligned}\right\} \tag{7}$$

It is true that

$$f(k) \geqslant 0, \tag{8}$$

$$f(k+1) - f(k) > 0. \tag{9}$$

The first is obvious; the second is easily proven by induction in k.

<u>Lemma 3.</u> It is true that (for $t \geq 1$, $l - 2t \geq 0$):

$$\mu(f(l) - f(l-t)) \geqslant f(l-t) - f(l-2t), \tag{$A_{l,t}$}$$

$$\mu(f(l+1) - f(l+1-t)) \geqslant f(l-t) - f(l-2t). \tag{$B_{l,t}$}$$

<u>Proof.</u> First we consider the case $l = 2t$. Then $(A_{l,t})$ has the form

$$\mu(f(2t) - f(t)) \geqslant f(t). \tag{A'_t}$$

Expressing $f(2t)$ in accordance with (7), we find an equivalent and obvious relationship (because $\mu \geq 2$):

$$(\mu^2 - \mu) f(t) + \mu f(t-1) \geqslant f(t).$$

In this case $(B_{l,t})$ takes the form

$$\mu(f(2t+1) - f(t+1)) \geqslant f(t). \tag{B'_t}$$

We shall prove (B'_t) by induction in t.

Obviously (B'_1) is true. Let $(B'_{t'})$ be true for all $t' < t$. We shall prove the validity of (B'_t).

a) Let t be even. Expressing $f(2t+1)$, $f(t+1)$, and $f(t)$ in accordance with (7), we arrive at a relationship equivalent to (B'_t):

$$\mu\left((\mu+1)f(t)-(\mu+1)f\left(\frac{t}{2}\right)\right)\geqslant\mu f\left(\frac{t}{2}\right)+f\left(\frac{t}{2}-1\right)+1. \tag{10}$$

Relationship (10) follows from:

$$\mu\left(\mu\left(f(t)-f\left(\frac{t}{2}\right)\right)\right)\geqslant\mu f\left(\frac{t}{2}\right), \tag{11}$$

$$\mu\left(f(t)-f\left(\frac{t}{2}\right)\right)\geqslant f\left(\frac{t}{2}-1\right)+1. \tag{12}$$

Equation (11) follows from the already proven relationship $(A'_{\frac{t}{2}})$, and (12) from $(A'_{\frac{t}{2}})$ and the inequality $f\left(\frac{t}{2}-1\right)+1\leqslant f\left(\frac{t}{2}\right)$ [see (9)].

b) Let t be odd.

Equation (B'_t) is equivalent to

$$\mu\left((\mu+1)f(t)-\mu f\left(\frac{t+1}{2}\right)-f\left(\frac{t+1}{2}-1\right)\right)\geqslant(\mu+1)f\left(\frac{t-1}{2}\right)+1. \tag{13}$$

Equation (13) follows from:

$$\mu\left(\mu\left(f(t)-f\left(\frac{t-1}{2}+1\right)\right)\right)\geqslant\mu f\left(\frac{t-1}{2}\right), \tag{14}$$

$$\mu\left(f(t)-f\left(\frac{t-1}{2}\right)\right)\geqslant f\left(\frac{t-1}{2}\right)+1. \tag{15}$$

Equation (14) is derived from $(B'_{\frac{t-1}{2}})$, which is true by assumption, and (15) is derived from $(A'_{\frac{t-1}{2}})$ and (9).

Now consider the case when $l-2t>0$. Equations $(A_{l,t})$ and $(B_{l,t})$ are proven by induction in $l-2t$ as (B'_t) was proven.

Equations $(A_{l,t})$ and $(B_{l,t})$ follow from the relationships shown in Table 1 and assumed by induction to be fulfilled.

TABLE 1

	$A_{l,t}$	$B_{l,t}$
l — even, t — even	$A_{l/2,\,t/2}$, $A_{(l/2)-1,\,t/2}$	$A_{l/2,\,t/2}$, $B_{(l/2)-1,\,t/2}$
l — odd, t — even	$A_{(l-1)/2,\,t/2}$	$B_{(l-1)/2,\,t/2}$, $A_{(l-1)/2,\,t/2}$
l — even, t — odd	$A_{l/2,\,(t+1)/2}$, $A_{(l/2)-2,\,(t-1)/2}$ *	$B_{(l-1)/2,\,(t-1)/2}$, $A_{l/2,\,(t+1)/2}$
l — odd, t — odd	$B_{(l-1)/2,\,(t-1)/2}$ **	$A_{(l+1)/2,\,(t+1)/2}$, $B_{((l-1)/2)-1,\,(t-1)/2}$

*Here we also use the inequality

$$\mu\left(f\left(\frac{l}{2}-\frac{t+1}{2}\right)-f\left(\frac{l}{2}-t\right)\right)\leqslant\mu\left(f\left(\frac{l}{2}-\frac{t+1}{2}\right)+f\left(\frac{l}{2}-(t+1)\right)\right).$$

**Terms must be grouped, with some of them transferred from one part of the inequality to the other.

Lemma 4. $$f(k)=\min_{\mathfrak{B}_k}\{f(k_1)+\ldots+f(k_{\mu+1})+1\}. \tag{16}$$

Proof. For k = 1, 2, Eq. (16) is verified directly (sets $\mathfrak{A}_1$ and $\mathfrak{A}_2$ contain one set each).

Now let k ≥ 3. We can set up an arbitrary set of numbers from $\mathfrak{B}_k$ such that:

$$\text{if}\quad k=2s+1,\ \text{then}\ k_1=s-t,\ k_2=k_3=\ldots=k_{\mu+1}=s+t;$$
$$\text{if}\quad k=2s+2,\ \text{then}\ k_1=s-t,\ k_2=k_3=\ldots=k_{\mu+1}=s+t+1.$$

In our case s ≥ 1 and s − t ≥ 0.

By virtue of (7), it is sufficient to show that for each s ≥ 1 and for each t ≥ 1 (for s − t ≥ 0) it is true that:

$$1+(\mu+1)f(s)\leqslant f(s-t)+\mu f(s+t)+1$$

and

$$1+\mu f(s+1)+f(s)\leqslant f(s-t)+\mu f(s+t+1)+1,$$

which are equivalent respectively to

$$\mu(f(s+t)-f(s))\geqslant f(s)-f(s-t),$$
$$\mu(f(s+t+1)-f(s+1))\geqslant f(s)-f(s-t).$$

Assuming s + t = l, we obtain the relationship ($A_{l,t}$) and ($B_{l,t}$) that were already proven.

Theorem. $$L(k)=f(k).$$

Proof. For k = 1 we have $L^\mu(1) = L_0^\mu(1) = f(1) = 1$. Let $L^\mu(k') = f(k')$ for all k' > k.

From Lemmas 2 and 4 and (4) it follows that:

$$L_0^\mu(k)=L^\mu(k)=1+\min_{\mathfrak{B}_k}\sum_{i=1}^{\mu+1}L_0^\mu(k_i)=1+\min_{\mathfrak{B}_k}(f(k_1)+\ldots+f(k_{\mu+1}))=f(k).$$

Note 2. The function $L^\mu(k)$ has an order $k^{\log_2(\mu+1)}$.

First we show by induction in m that $L^\mu(2^m-1)=\frac{(\mu+1)^m-1}{\mu}$. For m = 1 this equation is obvious. Let it be true for all m' < m that

$$L^\mu(2^{m'}-1)=\frac{(\mu+1)^{m'}-1}{\mu}.$$

From the theorem and (7) it follows that:

$$L^\mu(2^m-1)=1+(\mu+1)L^\mu(2^{m-1}-1)=\frac{(\mu+1)^m-1}{\mu}.$$

Further, let $2^{m-1}-1 < k < 2^m-1$; then from (3) it follows that

$$\frac{(\mu+1)^{m-1}-1}{\mu}\leqslant L^\mu(k)<\frac{(\mu+1)^m-1}{\mu},$$

or

$$\frac{(2^{m-1})^{\log_2(\mu+1)}-1}{\mu} < L^{\mu}(k) < \frac{(2^{m})^{\log_2(\mu+1)}-1}{\mu}.$$

Therefore the statement is proven.

Note 3. Obviously all the results of this analysis are also valid for the complexity of realization of the k-function by circuits of $\mathscr{E}_{\mu}$, elements realizing the function

$$h_{\tilde{\mu}}(x_1, \ldots, x_{\mu+1}) = \mathop{\&}_{i=1}^{\mu+1} (x_1 \vee x_2 \vee \ldots \vee x_{i-1} \vee x_{i+1} \vee \ldots \vee x_{\mu+1}),$$

doubled to $h_{\mu}(x_1, \ldots, x_{\mu+1})$, $\mu \geq 2$.

Addendum

Consider the application of these results to the "effective" design of a complexly realized function. Consider the equations in a base consisting of one function $h_{\mu}(x_1, .., x_{\mu+1})$.

The complexity of the equation $\mathfrak{A}$ is the number of base functions entering into $\mathfrak{A}$ For a function f, expressed by $h_{\mu}(x_1, \ldots, x_{\mu+1})$ the notation of the main text is preserved (by virtue of note 1), i.e., $L^{\mu}(f) = \min L^{\mu}(\mathfrak{A})$, where the minimum is taken over all equations $\mathfrak{A}$, realizing the given function f.

Let $f_n^{\mu}(x_1, \ldots, x_n)$ be a function that assumes the value 1 on sets $\tilde{\alpha}$ such that $l_0(\tilde{\alpha}) \leq [(n-1)/\mu]$, and is equal to 0 on the remaining sets.

The function $f_n^{\mu}(x_1, \ldots, x_n)$ can be defined differently as a symmetrical function with operating numbers $n - [(n-1)/\mu], n - [(n-1)/\mu] + 1, \ldots, n$. It is a monotonic symmetrical function.

It is easy to show that $f_n^{\mu}(x_1, \ldots, x_n)$ satisfies condition $\langle A^{\mu}\rangle$. Accordingly, it is expressed by $h_{\mu}(x_1, \ldots, x_{\mu+1})$. Furthermore, $f_n^{\mu}(x_1, \ldots, x_n)$ is an $[(n-1)/\mu]$-function. Therefore, by virtue of note 1

$$L^{\mu}(f_n^{\mu}) \geqslant L^{\mu}\left(\left[\frac{n-1}{\mu}\right]\right)$$

and by virtue of note 2 for sufficiently large n

$$L^{\mu}(f_n^{\mu}) \geqslant c_{\mu} n^{\log_2(\mu+1)},$$

where c_{μ} is a certain constant.

Thus, for any constant c, a base (incomplete) and a certain simply defined sequence of monotonic symmetrical functions $f_1(x_1), \ldots, f_n(x_1, \ldots, x_n)$ can be indicated such that the complexity of realization of the function f_n in this base has an order to less than n^c.

Literature Cited

1. G. I. Kirienko, "Self-correcting circuits," in: Problemy Kibernetiki, Vol. 12, Nauka, Moscow (1964).
2. S. V. Yablonskii, G. P. Gavrilov, and V. B. Kudryavtsev, Functions of Algebraic Logic and the Post Class, Nauka, Moscow (1965).